Maple V®

Example

EDITION

LIMITED WARRANTY AND DISCLAIMER OF LIABILITY

Maple V®
By Example

SECOND EDITION

Martha L. Abell
Department of Mathematics and Computer Science
Georgia Southern University
Statesboro, Georgia

James P. Braselton
Department of Mathematics and Computer Science
Georgia Southern University
Statesboro, Georgia

ACADEMIC PRESS
San Diego London Boston
New York Sydney Tokyo Toronto

This book is printed on acid-free paper. ∞

Academic Press
A Division of Harcourt Brace & Company
525 B Street, Suite 1900, San Diego, California 92101-4495, USA
http://www.apnet.com

Academic Press
24–28 Oval Road, London NW1 7DX, UK
http://www.hbuk.co.uk/ap/

Library of Congress Cataloging-in-Publication Data
Abell, Martha L., 1962–
Maple V by example / Martha L. Abell, James P. Braselton.—2nd ed.
p. cm.
Includes bibliographical references and index.
ISBN 0-12-041558-5 (book : alk. paper). —ISBN 0-12-041559-3 (CD-ROM)
1. Maple (Computer file) 2. Mathematics—Data processing.
I. Braselton, James P., 1965– . II. Title.
QA76.95.A212 1998
510′.285′53—dc21 98-38818
CIP

Printed in the United States of America
98 99 00 01 02 IP 9 8 7 6 5 4 3 2 1

Contents

Preface

Maple V By Example bridges the gap that exists between the very elementary handbooks available on Maple V and those reference books written for the advanced Maple V users. *Maple V By Example* is an appropriate reference for all users of Maple V and, in particular, for beginning users like students, instructors, engineers, business people, and other professionals first learning to use Maple V. *Maple V By Example* introduces the very basic commands and includes typical examples of applications of these commands. In addition, the text also includes commands useful in areas such as calculus, linear algebra, business mathematics, ordinary and partial differential equations, and graphics. In all cases, however, examples follow the introduction of new commands. Readers from the most elementary to advanced levels will find that the range of topics covered addresses their needs.

Taking advantage of Version 5 of Maple V, *Maple V By Example*, Second Edition, introduces the fundamental concepts of Maple V to solve typical problems of interest to students, instructors, scientists, and other professionals who use Maple V. Other features to help make *Maple V By Example*, Second Edition, as easy to use and as useful as possible include the following.

1. **Version 5 Compatibility.** All examples illustrated in Maple V By Example, Second Edition, were completed using Version 5 of Maple V. Although most computations can continue to be carried out with earlier versions of Maple V, like Versions 2, 3, and 4, we have taken advantage of the new features in Version 5, as much as possible.

2. **Applications.** New applications, many of which are documented by references, from a variety of fields, especially biology, physics, and engineering, are included throughout the text.
3. **Detailed Table of Contents.** The table of contents includes all chapter, section, and subsection headings. Along with the comprehensive index, we hope that users will be able to locate information quickly and easily.
4. **Additional examples.** We have considerably expanded the topics in Chapters 1 through 7. The results should be more useful to instructors, students, business people, engineers, and other professionals using Maple V on a variety of platforms. In addition, several sections have been added to help make locating information easier for the user.
5. **Comprehensive Index.** In the index, mathematical examples and applications are listed by topic, or name, as well as commands along with frequently used options: particular mathematical examples as well as examples illustrating how to use frequently used commands are easy to locate. In addition, commands in the index are cross-referenced with frequently used options. Functions available in the various packages are cross-referenced both by package and alphabetically.
6. **Compact Disc.** All Maple V input that appears in *Maple V By Example*, Second Edition, is included on the disk packaged with the text.

Of course, we must express our appreciate to those who assisted in this project. We would like to express appreciation to our editor, Charles B. Glaser, his assistant, Della Grayson, and our production editor, **, at Academic Press for providing a pleasant environment in which to work. In addition, Waterloo Maple Software, especially Ben Friedman, has been most helpful in providing us up-to-date information about Maple V. Finally, we thank those close to us, especially Imogene Abell, Lori Braselton, Ada Braselton, and Martha Braselton for enduring with us the pressures of meeting a deadline and for graciously accepting our demanding work schedules. We certainly could not have completed this task without their care and understanding.

M. L. Abell
J. P. Braselton
Statesboro, Georgia

CHAPTER 1

Getting Started

1.1 Introduction to Maple V

Maple V is one of the most powerful and reliable systems currently available for doing mathematics by computer. Editions of Maple and Maple V have been available for over ten years. The more than 100,000 users of Maple V consist of scientists, engineers, mathematicians, teachers, and students. Maple V is easy to use, accurate, and expandable. Although Maple V has modest minimum memory requirements, with the more than 2500 math routines available, Maple V can perform symbolic and numeric mathematics, generate two- and three-dimensional graphics and animations, display output using standard mathematical notation, and provide a high-level programming language similar to Pascal.

Maple V is available for nearly every operating system available including Macintosh, MS Windows, MS DOS, UNIX, VMS, NeXT, Ultrix, and UNICOS. Since Maple V worksheets share a common format, they can be taken from one platform to another.

A Note Regarding Different Versions of Maple

With the advent of Release 5 of Maple V, many new functions and features have been added to Maple V. Particular examples of features available with Release 5 but not some previous releases include: standard mathematical notation for output, contour and implicit plots, animation of graphics, and the interactive help browser. We encourage users of earlier releases to update to Release 5 as soon as they can. All examples in *Maple V By Example* were done using Release 5. In most cases, the same results will be obtained if you are using Release 4. Occasionally, however, particular features of Release 5 are illustrated in an example; of course, these features are not available with earlier releases. If you are using an earlier or later version of Maple, your results may not appear in a form identical to those found in this book. Some commands found in Release 5 are not available in earlier versions of Maple; in later versions some commands will certainly be changed, new commands will be added, and obsolete commands will be removed.

In this text, we assume that Maple V has been correctly installed on the computer you are using. If you need to install Maple V on your computer, please refer to the documentation that came with the Maple V software package.

1.2 Getting Started with Maple V

After the Maple V program has been properly installed, a user can access Maple V. Below we briefly describe methods of starting Maple V on several platforms.

Macintosh: In the same manner that folders are opened, the Maple application can be started by selecting the Maple icon, going to **File,** and selecting **Open,** or by simply clicking twice on the Maple icon. Of course, opening an existing Maple document also opens the Maple program.

Windows: To start Maple V for Windows, double-click on the Maple V for Windows application icon. This operation opens Maple V for Windows with a blank worksheet.

DOS: To run Maple, type the command **MAPLE** at the DOS prompt. After a few seconds, the screen will clear, the Maple logo will appear briefly, and a prompt will appear at the bottom of the screen above the status line. You will be in the Maple session, and you can now begin typing Maple commands.

UNIX: To start Maple V, type **xmaple** at the UNIX prompt. A new window will open, containing Maple V Release 2. You can now begin typing Maple commands.

If you are using a text-based interface (like UNIX), Maple V is started with the operating system command `maple` or `xmaple`. If you are using a notebook interface (like Macintosh, Windows, or NeXT), Maple V is started by selecting the Maple V icon and double-clicking or selecting the Maple V icon and selecting **Open** from the **File** menu.

Once Maple V has been started, computations can be carried out immediately. Maple V commands are typed to the right of the prompt; a semicolon is placed at the end of the command, and then it is evaluated by pressing **ENTER**. Generally, when a colon is placed at the end of the command, the resulting output is not displayed. Note that pressing **ENTER** evaluates commands and pressing **RETURN** yields a new line. Output is displayed below input. We illustrate some of the typical steps involved in working with Maple V in the calculations below. In each case, we type the command, place a semicolon at the end of the command, and press **ENTER**. Maple V evaluates the command, displays the result, and inserts a new prompt. For example, entering

```
> evalf(Pi,50);
  3.1415926535897932384626433832795028841971693993751
```

returns a 50-digit approximation of π.

The next calculation can then be typed and entered in the same manner as the first. For example, entering

```
> solve(x^3-2*x+1=0);
```

$$1, -\frac{1}{2}+\frac{1}{2}\sqrt{5}, -\frac{1}{2}+\frac{1}{2}\sqrt{5}$$

solves the equation $x^3 - 2x + 1 = 0$ for x. Subsequent calculations are entered in the same way. For example, entering

```
> plot({sin(x),2*cos(2*x)},x=0..3*Pi);
```

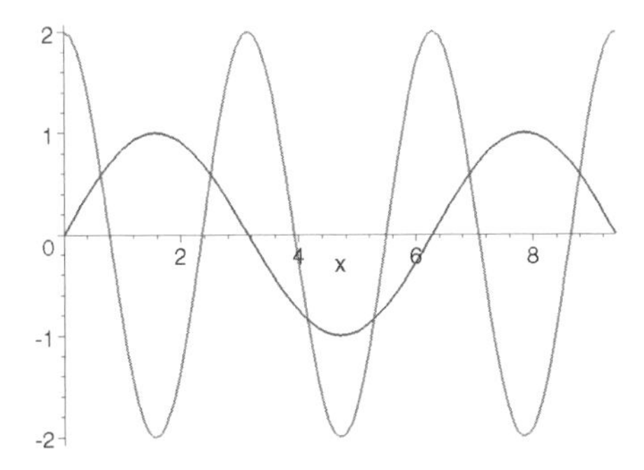

graphs the functions $\sin x$ and $2\cos 2x$ on the interval $[0, 3\pi]$. Similarly, entering

```
> plot3d(sin(x+cos(y)),x=0..4*Pi,y=0..4*Pi);
```

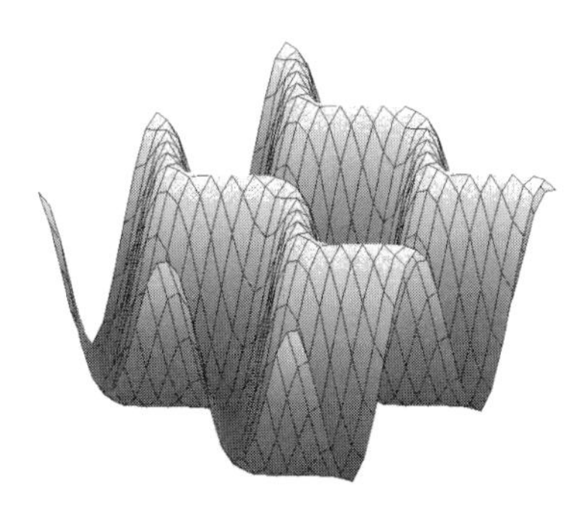

graphs the function $\sin(x + \cos y)$ on the rectangle $[0, 4\pi] \times [0, 4\pi]$.

Notice that you can change how your input and output appear in your Maple notebook by going to **Options** under the menu and selecting **Input Display ...**

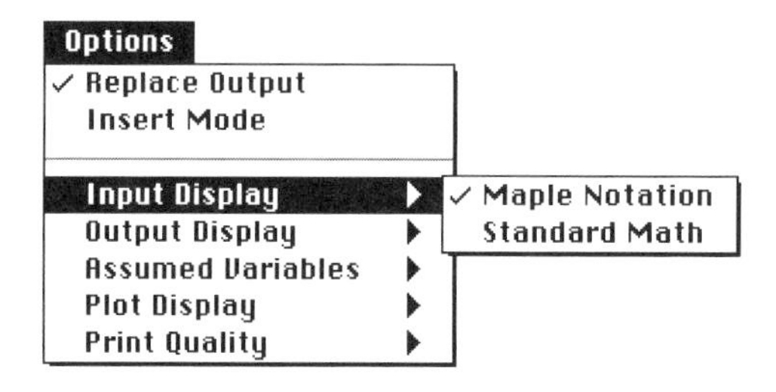

... or **Output Display**.

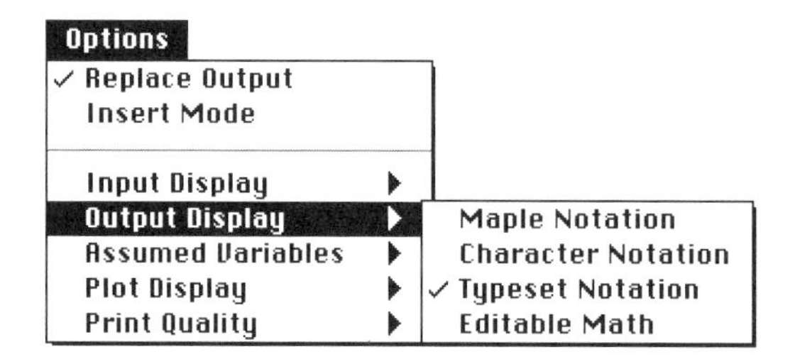

In this command, both the input and the ouput are in **Maple Notation**.

```
Int(x^2*sin(x)^2,x)=int(x^2*sin(x)^2,x);
       Int(x^2*sin(x)^2,x)=
   x^2*(-1/2*cos(x)*sin(x)+1/2*x)-
   1/2*x*cos(x)^2+1/4*cos(x)*sin(x)+1/4*x-1/3*x^3
```

In this case, the input is in **Maple Notation** and the output is in **Character Notation.**

```
Int(x^2*sin(x)^2,x)=int(x^2*sin(x)^2,x);
     /
    |   2       2       2
    |  x  sin(x)  dx = x  (- 1/2 cos(x) sin(x) + 1/2 x)
    |
   /
                            2                                    3
          - 1/2 x cos(x)  + 1/4 cos(x) sin(x) + 1/4 x - 1/3 x
```

Here, the output is in **Typeset Notation.**

```
Int(x^2*sin(x)^2,x)=int(x^2*sin(x)^2,x);
```

$$\int x^2 \sin(x)^2 dx =$$
$$x^2\left(-\frac{1}{2}\cos(x)\sin(x)+\frac{1}{2}x\right)-\frac{1}{2}x\cos(x)^2+\frac{1}{4}\cos(x)\sin(x)+\frac{1}{4}x-\frac{1}{3}x^3$$

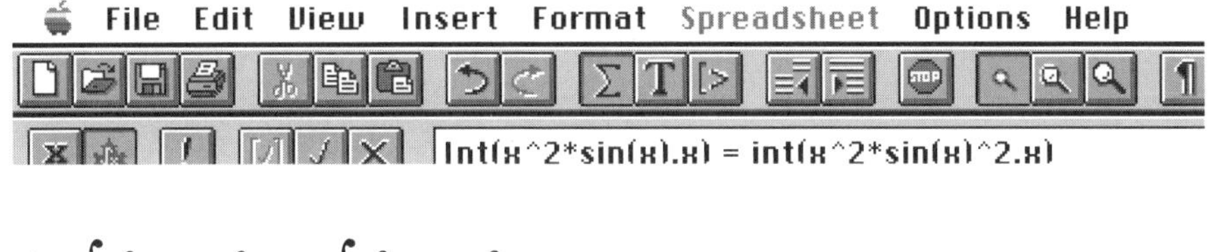

$$\int x^2 \sin(x)^2 dx = \int x^2 \sin(x)^2 dx$$
$$\int x^2 \sin(x)^2 dx =$$
$$x^2\left(-\frac{1}{2}\cos(x)\sin(x)+\frac{1}{2}x\right)-\frac{1}{2}x\cos(x)^2+\frac{1}{4}\cos(x)\sin(x)+\frac{1}{4}x-\frac{1}{3}x^3$$

Although **Editable Math** notation looks the same as **Typeset Notation**, portions of **Editable Math** notation can be selected, copied, and pasted elsewhere in your Maple notebook.

This book includes real input and output from Maple V Release 5. Appearances of input and output may vary depending on the version of Maple used, the fonts used to display input and ouput, the quality of the monitor, and the resolution and type of printer used to print the Maple worksheet; the results displayed on your computer may not be physically identical to those shown here.

Maple V sessions are terminated by entering `quit`, `done`, or `stop`. On several platforms with notebook interfaces (like Macintosh, Windows, and NeXT), Maple V sessions are ended by selecting **Quit** from the **File** menu, or by using the keyboard shortcut ⌘**Q**, as with other applications. They can be saved by referring to ⌘**S** from the File menu.

On these platforms, input and text regions in notebook interfaces can be edited. Editing input can create a notebook in which the mathematical output does not make sense in the sequence it appears. It is also possible to simply go into a notebook and alter input without doing any recalculation. This also creates misleading notebooks. Hence, common sense and caution should be used in the editing of input regions of notebooks. Recalculating all commands in the notebook will clarify any confusion.

Preview

In order for the Maple V user to take full advantage of the capabilities of this software, an understanding of its syntax is imperative. The goal of *Maple V By Example* is to introduce the reader to the Maple V commands and sequences of commands most frequently used by beginning users. Although the rules of Maple V syntax are far too numerous to list here, knowledge of the following five rules equips the beginner with the necessary tools to start using the Maple V program with little trouble.

Five Basic Rules of Maple V Syntax

1. The arguments of functions are given in parentheses (...).
2. A semicolon (;) or colon (:) must be included at the end of each command.
3. Multiplication is represented by a *
4. Powers are denoted by a ^.
5. If you get no response or an incorrect response, you may have entered or executed the command incorrectly. In some cases, the amount of memory allocated to Maple V can cause a crash; like people, Maple V is not perfect, and some errors can occur.

1.3 Loading Miscellaneous Library Functions and Packages

Maple V's modularity, which gives Maple a great deal of flexibility, helps minimize its memory requirements. Nevertheless, although Maple contains many built-in functions that are loaded immediately when called, some other functions are contained in packages that must be loaded separately. Other functions—miscellaneous library functions—must also be loaded separately.

Loading Miscellaneous Library Functions

Miscellaneous library commands must be loaded with the command `readlib(command)` before they are used during a Maple session. We show this with the following example by first listing the miscellaneous library functions.

We use Maple's help facility to view a list of the miscellaneous library functions. Entering the command

```
> ?index[misc]
```

displays a help window that lists the miscellaneous library functions:

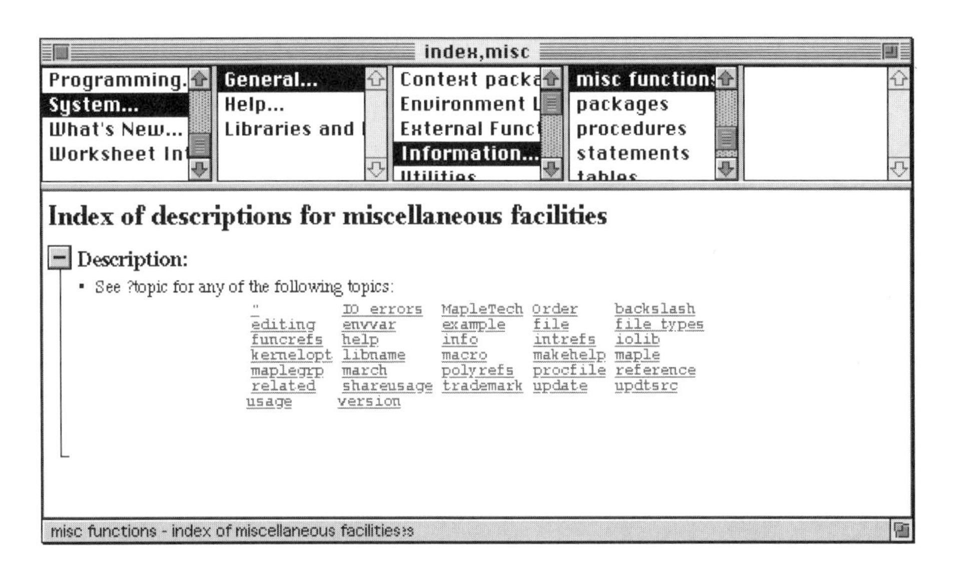

For example, if we let $f(z) = \dfrac{5z-2}{z^2-z}$, then, by the Residue theorem, the value of the integral $\int_C f(z)dz$, where C is the circle $|z| = 2$, is $2\pi i(res(f,0)+res(f,1))$, where $res(f,0)$ and $res(f,1)$ denote the residue of $f(z)$ at $z = 0$ and $z = 1$, respectively. We can use Maple to compute these values by loading the command residue, defining *f*, and then evaluating `residue (f,z=0)` and `residue (f,z=1)`, which compute the residue of $f(z)$ at $z = 0$ and $z = 1$, respectively.

We first load the command `residue` with

```
readlib(residue)
```

and then define *f*.

```
> readlib(residue):
  f:=(5*z-2)/(z^2-z);
```

$$f := \frac{5z-2}{z^2-z}$$

Next, we compute the desired residues.

```
> r0:=residue(f,z=0);
  r1:=residue(f,z=1);
```

$$r0 := 2$$

$$r1 := 3$$

Finally, we apply the Residue theorem to evaluate the integral.

```
> 2*Pi*I*(r0+r1);
```

$$10 I \pi$$

The following command calculates the residue of $\frac{\cot z}{z^4}$ if $z = 0$.

```
> residue(cot(z)/z^4,z=0);
```

$$\frac{-1}{45}$$

Loading Packages

In addition to the standard library functions and miscellaneous library functions, described above, a tremendous number of additional commands are available in various packages. We use Maple's help facility by entering `?packages` to view a list of the available packages. Entering the command

```
> ?index[packages]
```

displays the following window, which lists the available packages:

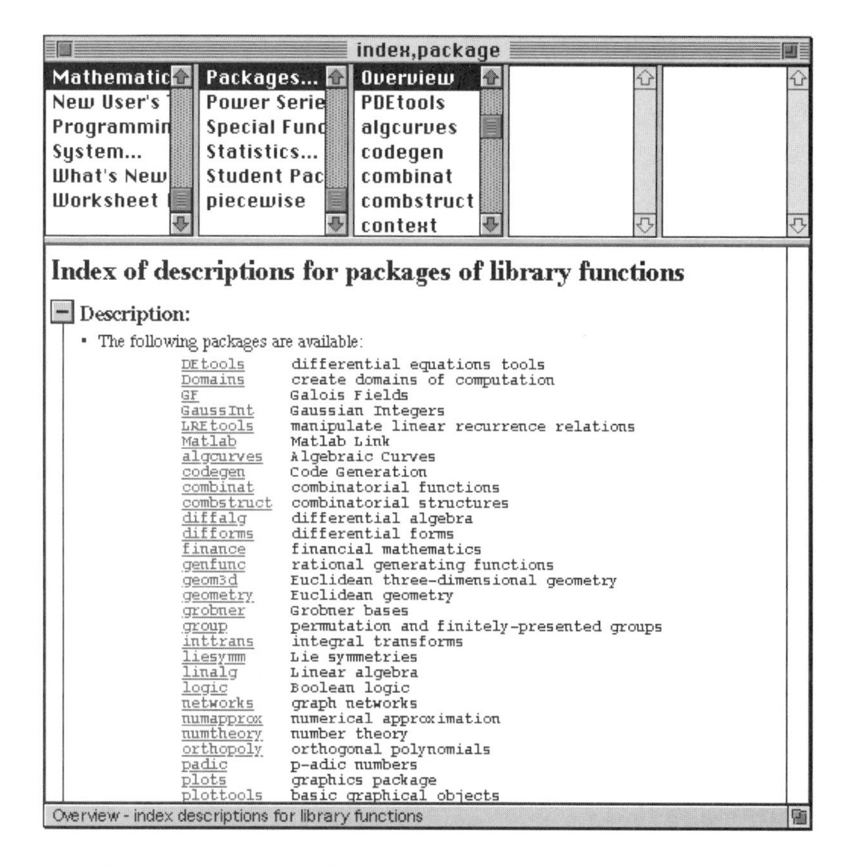

Information about particular packages is obtained with `?packagename`. For example, with the command `?student` we obtain information about the `student` calculus package, including a list of the commands contained in the package.

```
> ?student
```

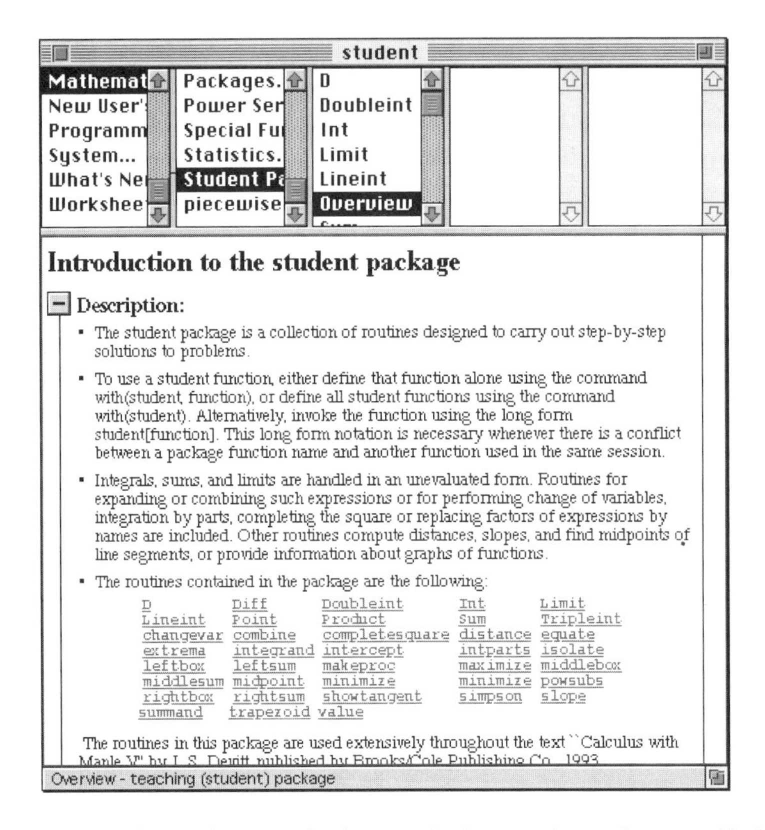

A typical example of a package is the linear algebra package that is called `linalg`. For example, to compute the determinant of the matrix $A = \begin{pmatrix} 10 & -6 & -9 \\ 6 & -5 & -7 \\ -10 & 9 & 12 \end{pmatrix}$, we must first define the matrix and then select a Maple command, or define a new one, which computes the determinant of a square matrix.

In this case, we first define A and then use the command `linalg[det]` to compute the determinant of A. Note that the command `det` is contained in the package `linalg`.

```
> A:=matrix([[10,-6,-9],[6,-5,-7],[-10,9,12]]);
```

$$A: = \begin{bmatrix} 10 & -6 & -9 \\ 6 & -5 & -7 \\ -10 & 9 & 12 \end{bmatrix}$$

```
> linalg[det](A);
```

6

However, we can use the `with` command to load a package. After a package has been loaded, subsequent calculations involving commands contained in the package can be entered directly.

Below, we use `with` to load the package `linalg`. The commands contained in the `linalg` package are displayed after the package is loaded. If a colon were used at the end of the command instead of a semicolon, then the package commands would be loaded but not listed.

```
> with(linalg);
```

[BlockDiagonal, GramSchmidt, JordanBlock, LUdecomp, QRdecomp, Wronskian, addcol, addrow, adj, adjoint, angle, augment, backsub, band, basis, bezout, blockmatrix, charmat, charpoly, cholesky, col, coldim, colspace, colspan, companion, concat, cond, copyinto, crossprod, curl, definite, delcols, delrows, det, diag, diverge, dotprod, eigenvals, eigenvalues, eigenvectors, eigenvects, entermatrix, equal, exponential, extend, ffgausselim, fibonacci, forwardsub, frobenius, gausselim, gaussjord, geneqns, genmatrix, grad, hadamard, hermite, hessian, hilbert, htranspose, ihermite, indexfunc, innerprod, intbasis, inverse, ismith, issimilar, iszero, jacobian, jordan, kernel, laplacian, leastsqrs, linsolve, matadd, matrix, minor, minpoly, mulcol, mulrow, multiply, norm, normalize, nullspace, orthog, permanent, pivot, potential, randmatrix, randvector, rank, ratform, row, rowdim, rowspace, rowspan, rref, scalarmul, singularvals, smith, stackmatrix, submatrix, subvector, sumbasis, swapcol, swaprow, sylvester, toeplitz, trace, transpose, vandermonde, vecpotent, vectdim, vector, wronskian]

We can then compute the determinant of A using the command `det` instead of `linalg[det]`.

```
> det(A);
```

6

1.4 Getting Help from Maple V

Help Commands

Becoming competent with Maple can take a serious investment of time. Beginning Maple users should try to view messages that result from syntax errors lightheartedly. Ideally, instead of becoming frustrated, they will find it challenging and fun to locate the source of errors. Frequently, Maple's error messages indicate where the error or errors may have occurred. In this process, users will naturally become more proficient with Maple.

One way to obtain information about commands and functions is the commands `?` and `help`. In general, `?f and help(f)` give information on the Maple function `f`. This information appears in a separate window.

EXAMPLE 1: Use ? to obtain information about the command help.

SOLUTION:

```
> ?help
```

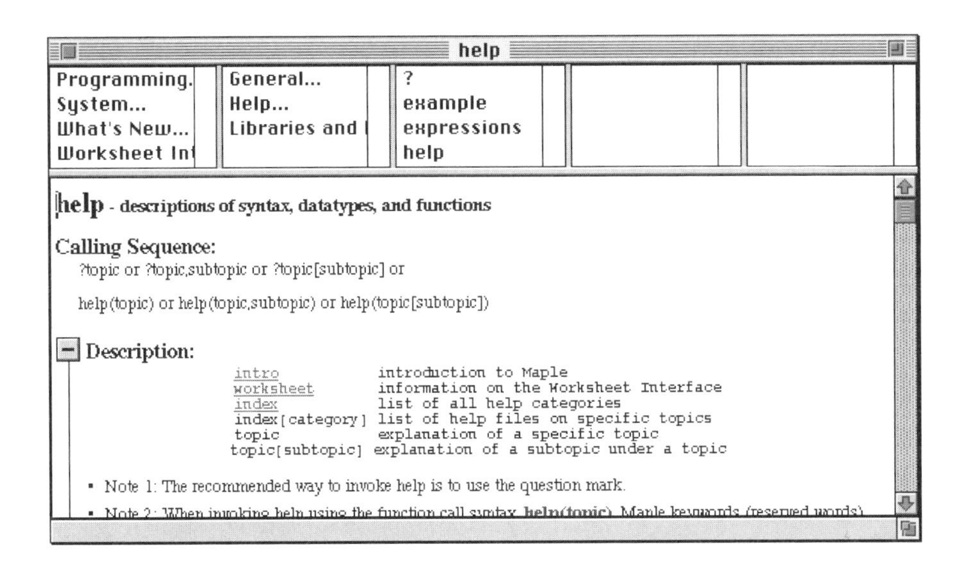

■

Alternatively, Maple's help facilty can be accessed from the Maple menu. In this case, we select **Using Help**,

which presents a window that describes Maple's online help facility.

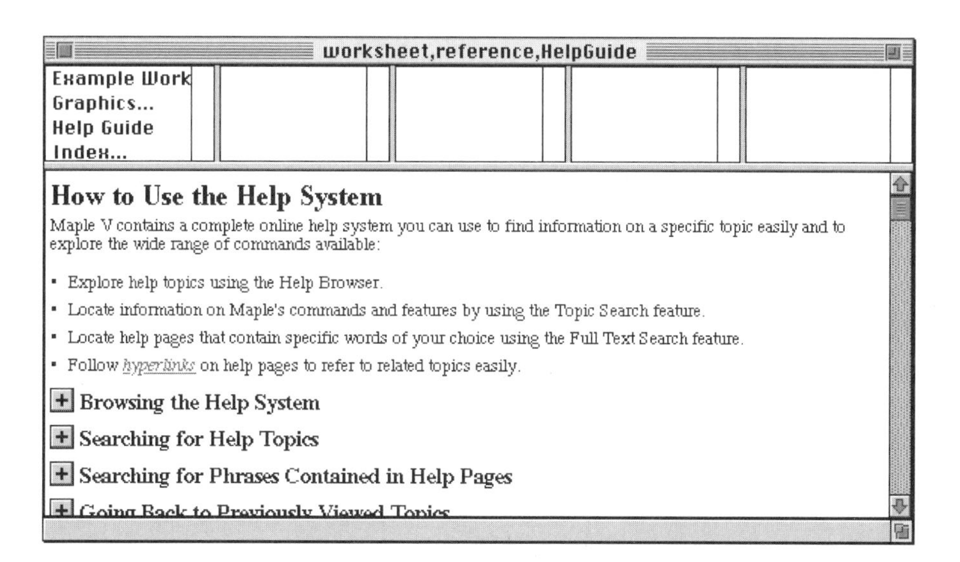

Additional Ways of Obtaining Help from Maple V

The Maple Menu offers other ways of obtaining help. For example, if a command is selected, and then **Help for Context** is selected from the **Windows** menu, Maple displays the help window for that command. This is illustrated below with the `solve` command. We begin by typing and selecting the `solve` command.

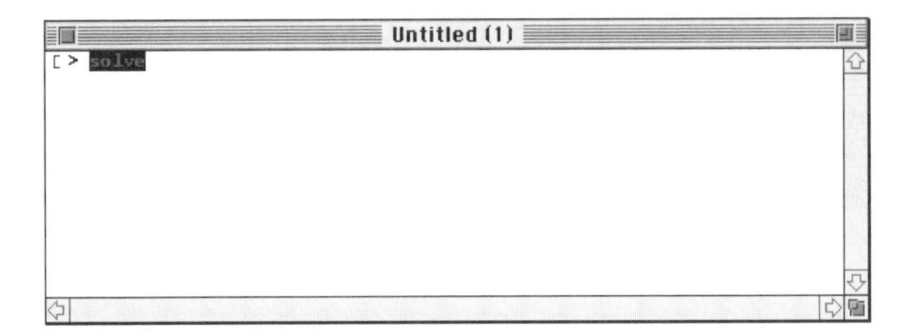

We then select **Help for Context** from the **Windows** menu. Notice that in this case the word **Context** is replaced by the word **solve**.

This is the resulting help window:

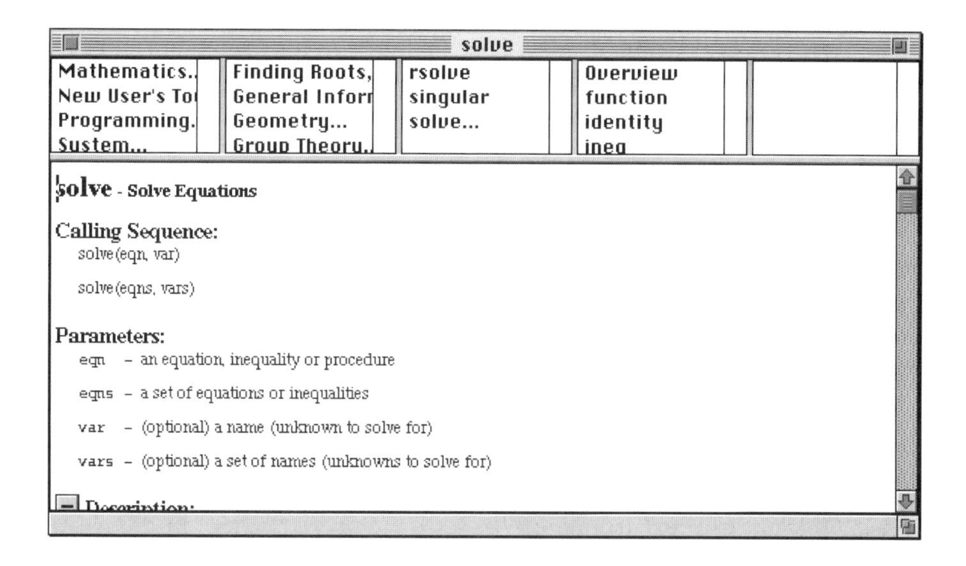

In this particular case, we wish to solve the literal equation $v = \frac{1}{3}\pi h(r^2 + rR + R^2)$ for R. We scroll through the help window and locate an example similar to the one we wish to solve. We then complete the typing of the command and enter the result. The output corresponds to the solution of solving $v = \frac{1}{3}\pi h(r^2 + rR + R^2)$ for R.

```
> solve(v=1/3*Pi*h*(r^2+r*R+R^2),R);
```

$$\frac{1}{2}\,\frac{-\pi h r + \sqrt{-3\pi^2 h^2 r^2 + 12\pi h r v}}{\pi h},\ \frac{1}{2}\,\frac{-\pi h r - \sqrt{-3\pi^2 h^2 r^2 + 12\pi h v}}{\pi h}$$

Assistance can also be obtained by selecting **Topic Search...** or **Full Text Search...** from the **Help** menu.

Selecting **Topic Search...** yields the following help window:

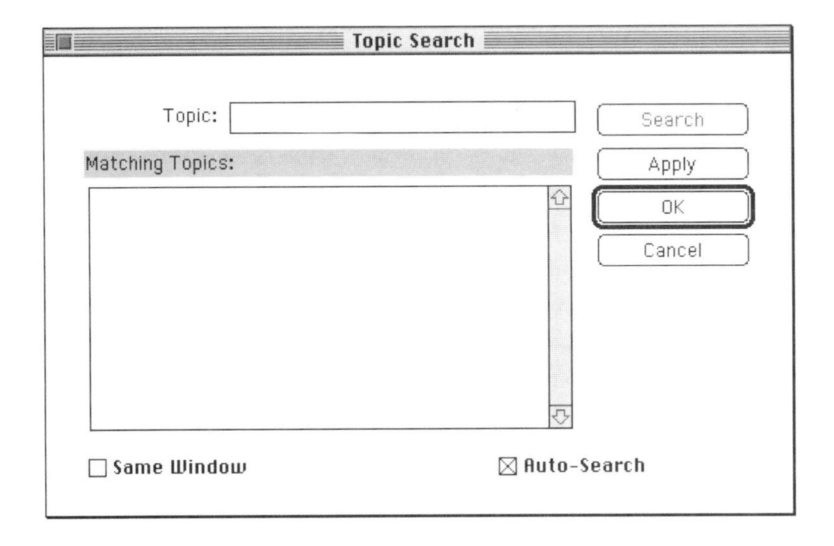

CHAPTER 

Numerical Operations on Numbers, Expressions, and Functions

2.1 Numerical Calculations and Built-in Functions

Numerical Calculations

The basic arithmetic operations (addition, subtraction, multiplication, and division) are performed in the natural way with Maple. Whenever possible, Maple gives an exact answer and reduces fractions.

"a plus b" is entered as a+b;

"a minus b" is entered as a–b;

"a times b" is entered as a*b; and

"a divided by b" is entered as a/b. Generally, when a and b are numbers, a/b results in the reduced fraction.

EXAMPLE 1: Calculate (a) 121 + 542; (b) 3231 – 9876; (c) (-23)(76); (d) (22341)(832748)(387281); (e) $\frac{467}{31}$; and (f) $\frac{12315}{35}$.

SOLUTION: These calculations are carried out below. Notice that in (f), Maple simplifies the quotient since the numerator and denominator have a common factor of 5. In each case, the input is typed to the right of the prompt; a semicolon is placed at the end of the command, and then it is evaluated by pressing **Enter**. Note that pressing **Return** gives a new line; pressing **Enter** causes Maple to evaluate the input.

```
> 121+542;
```

663

```
> 3231-9876;
```

-6645

```
> -23*76;
```

-1748

```
> 22361*832748*387281;
```

7211589719761868

```
> 467/31;
```

$$\frac{467}{31}$$

```
> 12315/35;
```

$$\frac{2463}{7}$$

■

Maple performs the order of operations in the usual manner.

EXAMPLE 2: Compare the results of entering `3+8*4+16/2` and `((3+8)*4+16)/2`.

SOLUTION: In the first expression, Maple first calculates the product 8*4 and the quotient 16/2, and then performs the operation of addition. In the second expression, the grouping of the parentheses is following. Hence, 3+8 is calculated, multiplied by 4, added to 16, and then divided by 2.

```
> 3+8*4+16/2;
                              43
> ((3+8)*4+16)/2;
                              30
```

■

The expression "a raised to the power b" is entered as `a^b`. The term $a^{n/m} = \sqrt[m]{a^n} = (\sqrt[m]{a})^n$ is entered as `a^(n/m)`. For $b = 1/2$, the command `sqrt(a)` can be used instead. Usually, the result is returned in unevaluated form but `evalf` can be used to obtain numerical approximations to virtually any degree of accuracy. At other times, `simplify` can be used to produce the expected result.

EXAMPLE 3: Calculate (a) $(-5)^{121}$, (b) $(-5)^{1/9}$, and (c) $8^{2/3}$.

SOLUTION: Maple yields a numerical value for $(-5)^{121}$. Notice that "\" in this result indicates that the output continues on the next line. The expression that involves a fractional exponent is returned in unevaluated form.

```
> (-5)^121;
  -3761581922631320025499956919111186169019729781670 68\
                 006882800546009093523025512695312 5
```

```
> (-5)^(1/9);
```

$$(-5)^{\frac{1}{9}}$$

A numerical approximation is obtained with `evalf`. Notice that Maple returns the "principal root," which is an imaginary number.

```
> evalf((-5)^(1/9));
                1.123696816 + .4089921933 I
```

Maple does not automatically simplify $8^{2/3} = (\sqrt[3]{8})^2 = 4$.

```
> 8^(2/3);
```

$$8^{\left(\frac{2}{3}\right)}$$

We use `simplify` to obtain 4 as the result.

```
> simplify(8^(2/3));
```

$$4$$

■

Often, a numerical approximation of a result is either more meaningful or more desirable than an exact or unevaluated `result`. The command `evalf(expr)` is used to obtain a numerical approximation of the expression `expr`. With `evalf(expr,n)`, Maple yields a numerical approximation of expr to `n` digits of precision, whenever possible.

EXAMPLE 4: Use Maple to obtain an approximate value of each of the following: (a) $(-5)^{121}$ and (b) $\sqrt{233}$.

SOLUTION: (a) The approximate value of $(-5)^{121}$ is given in scientific notation with `evalf`. (b) `evalf` is used to yield the approximate value of $\sqrt{233}$ which is first returned in unevaluated form.

```
> evalf((-5)^121);
```

$$-.3761581923\ 10^{85}$$

```
> sqrt(233);
```

$$\sqrt{233}$$

```
> evalf(sqrt(233),20);
```

$$15.264337522473748025$$

■

When odd roots of negative numbers are computed, Maple's results are surprising to the novice. Namely, Maple returns a complex number. We will see that this has important consequences for graphing certain functions.

EXAMPLE 5: Calculate $\left(-\frac{27}{64}\right)^{2/3}$.

SOLUTION: Maple does not automatically simplify $\left(-\frac{27}{64}\right)^{2/3}$.

```
> (-27/64)^(2/3);
```

$$\frac{1}{64}(-27)^{\left(\frac{2}{3}\right)}64^{\left(\frac{1}{3}\right)}$$

However, when we use `simplify`, Maple returns the simplified version of the principal root of $\left(-\frac{27}{64}\right)^{2/3}$.

```
> simplify((-27/64)^(2/3));
```

$$\frac{9}{64}(1+I\sqrt{3})^2$$

An approximation of this imaginary number is obtained with `evalf`.

```
> evalf((-27/64)^(2/3));
```

```
-.2812500000 + .4871392897 I
```

To obtain the result

$$\left(-\frac{27}{64}\right)^{2/3} = \left(\sqrt[3]{-\frac{37}{64}}\right)^2 = \left(-\frac{3}{4}\right)^2 = \frac{9}{16},$$

which would be expected by most algrebra and calculus students, we use `surd`. For negative real x and odd n, `surd(x^n,n)` returns `x`. Thus, `surd(-8,3)` returns $\sqrt[3]{-8} = -2$,

```
> surd(-27/64,3);
```

$$\frac{-3}{4}$$

returns $\sqrt[3]{-27/64} = -3/4$, and

```
> surd(-27/64,3)^2;
```

$$\frac{9}{16}$$

returns $\sqrt[3]{-27/64} = 9/16$.

■

Built-in Constants

Maple has built-in definitions of many commonly used constants. In particular, $e \approx 2.71828$ is denoted by `exp(1)`, $\pi \approx 3.14159$ is denoted by `Pi`, and $i = \sqrt{-1}$ is denoted by `I`.

Usually, Maple performs complex arithmetic automatically. If you are using Release 5, `evalc(expr)` evaluates and simplifies the complex expression `expr` and attempts to write `expr` as a sum of its real and imaginary components.

In many cases, expressions must be expanded. The command `expand(expr)` accomplishes this primarily by distributing products over sums or by making use of the properties of most mathematical functions.

EXAMPLE 6: Approximate e to 50 digits of precision and π to 25 digits of precision. Also, calculate $\sqrt{-9}$, expand $(1 - i^4)$, and write $\frac{3+i}{4-i}$ in standard form.

SOLUTION: First, `evalf` is used to approximate `E` and `Pi` to 50 and 25 digits of precision, respectively. Then, `sqrt(-9)` yields the imaginary number `3I`.

```
> evalf(exp(1),50);
```

2.7182818284590452353602874713526624977572470937

```
> evalf(Pi,25);
```

3.141592653589793238462643

```
> sqrt(-9);
```

$$3I$$

```
> (1-I)^4;
```

$$-4$$

```
> (3+I)/(4-I);
```

$$\frac{11}{17} + \frac{7}{17}I$$

■

Built-in Functions

Maple recognizes numerous built-in functions. These include the exponential function, `exp(x);` the natural logarithm, `ln(x);` the absolute value function, `abs(x);` the trigonometric functions `sin(x)`, `cos(x)`, `tan(x)`, `sec(x)`, `csc(x)`, and `cot(x);` and the inverse trigonometric functions `arcsin(x)`, `arccos(x)`, `arctan(x)`, `arcsec(x)`, `arccsc(x)`, and `arccot(x)`.

Several examples of these functions are given next. Notice that Maple returns an exact value unless otherwise specified with `evalf` or `evalc`. You can obtain a complete list of all initially known built-in functions by entering `?inifcn`.

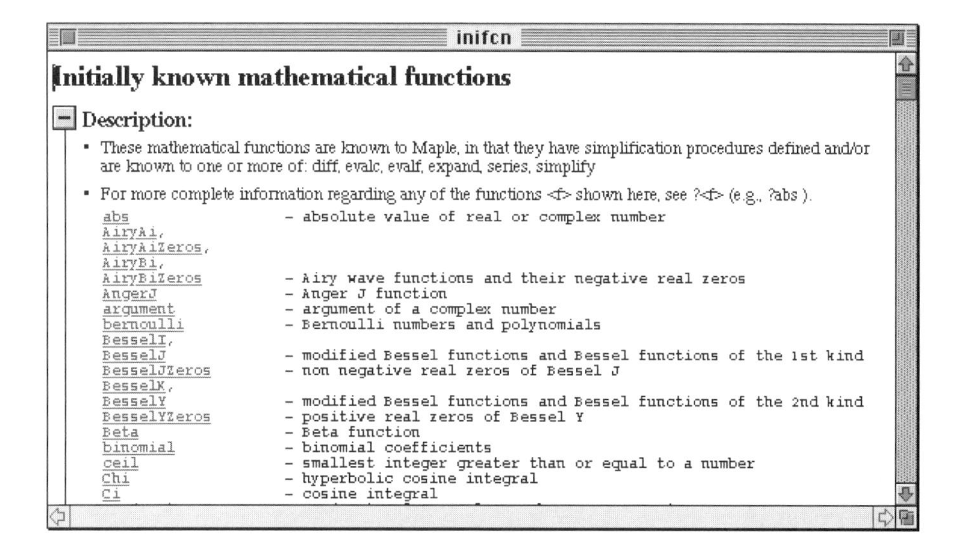

inifcn

Initially known mathematical functions

Description:

- These mathematical functions are known to Maple, in that they have simplification procedures defined and/or are known to one or more of: diff, evalc, evalf, expand, series, simplify
- For more complete information regarding any of the functions <f> shown here, see ?<f> (e.g., ?abs).

```
abs            - absolute value of real or complex number
AiryAi,
AiryAiZeros,
AiryBi,
AiryBiZeros    - Airy wave functions and their negative real zeros
AngerJ         - Anger J function
argument       - argument of a complex number
bernoulli      - Bernoulli numbers and polynomials
BesselI,
BesselJ        - modified Bessel functions and Bessel functions of the 1st kind
BesselJZeros   - non negative real zeros of Bessel J
BesselK,
BesselY        - modified Bessel functions and Bessel functions of the 2nd kind
BesselYZeros   - positive real zeros of Bessel Y
Beta           - Beta function
binomial       - binomial coefficients
ceil           - smallest integer greater than or equal to a number
Chi            - hyperbolic cosine integral
Ci             - cosine integral
```

EXAMPLE 7: Approximate e^{-5}, and obtain exact values of $|-5|$ and $|14|$.

SOLUTION: Entering

```
> evalf(exp(-5));
```

$$.006737946999$$

computes an approximation of $e^{-5} = 1/e^{5}$. Entering

```
> abs(-5);
```

$$5$$

computes $|-5|$ and entering

```
> abs(14);
```

14

computes $|14|$.

■

Real-valued functions of a single variable are graphed with `plot`, which we will discuss in more detail in Section 2.3. For now, we remark that the command

```
plot(f(x),x=a..b)
```

graphs the function $y = f(x)$ on the interval $[a, b]$. Thus, entering

```
> plot(exp(x),x=-2..2);
```

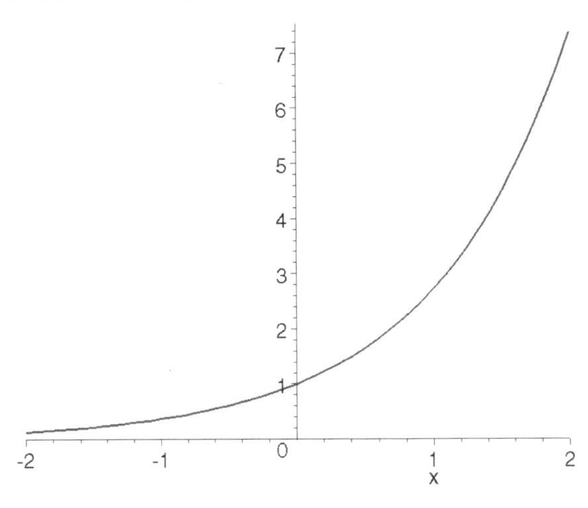

graphs $y = e^x$ on the interval $[-2, 2]$.

In addition to finding real numbers, the absolute value function `abs(x)` can be used to find the absolute value (or modulus) of the complex number $a + bi$ where $|a + bi| = \sqrt{a^2 + b^2}$.

EXAMPLE 8: Calculate $|3 - 4i|$ and $\left|\frac{3 + 2i}{2 - 9i}\right|$.

SOLUTION: The value of 5 is obtained for $|3-4i|$; the modulus of the quotient, is returned by entering abs((3+2*I)/(2-9*I)). Hence, $\left|\frac{3+2i}{2-9i}\right| = \frac{1}{85}\sqrt{1105}$.

```
> abs(3-4*I);
```

$$5$$

```
> abs((3+2*I)/(2-9*I));
```

$$\frac{1}{85}\sqrt{1105}$$

■

Several examples of the natural logarithm and the exponential functions are given next. Notice that Maple recognizes the properties associated with these functions and simplifies expressions accordingly.

EXAMPLE 9: Calculate: (a) $\ln e$; (b) $\ln e^3$; (c) $e^{\ln\pi}$

SOLUTION: Maple understands that ln(x) and exp(x) are inverses of each other. Hence, $\ln e = \ln(\exp(1))$ returns 1. In a similar fashion, $e^{\ln\pi}$ yields π.

```
> ln(exp(1));
```

$$1$$

```
> ln(exp(3));
```

$$3$$

```
> exp(ln(Pi));
```

$$\pi$$

Because $y = \ln x$ and $y = e^x$ are inverse functions, their graphs are symmetric about the line $y = x$. As illustrated with the following plot command, several functions can be graphed together with plot.

```
> plot([exp(x),ln(x),x],x=-2..2,-2..2);
```

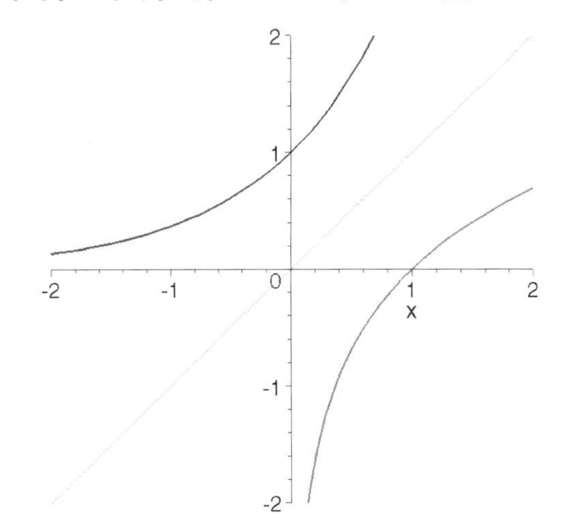

■

Several calculations that involve trigonometric functions are demonstrated in the following examples. Notice that an exact value is returned unless `evalf` is used with the calculation.

EXAMPLE 10: Evaluate (a) $\cos(\pi/4)$; (b) $\sin(\pi/3)$; (c) $\tan(3\pi/4)$; (d) $\cos(\pi/12)$; (e) $\cos(\pi/18)$; (f) $\sin(-9\pi/4)$.

SOLUTION: In the first five cases in which the value is well known, Maple returns an exact value. In the last case, however, `evalf` must be used to obtain a numerical approximation of the results.

```
> cos(Pi/4);
```

$$\frac{1}{2}\sqrt{2}$$

```
> sin(Pi/3);
```

$$\frac{1}{2}\sqrt{3}$$

```
> tan(3*Pi/4);
```

$$-1$$

```
> cos(Pi/12);
```

$$\frac{1}{4}\sqrt{6}\left(1+\frac{1}{3}\sqrt{3}\right)$$

```
> sin(-9*Pi/8);
```

$$\frac{1}{2}\sqrt{2-\sqrt{2}}$$

```
> cos(Pi/18);
```

$$\cos\left(\frac{1}{18}\pi\right)$$

```
> evalf(cos(Pi/18));
```

.9848077530

The trigonometric functions $\sin x$, $\cos x$, and $\tan x$ are related to $\csc x$, $\sec x$, and $\cot x$ by $\sin x = 1/\csc x$, $\cos x = 1/\sec x$, and $\tan x = 1/\cot x$.

```
> plot([sin(x),csc(x)],x=0..2*Pi,-Pi..Pi);
  plot([cos(x),sec(x)],x=0..2*Pi,-Pi..Pi);
  plot([tan(x),cot(x)],x=0..2*Pi,-Pi..Pi);
```

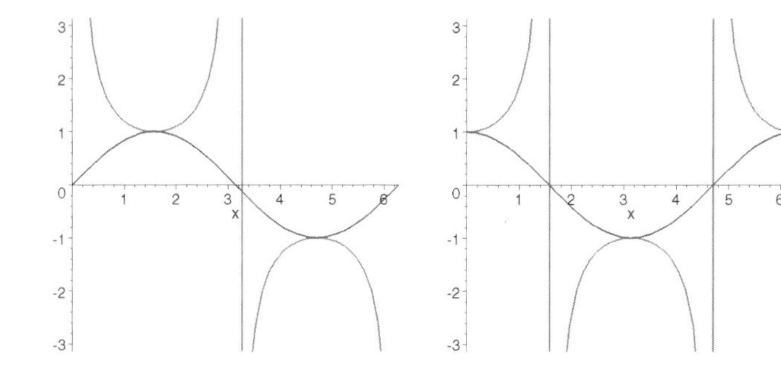
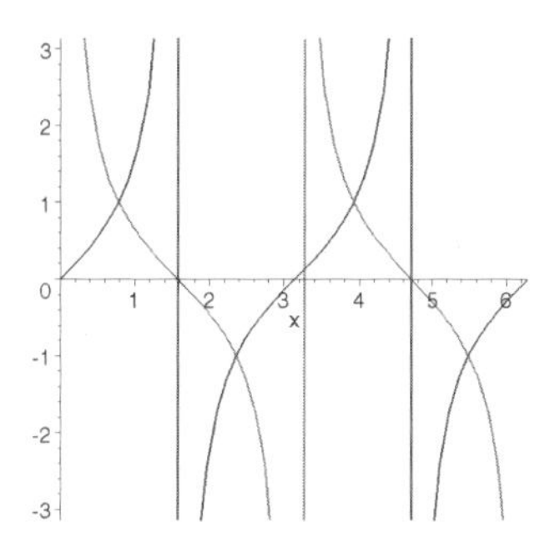

■

Remember that there are certain ambiguities in traditional mathematical notation. For example, the expression $\sin^2(\pi/6)$ is usually interpreted to mean "compute $\sin(\pi/6)$ and square the result." That is: $\sin^2(\pi/6) = (\sin(\pi/6))^2$. The symbol sin is *not* being squared; the number $\sin(\pi/6)$ *is* squared. With Maple, we must be especially careful and follow the standard order of operations exactly. We see that entering

```
> sin(Pi/6)^2;
```

$$\frac{1}{4}$$

computes $\sin^2(\pi/6) = (\sin(\pi/6))^2$, while entering

```
> sin^2(Pi/6);
```

$$\sin^2$$

raises the symbol sin to the power 2.

Commands involving the inverse trigonometric functions are similar to those demonstrated for the trigonometric functions.

EXAMPLE 11: Evaluate: (a) $\cos^{-1}(1/2)$; (b) $\sin^{-1}(-1)$; (c) $\tan^{-1}(1)$.

SOLUTION: The exact value of each of these three expressions is given next.

```
> arccos(1/2);
```

$$\frac{1}{3}\pi$$

```
> arcsin(-1);
```

$$-\frac{1}{2}\pi$$

```
> arctan(1);
```

$$\frac{1}{4}\pi$$

■

If exact values cannot be obtained, `evalf` must used to approximate the value. Notice that `(evalf@func)(x)` performs the same operation as `evalf(func(x))`. The symbol `@` represents the composition operator $(f \circ g)(x) = f(g(x))$.

EXAMPLE 12: Approximate (a) $\sin^{-1}(1/3)$; (b) $\cos^{-1}(2/3)$; (c) $\tan^{-1}(100)$.

SOLUTION: The numerical approximation of each of these expressions is obtained with `evalf`.

```
> evalf(arcsin(1/3));
                         .3398369094

> (evalf@arccos)(2/3);
                         .8410686705

> (evalf@arctan)(100);
                         1.560796660
```

■

Because the trigonometric functions and their inverse are inverse functions, their graphs are symmetric about the line $y = x$. In the following, we generate a graph of $y = \cos x$, $0 \le x \le \pi$, and name the result `p1`. The option `color=black` included in the plot specifies that the resulting graph is displayed in black; `linestyle=2` specifies that the graph be dotted. In `p2`, we generate a graph of $y = \cos^{-1} x$, $-1 \le x \le 1$, and in `p3` we generate a graph of gray, dashed graph of $y = x$. All three graphs are displayed together using `display`, which we load by entering `with(plots)` because the `display` command is not automatically loaded when we start a Maple session.

```
> with(plots):
  p1:=plot(cos(x),x=0..Pi,color=black,linestyle=2):
  p2:=plot(arccos(x),x=-1..1,color=black):
  p3:=plot(x,x=-1..Pi,color=gray,linestyle=3):
  display({p1,p2,p3});
```

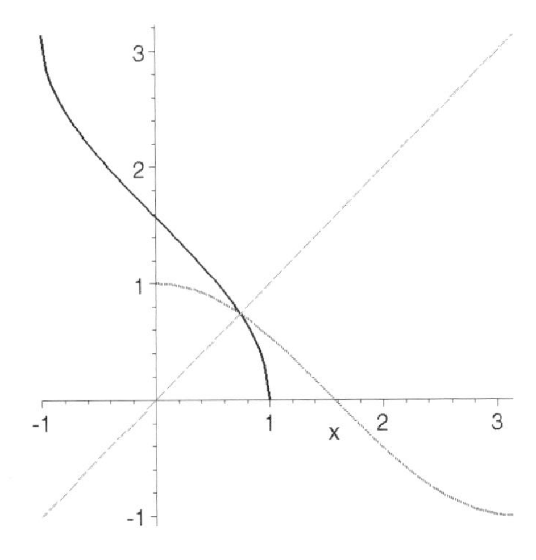

In the same way, we generate graphs of $y = \sin x$ and $y = \sin^{-1} x$.

```
> p1:=plot(sin(x),x=-Pi/2..Pi/2,color=black,linestyle=2):
  p2:=plot(arcsin(x),x=-1..1,color=black):
  p3:=plot(x,x=-Pi/2..Pi/2,color=gray,linestyle=3):
  display({p1,p2,p3});
```

Because the range of $y = \tan x$, $-\pi/2 < x < \pi/2$, is $-\infty < y < \infty$ and the domain of $y = \tan^{-1} x$ is $-\infty < x < \infty$ we are more careful in generating these graphs. Including `-10..10` in the plot command for `p1` specifies that the range of y-values displayed correspond to $-10 < y < 10$.

```
> p1:=plot(tan(x),x=-Pi/2..Pi/2,-10..10,color=black,
      linestyle=2):
  p2:=plot(arctan(x),x=-10..10,color=black):
  p3:=plot(x,x=-10..10,color=gray,linestyle=3):
  display({p1,p2,p3});
```

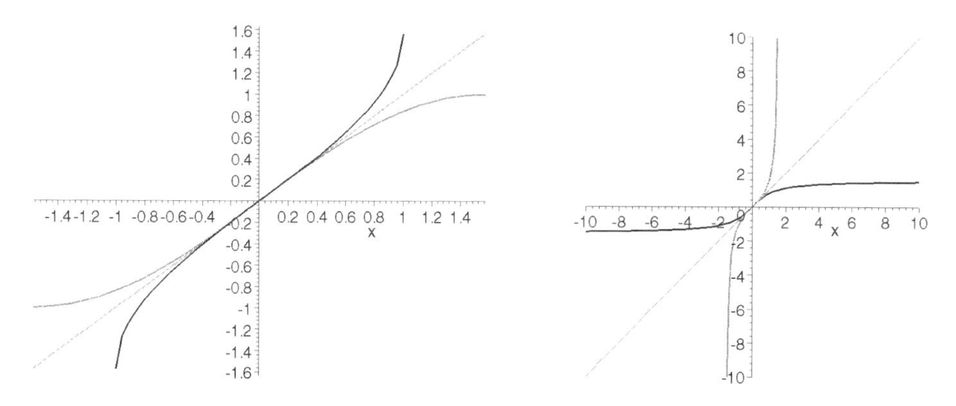

The hyperbolic trigonometric functions and their inverses are computed in the same way as those above. `evalf` is used to obtain an approximation, when necessary.

Hyperbolic Trigonometric Functions and Their Inverses

Function	Inverse
`sinh(x)`	`cosh(x)`
`tanh(x)`	`arcsinh(x)`
`arccosh(x)`	`arctanh(x)`

EXAMPLE 13: Compute $\sinh 5$, $\cosh(-5)$, and $\tanh 1$. Graph each of the functions $y = \sinh x$, $y = \cosh x$, and $y = \tanh x$ together with their inverse.

SOLUTION: We first compute the indicated values.

```
> evalf(sinh(5));
                                74.20321058
> evalf(cosh(-5));
                                74.20994852
> evalf(tanh(1));
                                .7615941560
```

Next, we use `plot` to graph each function together with its inverse.

```
> plot([sinh(x),arcsinh(x)],x=-10..10,-10..10,
   color=[black,gray]);
  plot([cosh(x),arccosh(x)],x=-2..5,-2..5,
   color=[black,gray]);
  plot([tanh(x),arctanh(x)],x=-2..2,-2..2,
   color=[black,gray]);
```

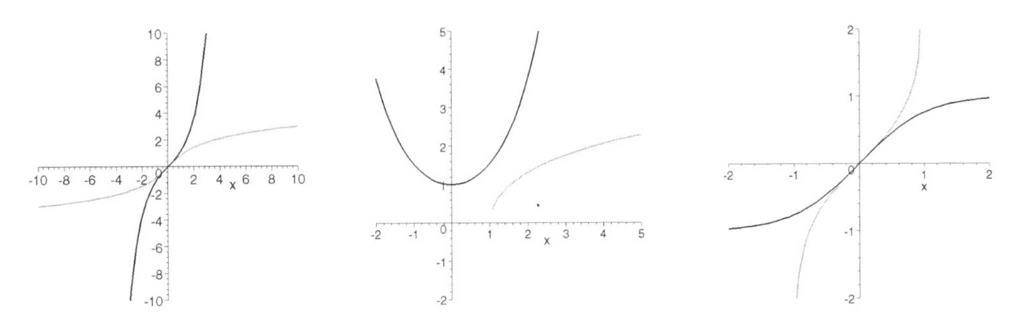

■

A Word of Caution

Maple is very sensitive about syntax. For example, if commands are misspelled or if a semi-colon or colon are omitted from the end of a command, errors result. For example, entering

```
> Sin(Pi/2);
```

$$\mathrm{Sin}\left(\frac{1}{2}\pi\right)$$

does not return 1 since `Sin` is used instead of `sin`.

Similarly, Maple returns a prompt when we enter

```
> cos(Pi)
>
```

but if we type a semicolon and then press **ENTER**, the desired result is obtained. If a colon is included at the end of a command, the resulting output is *not* displayed.

```
> cos(Pi)
> ;
```

$$-1$$

2.2 Expressions and Functions

Basic Algebraic Operations on Expressions

Maple performs standard algebraic operations on mathematical expressions. For example, the command `factor(expr)` factors the polynomial given in `expr`; `expand(expr)` distributes products over sums in expr; and `simplify(expr)` writes `expr` as a single fraction when `expr` represents the sum of fractions. Many built-in functions, like `expand` and `factor`, have corresponding *inert* functions, like `Expand` and `Factor`. Inert functions work as "place holders" for the corresponding noninert function: They remain unevaluated until commands, like `evalf`, or `value`, are used to evaluate them.

For basic information about any of these commands, enter `?command` as we do here for `factor`.

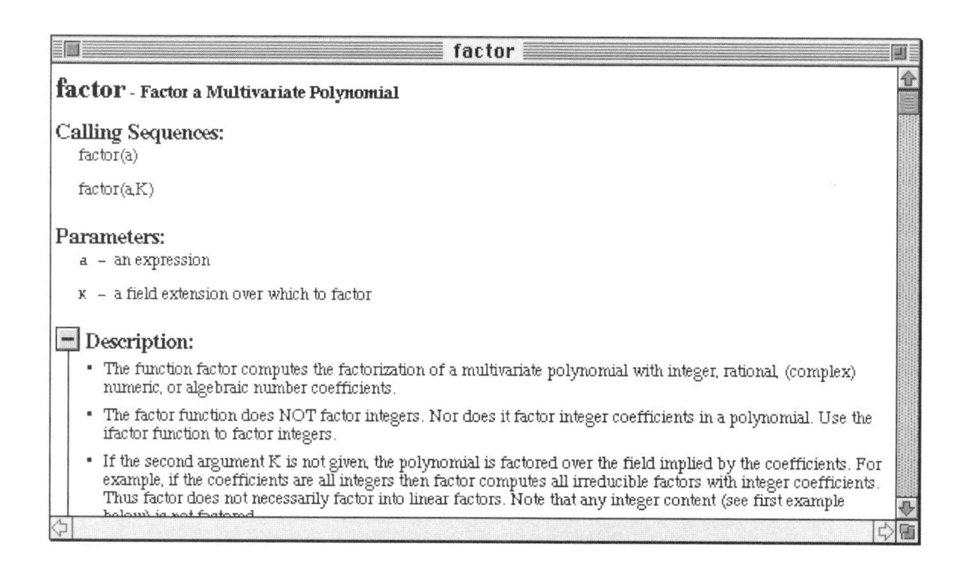

factor

factor - Factor a Multivariate Polynomial

Calling Sequences:

factor(a)

factor(a,K)

Parameters:

a - an expression

K - a field extension over which to factor

Description:

- The function factor computes the factorization of a multivariate polynomial with integer, rational, (complex) numeric, or algebraic number coefficients.
- The factor function does NOT factor integers. Nor does it factor integer coefficients in a polynomial. Use the ifactor function to factor integers.
- If the second argument K is not given, the polynomial is factored over the field implied by the coefficients. For example, if the coefficients are all integers then factor computes all irreducible factors with integer coefficients. Thus factor does not necessarily factor into linear factors. Note that any integer content (see first example

When entering expressions, be sure to include a * between variables to denote multiplication.

EXAMPLE 1: (a) Factor the polynomial $12x^2 + 27xy - 84y^2$; (b) expand the expression $(x + y)^2(3x - y)^3$; and (c) write the sum $\frac{2}{x^2} - \frac{x^2}{2}$ as a single fraction.

SOLUTION: The first command indicates that

$$12x^2 + 27xy - 84y^2 = 3(x + 4y)(4x - 7y),$$

the second computes the product $(x + y)^2(3x - y)^3$, and the third expresses $\frac{2}{x^2} - \frac{x^2}{2}$ as a single fraction.

```
> factor(12*x^2+27*x*y-84*y^2);
```

$$3(x + 4y)(4x - 7y)$$

```
> expand((x+y)^2*(3*x-y)^3);
```

$$27x^5 + 27x^4y - 18x^3y^2 - 10x^2y^3 + 7xy^4 - y^5$$

```
> simplify(2/x^2-x^2/2);
```

$$-\frac{1}{2}\frac{-4 + x^4}{x^2}$$

■

The command convert(fraction,parfrac,x) computes the partial fraction decomposition of the rational function fraction of the variable x. Similarly, normal(fraction) factors the numerator and denominator of the rational function fraction and reduces fraction to lowest terms.

EXAMPLE 2: Determine the partial fraction decomposition of the rational function $\frac{1}{(x - 3)(x - 1)}$. Simplify the expression $\frac{x^2 - 1}{x^2 - 2x + 1}$.

SOLUTION: convert is used to show that $\frac{1}{(x-3)(x-1)} = \frac{1}{2(x-3)} - \frac{1}{2(x-1)}$. Then, normal is used to find that $\frac{x^2-1}{x^2-2x+1} = \frac{x+1}{x-1}$.

```
> convert(1/((x-3)*(x-1)),parfrac,x);
```

$$\frac{1}{2}\frac{1}{x-3} - \frac{1}{2}\frac{1}{x-1}$$

```
> normal((x^2-1)/(x^2-2*x+1));
```

$$\frac{x+1}{x-1}$$

Other Maple functions for manipulation fractions include numer(fraction), which yields the numerator of fraction, and denom(fraction), which yields the denominator of fraction.

```
> ?numer
```

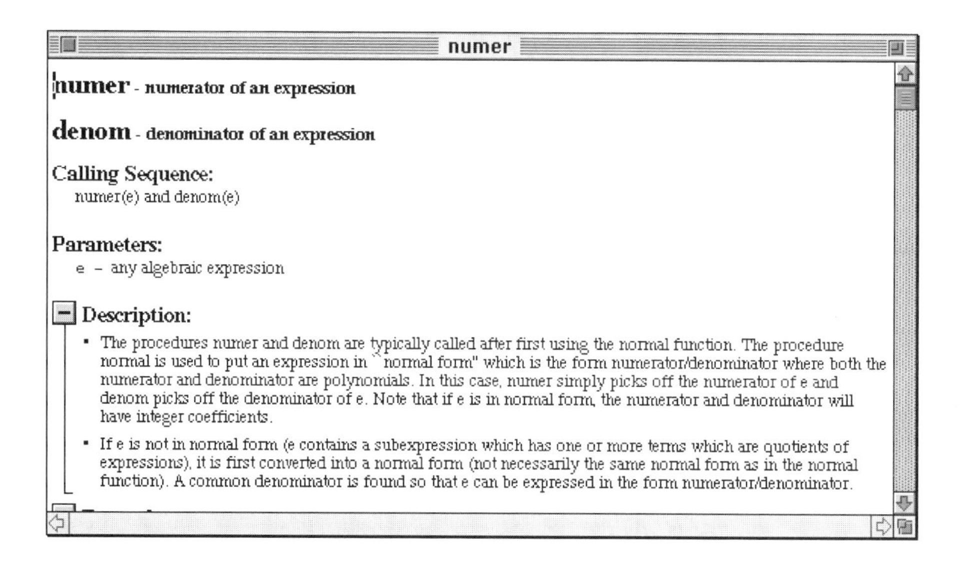
numer

numer - numerator of an expression

denom - denominator of an expression

Calling Sequence:
numer(e) and denom(e)

Parameters:
e - any algebraic expression

Description:

- The procedures numer and denom are typically called after first using the normal function. The procedure normal is used to put an expression in "normal form" which is the form numerator/denominator where both the numerator and denominator are polynomials. In this case, numer simply picks off the numerator of e and denom picks off the denominator of e. Note that if e is in normal form, the numerator and denominator will have integer coefficients.
- If e is not in normal form (e contains a subexpression which has one or more terms which are quotients of expressions), it is first converted into a normal form (not necessarily the same normal form as in the normal function). A common denominator is found so that e can be expressed in the form numerator/denominator.

EXAMPLE 3: Given the rational expression $\frac{x^3+2x^2-x-2}{x^3+x^2-4x-4}$, (a) factor both the numerator and denominator; (b) reduce $\frac{x^3+2x^2-x-2}{x^3+x^2-4x-4}$ to lowest terms; and (c) find the partial fraction decomposition of $\frac{x^3+2x^2-x-2}{x^3+x^2-4x-4}$.

SOLUTION: The numerator of a fraction is extracted with `numer`. We then use `factor` together with `%`, which is used to refer to the most recent output, to factor the result of executing the `numer` command.

```
> numer((x^3+2*x^2-x-2)/(x^3+x^2-4*x-4));
```

$$x^2 + 2x^2 - x - 2$$

```
> factor(%);
```

$$(x-1)(x+2)(x+1)$$

Similarly, we use `denom` to extract the denominator of the fraction. Again, `factor` together with `%` is used to factor the previous result, which corresponds to the denominator of the fraction.

```
> denom((x^3+2*x^2-x-2)/(x^3+x^2-4*x-4));
```

$$x^3 + x^2 - 4x - 4$$

```
> factor(%);
```

$$(x-2)(x+2)(x+1)$$

`normal` is used to reduce the fraction to lowest terms.

```
> normal((x^3+2*x^2-x-2)/(x^3+x^2-4*x-4));
```

$$\frac{x-1}{x-2}$$

Finally, convert together with parfrac is used tofind the partial fraction decomposition.

```
> convert((x^3+2*x^2-x-2)/(x^3+x^2-4*x-4),parfrac,x);
```

$$1+\frac{1}{x-2}$$

■

Naming and Evaluating Expressions

In Maple, mathematical objects can be named. Naming objects is convenient: We can avoid typing the same mathematical expression repeatedly (as we did in Example 3), and named expressions can be referenced throughout a notebook or Maple session. Every Maple object can be named; even graphics and functions can be named with Maple.

An expression is named by using a colon (:) followed by a single equals sign (=). For example, entering

```
> graph=plot(sin(x),x=0..2*Pi):
```

names the graph of $y = \sin x$ on the interval $[0,2\,\pi]$. (Note that because we include a colon at the end of this command, Maple does not return any output.)

Expressions can be evaluated with the command subs(x=x0,expr), which substitutes the value of x0 for x in the expression expr.

EXAMPLE 4: Evaluate $\frac{x^3+2x^2-x-2}{x^3+x^2-4x-4}$ if $x = 4$ and if $x = -3$.

SOLUTION: To avoid retyping $\frac{x^3+2x^2-x-2}{x^3+x^2-4x-4}$, we define f to be $\frac{x^3+2x^2-x-2}{x^3+x^2-4x-4}$. (Of course, you can simply copy and paste the expression from Example 3 if you do not want to name it or retype it.)

```
> f:=(x^3+2*x^2-x-2)/(x^3+x^2-4*x-4);
```

$$f := \frac{x^3+2\,x^2-x-2}{x^3+x^2-4\,x-4}$$

subs is used to evaluate $f(x)$ if $x = 4$ and then if $x = -3$.

```
> subs(x=4,f);
```

$$\frac{3}{2}$$

```
> subs(x=-3,f);
```

$$\frac{4}{5}$$

The eval command is closely related to the subs command. Entering

```
> eval(f,x=1/2);
```

$$\frac{1}{3}$$

■

Defining and Evaluating Functions

It is important to remember that functions, expressions, and graphics can be named anything that is not the name of a built-in Maple function or command. Since definitions of functions are frequently modified, we introduce the command that is used to clear a definition. This command `f:='f'` clears all definitions of `f`. Consequently, we are certain to avoid any ambiguity when we create a new definition. The function `f(x)=expr` can be defined with Maple as `f:=x->expr`. Hence, a function such as $f(x) = x^4$ is defined with

```
f:=x->x^4.
```

Similarly, $h(x) = x - \cos x$ is defined by

```
h:=x->x-cos(x).
```

An alternative method is to define functions as procedures with `proc`. `f:=proc(x) x^4 end` and `h:=proc(x) x-cos(x) end`, which, respectively, produce the same results as above. After a function is defined, then it can be evaluated easily. For example, the function `f` is evaluated at `x=x0` with the command `f(x0)`.

EXAMPLE 5: Define the functions $f(x) = x^2$, $g(x) = \sqrt{x}$, and $h(x) = x + \sin x$. Evaluate $f(2)$, $g(4)$, and $h(\pi/2)$.

SOLUTION: The functions *f*, *g*, and *h* are defined after we have cleared all function names on the first line. A colon (:) causes the output resulting from a command to be suppressed. Hence, the definition of h is not displayed.

```
> f:='f':g:='g':h:='h':
  f:=x->x^2;
```

$$f := x \to x^2$$

```
> g:=x->sqrt(x);
```

$$g := \text{sqrt}$$

```
> h:=x->x+sin(x):
```

The functions are evaluated at the indicated values to yield 4, 2, and 1 + π/2, respectively.

```
> f(2);
```

$$4$$

```
> g(4);
```

$$2$$

```
> h(Pi/2);
```

$$\frac{1}{2}\pi + 1$$

■

Maple can symbolically evaluate and manipulate functions as well.

EXAMPLE 6: Let $f(t) = 4 - t^2$. (a) Calculate $f(a - b^2)$. (b) Calculate and expand $f(a - b^2)$. (c) Compute $\frac{f(x+h) - f(x)}{h}$. (d) Compute and simplify $\frac{f(x+h) - f(x)}{h}$.

SOLUTION: First, we define *f* after clearing any previous definitions of *f*. Note that if we include a colon at the end of a command, the resulting output is not displayed.

```
> f:='f':
  f:=t->4-t^2:
```

We evaluate functions when the argument consists of symbols other than numbers in the same way as we evaluate functions when the argument consists of numbers in the functions' domain.

Entering

```
> f(a-b^2);
```

$$4-(a-b^2)^2$$

calculates $f(a-b^2)$; entering

```
> expand(f(a-b^2));
```

$$4-a^2+2\,a\,b^2-b^4$$

computes and expands $f(a-b^2)$; entering

```
> (f(x+h)-f(x))/h;
```

$$\frac{-(x+h)^2+x^2}{h}$$

computes, but does not simplify, $\frac{f(x+h)-f(x)}{h}$; and entering

```
> simplify((f(x+h)-f(x))/h);
```

$$-2\,x-h$$

computes and simplifies $\frac{f(x+h)-f(x)}{h}$.

■

Many different types of functions can be defined using Maple, including piecewise-defined functions, functions of more than one variable, and vector-valued functions. We will discuss additional ways of defining functions as needed, throughout the text. However, Maple's extensive programming language allows a great deal of flexibility in defining functions, many of which are beyond the scope of this text.

EXAMPLE 7: Define the function $f(x,y) = 1 - \sin(x^2 + y^2)$, and evaluate $f(1,2)$, $f(2\sqrt{\pi}, 3\sqrt{\pi}/2)$, $f(0,a)$, and $f(a^2 - b^2, b^2 - a^2)$.

SOLUTION: First, the function f is cleared and defined. The exact value is then returned for each of the following expressions.

```
> f:='f':
  f:=(x,y)->1-sin(x^2+y^2);
```

$$f := (x, y) \rightarrow 1 - \sin(x^2 + y^2)$$

```
> f(1,2);
```

$$1 - \sin(5)$$

```
> f(2*sqrt(Pi),3*sqrt(Pi)/2);
```

$$1 - \frac{1}{2}\sqrt{2}$$

```
> f(0,a);
```

$$1 - \sin(a^2)$$

```
> f(a^2-b^2,b^2-a^2);
```

$$1 - \sin((a^2 - b^2)^2 + (b^2 - a^2)^2)$$

■

Vector-valued functions can be defined with Maple as well. For example, $f(t) = \langle x(t), y(t) \rangle$ is defined with `f:='f':f:=x->[x(t),y(t)]`. Functions of higher dimension are defined in a similar manner.

EXAMPLE 8: Define **g** to be the vector-valued function $\mathbf{g}(t) = \langle t^2, 1 - t^2 \rangle$. Calculate $\mathbf{g}(1)$ and $\mathbf{g}(\sin b)$.

SOLUTION: After **g** is defined, the exact values of **g**(1) and **g**(sin*b*) are computed.

```
> g:='g':
  g:=t->[t^2,1-t^2];
```

$$g := t \to [t^2, 1 - t^2]$$

```
> g(1);
```

$$[1, 0]$$

```
> g(sin(b));
```

$$[\sin(b)^2, 1 - \sin(b)^2]$$

■

Composition of Functions

Maple can easily perform the calculation $(f \circ g)(x) = f(g(x))$ by entering `f(g(x))`. However, when several different functions are composed or a function is repeatedly composed with itself, the commands `@` and `@@` are provided:

(1) `(f1@f2@f3@...@fn)(x)` computes the composition $(f_1 \circ f_2 \circ \ldots \circ f_n)(x) = f_1(f_2 \ldots (f_n(x)))$.

(2) `(f@@n)(x)` computes the composition $\underbrace{(f \circ f \circ \ldots \circ f)}_{n \text{ times}}(x) = \underbrace{f(f \ldots f(x))}_{n \text{ times}}$, where f is a function, n is a positive integer, and x is an expression.

EXAMPLE 9: Let $f(x) = x^2 + x$, $g(x) = x^3 + 1$, and $k(x) = \sin x + \cos x$. Compute (a) $(f \circ g)(x) = f(g(x))$; (b) $(g \circ f)(x-1) = g(f(x-1))$; (c) $(f \circ k)(\pi/3) = f(k(\pi/3))$; (d) $f(\sin x)$; (e) $(f \circ k)(x) = f(k(x))$; and (f) $f(\sin(x + iy))$.

SOLUTION: We begin by clearing all prior definitions of *f*, *g*, and *k*, if any, and then defining *f*, *g*, and *k*. Generally, Maple displays output for *each* command as it is generated unless a colon (:) is included at the end of the command. Thus, in this case, the formulas for `f(x)`, `g(x)`, and `h(x)` are not displayed because a colon is placed at the end of each command.

```
> f:='f':h:='h':
  f:=x->x^2+x:
```

```
g:=x->x^3+1:
k:=x->sin(x)+cos(x):
```

Notice that both `f(g(x))` and `(f@g)(x)` compute $(f \circ g)(x) = f(g(x))$.

```
> f(g(x));
```

$$(x^3+1)^2+x^3+1$$

```
> (f@g)(x);
```

$$(x^3+1)^2+x^3+1$$

Generally, Maple does not automatically expand and/or simplify results.

```
> (g@f)(x-1);
```

$$((x-1)^2+x-1)^3+1$$

```
> (f@k)(Pi/3);
```

$$\left(\frac{1}{2}\sqrt{3}+\frac{1}{2}\right)^2+\frac{1}{2}\sqrt{3}+\frac{1}{2}$$

Entering `f(sin(x))` also computes $f(\sin x)$.

```
> (f@sin)(x);
```

$$\sin(x)^2+\sin(x)$$

When we use `simplify`, elementary trigonometric identities are applied to simplify the result.

```
> (f@k)(x);
```

$$(\sin(x)+\cos(x))^2+\sin(x)+\cos(x)$$

```
> simplify((f@k)(x));
```

$$2\sin(x)\cos(x)+\sin(x)+\cos(x)+1$$

On the other hand, when we use `expand`, expressions are multiplied out but no trigonometric identities are applied to simplify the result.

```
> expand((f@k)(x));
```

$$\sin(x)^2 + 2\sin(x)\cos(x) + \cos(x)^2 + \sin(x) + \cos(x)$$

When we use `combine`, trigonometric identities are applied to simplify the result. In addition, products of trigonometric functions are converted to single functions.

```
> combine((f@k)(x));
```

$$1 + \sin(2x) + \sin(x) + \cos(x)$$

The imaginary number $i = \sqrt{-1}$ is denoted by `I`.

```
> (f@sin)(x+I*y);
```

$$\sin(x + I\,y)^2 + \sin(x + I\,y)$$

■

EXAMPLE 10: Let $f(x) = x^2 + x$. Compute: (a) $(f \circ f \circ f)(x) = f(f(f(x)))$, (b) $t(x) = \sin(\sin(\sin(\sin(\sin(\sin x)))))$, and (c) $g(x) = \underbrace{\sin \circ \sin \circ \ldots \sin}_{50\text{ times}}(x)$.

SOLUTION: After clearing all prior definitions of f, we define f and then use `@@` to compose f with itself 3 times.

```
> f:='f':
> f:=x->x^2+x:
> (f@@3)(x);
```

$$((x^2 + x)^2 + x^2 + x)^2 + (x^2 + x)^2 + x^2 + x$$

We use `expand` to simplify the result. Remember that the `%` operator is used to refer to the most recent result. (Note that using `simplify` returns the same result.)

```
> expand(%);
```

$$x^8 + 4x^7 + 8x^6 + 10x^5 + 9x^4 + 6x^3 + 3x^2 + x$$

Similarly, we use `@@` to compose $\sin x$ with itself 6 and then 50 times. The result is displayed in a more traditional form if we use `expand`.

```
> (sin@@6)(x);
```

$$(\sin^{(6)})(x)$$

```
> expand(%);
```

$$\sin(\sin(\sin(\sin(\sin(\sin(x))))))$$

```
> (sin@@50)(x);
```

$$(\sin^{(50)})(x)$$

We use `plot` to compare the graphs of these two functions to the graph of $y = \sin x$. Note that $y = \sin x$ and $y = \sin^{(50)} x$ are in black; $y = \sin^{(6)} x$ is in gray.

```
> plot({sin(x),(sin@@6)(x),(sin@@50)(x)},
   x=0..2*Pi,color=[black,gray,black]);
```

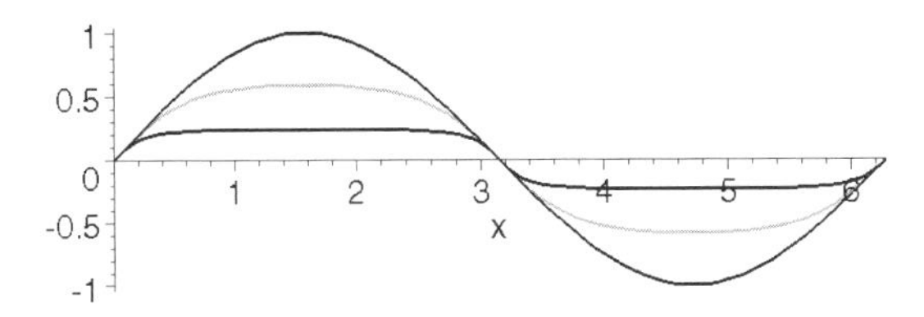

■

2.3 Graphing Functions and Expressions

One of the best features of Maple is its graphics capabilities. In this section, we discuss methods of graphing functions, expressions, and equations, and several of the options available to help graph functions. We will use the techniques discussed throughout the remainder of the text. Other graphics topics are discussed throughout the text as well as in Chapter 8.

Graphing Functions of a Single Variable

The command used to graph real-valued functions of a single variable is `plot`. The command

```
plot(f(x),x=a..b).
```

graphs the function $f(x)$ on the interval $[a, b]$.

Maple returns information about the basic syntax of the `plot` command with `?plot`.

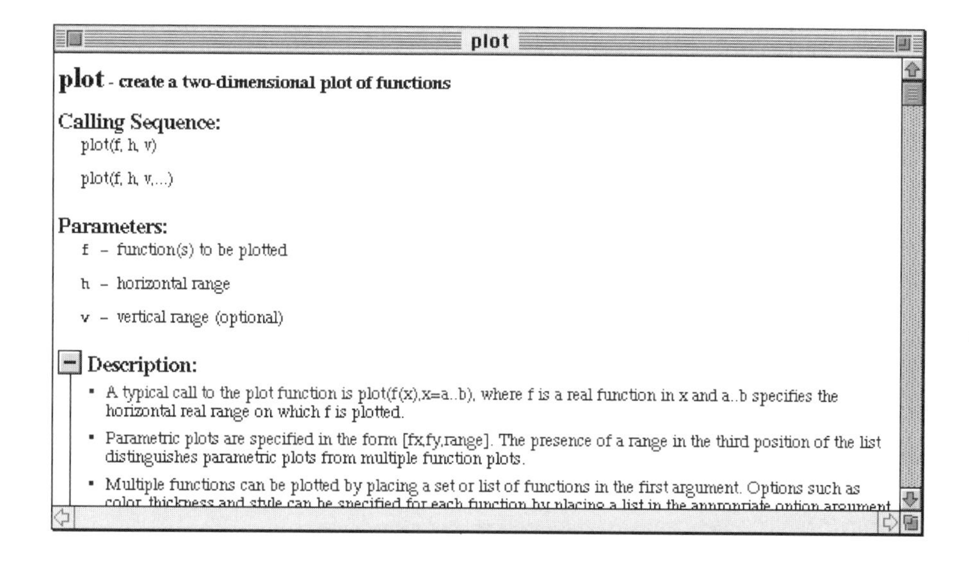

plot

plot - create a two-dimensional plot of functions

Calling Sequence:

plot(f, h, v)

plot(f, h, v,...)

Parameters:

f - function(s) to be plotted

h - horizontal range

v - vertical range (optional)

Description:

- A typical call to the plot function is plot(f(x),x=a..b), where f is a real function in x and a..b specifies the horizontal real range on which f is plotted.
- Parametric plots are specified in the form [fx,fy,range]. The presence of a range in the third position of the list distinguishes parametric plots from multiple function plots.
- Multiple functions can be plotted by placing a set or list of functions in the first argument. Options such as

---

EXAMPLE 1: Let $f(x) = 4x^3 + 6x^2 - 9x + 2$. Graph $f(x)$ on the interval $[-3, 2]$.

SOLUTION: After clearing all prior definitions of f, if any, we define f and then use `plot` to graph f on the interval $[-3, 2]$.

```
> f:='f':
  f:=4*x^3+6*x^2-9*x+2:
  plot(f(x),x=-3..2);
```

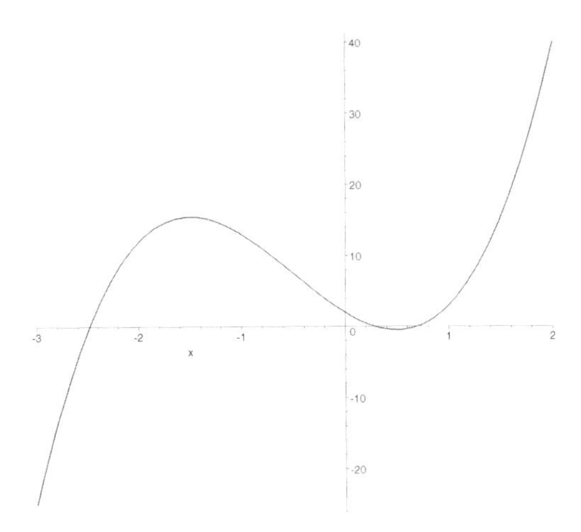

Notice that the graph is not drawn to scale. Maple "chooses" the scale so that the graph appears "reasonable" to us. Maple redraws the graph to scale if you select **Projection** from the menu, followed by **Constrained**.

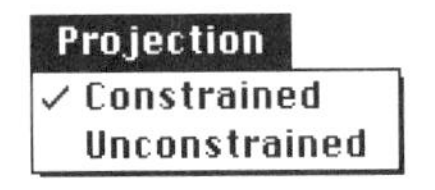

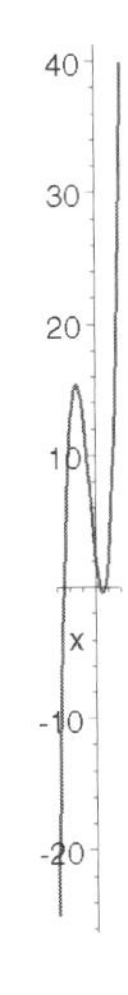

■

To plot the graph of $f(x)$ in different colors, the command is

```
plot(f(x),x=a..b,color=n)
```

where `n` is one of the predefined colors aquamarine, black, blue, navy, coral, cyan, brown, gold, green, gray, khaki, magenta, maroon, orange, pink, plum, red, sienna, tan, turquoise, violet, wheat, white, or yellow. In addition, Maple allows us to define our own colors.

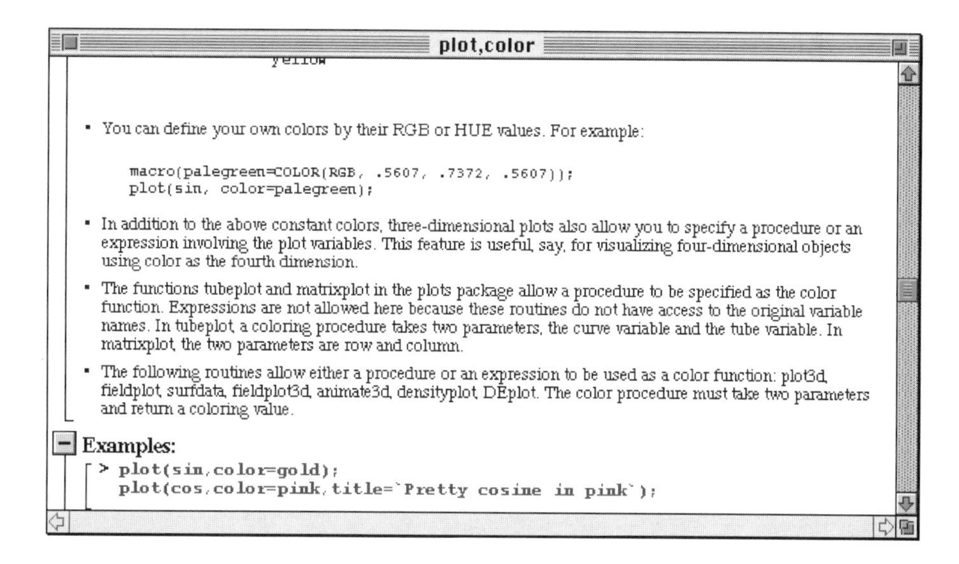

plot,color

yellow

- You can define your own colors by their RGB or HUE values. For example:

```
macro(palegreen=COLOR(RGB, .5607, .7372, .5607));
plot(sin, color=palegreen);
```

- In addition to the above constant colors, three-dimensional plots also allow you to specify a procedure or an expression involving the plot variables. This feature is useful, say, for visualizing four-dimensional objects using color as the fourth dimension.
- The functions tubeplot and matrixplot in the plots package allow a procedure to be specified as the color function. Expressions are not allowed here because these routines do not have access to the original variable names. In tubeplot, a coloring procedure takes two parameters, the curve variable and the tube variable. In matrixplot, the two parameters are row and column.
- The following routines allow either a procedure or an expression to be used as a color function: plot3d, fieldplot, surfdata, fieldplot3d, animate3d, densityplot, DEplot. The color procedure must take two parameters and return a coloring value.

Examples:

```
> plot(sin,color=gold);
  plot(cos,color=pink,title=`Pretty cosine in pink`);
```

A dashed graph can be generated by using the option `linestyle=n`: if `n=1`, the line is solid.; `n=2`, the style is dot; `n=3` gives a dashed graph; and `n=4` gives dash-dot.

One of the more useful `plot` command options is that of `title` which allows for a label to be placed on the graph. This option is used in the following way:

```
plot(f(x),x=a..b,title=`plotlabel`)
```

where `plotlabel` represents the expression to be placed at the top of the graph.
Additional `plot` options include:

1. `coords=system`

   This uses the coordinate system indicated by the setting of `system` to produce the graph. The default setting of this option is `cartesian`. For polar plots, the command is entered in the form:

   ```
   plot([r(theta),theta,theta=t0..t1],coords=polar)
   ```

   or

   ```
   plot(f(theta),theta=t0..t1,coords=polar)
   ```

   in which the function $r=f(\theta)$ is assumed.

2. `numpoints=n`

   The value of `n` specifies the number of sample points to be used to graph the function. The default setting of this option is 49. An adaptive method is used to select sample points after an initial selection of points is made uniformly over the domain indicated in the plot command. Other points are selected so that curves do not resemble lines when they are plotted.

3. `resolution=n`

   This setting indicates the horizontal resolution of the display in pixels. The default setting is 200. The value of `n` specifies when the adaptive method used for plotting is terminated. Hence, higher values of `n` lead to more sample points which, in turn, leads to a smoother graph. This option is particularly useful for functions that are not smooth.

4. `xtickmarks=n`

   Th e value of `n` indicates the number of divisions to place along the horizontal axis.

5. `ytickmarks=n`

   The value of `n` indicates the number of divisions to place along the vertical axis.

6. `view=[xmin..xmax,ymin..ymax]`

   specifies that the axes displayed correspond to the rectangle [`xmin`,`xmax`] × [`ymin`,`ymax`].

   A complete description of all options is obtained by entering `?plot[options]`.

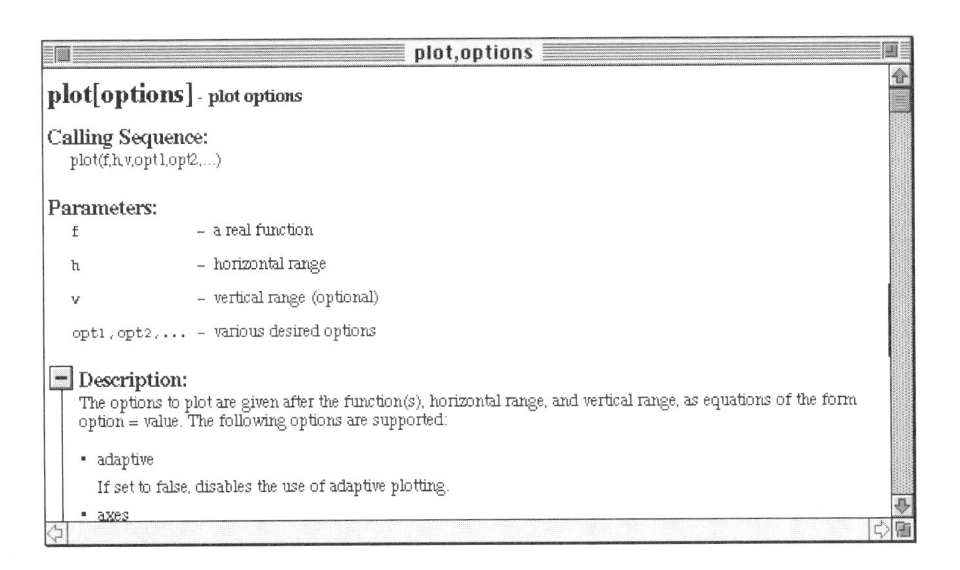

Global options for two-dimensional graphics can be set using the `setoptions` command, which is contained in the `plots` package. For example, entering

```
with(plots);
```

loads the `plots` package, and then entering

```
setoptions(axes=BOXED);
```

instructs Maple to place a box around all subsequently generated two-dimensional graphics during the Maple session.

Graphing Several Functions

The `plot` command can be used to graph several functions simultaneously. To display the graphs of the functions $f(x)$, $g(x)$, and $h(x)$ on the interval $[a,b]$ on the same axes, enter commands of the form `plot({f(x),g(x),h(x)},x=a..b,options)` or `plot([f(x), g(x),h(x)],x=a..b,options)`. This command can be generalized to include more than three functions.

EXAMPLE 2: Graph the functions $\sin x$, $\sin 2x$, and $\sin(x/2)$ on the interval $[0,4\pi]$. Display all three graphs on the same axes.

SOLUTION: With `macro`, we define new colors that correspond to shades of gray named `gray3` and `gray5`. Then, we use `plot` to graph the functions on the interval $[0,4\pi]$. The option `color=[black,gray3,gray5]` indicates that $\sin x$ is to be graphed in black, $\sin 2x$ in `gray3`, and $\sin(x/2)$ in `gray5`; `linestyle=[1,3,4]` indicates that $\sin x$ is solid, $\sin 2x$ dashed, and $\sin(x/2)$ dashed-dot; `ytickmarks=[-1,1]` indicates that tick marks are to be placed at $y = 1$ and $y = -1$; and `view=[0..4*Pi,-2*Pi..2*Pi]` indicates that the viewing window displayed correspond to $0 \le x \le 4\pi$ and $-2\pi \le y \le 2\pi$.

```
> macro(gray3=COLOR(RGB, .3, .3, .3));
  macro(gray5=COLOR(RGB, .5, .5, .5));

> plot({sin(x),sin(2*x),sin(x/2)},x=0..4*Pi,
   color=[black,gray3,gray5],linestyle=[1,3,4],
   ytickmarks=[-1,1],view=[0..4*Pi,-2*Pi..2*Pi]);
```

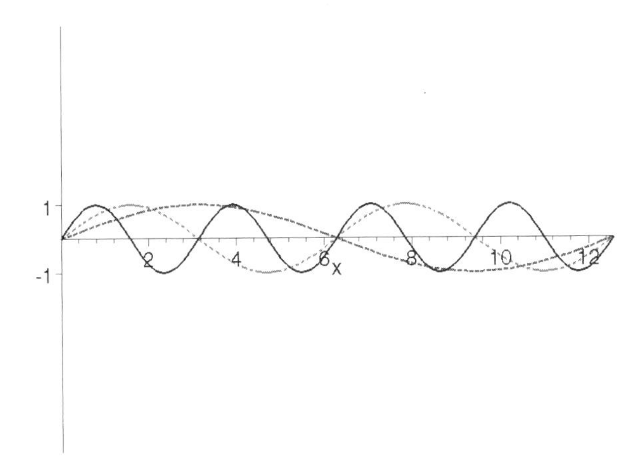

■

Graphs of functions, like expressions, can be named. This is particularly useful when one needs to refer to the graph of particular functions repeatedly or to display several graphs on the same axes. The command that is used to display several graphs on the same axes is `display`, which is located in the `plots` package. Hence, the command `with(plots)` must be entered before the `display` command is used, or the `display` command must be entered in the form `plots[display]`.

EXAMPLE 3: Graph the circle $x^2 - 4x + y^2 - 2y = 4$.

SOLUTION: We find the center and radius of the circle $x^2 - 4x + y^2 - 2y = 4$ by completing the square and obtain the equation $(x - 2)^2 + (y - 1)^2 = 3^2$. Thus, the center is (2, 1) and the radius is 3. Solving this equation for y results in

$$y = 1 \mp \sqrt{9-(x-2)^2}.$$

Thus, a function describing the top half of the circle is given by $y_1(x) = 1 + \sqrt{9-(x-2)^2}$, while a function describing the bottom half is given by $y_2(x) = 1 - \sqrt{9-(x-2)^2}$. We define `y1` and `y2` to be the functions describing the top and bottom half of the circle, respectively. We then use `plot` to graph `y1` and `y2` on the interval [–1,5], naming the resulting graphs `p1` and `p2`, respectively. Neither graph is displayed because we include a colon at the end of each `plot` command.

```
> y1:=1+sqrt(9-(x-2)^2):
  y2:=1-sqrt(9-(x-2)^2):
> p1:=plot(y1,x=-1..5):
  p2:=plot(y2,x=-1..5):
```

In order to use `display`, we load the `plots` package. Note that `display` appears in the list of commands located in this package.

```
> with(plots);
```

[*animate, animate3d, animatecurve, changecoords, complexplot, complexplot3d, conformal, contourplot, contourplot3d, coordplot, coordplot3d, cylinderplot, densityplot, display, display3d, fieldplot, fieldplot3d, gradplot, gradplot3d, implicitplot, implicitplot3d, inequal, listcontplot, listcontplot3d, listdensityplot, listplot, listplot3d, loglogplot, logplot, matrixplot, odeplot, pareto, pointplot, pointplot3d, polarplot, polygonplot, polygonplot3d, polyhedra_supported, polyhedraplot, replot, rootlocus, semilogplot, setoptions, setoptions3d, spacecurve, sparsematrixplot, sphereplot, surfdata, textplot, textplot3d, tubeplot*]

We then use `display` **to show both graphs together.**

```
> display({p1,p2});
```

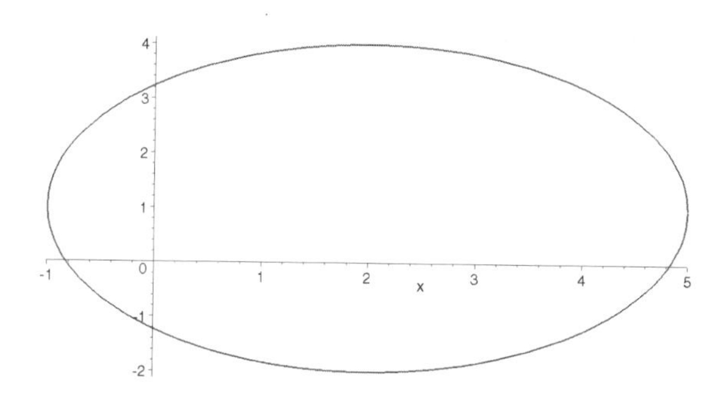

When we go to the menu and change **Projection** to **Constrained**, the resulting displayed graphics object looks like a circle.

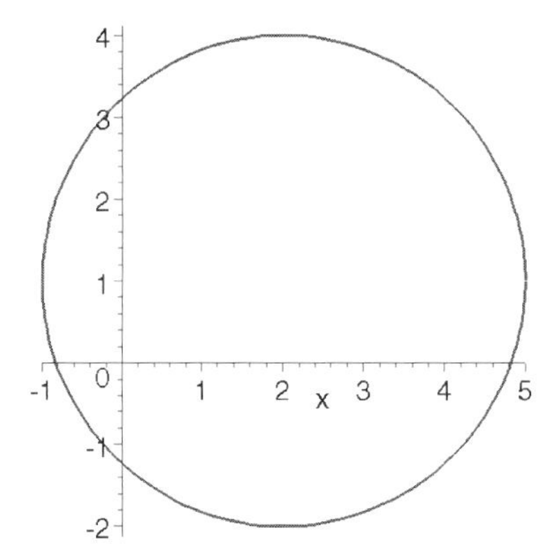

■

Piecewise-Defined Functions

Piecewise-defined functions may also be defined with Maple using the `piecewise` command.

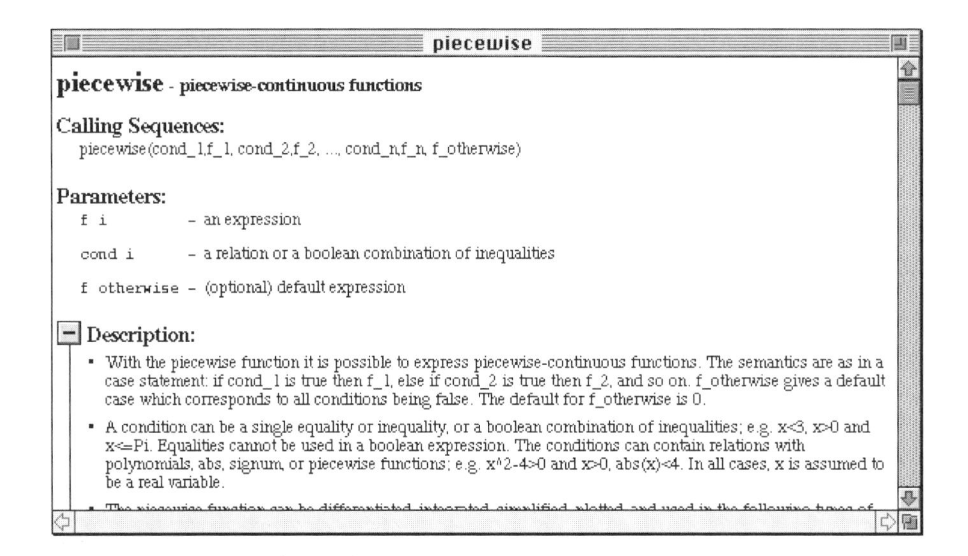
piecewise

piecewise - piecewise-continuous functions

Calling Sequences:

piecewise(cond_1,f_1, cond_2,f_2, ..., cond_n,f_n, f_otherwise)

Parameters:

f i - an expression

cond i - a relation or a boolean combination of inequalities

f otherwise - (optional) default expression

Description:

- With the piecewise function it is possible to express piecewise-continuous functions. The semantics are as in a case statement: if cond_1 is true then f_1, else if cond_2 is true then f_2, and so on. f_otherwise gives a default case which corresponds to all conditions being false. The default for f_otherwise is 0.
- A condition can be a single equality or inequality, or a boolean combination of inequalities; e.g. x<3, x>0 and x<=Pi. Equalities cannot be used in a boolean expression. The conditions can contain relations with polynomials, abs, signum, or piecewise functions; e.g. x^2-4>0 and x>0, abs(x)<4. In all cases, x is assumed to be a real variable.

EXAMPLE 4: If $f(x) = \begin{cases} x^2+1, x \geq 0 \\ -x-1, x<0 \end{cases}$, graph f on the interval $[-3,3]$. Calculate $f(7)$, $f(-10)$, and $f(a)$.

SOLUTION: We define f using `piecewise` and then graph f on the interval $[-3,3]$ with `plot`.

```
> f:=x->piecewise(x>=0,x^2+1,x<0,-x-1);
```

$$f := x \rightarrow \text{piecewise}(\,0 \le x, x^2 + 1, x < 0, -x - 1\,)$$

```
> plot(f(x),x=-3..3);
```

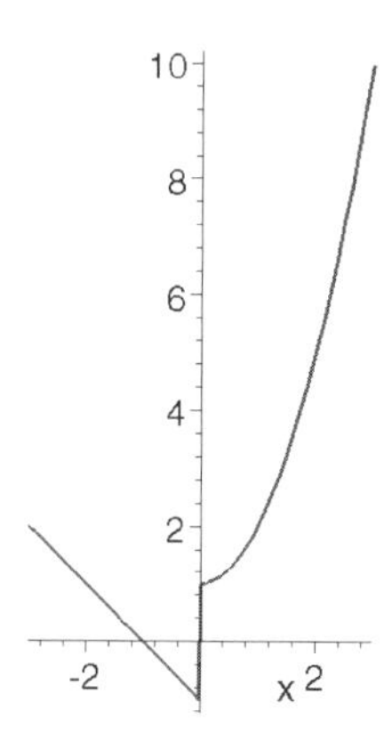

We then compute $f(7)$ and $f(-10)$.

```
> f(7);
```

$$50$$

```
> f(-10);
```

$$9$$

However, Maple does not return a value for $f(a)$ because Maple does not know which condition a satisfies.

```
> f(a);
```

$$\begin{cases} a^2 + 1 & 0 \le a \\ -a - 1 & a < 0 \end{cases}$$

However, Maple's `assume` facilities lets us make assumptions regarding variables. In this case, we use `assume` to instruct Maple to assume that a is greater than 0. With this assumption, Maple returns $f(a) = a^2 + 1$. (Note that Maple uses a ~ to indicate that assumptions have been made about a variable.)

```
> assume(a>0);
```

```
f(a);
```

$$a\sim^2 + 1$$

■

An alternative method to defining piecewise-defined functions is to use the `Heaviside` function or to define them as procedures using `proc`. If you use `proc`, the definition of these segments involves an if-then-else statement that is entered in the form `if (expr) then (statseq1) else (statseq2) fi` where `expr` represents the conditional expression, and `statseq1` and `statseq2` represent sequences of statements.

Thus, entering

```
> f:='f':
  f:=proc(x) if x>=0 then x^2+1 else -x-1 fi end:
```

defines $f(x) = \begin{cases} x^2+1, x \ge 0 \\ -x-1, x < 0 \end{cases}$. Because we have defined f as a procedure, we must delay the evaluation of f using

```
'f(x)'
```

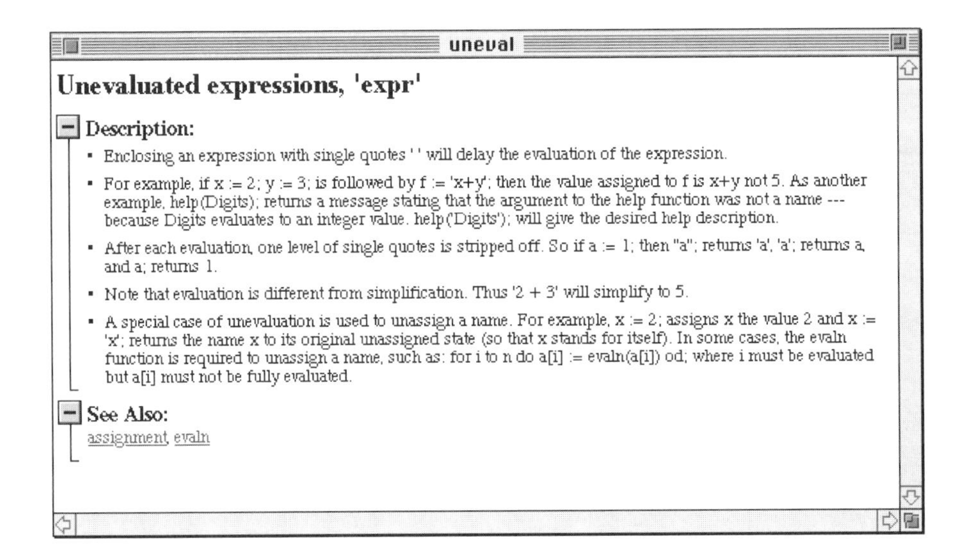
uneval

Unevaluated expressions, 'expr'

Description:

- Enclosing an expression with single quotes '' will delay the evaluation of the expression.
- For example, if x := 2; y := 3; is followed by f := 'x+y'; then the value assigned to f is x+y not 5. As another example, help(Digits); returns a message stating that the argument to the help function was not a name --- because Digits evaluates to an integer value. help('Digits'); will give the desired help description.
- After each evaluation, one level of single quotes is stripped off. So if a := 1; then ''a''; returns 'a', 'a'; returns a, and a; returns 1.
- Note that evaluation is different from simplification. Thus '2 + 3' will simplify to 5.
- A special case of unevaluation is used to unassign a name. For example, x := 2; assigns x the value 2 and x := 'x'; returns the name x to its original unassigned state (so that x stands for itself). In some cases, the evaln function is required to unassign a name, such as: for i to n do a[i] := evaln(a[i]) od; where i must be evaluated but a[i] must not be fully evaluated.

See Also:

assignment, evaln

when graphing f with `plot`, as indicated in the following `plot` command.

```
> plot('f(x)',x=-3..3);
```

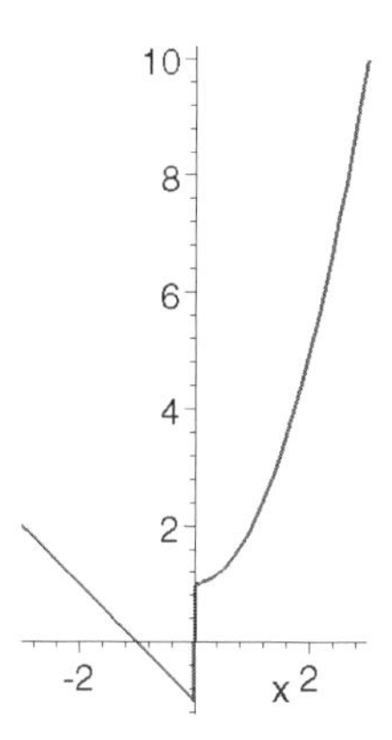

EXAMPLE 5: Graph $h(x) = \begin{cases} 6+2x, x \le -2 \\ x^2, -2 < x \le 2 \\ 11-3x, x > 2 \end{cases}$ on the interval [–3,3].

SOLUTION: The three pieces of the function are cleared and defined as `h1`, `h2`, and `h3`. Each piece is then defined with `proc` and depends only on the variable x. Since the `else` statement within each definition assigns the value of 0 if x is not contained in the associated interval, `h` can be expressed as the sum of `h1`, `h2`, and `h3`. After `h` is defined, it is plotted on the interval [–3,3] using the option `numpoints=100` so that we obtain a smoother graph. Note that h could have been defined using a single `if` statement by taking advantage of the `elif` command.

```
> h:='h':h1:='h1':h2:='h2':h2:='h3':
  h1:=proc(x) if x < -2 then 6+2*x else 0 fi; end:
  h2:=proc(x)
       if x>-2 and x<2 then x^2 else 0 fi;
    end:
  h3:=proc(x) if x>2 then 11-3*x else 0 fi; end:
  h:=x->h1(x)+h2(x)+h3(x):
> plot('h(x)',x=-3..3);
```

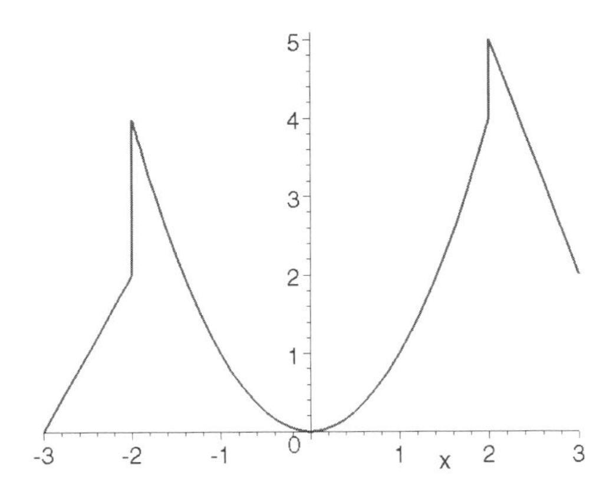

■

The **unit step function**, `Heaviside(x)`, is defined by

$$\boldsymbol{U}(x) = \begin{cases} 1, x > 0 \\ 0, x < 0 \end{cases}$$

Thus, a piecewise-defined function like

$$f(x) = \begin{cases} g(x), 0 < x < a \\ h(x), x > a \end{cases}$$

can be defined in terms of the unit step function with $f(x) = g(x)\,[1 - \boldsymbol{U}(x - a)] + h(x)\boldsymbol{U}(x - a)$. Defining functions using `Heaviside` is discussed in more detail in Chapter 6.

Functions can also be defined recursively. (Functions of this type are useful in the study of Fourier series.) Before illustrating how to enter recursively defined functions, we introduce the following commands:

`trunc(x)` returns the largest integer smaller than x if x is positive and the smallest integer larger than x if x is negative.

`frac(x)` returns x modulo 1 (the decimal part of x).

`round(x)` rounds x to the nearest integer.

`floor(x)` returns the greater integer less than or equal to x.

`ceil(x)` returns the smallest integer greater than or equal to x.

```
> trunc(2.9);
```

$$2$$

```
> frac(2.9);
```

$$.9$$

```
> round(2.9);
```

$$3$$

```
> floor(2.9);
```

$$2$$

```
> ceil(2.9);
```

$$3$$

Recursively defined functions are illustrated in the following two examples: If $f(x)$ is defined on the interval $[a,b]$, then the periodic extension of f with period p is defined with the following procedure:

```
g:=proc(x)
      local a,b;
      a:=trunc(x) mod p;
      b:=frac(x);
      f(a+b);
   end:
```

EXAMPLE 6: Let $f(x) = \begin{cases} 1, 0 \le x < 1 \\ -1, 1 \le x < 2 \end{cases}$. If $x \ge 2$, define $f(x)$ recursively by $f(x) = f(x-2)$. Graph f on the interval $[0,6]$.

SOLUTION: We begin by clearing the definitions of the two pieces of `f`, `f1` and `f2`, as well as `f` and `g`. In this example, we will use `f` to represent the function defined on $0 \le x < 2$ and `g` to represent the periodic extension $f(x) = f(x-2)$. As in the previous example, we use `proc` to define the two segments of `f` and then define `f` as the sum of `f1` and `f2`. This function is then plotted on the interval $[-1,3]$. Notice that the function is 0 outside of the interval $0 \le x \le 2$.

```
> f1:='f1':f2:='f2':f:='f':g:='g':
  f1:=proc(x) if x>=0 and x<1 then 1 else 0 fi;
   end:
  f2:=proc(x) if x>=1 and x<=2 then -1 else 0 fi;
   end:
  f:=x->f1(x)+f2(x):
> plot(f,-1..3,-2..2,numpoints=200);
```

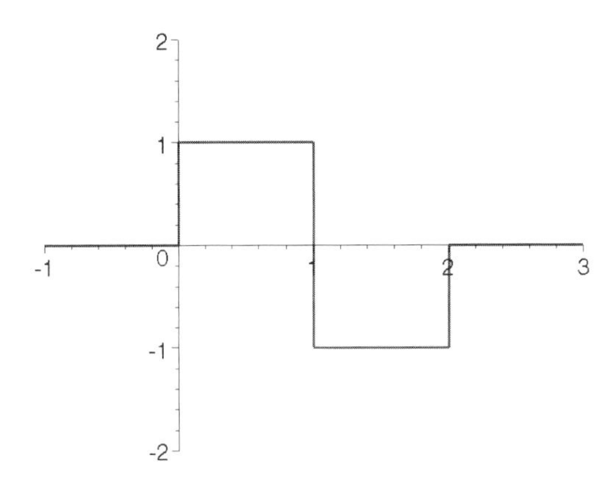

The periodic extension is then defined with `proc`. Given a value of `x`, this procedure determines the integer part of `x` with `trunc(x)` and the fractional part of `x` with `frac(x)`. Hence, the sum of `a=trunc(x) mod 2` and `b=frac(x)` yields the corresponding value on the interval [0,2]. Therefore, `f(a+b)` yields the appropriate functional value at `x`. Note that for a function of period p, the command `a=trunc(x) mod p` would be used instead of the one used in this example. After defining the periodic extension in `g`, we plot `g` on [0,6].

```
> g:=proc(x)
      local a,b;
      a:=trunc(x) mod 2;
      b:=frac(x);
      f(a+b);
    end:
```

```
> plot(g,0..6,-2..2,numpoints=200);
```

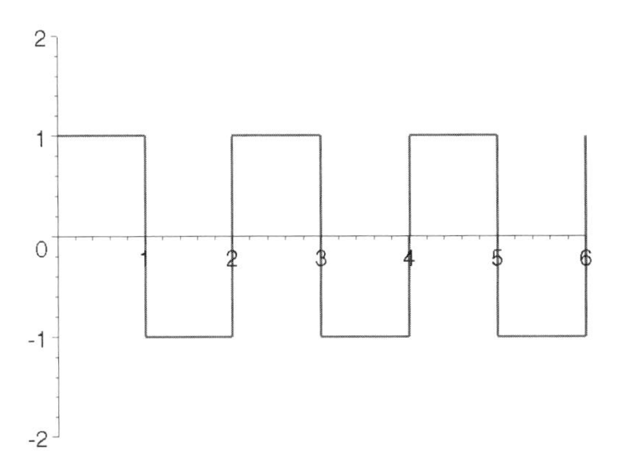

■

Our last example illustrates how to define piecewise-defined functions taking advantage of `elif` instead of using several `if-then` statements.

EXAMPLE 7: Let $g(x)$ be the periodic extension of the function

$$f(x) = \begin{cases} x, 0 \le x < 1 \\ 1, 1 \le x < 2 \\ 3 - x, 2 \le x < 3 \end{cases}$$. Graph g on the interval [0,12].

SOLUTION: We begin by defining f. As in the previous examples, we must use operator notation when graphing f.

```
> f:='f':
  f:=proc(x)
      if 0<=x and x<1 then x
         elif 1<=x and x<2 then 1
         elif 2<=x and x<3 then 3-x fi
      end:
> plot(f,0..3);
```

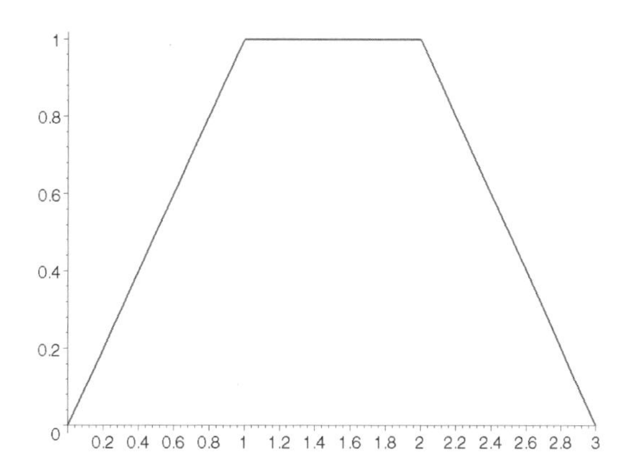

Then, in the same manner as in the previous example, we define and graph g.

```
> g:='g':
  g:=proc(x)
      local a,b;
```

```
        a:=trunc(x) mod 3;
        b:=frac(x);
        f(a+b);
    end:
> plot(g,0..12);
```

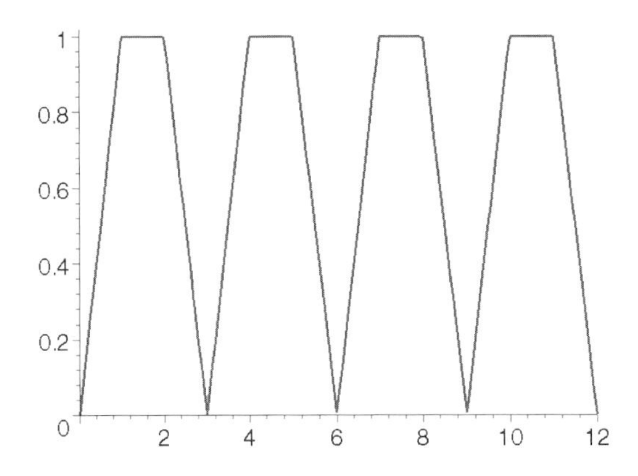

■

Polar Plots

Graphs of $r = f(\theta)$, $a \leq \theta \leq b$, in polar coordinates are generated by including the option `coords=polar` in the plot command: `plot(f(theta),theta=a..b,coords=polar)`.

EXAMPLE 8: Graph $r = 1 + 2\sin\theta$, $r = 1$, and $r = \cos 3\theta$ on the same axes.

SOLUTION: We include the option `coords=polar` in the `plot` command.

```
> plot({1+2*sin(theta),1,cos(3*theta)},theta=0..2*Pi,
   coords=polar,color=black,linestyle=[1,3,4]);
```

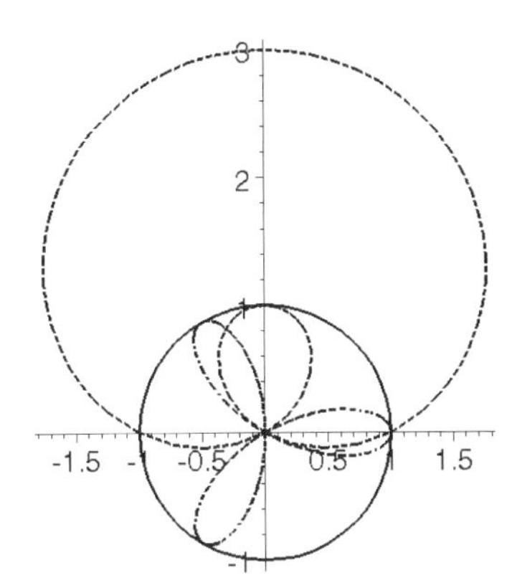

■

In addition to using `display` to show sets of graphs together, we can also use `display` to display arrays of graphics objects.

EXAMPLE 9: Graph (a) $r = \sin(8\theta/7), 0 \le \theta \le 14\pi$; (b) $r = \theta\cos\theta$, $-19\pi/2 \le \theta \le 19\pi/2$; (c) ("The Butterfly") $0 \le \theta \le 24\pi$, $r = e^{\cos\theta} - 2\cos 4\theta + \sin^5(\theta/12)$; and (d) $r = \cos\theta + \cos^3(3\theta/2), 0 \le \theta \le 4\pi$.

SOLUTION: We will display the graphs of these four equations as a 2×2 array with display. We declare `A` to be a 2×2 array. (Note that you do not need to reload the **plots** package if you have already loaded it during your current Maple session.)

```
> with(plots):
  A:=array(1..2,1..2):
```

We then use `plot` together with the option `coords=polar` to graph each polar equation.

```
> A[1,1]:=plot(sin(8*theta/7),theta=0..14*Pi,
   coords=polar,color=black):
> A[1,2]:=plot(theta*cos(theta),theta=-19*Pi/2..19*Pi/2,
   coords=polar,color=black):
```

Using standard mathematical notation, we know that $\sin^5(\theta/12) = (\sin(\theta/12))^5$. However, with Maple, be sure you use the form `sin(θ/12)^5`, not `sin^5(θ/12)`, which Maple will not interpret in the way intended.

```
> A[2,1]:=plot(exp(cos(theta))-2*cos(4*theta)+
      sin(theta/12)^5,
   theta=0..24*Pi,coords=polar,color=black):

> A[2,2]:=plot(cos(theta)+cos(3*theta/2)^3,theta=0..4*Pi,
   coords=polar,color=black):
```

Finally, we use `display` to display the four graphs in `A`. `A[1,1]` and `A[1,2]` are shown in the first row; `A[2,1]` and `A[2,2]` are shown in the second.

```
> display(A);
```

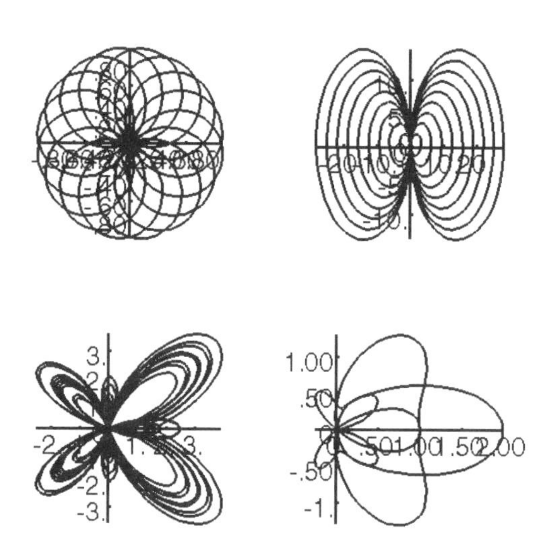

■

Parametric Plots

Parametric plots are generated with `plot`. The graph of $\begin{cases} x = x(t) \\ y = y(t) \end{cases}, a \le t \le b$, is generated with

`plot([x(t),y(t),t=a..b],horzl,vert,opts)`

where `horzl` and `vert` represent the horizontal and vertical ranges used in the display of the graph, respectively. Both of these settings are made in the form `min..max`.

EXAMPLE 10: Graph the circle $x^2 - 4x + y^2 = 4$.

SOLUTION: In Example 3, we saw the equation $x^2 - 4x + y^2 = 4$ is equivalent to the equation $(x-2)^2 + (y-1)^2 = 3^2$. Thus, the center of the circle $x^2 - 4x + y^2 = 4$ is $(2,1)$ and the radius is 3. Parametric equations of the circle are then given by $\begin{cases} x = 2 + 3\cos t \\ y = 1 + 3\sin t \end{cases}$, $0 \le t \le 2\pi$. We use `plot` to graph the circle.

```
> plot([2+3*cos(t),1+3*sin(t),t=0..2*Pi]);
```

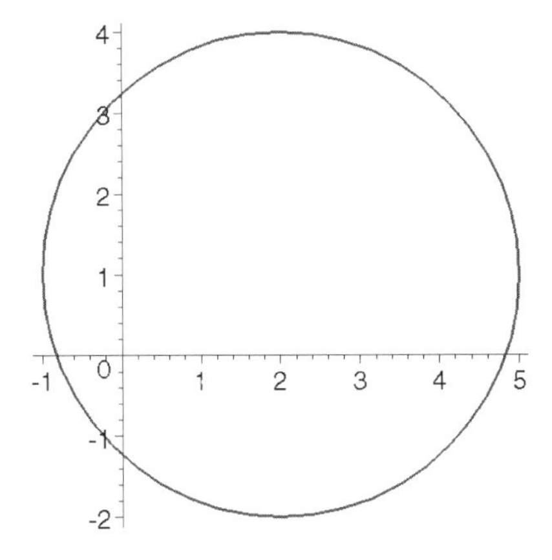

■

EXAMPLE 11: For $a < b$, the **Prolate Cycloid** is the graph of the parametric equations $\begin{cases} x = at - b\sin t \\ y = a - b\cos t \end{cases}$. The **Folium of Descartes** has parametrization $\begin{cases} x = 3at/(1+t^3) \\ y = 3at^2/(1+t^3) \end{cases}$. Graph the Prolate Cycloid and the Folium of Descartes if $a = 1$ and $b = 2$.

SOLUTION: We define x and y and then use `plot` to graph the Prolate Cycloid.

```
> x:=t->t-2*sin(t):
```

```
y:=t->1-2*cos(t):
plot([x(t),y(t),t=0..8*Pi]);
```

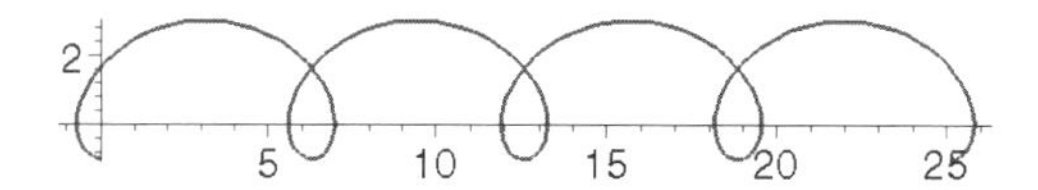

Similarly, we use `plot` to graph the Folium of Descartes. In this case, we include `-4..5,-4..5` to indicate that the x and y-axes displayed each correspond to the interval $[-4,5]$.

```
> x:='x':y:='y':
  x:=t->3*t/(1+t^3):
  y:=t->3*t^2/(1+t^3):
  plot([x(t),y(t),t=-10..10],-4..5,-4..5);
```

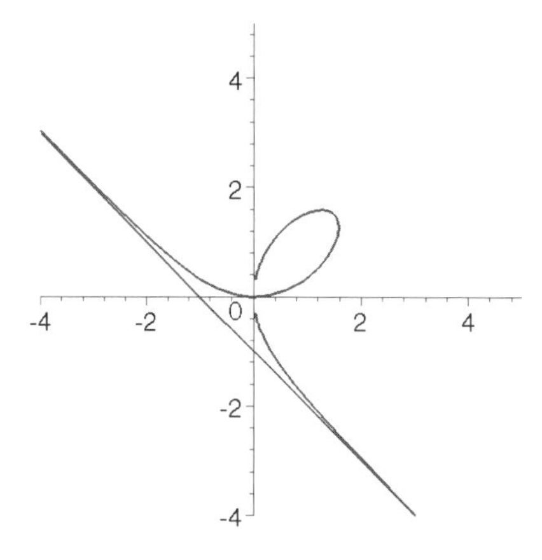

■

Polar plots are given with the option setting `coords=polar`, which assumes a function of the form `{r(t),theta(t)}`. Hence, the `plot` command should be entered as

```
plot([r(t),theta(t),t=a..b],coords=polar)
```

for polar plots.

EXAMPLE 12: Graph the polar parametric equations $\begin{cases} r = \cos t \\ \theta = \sin t \end{cases}, 0 \le t \le 2\pi$.

SOLUTION: The graph is generated using `plot` together with the options `coords=polar`.

```
> plot([cos(t),sin(t),t=0..2*Pi],coords=polar);
```

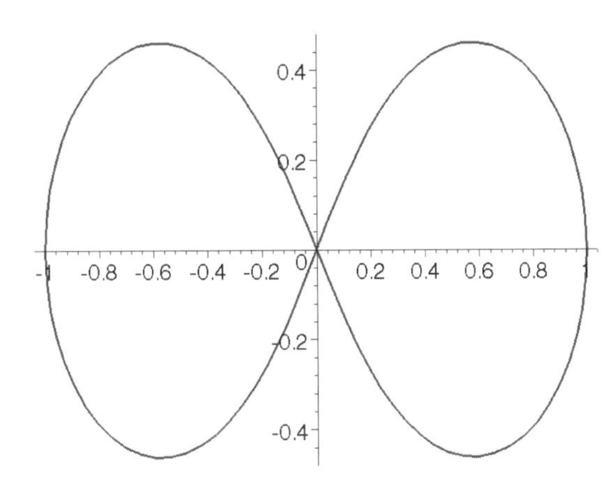

■

Three-Dimensional Graphics

As was mentioned in Chapter 1, functions of more than one variable can be graphed with Maple. Of particular interest are functions of two variables. The command that plots the graph of the function $f(x,y)$, $a \le x \le b$, $c \le y \le d$, is

```
plot3d(f(x,y),x=a..b,y=c..d).
```

You can use `?plot3d` or `?plot3d[options]` to obtain information about the `plot3d` command.

EXAMPLE 13: Graph $f(x, y) = xye^{-x^2-y^2}$ for $-2 \le x \le 2$ and $-2 \le y \le 2$.

SOLUTION: After clearing all prior definitions of f, if any, we define f and then use `plot3d` to graph f on the rectangle $[-2,2] \times [-2,2]$.

```
> f:='f':
  f:=(x,y)->x*y*exp(-x^2-y^2);
```

$$f:=(x,y) \rightarrow x y e^{(-x^2-y^2)}$$

```
> plot3d(f(x,y),x=-2..2,y=-2..2);
```

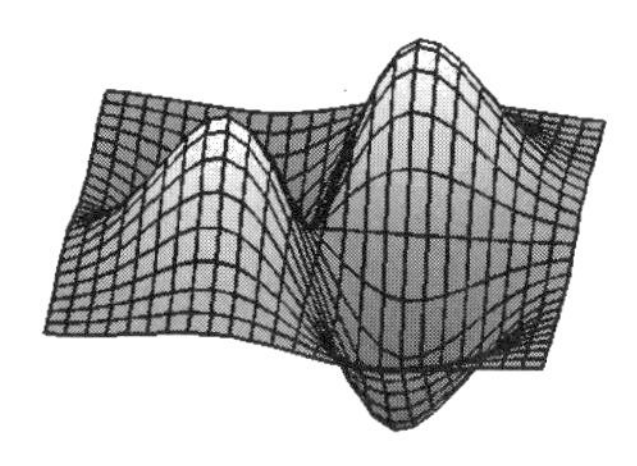

■

The various options available with the `plot3d` command are obtained by entering `?plot3d[options]`.

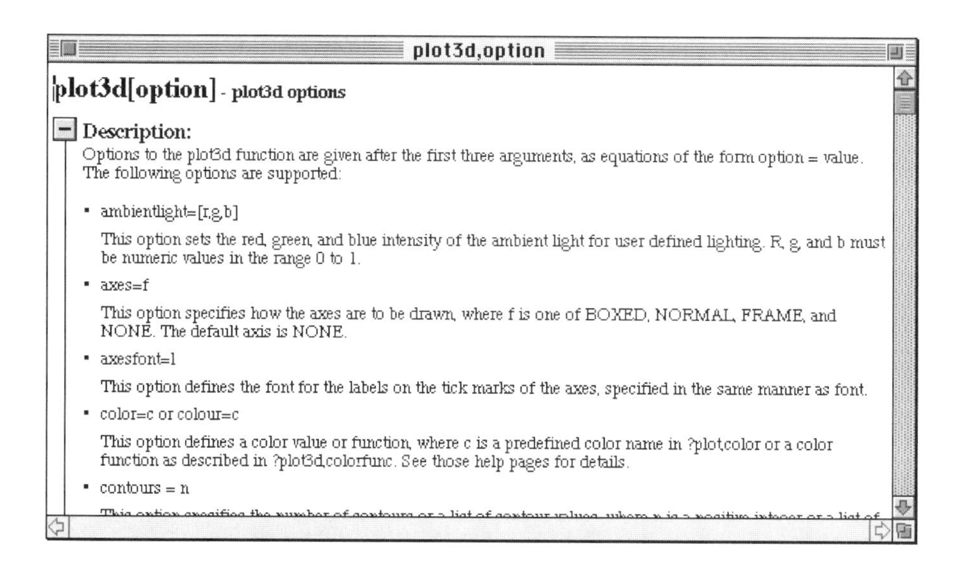

plot3d,option

plot3d[option] - plot3d options

Description:

Options to the plot3d function are given after the first three arguments, as equations of the form option = value. The following options are supported:

- ambientlight=[r,g,b]

  This option sets the red, green, and blue intensity of the ambient light for user defined lighting. R, g, and b must be numeric values in the range 0 to 1.

- axes=f

  This option specifies how the axes are to be drawn, where f is one of BOXED, NORMAL, FRAME, and NONE. The default axis is NONE.

- axesfont=l

  This option defines the font for the labels on the tick marks of the axes, specified in the same manner as font.

- color=c or colour=c

  This option defines a color value or function, where c is a predefined color name in ?plot,color or a color function as described in ?plot3d,colorfunc. See those help pages for details.

- contours = n

Global options for three-dimensional graphics can be set using the `setoptions3d` command, which is contained in the `plots` package. For example, entering

```
with(plots);
```

loads the `plots` package, and then entering

```
setoptions3d(axes=BOXED):
```

instructs Maple to place a box around all subsequently generated three-dimensional graphics during the Maple session.

In addition, note that many options can be accessed directly from the Maple menu or the contol strip

once a graphic has been selected, as illustrated in the following example.

EXAMPLE 14: Graph

$$f(x,y) = (x-1)^2 e^{-x^2-(y+1)^2} + \left(-\frac{2}{3}x + \frac{10}{3}x^3 + \frac{10}{3}y^5\right)e^{-x^2-y^2} - \frac{1}{9}e^{-(x+1)^2-y^2}$$

on the rectangle $[-3,3] \times [-3,3]$.

SOLUTION: After clearing all prior definitions of *f*, we define *f*

```
> f:='f':
  f:=(x,y)->(x-1)^2*exp(-x^2-(y+1)^2)+
   (-2/3*x+10/3*x^3+10/3*y^5)*exp(-x^2-y^2)-
   10/3*exp(-(x+1)^2-y^2);
```

$$f := (x,y) \to (x-1)^2 \mathbf{e}^{(-x^2-(y+1)^2)} + \left(-\frac{2}{3}x + \frac{10}{3}x^3 + \frac{10}{3}y^5\right)\mathbf{e}^{(-x^2-y^2)} - \frac{10}{3}\mathbf{e}^{(-(x+1)^2-y^2)}$$

and then use `plot3d` together with the option `grid=[40,40]` to graph *f* on the rectangle $[-3,3] \times [-3,3]$.

```
> plot3d(f(x,y),x=-3..3,y=-3..3,grid=[40,40]);
```

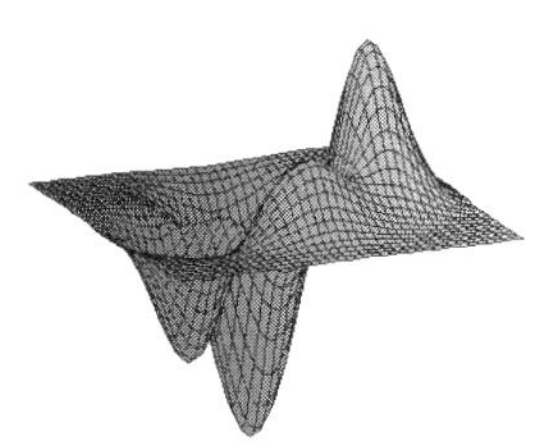

Once you have generated a graphic, you can resize it and adjust its orientation. First, select the graphic. By clicking and dragging on the toggle bars, you can resize the graphic.

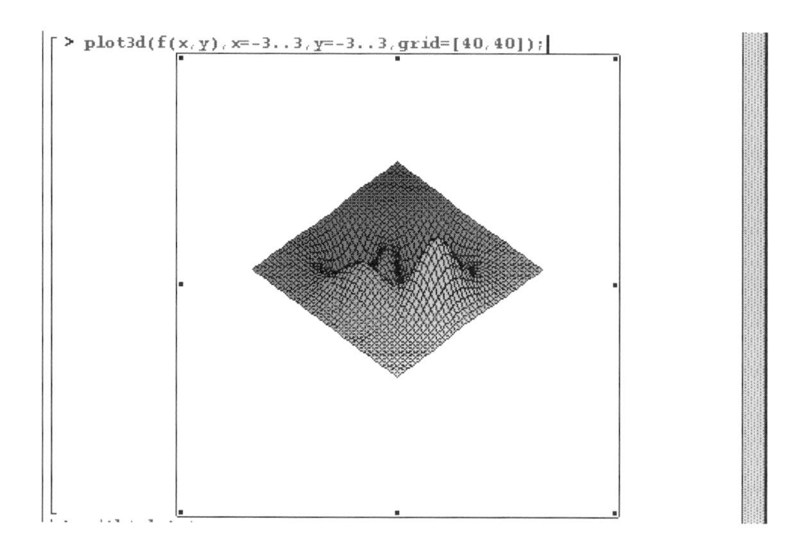

You can click and drag the image directly to change the orientation of the graphic.

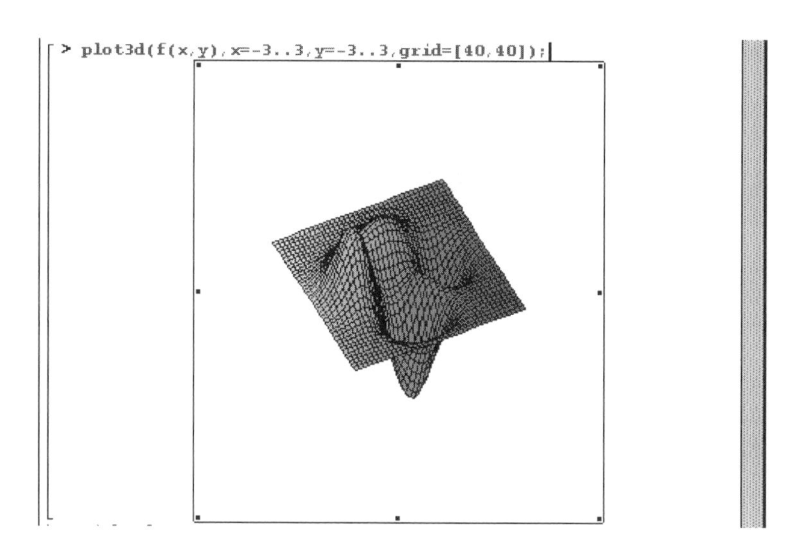

In addition, you can invoke various options directly from the control strip or the menu. From **Projection**, we select **Far Perspective** and **Constrained**.

Projection
No Perspective
Near Perspective
Medium Perspective
✓ Far Perspective

✓ Constrained
Unconstrained

From **Axes**, we select **Framed**.

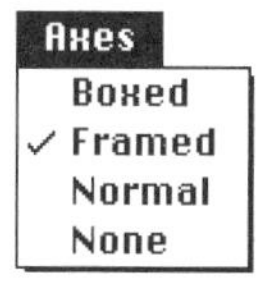

From **Color**, we choose to display the graph in various shades of gray.

Color

- XYZ
- XY
- Z
- Z (Hue)
- ✓ Z (Grayscale)
- No Coloring

- ✓ No Lighting
- User Lighting
- Light Scheme 1
- Light Scheme 2
- Light Scheme 3
- Light Scheme 4

From **Style**, we select **Patch and contour**.

Style

- Patch
- Patch w/o grid
- ✓ Patch and contour
- Hidden line
- Contour
- Wireframe
- Point

- Symbol ▶
- Line Style ▶
- Line Width ▶
- Grid Style ▶

Maple displays the results of these changes immediately.

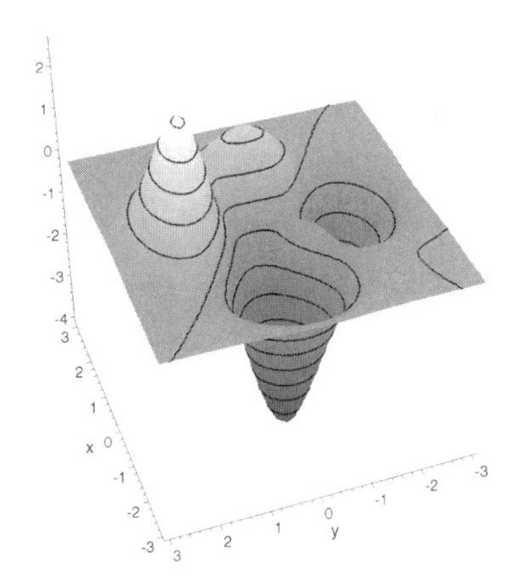

■

You can graph several three-dimensional images together with plot3d in the same way that you can graph several two-dimensional images together with plot. Also, the command display3d, contained in the plots package, can be used to display several three-dimensional graphs together in the same way that the command display can be used to display several two-dimensional graphs together.

EXAMPLE 15: Graph $z = 2x^2 + y^2$ and $z = 8 - x^2 - 2y^2$ together on the rectangle $[-3,3] \times [-3,3]$.

SOLUTION: We use plot3d to graph these two functions together on the rectangle $[-3,3] \times [-3,3]$. We also illustrate the options axes=BOXED and gridstyle= triangular.

```
> plot3d({2*x^2+y^2,8-x^2-2*y^2},
  x=-3..3,y=-3..3,axes=BOXED,gridstyle=triangular);
```

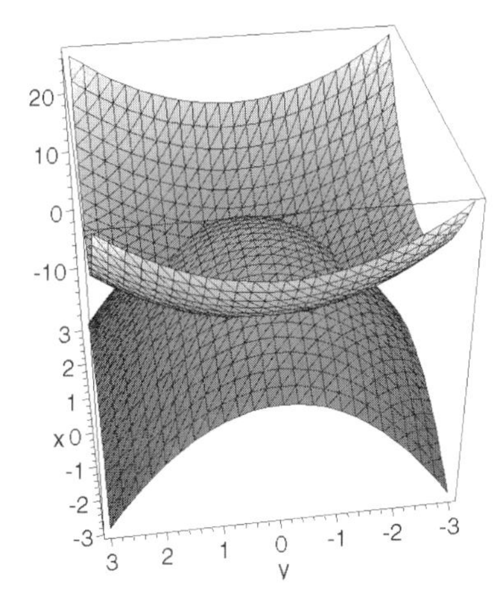

■

EXAMPLE 16: Display the graphs of $f(x, y) = |x||y|$ and $g(x, y) = -|x||y|$ on the rectangle $[-1,1] \times [-1,1]$ together.

SOLUTION: After loading the `plots` package we use `plot3d` to graph $f(x, y) = |x||y|$ and $g(x, y) = -|x||y|$. Be sure to include a colon (:) at the end of the commands defining `POne` and `PTwo`.

```
> with(plots):
  f:=(x,y)->abs(x)*abs(y):
  g:=(x,y)->-abs(x)*abs(y):
> POne:=plot3d(f(x,y),x=-1..1,y=-1..1,grid=[40,40]):
  PTwo:=plot3d(g(x,y),x=-1..1,y=-1..1,grid=[40,40]):
```

Then, `display3d` is used to display both graphs together. To display only the graph of $f(x, y) = |x||y|$, enter the command

`display3d({POne});`

enter `display3d({PTwo})` to display the graph of $g(x, y) = -|x||y|$.

```
> display({POne,PTwo},axes=BOXED);
```

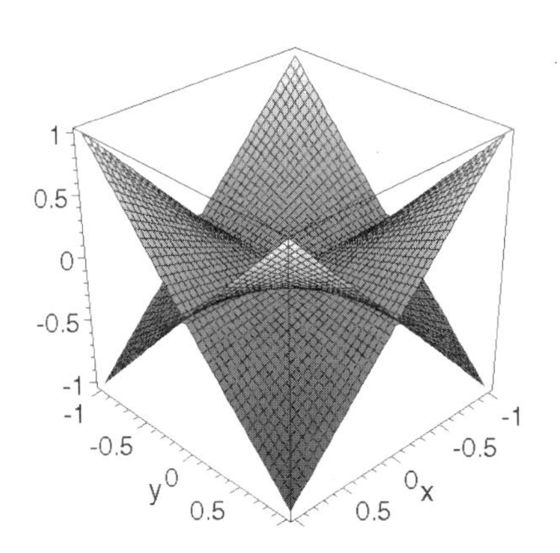

■

Graphing Level Curves of Functions of Two Variables

The **level curves** of the function $f(x,y)$ are curves in the xy-plane that satisfy the equation $f(x,y) = C$, where C is a constant. Maple graphs several of the level curves of the function $f(x,y)$ for $a \le x \le b$ and $c \le y \le d$ with the command

`contourplot(f(x,y),x=xmin..xmax,y=ymin..ymax).`

As with command display3d, contourplot is contained in the plots package so this package must be loaded by first entering with(plots) before using contourplot or the command should be entered in the form plots[contourplot](...).

EXAMPLE 17: Graph several level curves of $g(x,y) = x\sin y + y\sin x$ on the rectangle $[0,5\pi] \times [0,5\pi]$.

SOLUTION: We use contourplot to generate various level curves of g on the rectangle $[0,5\pi] \times [0,5\pi]$.

```
> with(plots):
  g:=x*sin(y)+y*sin(x):
  contourplot(g(x,y),x=0..5*Pi,y=0..5*Pi);
```

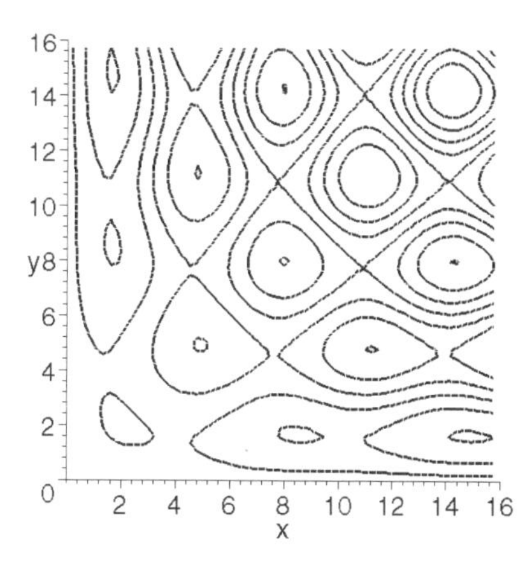

■

Equations are graphed with the implicitplot command, also contained in the plots package. The command

```
implicitplot(equation,x=a..b,y=c..d)
```

graphs the equation equation, with variables x and y, on the rectangle $[a,b] \times [c,d]$.

EXAMPLE 18: (a) Graph several level curves of $f(x,y) = x^2 - 4x + y^2 - 2y + 5$ on the rectangle $[-2,6] \times [-3,5]$. (b) Graph the circle $x^2 - 4x + y^2 - 2y + 5 = 9$.

SOLUTION: After defining f, we use `contourplot` along with the options `grid` and `axes`, `color`, and `contours` to graph several level curves.

```
> f:='f':
  f:=(x,y)->x^2-4*x+y^2-2*y+5:
  contourplot(f(x,y),x=-2..6,y=-3..5,
  grid=[40,40],axes=NORMAL,color=black,contours=20);
```

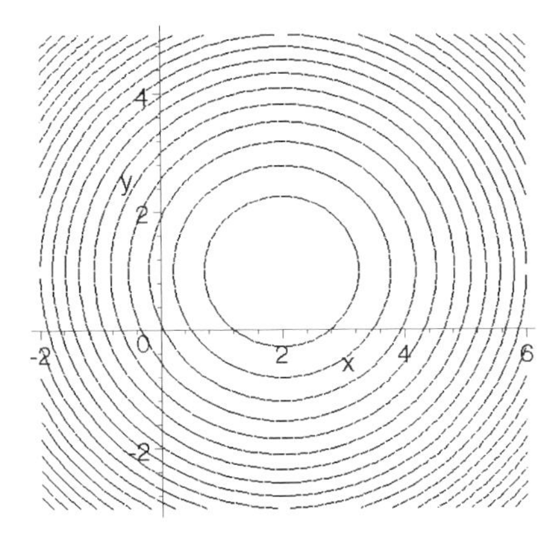

To graph the circle $x^2 - 4x + y^2 - 2y + 5 = 9$, which is the same as the circle $(x-2)^2 + (y-1)^2 = 3^2$ considered in previous examples, we note that the graph of $x^2 - 4x + y^2 - 2y + 5 = 9$ is the level curve of $f(x,y)$ corresponding to 9 defined next as `Eq1` and graphed with `implicitplot`.

```
> Eq1:='Eq1':
  Eq1:=f(x,y)=9:
> implicitplot(Eq1,x=-2..6,y=-3..5,color=black);
```

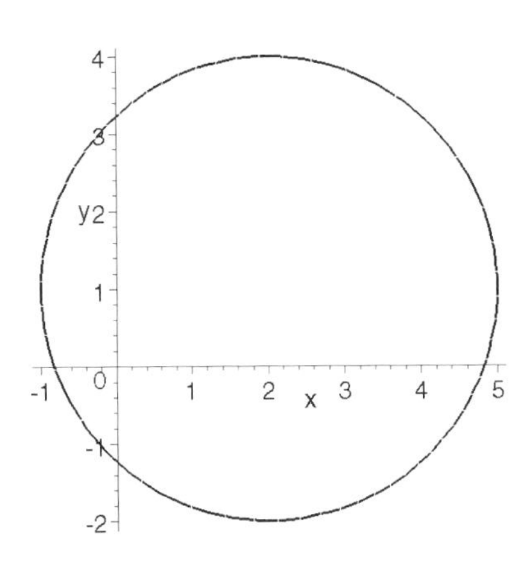

Note that the same graph is obtained by entering

```
contourplot(f(x,y),x=-2..6,y=-3..5,color=black,contours=[9]);.
```

■

`contourplot` can also help us investigate the behavior of some functions near particular points.

EXAMPLE 19: Graph h and several level curves of h on the rectangle $[-2,2]\times[-2,2]$ if $h(x,y) = \dfrac{x^2-y^2}{x^2+y^2}$.

SOLUTION: In the following graphs, Maple does not compute `h[0,0]`, and thus no error messages are generated, even though h is undefined if $x = 0$ and $y = 0$. In the following code, we use `plot3d` to graph h and `contourplot` to graph several level curves of h.

```
> h:='h':
  h:=(x,y)->(x^2-y^2)/(x^2+y^2):
  plot3d(h(x,y),x=-2..2,y=-2..2,
      grid=[30,30],axes=BOXED);
  contourplot(h(x,y),x=-2..2,y=-2..2,
  grid=[30,30],axes=NORMAL);
```

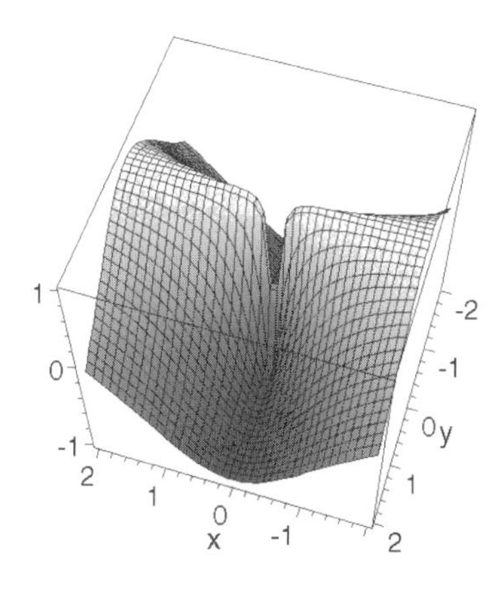

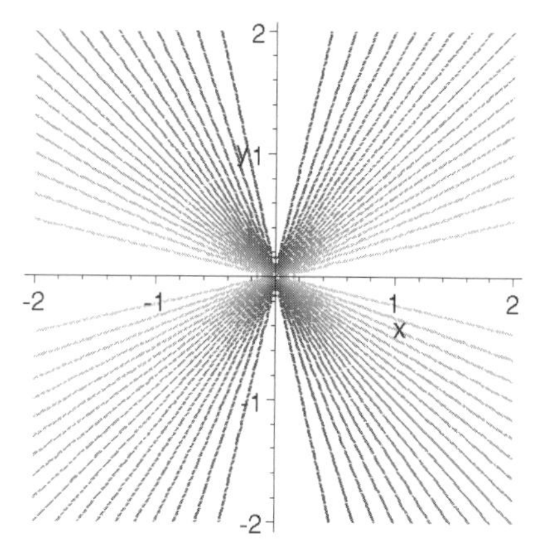

From the graph on the left, we see that h behaves strangely near (0,0). In fact, if h had been graphed on a different rectangle, Maple might have sampled (0,0) and subsequently might have displayed an error message. In any case, the resulting graph would have been sufficiently accurate for our purposes. From the graph on the right, we see that all contours are approaching (0,0). In fact, near (0,0), h attains every value between –1 and 1, and in calculus we show that $\lim_{(x, y) \to (0, 0)} h(x, y)$ does not exist.

■

Graphing Parametric Curves and Surfaces in Space

`spacecurve` is used to graph parametric curves and surfaces in space. The command

```
spacecurve([x(t),y(t),z(t)],t=a..b)
```

generates the three-dimensional curve $\begin{cases} x = x(t) \\ y = y(t), (a \le t \le b), \\ z = z(t) \end{cases}$ and the command

```
plot3d[[x(u,v),y(u,v),z(u,v)],u=a..b,v=c..d]
```

plots the surface $\begin{cases} x = x(u, v) \\ y = y(u, v), (a \le u \le b), (c \le v \le d). \\ z = z(u, v) \end{cases}$ As with `contourplot` and `implicit-plot`, `spacecurve` is also contained in the `plots` package so must be loaded first by entering `with(plots)` or entering the command in the form `plots[spacecurve](...)`.

EXAMPLE 20: Compare the graphs of $\begin{cases} x = \cos 2t \\ y = \sin 2t, 0 \le t \le 8\pi \\ z = t/5 \end{cases}$ and $\begin{cases} x = t\cos 2t \\ y = t\sin 2t, 0 \le t \le 8\pi. \\ z = t/5 \end{cases}$

SOLUTION: In each case, we use `spacecurve`. Including the option `numpoints=240` increases the number of sample points and helps assure that the resulting curves appear smooth.

```
> spacecurve([cos(2*t),sin(2*t),t/5],t=0..8*Pi,
```

```
    numpoints=240,axes=NORMAL,color=black);
```

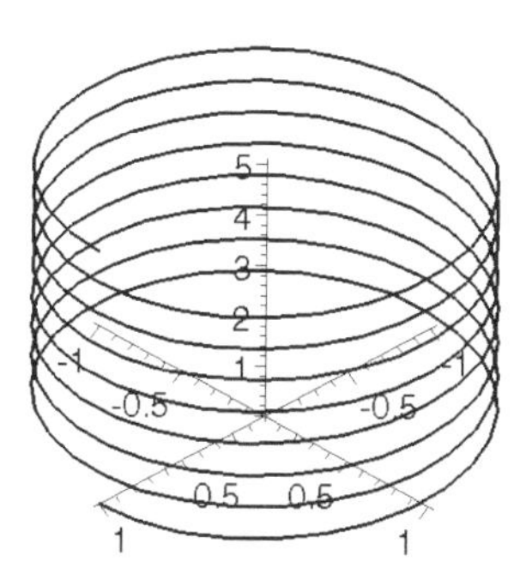


```
> x:=t->t*cos(2*t):
  y:=t->t*sin(2*t):
  z:=t->t/5:
  spacecurve([x(t),y(t),z(t)],t=0..8*Pi,
     numpoints=240,axes=NORMAL,color=black);
```

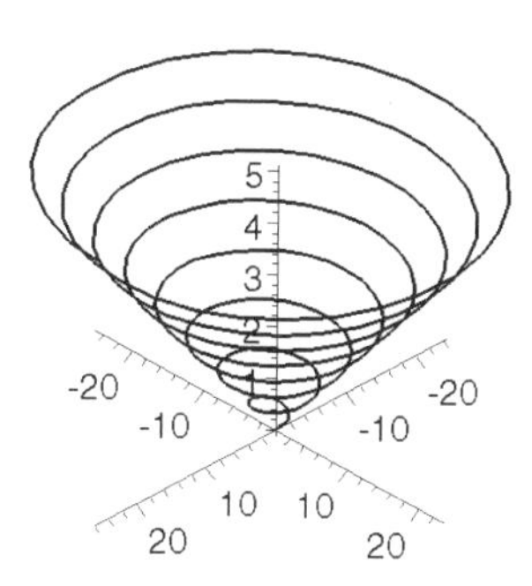

■

The intersection of a plane and a surface is called the **trace** of the surface. We can use `spacecurve` to help us visualize the traces of some surfaces.

EXAMPLE 21: Let $g(x,y) = \cos(x + \sin y)$. Graph the intersection of the graph of g with the planes (a) $x = 5$; (b) $y = 6$; (c) $x = y$; and (d) $y = 4\pi - x$.

SOLUTION: We begin by using `plot3d` to graph g on the rectangle $[0,4\pi]\times[0,4\pi]$.

```
> g:='g':
  g:=(x,y)->cos(x+sin(y)):
  plot3d(g(x,y),x=0..4*Pi,y=0..4*Pi,
     axes=FRAME,grid=[40,40],shading=NONE,
     gridstyle=triangular);
```

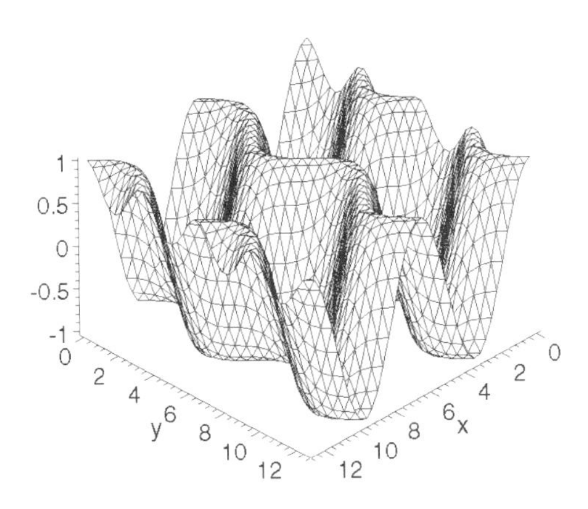

For (a), the intersection of the plane $x=5$ and the xy-plane is the line $x=5$, which has parametrization $\begin{cases} x=5 \\ y=t \\ z=0 \end{cases}$. The intersection of the plane $x=5$ and the graph of g is the set of points on the graph of g above the line $x=5$ that has parametrization $\begin{cases} x=5 \\ y=t \\ z=g(5,t) \end{cases}$. The lines $\begin{cases} x=5 \\ y=t \\ z=0 \end{cases}$ and $\begin{cases} x=5 \\ y=t \\ z=g(5,t) \end{cases}$ are graphed using `spacecurve` for $0\le t\le 4\pi$ in `SC1` and `SC2`, respectively. Similarly, for (b), the intersection of the plane $y=6$ and the xy-plane is the line $y=6$ with parametrization $\begin{cases} x=t \\ y=6 \\ z=0 \end{cases}$; the intersection of the plane $y=6$ and the graph of g is the set of points on the graph of g above the line $y=6$ which has parametrization $\begin{cases} x=t \\ y=6 \\ z=g(t,6) \end{cases}$. These two curves are graphed in `SC3` and `SC4`. Finally, `display3d` is used to display the graphs `SC1`, `SC2`, `SC3`, and `SC4`. As with the commands `plot3d` and `spacecurve`, `display3d` is included in the `plots` package.

```
> SC1:=spacecurve([5,t,0],t=0..4*Pi):
  SC2:=spacecurve([5,t,g(5,t)],t=0..4*Pi):
  SC3:=spacecurve([t,6,0],t=0..4*Pi):
  SC4:=spacecurve([t,6,g(t,6)],t=0..4*Pi):
> display3d([SC1,SC2,SC3,SC4],axes=BOXED);
```

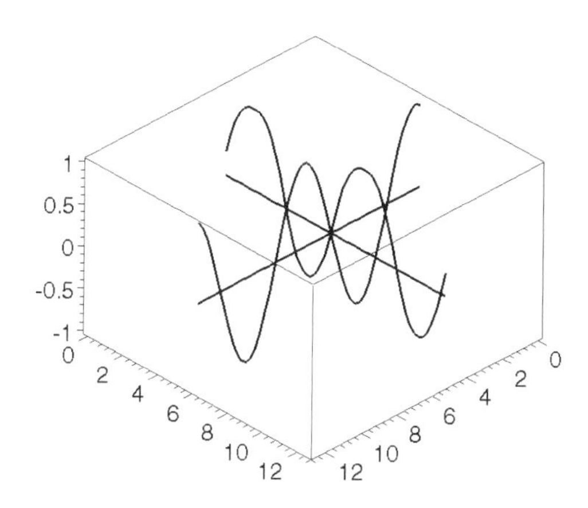

For (c) and (d) we proceed in the same manner as in (a) and (b). The line of intersection of the plane $x = y$ with the xy-plane is the line $x = y$ with parametrization $\begin{cases} x = t \\ y = t \\ z = 0 \end{cases}$; the intersection of the plane $x = y$ with g has parametrization $\begin{cases} x = t \\ y = t \\ z = g(t,t) \end{cases}$.
Similarly, the line of intersection of the plane $y = 4\pi - x$ with the xy-plane is the line $y = 4\pi - x$ with parametrization $\begin{cases} x = t \\ y = 4\pi - t \\ z = 0 \end{cases}$; the intersection of the plane $y = 4\pi - x$

with g has parametrization $\begin{cases} x = t \\ y = 4\pi - t \\ z = g(t, 4\pi - t) \end{cases}$. All four curves are graphed with `spacecurve`.

```
> spacecurve({[t,t,0],[t,t,g(t,t)],
   [t,4*Pi-t,0],[t,4*Pi-t,g(t,4*Pi-t)]},
   t=0..4*Pi,axes=FRAME,color=black);
```

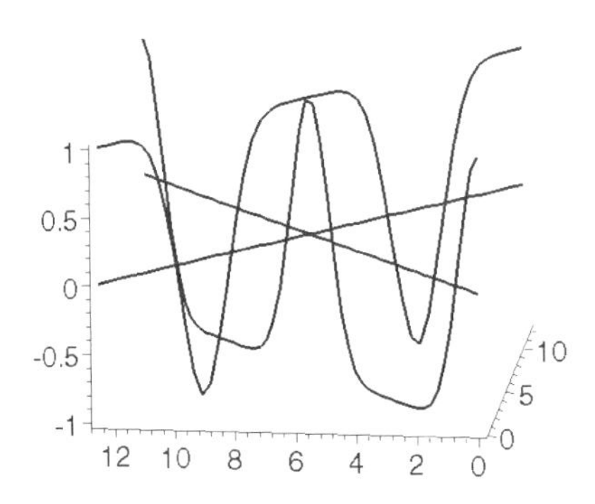

■

In Chapter 3, we will use the **Method of Lagrange Multipliers** to solve problems of the form "Find the minimum and maximum values of $f(x,y)$ subject to the constraint $f(x,y) = C$, C a constant." To see that there exist maximum and minimum values subject to the constraint, we can graph $f(x,y)$ for points (x,y) on the graph of $g(x,y) = C$ if we know a parametrization of the equation $g(x,y) = C$ with `spacecurve`.

EXAMPLE 22: Graph $f(x,y)$ for points (x,y) on the circle $x^2 + y^2 = 1$ if $f(x,y) = x^3\sin 4y + y^2\cos 3x$.

SOLUTION: A parametrization of the circle $x^2 + y^2 = 1$ is given by $\begin{cases} x = \cos t \\ y = \sin t \\ z = 0 \end{cases}$, $0 \le t \le 2\pi$. Thus, a graph of $f(x,y)$ for points (x,y) on the circle is obtained by graphing $\begin{cases} x = \cos t \\ y = \sin t \\ z = f(\cos t, \sin t) \end{cases}$, $0 \le t \le 2\pi$. We use `spacecurve` to graph these two curves together.

```
> f:='f':
  f:=(x,y)->x^3*sin(4*y)+y^2*cos(3*x):
  spacecurve({[cos(t),sin(t),0],
      [cos(t),sin(t),f(cos(t),sin(t))]},
        t=0..2*Pi,axes=NORMAL,numpoints=150);
```

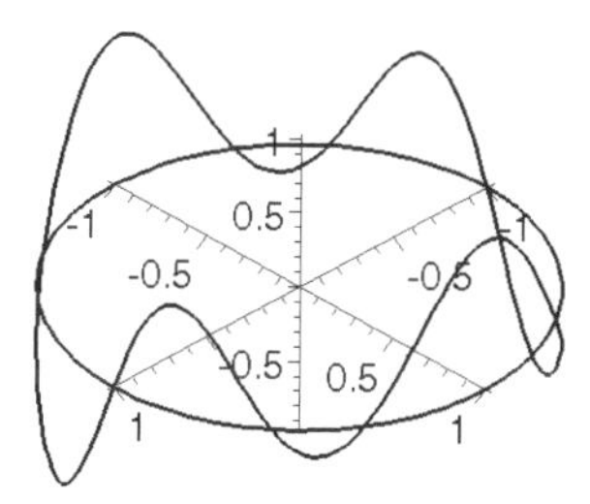

■

`plot3d` can also be used to graph parametric equations of surfaces.

EXAMPLE 23: The quadric surfaces are the three-dimensional objects corresponding to the conic sections in two dimensions. A **quadric surface** is a graph of

$$Ax^2 + By^2 + Cz^2 + Dxy + Exz + Fyz + Gx + Hy + Iz + J = 0.$$

The intersection of a plane and a quadric surface is a conic section. Several of the basic quadric surfaces, in standard form, and a parametrization of the surface are listed in the following table.

Use `plot3d` to graph the ellipsoid with equation $\frac{1}{16}x^2 + \frac{1}{4}y^2 + z^2 = 1$ and the hyperboloid of one sheet with equation $\frac{1}{16}x^2 + \frac{1}{4}y^2 - z^2 = 1$.

Name	Parametric Equations
Ellipsoid $\frac{x^2}{a^2}+\frac{y^2}{b^2}+\frac{z^2}{c^2}=1$	$\begin{cases} x = a\cos t\cos r \\ y = b\cos t\sin r \\ z = c\sin t \end{cases}$, $-\pi/2 \le t \le \pi/2$ and $-\pi \le r \le \pi$
Hyperboloid of One Sheet $\frac{x^2}{a^2}+\frac{y^2}{b^2}-\frac{z^2}{c^2}=1$	$\begin{cases} x = a\sec t\cos r \\ y = b\sec t\sin r \\ z = c\tan t \end{cases}$, $-\pi/2 \le t \le \pi/2$ and $-\pi \le r \le \pi$
Hyperboloid of Two Sheets $\frac{x^2}{a^2}-\frac{y^2}{b^2}-\frac{z^2}{c^2}=1$	$\begin{cases} x = a\sec t \\ y = b\tan t\cos r \\ z = c\tan t\sin r \end{cases}$, $-\pi/2 \le t \le \pi/2$ or $\pi/2 < t < 3\pi/2$ and $-\pi \le r \le \pi$

SOLUTION: A parametrization of the ellipsoid with equation $\frac{1}{16}x^2+\frac{1}{4}y^2+z^2=1$ is given by $\begin{cases} x = 4\sec t\cos r \\ y = 2\sec t\sin r \\ z = \tan t \end{cases}$, $-\pi/2 \le t \le \pi/2$ and $-\pi \le r \le \pi$, which is graphed with `plot3d`.

```
> x:='x':y:='y':z:='z':
  x:=(t,r)->4*cos(t)*cos(r):
  y:=(t,r)->2*cos(t)*sin(r):
  z:=(t,r)->sin(t):
  plot3d([x(t,r),y(t,r),z(t,r)],
    t=-Pi/2..Pi/2,r=-Pi..Pi,axes=BOXED);
```

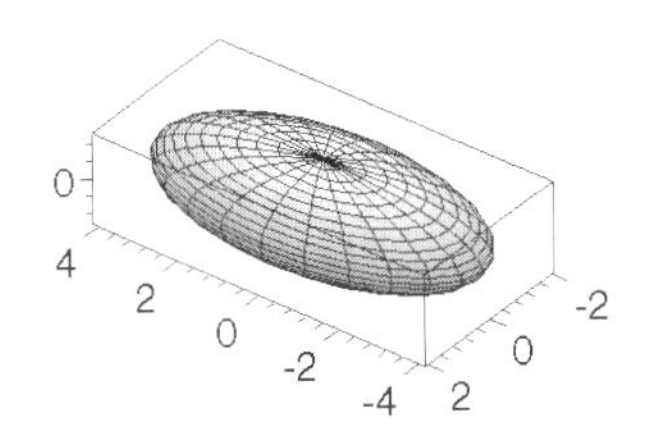

A parametrization of the hyperboloid of one sheet with equation $\frac{1}{16}x^2 + \frac{1}{4}y^2 - z^2 = 1$ is given by $\begin{cases} x = 4\sec t \cos r \\ y = 2\sec t \sin r \\ z = \tan t \end{cases}$, $-\pi/2 \le t \le \pi/2$ and $-\pi \le r \le \pi$. Because $\sec t$ and $\tan t$ are undefined if $t = \pm\pi/2$, we use `plot3d` to graph these parametric equations on a subinterval of $[-\pi/2, \pi/2]$, $[-\pi/3, \pi/3]$.

```
> x:='x':y:='y':z:='z':
  x:=(t,r)->4*sec(t)*cos(r):
  y:=(t,r)->2*sec(t)*sin(r):
  z:=(t,r)->tan(t):
  plot3d([x(t,r),y(t,r),z(t,r)],
     t=-Pi/3..Pi/3,r=-Pi..Pi,axes=BOXED);
```

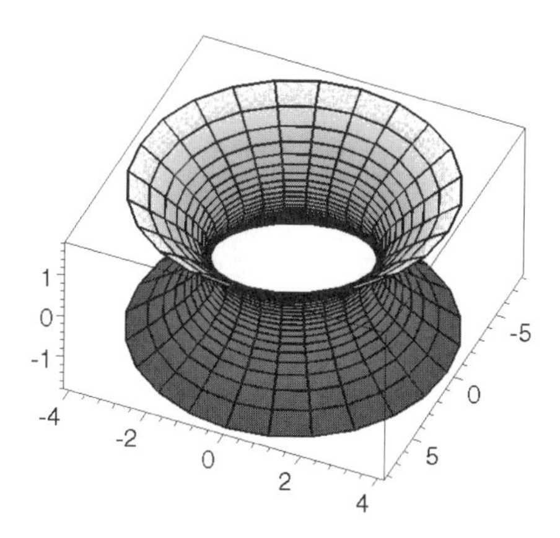

■

The command `implicitplot3d` can be used to graph equations in three variables in the same way that `implicitplot` can be used to graph equations in two variables. As with the command `implicitplot`, `implicitplot3d` is also contained in the `plots` package.

EXAMPLE 24: Graph the hyperboloid of two sheets $\frac{1}{16}x^2 - \frac{1}{4}y^2 - z^2 = 1$.

SOLUTION: We use `implicitplot3d` to graph $\frac{1}{16}x^2 - \frac{1}{4}y^2 - z^2 = 1$.

```
> with(plots):
  implicitplot3d(x^2/16-y^2/4-z^2=1,
    x=-10..10,y=-8..8,z=-5..5,axes=BOXED,
        grid=[20,20,20]);
```

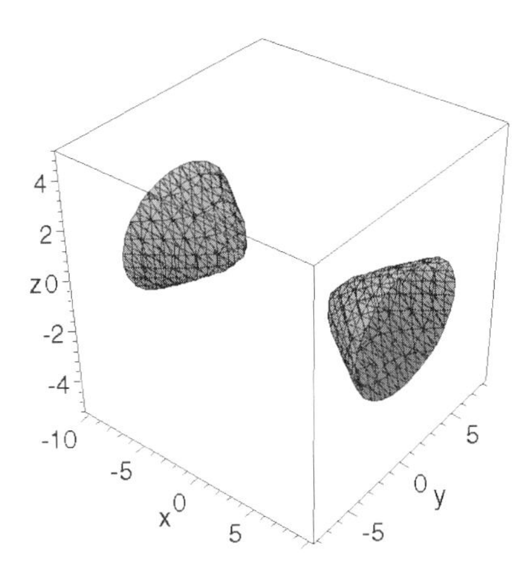

■

A Word of Caution

As indicated in the previous examples, many of Maple's graphics functions are contained in the `plots` package. Before using any of the following commands, be sure to load the `plots` package by entering `with(plots)`.

```
animate            animate3d        conformal
contourplot        cylinderplot     densityplot
display            display3d        fieldplot
fieldplot3d        gradplot         gradplot3d
implicitplot       implicitplot3d   loglogplot
logplot            matrixplot       odeplot
pointplot          polarplot        polygonplot
polygonplot3d      polyhedraplot    replot
setoptions         setoptions3d     spacecurve
sparsematrixplot   sphereplot       surfdata
textplot           textplot3d       tubeplot
```

2.4 Exact and Approximate Solutions of Equations

Exact Solutions of Equations

Maple can solve many equations exactly. For example, it can find exact solutions to systems of equations and exact solutions to polynomial equations of degree four or less. The command

```
solve(lhs=rhs,var)
```

solves the equation `lhs=rhs` for the variable `var`. If there is only one variable, or unknown, in the indicated equation, then the command `solve(lhs=rhs)` yields the same result. Enter `?solve` to obtain detailed information regarding `solve`.

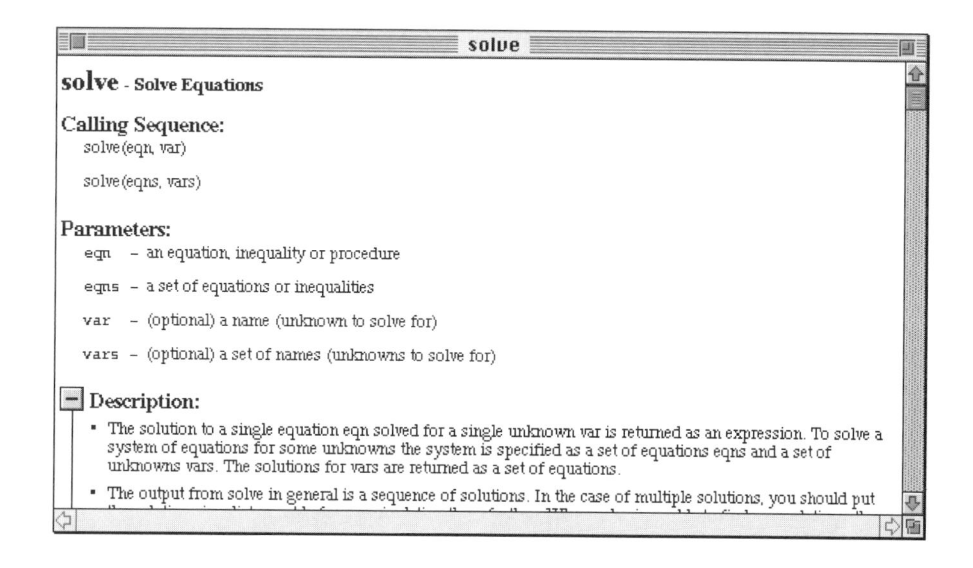

EXAMPLE 1: Solve the equations $3x + 7 = 4$, $\frac{x^2 - 1}{x - 1} = 0$, and $x^3 + x^2 + x + 1 = 0$.

SOLUTION: Because x is the only unknown in the equation $3x + 7 = 4$, Maple understands to solve for x even though the variable is not specified in the `solve` command. Next, the root of the rational function is found to be $x = -1$ with `solve`. Finally, the three roots $x = -1$, i, and $-i$ of the polynomial $x^3 + x^2 + x + 1$ are determined. Again, the variable does not have to be specified since there is only one unknown.

```
> solve(3*x+7=4);
```

$$-1$$

```
> solve((x^2-1)/(x-1)=0,x);
```

$$-1$$

```
> solve(x^3+x^2+x+1=0);
```

$$-1, I, -I$$

■

EXAMPLE 2: Find a solution of $\sin^2 x - 2\sin x - 3 = 0$.

SOLUTION: Since this equation depends only on the variable x, we find the two roots $x = \sin^{-1}3$ and $x = -\pi/2$ with `solve`. Note, of course, that the first root, $x = \sin^{-1}3$, does not make any sense (in the context of the problem) because the domain of the arcsine function is $[-1,1]$.

```
> solve(sin(x)^2-2*sin(x)-3=0);
```

$$-\frac{1}{2}\pi, \arcsin(3),$$

When we set `_EnvAllSolutions:=true`, Maple succeeds in finding all solutions to the equation. The result indicates that all *real* solutions to the equation are $x = -\pi/2 + 2n\pi$, n any integer.

```
> _EnvAllSolutions:=true:
  solve(sin(x)^2-2*sin(x)-3=0);
```

$$-\frac{1}{2}\pi + 2\pi\_Z1\sim,\ \arcsin(3) - 2\arcsin(3)\_B1 + 2\pi\_Z1\sim + \pi\_B1\sim$$

■

We can also use `solve` to find the solutions, if any, of various types of systems of equations. For example,

```
solve[{lhs1=rhs1,lhs2=rhs2},{x,y}]
```

solves a system of two equations for x and y, while entering

```
solve[{lhs1=rhs1,lhs2=rhs2}]
```

attempts to solve the system of equations for all unknowns. In general, `solve` can find the solutions to a system of linear equations. In fact, if the systems to be solved are inconsistent or dependent, Maple's output will tell you so.

EXAMPLE 3: Solve each system: (a) $\begin{cases} 3x - y = 4 \\ x + y = 2 \end{cases}$; (b) $\begin{cases} 2x - 3y + 4z = 2 \\ 3x - 2y + z = 0 \\ x + y - z = 0 \end{cases}$;

(c) $\begin{cases} 2x - 2y - 2z = -2 \\ -x + y + 3z = 0 \\ -3x + 3y - 2z = 1 \end{cases}$; and (d) $\begin{cases} -2x + 2y - 2z = -2 \\ 3x - 2y + 2z = 2 \\ x + 3y - 3z = -3 \end{cases}$.

SOLUTION: In each case we use `solve` to solve the given system.

(a) For this system, the result of entering

```
> solve({3*x-y=4,x+y=2});
```

$$\{y = \frac{1}{2}, x = \frac{3}{2}\}$$

means that the solution of $\begin{cases} 3x - y = 4 \\ x + y = 2 \end{cases}$ is $(x,y) = (3/2,1/2)$.

(b) We can verify that the results returned by Maple are correct. First, we name the system of equations `sys`, and then we use `solve` to solve the system of equations, naming the result `sols`.

```
> sys:={2*x-3*y+4*z=2,3*x-2*y+z=0,x+y-z=1}:
> sols:=solve(sys);
```

$$sols := \left\{x = \frac{7}{10}, z = \frac{3}{2}, y = \frac{9}{5}\right\}$$

We verify the result by substituting the values obtained with `solve` back into `sys` with

```
> subs(sols,sys);
```

$$\{1 = 1, 0 = 0, 2 = 2\}$$

(c) When we use `solve` to solve this system, Maple returns nothing, which indicates that the system has no solution; the system is inconsistent.

```
> solve({2*x-2*y-2*z=-2,-x+y+3*z=0,-3*x+3*y-2*z=1},
  {x,y,z});
```

(d) On the other hand, when we use solve to solve this system, Maple's result indicates that the system has infinitely many solutions. That is, all ordered triples of the form $\{(0, z-1, z) | z \text{ real}\}$ are solutions of the system.

```
> solve({-2*x+2*y-2*z=-2,3*x-2*y+2*z=2,x+3*y-3*z=-3});
```

$$\{z = z, y = z - 1, x = 0\}$$

■

We can often use solve to find solutions of a nonlinear system of equations as well.

EXAMPLE 4: Solve the systems (a) $\begin{cases} 4x^2 + y^2 = 4 \\ x^2 + 4y^2 = 4 \end{cases}$ and (b) $\begin{cases} \dfrac{x^2}{a^2} + \dfrac{y^2}{b^2} = 1 \\ y = mx \end{cases}$ (a, b greater than zero) for x and y.

SOLUTION: The graphs of the equations $4x^2 + y^2 = 4$ and $x^2 + 4y^2 = 4$ are both ellipses. We use contourplot to graph each equation. The solutions to the system $\begin{cases} 4x^2 + y^2 = 4 \\ x^2 + 4y^2 = 4 \end{cases}$ correspond to the intersection points of the two graphs.

```
> with(plots):
  cp1:=contourplot(4*x^2+y^2,x=-4..4,y=-4..4,
     contours=[4],color=black):
  cp2:=contourplot(x^2+4*y^2,x=-4..4,y=-4..4,
     contours=[4],color=black):
  display({cp1,cp2});
```

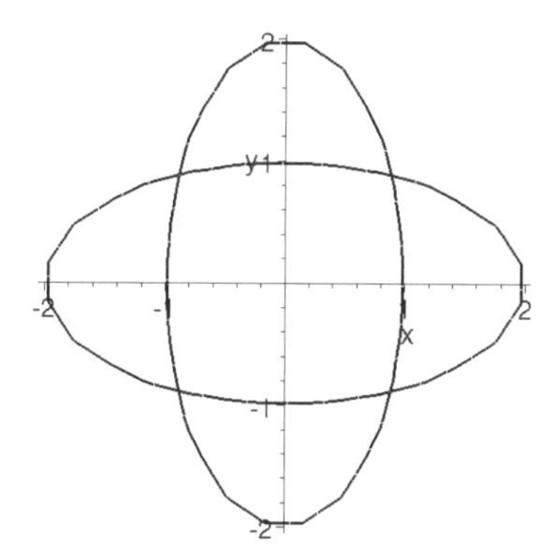

Now we use `solve` to find the solutions of the system.

```
> sola:=solve({4*x^2+y^2=4,x^2+4*y^2=4});
```

$$sola := \{ y = 2 \operatorname{RootOf}(5\_Z^2 - 1), x = 2 \operatorname{RootOf}(5\_Z^2 - 1) \}$$

Notice that the solution is given in terms of `RootOf`. The exact values are obtained with `allvalues`.

```
> allvalues(sola);
```

$$\{ y = \frac{2}{5}\sqrt{5}, x = \frac{2}{5}\sqrt{5} \}, \{ y = -\frac{2}{5}\sqrt{5}, x = -\frac{2}{5}\sqrt{5} \}$$

For (b), we also use `solve` to find the solutions of the system. However, because the unknowns in the equations are *a*, *b*, *m*, *x*, and *y*, we must specify that we want to solve for x and y in the `solve` command. Once again we use `allvalues` to obtain the exact value of the solutions.

```
> solb:=solve({x^2/a^2+y^2/b^2=1,y=m*x},{x,y});
```

$$solb := \{ y = m \operatorname{RootOf}((b^2 + m^2 a^2)\_Z^2 - 1)\, b\, a,$$
$$x = \operatorname{RootOf}((b^2 + m^2 a^2)\_Z^2 - 1)\, b\, a \}$$

```
> allvalues(solb);
```

$$\left\{ x = \frac{ba}{\sqrt{\%1}}, y = \frac{mba}{\sqrt{\%1}} \right\}, \left\{ y = -\frac{mba}{\sqrt{\%1}}, x = -\frac{ba}{\sqrt{\%1}} \right\}$$

$$\%1 := b^2 + m^2 a^2$$

■

Although Maple can find the exact solution to every polynomial equation of degree four or less, exact solutions to some equations may not be meaningful. In those cases, Maple can provide approximations of the exact solutions using `evalf`.

EXAMPLE 5: Approximate the solutions to the equations (a) $x^4 - 2x^2 = 1 - x$; and (b) $1 - x^2 = x^3$.

SOLUTION: Each of these is a polynomial equation with degree less than five, so `solve` will find the exact solutions of each equation. However, the solutions are quite complicated, so we use `evalf` to obtain approximate solutions of all solutions to each equation.

For (a), entering

```
> evalf(allvalues(solve(x^4-2*x^2=1-x)));
```

$$.1827773505 + .6333971206\,I,\ .1827773505 - .6333971206\,I,\ 1.345089395,\ -1.710644096$$

first finds the exact solutions of the equation $x^4 - 2x^2 = 1 - x$ and then computes approximations of those solutions. For (b), entering

```
> solb:=solve(1-x^2=x^3,x);
```

$$solb := \frac{1}{6}\%1^{\left(\frac{1}{3}\right)} + \frac{2}{3}\frac{1}{\%1^{\left(\frac{1}{3}\right)}} - \frac{1}{3},$$

$$-\frac{1}{12}\%1^{\left(\frac{1}{3}\right)} - \frac{1}{3}\frac{1}{\%1^{\left(\frac{1}{3}\right)}} - \frac{1}{3} + \frac{1}{2}I\sqrt{3}\left(\frac{1}{6}\%1^{\left(\frac{1}{3}\right)} - \frac{2}{3}\frac{1}{\%1^{\left(\frac{1}{3}\right)}}\right),$$

$$-\frac{1}{12}\%1^{\left(\frac{1}{3}\right)} - \frac{1}{3}\frac{1}{\%1^{\left(\frac{1}{3}\right)}} - \frac{1}{3} - \frac{1}{2}I\sqrt{3}\left(\frac{1}{6}\%1^{\left(\frac{1}{3}\right)} - \frac{2}{3}\frac{1}{\%1^{\left(\frac{1}{3}\right)}}\right),$$

$$\%1 := 100 + 12\sqrt{69}$$

first finds the exact solutions of the equation $1 - x^2 = x^3$ and names the result `solb`. Then, entering

```
> evalf(solb);
```

$$.7548776667, -.8774388331 + .7448617670\,I, -.8774388331 - .7448617670\,I$$

computes approximations of those solutions.

■

Maple can also solve equations involving more than one variable for one of the variables in terms of the other variables.

EXAMPLE 6: (a) Solve the equation $v = \pi r^2/h$ for h. (b) Solve the equation $a^2 + b^2 = c^2$ for c.

SOLUTION: These equations involve more than one unknown, so we must specify the variable for which we are solving. Thus, entering

```
> solve(v=Pi*r^2/h,h);
```

$$\frac{\pi\, r^2}{v}$$

solves the equation $v = \pi r^2/h$ for h. Note that if we had wanted to solve for r instead, then we would have entered `solve(v=π r^2*h,r)`. Similarly, entering

```
> solve(a^2+b^2=c^2,a);
```

$$\sqrt{-b^2 + c^2}, -\sqrt{-b^2 + c^2}$$

solves the equation $a^2 + b^2 = c^2$ for a.

■

Numerical Approximation of Solutions of Equations

When solving an equation is either impractical or impossible, Maple provides the function `fsolve` to approximate the roots of equations. In the previous examples above, Maple first computed the solutions of the equations exactly before approximating their value with `evalf`. However, `fsolve` uses a numerical method to approximate the roots. This command has several different forms depending on the equation. We first consider a polynomial equation. In this case, `fsolve(lhs=rhs)` or `fsolve(lhs=rhs,var)` attempts to compute all real roots of the polynomial equation `lhs=rhs`. To compute all roots (including complex roots), the `complex` option must be indicated.

EXAMPLE 7: Approximate the solutions of $x^5+x^4-4x^3+2x^2-7=0$.

SOLUTION: Approximations of all the real roots are determined with the first command. Then, with the `complex` option setting, the real and complex roots are given.

```
> fsolve(x^5+x^4-4*x^3+2*x^2-3*x-7=0);
```

$$-2.744632420, -.8808584760, 1.796450526$$

```
> fsolve(x^5+x^4-4*x^3+2*x^2-3*x-7=0,
  x,complex);
```

$$-2.744632420, -.8808584760, .4145201849-1.199959840I,$$
$$.4145201849+1.199959840\,I, 1.796450526$$

■

For a general equation, `fsolve` attempts to determine a single real root.

EXAMPLE 8: Approximate the positive solutions of the equation $\cos x - x = 0$.

SOLUTION: Note that because $\cos x \leq 1$ for all values of x, $\cos x < x$ if $x > 1$. Thus, all positive solutions of the equation $\cos x - x = 0$, if any, must be contained in the interval [0,1]. We begin by using `plot` to graph the function $\cos x - x$ and see that the equation $\cos x - x = 0$ has only one real solution.

```
> plot(cos(x)-x,x=0..1,color=black);
```

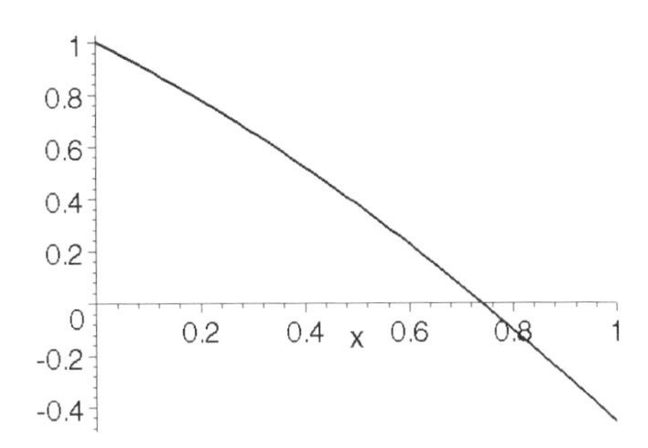

Because this equation has only one real solution, `fsolve` finds the approximate solution with the following command:

```
> fsolve(cos(x)-x=0);
```

                    .7390851332

■

`fsolve` can also be used to approximate solutions to systems of equations.

EXAMPLE 9: Approximate the solutions to the system of equationsy

$$\begin{cases} x^2 + 4xy + y^2 = 4 \\ 5x^2 - 4xy + 2y^2 = 8 \end{cases}$$

SOLUTION: We begin by using `contourplot` to graph each equation. From the resulting graph, we see that $x^2 + 4xy + y^2 = 4$ is a hyperbola, $5x^2 - 4xy + 2y^2 = 8$ is an ellipse, and there are four solutions to the system of equations.

```
> with(plots):
  cp1:=contourplot(x^2+4*x*y+y^2,x=-4..4,y=-4..4,
    contours=[4],grid=[30,30],color=black):
  cp2:=contourplot(5*x^2-4*x*y+2*y^2,x=-4..4,y=-4..4,
    contours=[8],grid=[30,30],color=black,linestyle=4):
  display({cp1,cp2});
```

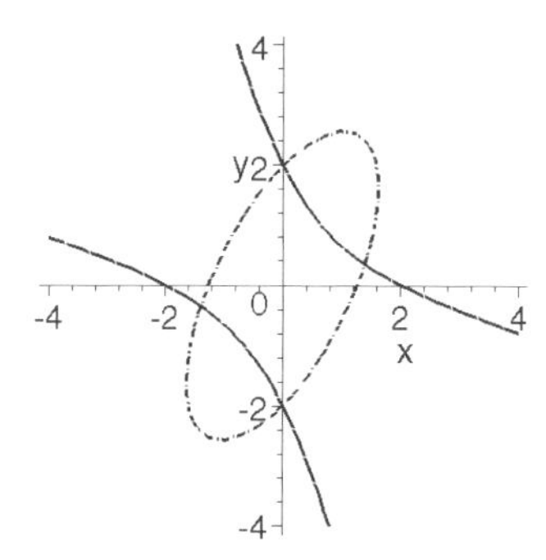

Now, we use `fsolve` to approximate the solutions to the equation. Our first attempt finds the solution $x = 0$, $y = -2$. To find the remaining solutions, we must specify the interval containing the solution.

```
> fsolve({x^2+4*x*y+y^2=4,5*x^2-4*x*y+2*y^2=8});
```

$$\{x = 0, y = -2.000000000\}$$

Thus, entering

```
> fsolve({x^2+4*x*y+y^2=4,5*x^2-4*x*y+2*y^2=8},
   {x=-2..0,y=-1..0});
```

$$\{x = -1.392621248, y = -.3481553119\}$$

searches for a solution that satisfies $-2 \le x \le 0$ and $-1 \le y \le 0$; entering

```
> fsolve({x^2+4*x*y+y^2=4,5*x^2-4*x*y+2*y^2=8},{x,y},
   {x=-1..1,y=-3..-1});
```

$$\{x = 0, y = -2.000000000\}$$

searches for a solution that satisfies $-1 \le x \le 1$ and $-3 \le y \le -1$; and entering

```
> fsolve({x^2+4*x*y+y^2=4,5*x^2-4*x*y+2*y^2=8},{x,y},
   {x=0..2,y=0..1});
```

$$\{x = 1.392621248, y = .3481553119\}$$

searches for a solution that satisfies $0 \le x \le 2$ and $0 \le y \le 1$.

■

Application: Intersection Points of Graphs of Functions

In several later examples, we will need to locate the intersection points of graphs of functions. Here we discuss several methods to locate the intersection points of graphs of functions.

EXAMPLE 10: Find the point(s) where the graphs of $h(x) = -3x^2 + 12x - 5$ and $k(x) = 2x^2 - 4x - 3$ intersect.

SOLUTION: Although it is not necessary to do this to solve the problem, we first graph h and k and see that these functions intersect twice.

```
> h:='h':k:='k':
  h:=x->-3*x^2+12*x-5:
  k:=x->2*x^2-4*x-3:
  plot({h(x),k(x)},x=-1..5,color=[black,gray]);
```

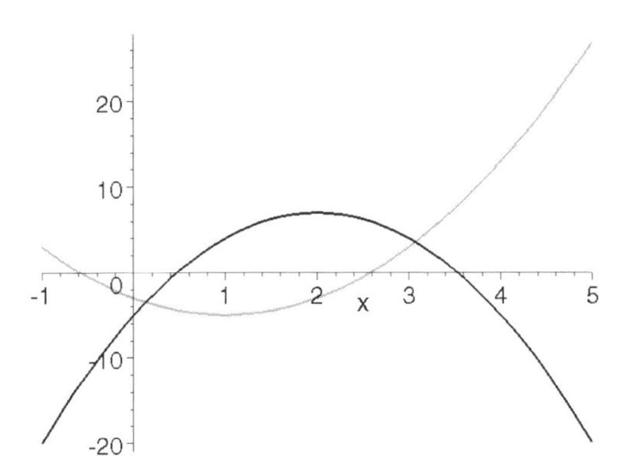

The x-coordinates of the intersection points satisfy the equation $h(x) = k(x)$. Consequently, to locate the intersection points, we must solve the quadratic equation $h(x) = k(x)$, which Maple can solve exactly because it is a polynomial equation with degree less than five.

```
> xcoords:=solve(h(x)=k(x));
```

$$xcoords := \frac{8}{5} + \frac{3}{5}\sqrt{6}, \frac{8}{5} - \frac{3}{5}\sqrt{6}$$

We use `evalf` to obtain an approximation of these numbers as well.

```
> evalf(xcoords);
```

3.069693846,.13036154

The y-coordinates of the intersection points are found by evaluating $h(x)$ (or $k(x)$) at the values of x obtained in `xcoords`.

```
> simplify(h(xcoords[1]));
```

$$\frac{1}{25}+\frac{36}{25}\sqrt{6}$$

```
> simplify(h(xcoords[2]));
```

$$\frac{1}{25}-\frac{36}{25}\sqrt{6}$$

Thus, we conclude that the graphs of h and k intersect at

$$\left(\frac{1}{5}(8-3\sqrt{6}), \frac{1}{25}(1-36\sqrt{6})\right) \approx (0.130306, -3.48727)$$

and

$$\left(\frac{1}{5}(8+3\sqrt{6}), \frac{1}{25}(1+36\sqrt{6})\right) \approx (3.06969, 3.56727).$$

■

If the equation involves functions other than polynomials and cannot be solved with `solve`, we must use the `interval` option with `fsolve`. The command `fsolve(lhs=rhs,a..b)` or `fsolve(lhs=rhs,x,a..b)` searches for a root of the equation `lhs=rhs` contained in the interval (a,b). This command can be entered as `fsolve(lhs=rhs,x=a..b)` or `fsolve(lhs=rhs,x,x=a..b)` as well. If the equation or system of equations involves more than one variable, then the option setting `{x=a..b,y=c..d,...}` is used to find a solution on the intersection of these open intervals. The interval option is demonstrated below.

EXAMPLE 11: Approximate the points where the graphs of $f(x) = e^{-(x/4)^2}\cos(x/\pi)$ and $g(x) = \frac{5}{4}+\sin x^{3/2}$ intersect.

SOLUTION: Notice that the x-coordinates of the intersection points satisfy the equation $f(x) = g(x)$. Consequently, to locate the points of intersection, it is sufficient to solve the equation $f(x) = g(x)$. Since this problem does not involve polynomials, we first graph f and g and notice that they intersect twice. In order to approximate the intersection points, we will use `fsolve` with the `interval` option described above. The two roots appear to be on the interval (2,3). Hence, we use the `fsolve` `interval` option setting `2..3` which yields the root $x \approx 2.541050127$. To obtain the second root, which appears to be near 3, we use the interval `2.75..3` which gives us the desired root $x \approx 2.974603408$.

```
> f:='f':g:='g':
  f:=x->exp(-(x/4)^2)*cos(x/Pi):
  g:=x->sin(x^(3/2))+5/4:
  plot({f(x),g(x)},x=0..5,
     color=[black,black],linestyle=[1,4]);
```

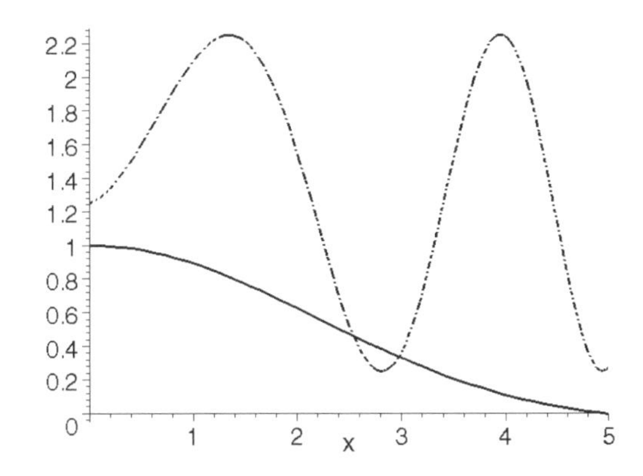


```
> fsolve(f(x)=g(x),x,2..3);
```

2.541050127

```
> fsolve(f(x)=g(x),x,2.75..3);
```

2.974603408

We conclude that one intersection point is approximately (2.54105, 0.461103), and the other intersection point is approximately (2.9746, 0.336066).

■

2.5 Simplifying Expressions with Trigonometric Terms

Several commands are useful in working with expressions involving trigonometric functions. For example, Maple can be used to test whether or not expressions are equivalent. This is done with the command `testeq(expr1=expr2)`, where the expressions can involve rational constants, `I`, combined algebraic operations, exponentials, and trigonometric functions. The result of `testeq` is `true` or `false`. We illustrate with the example below.

EXAMPLE 1: Determine if the following are trigonometric identities: (a) $\cos(x+y) = \cos x \cos y - \sin x \sin y$ and (b) $\sin 2x = 2\sin x$.

SOLUTION: Since the output of the first command is `true`, the equation in (a) is an identity. However, that in (b) is not an identity since the result is `false`.

```
> testeq(cos(x+y)=cos(x)*cos(y)-sin(x)*sin(y));
  testeq(sin(2*x)=2*sin(x));
```

```
true
false
```

■

Expressions involving trigonometric functions can be simplified with the command `simplify(expr,trig)`. This command uses the identity $\sin^2 x + \cos^2 x = 1$ to simplify these expressions and uses $\cosh^2 x - \sinh^2 x = 1$ to simplify expressions involving hyperbolic trigonometric functions.

EXAMPLE 2: Simplify the trigonometric expressions: (a) $\sin^3 x + \cos^3 x$ and (b) $\dfrac{\sec^4 x - 1}{\tan^2 x}$.

SOLUTION: These two expressions are simplified with `simplify` with the `trig` option setting.

```
> simplify(sin(x)^3+cos(x)^3,trig);
```

$$\cos(x)^3 + \sin(x) - \sin(x)\cos(x)^2$$

```
> simplify((sec(x)^4-1)/tan(x)^2,trig);
```

$$\frac{\cos(x)^2+1}{\cos(x)^2}$$

■

Maple provides the command `combine(expr,trig)` to combine powers and products of trigonometric terms into a sum of trigonometric terms. This is done by applying the identities $\sin x \sin y = \frac{1}{2}(\cos(x-y) - \cos(x+y))$, $\sin x \cos y = \frac{1}{2}(\sin(x-y) - \sin(x+y))$, and $\cos x \cos y = \frac{1}{2}(\cos(x-y) + \cos(x+y))$, until the desired result is obtained. We use this command to simplify the expression considered in Example 2 to illustrate the difference in `simplify(expr,trig)` and `combine(expr,trig)`.

EXAMPLE 3: Transform the expression $\sin^3 x + \cos^3 x$ into a sum of trigonometric terms.

SOLUTION: This transformation is carried out with the following command.

```
> combine(sin(x)^3+cos(x)^3,trig);
```

$$-\frac{1}{4}\sin(3x) + \frac{3}{4}\sin(x) + \frac{1}{4}\cos(3x) + \frac{3}{4}\cos(x)$$

■

CHAPTER 3

Calculus

Chapter 3 introduces Maple's built-in calculus commands. The examples used to illustrate the various commands are similar to examples routinely done in first-year calculus courses.

3.1 Computing Limits

One of the first topics discussed in calculus is limits. Maple can be used to investigate limits graphically and numerically. In addition, Maple uses the command

```
limit(f(x),x=a)
```

to find $\lim_{x \to a} f(x)$: the limit of `f(x)` as `x` approaches the value `a`, where `a` can be a finite number, positive infinity (`infinity`), or negative infinity (`-infinity`).

EXAMPLE 1: Use a graph and table of values to investigate $\lim_{x \to 0} \frac{\sin 3x}{x}$.

SOLUTION: We clear all prior definitions of f, define $f(x) = \frac{\sin 3x}{x}$, and then graph f on the interval $[-\pi, \pi]$ with `plot`.

```
> f:='f':
  f:=x->sin(3*x)/x:
  plot(f(x),x=-Pi..Pi,color=black);
```

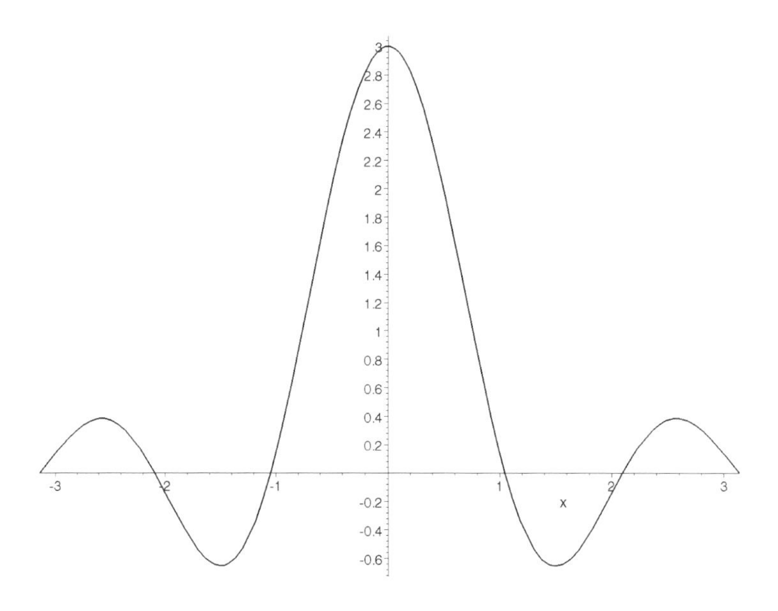

From the graph we might, correctly, conclude that $\lim_{x\to 0} \frac{\sin 3x}{x} = 3$. Further evidence that $\lim_{x\to 0} \frac{\sin 3x}{x} = 3$ can be obtained by computing the values of $f(x)$ for values of x "near" 0. In the following, we use `rand` to define `xvals` to be a table of 6 "random" real numbers. The first number in `xvals` is between –1 and 1, the second between $-1/10$ and $1/10$, and so on. (Note that because we are generating "random" numbers, your results will almost certainly differ from those obtained here.)

```
> xvals:=[seq((-1)^rand()*rand()*10.^(-12-n),n=0..5)];
```

$xvals :=$

$[.6283634430, .0005862664913, .006438424438, -.0006720753584, -.00009916381555, -.7428672314\,10^{-5}]$

We then use `map` to compute the value of $f(x)$ for each x in `xvals`.

```
> fvals:=map(f,xvals);
```

$fvals := [\,1.513478994, 2.999998453, 2.999813463, 2.999997967, 2.999999956, 3.000000000\,]$

From these values, we might again correctly deduce that $\lim_{x \to 0} \frac{\sin 3x}{x} = 3$. Of course, these results do not prove that $\lim_{x \to 0} \frac{\sin 3x}{x} = 3$, but they are helpful in convincing us that $\lim_{x \to 0} \frac{\sin 3x}{x} = 3$.

■

Computing Limits

Some limits involving rational functions can be computed by factoring the numerator and denominator.

EXAMPLE 2: Compute $\lim_{x \to -9/2} \frac{2x^2 + 25x + 72}{72 - 47x - 14x^2}$.

SOLUTION: We define `frac1` to be the rational expression $\frac{2x^2 + 25x + 72}{72 - 47x - 14x^2}$. We then attempt to compute the value of $\frac{2x^2 + 25x + 72}{72 - 47x - 14x^2}$ if $x = -9/2$ but see that it is undefined.

```
> frac1:=(2*x^2+25*x+72)/(72-47*x-14*x^2):
  eval(frac1,x=-9/2);
```

```
Error, division by zero
```

Factoring the numerator and denominator with `factor`, `numer`, and `denom`, we see that

$$\lim_{x \to -9/2} \frac{2x^2 + 25x + 72}{72 - 47x - 14x^2} = \lim_{x \to -9/2} \frac{(x+8)(2x+9)}{(8-7x)(2x+9)} = \lim_{x \to -9/2} \frac{x+8}{8-7x}.$$

The fraction $\frac{x+8}{8-7x}$ is named `frac2` and the limit is evaluated by computing the value of `frac2` if $x = -9/2$.

```
> factor(numer(frac1));
```

$$-(x+8)(2x+9)$$

```
> factor(denom(frac1));
```

$$(2x+9)(7x-8)$$

```
> frac2:=simplify(frac1);
```

$$frac2 := -\frac{x+8}{7x-8}$$

```
> eval(frac2,x=-9/2);
```

$$\frac{7}{79}$$

We conclude that $\lim\limits_{x \to -9/2} \frac{2x^2+25x+72}{72-47x-14x^2} = \frac{7}{79}$.

■

We can also use the `limit` command to evaluate frequently encountered limits.

EXAMPLE 3: Calculate the indicated limits.
(a) $\lim\limits_{x \to -5/3} \frac{3x^2-7x-20}{21x^2+14x-35}$; b) $\lim\limits_{x \to 0} \frac{\sin x}{x}$;
(c) $\lim\limits_{x \to +\infty} \frac{50x^2+95x+24}{20x^2+77x+72}$; and (d) $\lim\limits_{x \to -\infty} \frac{1+4x-16x^2-64x^3}{20x^2+13x+2}$.

SOLUTION: In each case, we use `limit` to evaluate the indicated limit. Entering

```
> limit((3*x^2-7*x-20)/(21*x^2+14*x-35),x=-5/3);
```

$$\frac{17}{56}$$

computes $\lim\limits_{x \to -5/3} \frac{3x^2-7x-20}{21x^2+14x-35}$; $= \frac{17}{56}$; entering

```
> limit(sin(x)/x,x=0);
```

$$1$$

computes $\lim_{x \to 0} \frac{\sin x}{x} = 1$; entering

```
> limit((50*x^2+95*x+24)/(20*x^2+77*x+72),x=infinity);
```

$$\frac{5}{2}$$

computes $\lim_{x \to +\infty} \frac{50x^2 + 95x + 24}{20x^2 + 77x + 72} = \frac{5}{2}$; and entering

```
> limit((1+4*x-16*x^2-64*x^3)/(20*x^2+13*x+2),
   x=-infinity);
```

$$\infty$$

computes $\lim_{x \to -\infty} \frac{1 + 4x - 16x^2 - 64x^3}{20x^2 + 13x + 2} = +\infty$.
■

In differential calculus, we learn that the **derivative** of f at x is given by

$$f'(x) = \lim_{h \to 0} \frac{f(x+h) - f(x)}{h},$$

provided the limit exists. The `limit` can be used along with `simplify` to compute the derivative of a function using the definition of the derivative.

EXAMPLE 4: Compute and simplify (a) $\frac{g(x+h) - g(x)}{h}$ and (b) $\lim_{h \to 0} \frac{g(x+h) - g(x)}{h}$ if $g(x) = x^3 - 3x^2 + x + 1$.

SOLUTION: We proceed by clearing prior definitions of g, defining g, and then computing and simplifying $\frac{g(x+h) - g(x)}{h}$.

```
> g:='g':
  g:=x->x^3-x^2+x+1:
```

```
> stepone:=simplify((g(x+h)-g(x))/h);
```

$$stepone := 3x^2 + 3xh + h^2 - 2x - h + 1$$

Although the limit $\lim_{h\to 0} \frac{g(x+h)-g(x)}{h}$ could be computed directly with the command `limit(stepone,h=0)`, we use the inert form of the `limit` command, `Limit`, and then use `value` to compute the value of the limit. The result, $g'(x)$, is named `dg`..

```
> steptwo:=Limit(stepone,h=0);
```

$$steptwo := \lim_{h\to 0} 3x^2 + 3xh + h^2 - 2x - h + 1$$

```
> dg:=value(steptwo);
```

$$dg := 3x^2 + 1 - 2x$$

Last, we use `plot` to graph $g(x)$ and $g'(x)$. The graph of `dg` ($g'(x)$) is dashed; the graph of $g(x)$ is in black.

```
> plot({g(x),dg},x=-2..4,y=-2..4,
        color=[black,black],linestyle=[1,4]);
```

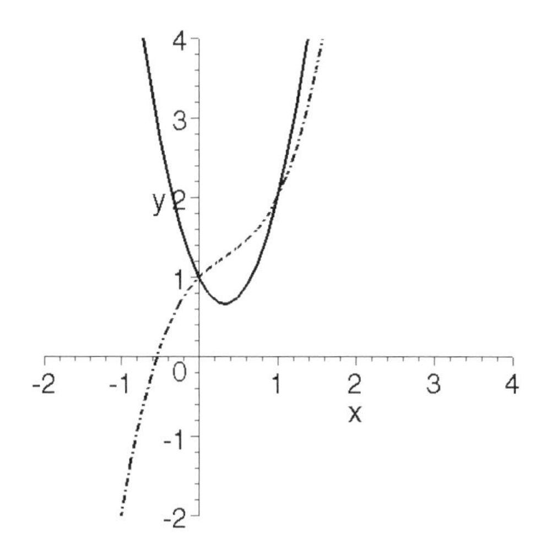

■

The following example illustrates how several Maple commands can be combined in a single statement to obtain the desired result.

EXAMPLE 5: Compute and simplify (a) $\frac{f(x+h)-f(x)}{h}$ and (b) $\lim_{h\to 0}\frac{f(x+h)-f(x)}{h}$ if $f(x)=\frac{1}{\sqrt{x}}+\sqrt{x}$.

SOLUTION: In the same manner as in Example 4, we begin by defining *f*:

```
> f:='f':
  f:=x->1/sqrt(x)+sqrt(x):
```

and then computing $\frac{f(x+h)-f(x)}{h}$ and naming the resulting output `stepone`:

```
> stepone:=simplify((f(x+h)-f(x))/h);
```

$$stepone := -\frac{-\sqrt{x}-x^{(3/2)}-\sqrt{x}\,h+\sqrt{x+h}+\sqrt{x+h}\,x}{\sqrt{x+h}\,\sqrt{x}\,h}$$

and finally computing $\lim_{h\to 0}\frac{f(x+h)-f(x)}{h}$.

```
> df:=limit(stepone,h=0);
```

$$df := \frac{1}{2}\frac{x-1}{x^{(3/2)}}$$

Last, we use `plot` to graph $f(x)$ = 1and $f'(x)$ (`df`). To see that *f* has a minimum at $x = 1$, we use the option `y=-4..4` to indicate that the range displayed (the vertical axis) corresponds to the interval [-4,4].

```
> plot({f(x),df},x=0..8,y=-4..4,
   color=[black,black],linestyle=[1,4]);
```

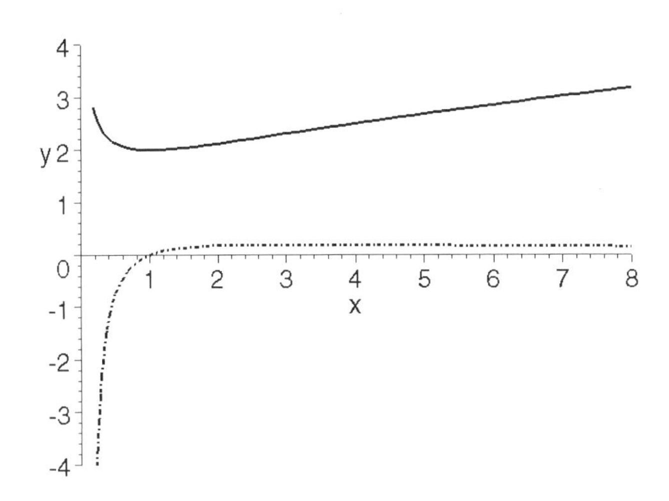

■

One-Sided Limits

In some cases, Maple can compute certain one-sided limits. The command

```
limit(f(x),x=a,left)
```

attempts to compute $\lim_{x \to a^-} f(x)$ while

```
limit(f(x),x=a,right)
```

attempts to compute $\lim_{x \to a^+} f(x)$.

EXAMPLE 6: Compute (a) $\lim_{x \to 0^-} \frac{1}{x}$; (b) $\lim_{x \to 0^+} \frac{1}{x}$; (c) $\lim_{x \to 0^+} \frac{|x|}{x}$; and (d) $\lim_{x \to 0^-} \frac{|x|}{x}$.

SOLUTION: Entering

```
> limit(1/x,x=0,left);
```

$$-\infty$$

```
> limit(1/x,x=0,right);
```

$$\infty$$

Even though $\lim_{x \to 0} \frac{|x|}{x}$ does not exist, both $\lim_{x \to 0^-} \frac{|x|}{x}$ and $\lim_{x \to 0^+} \frac{|x|}{x}$ do exist. The `right` and `left` settings are used to calculate the correct values for (c) and (d), respectively.

```
> limit(abs(x)/x,x=0);
```

undefined

Thus, entering

```
> limit(abs(x)/x,x=0,right);
```

1

computes $\lim_{x \to 0^+} \frac{|x|}{x}$; and entering

```
> limit(abs(x)/x,x=0,left);
```

-1

computes $\lim_{x \to 0^-} \frac{|x|}{x}$

■

3.2 Differential Calculus

The **derivative** of $y = f(x)$ is

$$f'(x) = \lim_{h \to 0} \frac{f(x+h) - f(x)}{h},$$

provided this limit exists. If $f'(x)$ exists for all values of x in (a,b) then f is said to be **differentiable** on (a,b). One geometric interpretation of $f'(x)$ is that $f'(x)$ is the slope of the line tangent to the graph of f at the point $(x, f(x))$. Recall that in Section 3.1, we used limiting process to compute $f'(x)$ for several functions. A typical example in calculus texts, when the derivative is introduced, involves choosing a value of x_0 and then sketching the graph of the secant line passing through the points $(x_0, f(x_0))$ and $(x_0 + h, f(x_0 + h))$ for "small" values of h

to show that as h approaches 0, the secant line approaches the tangent line. An equation of the secant line passing through the points $(x_0, f(x_0))$ and $(x_0 + h, f(x_0 + h))$ is given by

$$y - f(x_0) = \frac{f(x_0 + h) - f(x_0)}{h}(x - x_0)$$

$$y = \frac{f(x_0 + h) - f(x_0)}{h}(x - x_0) + f(x_0).$$

EXAMPLE 1: Let $f(x) = e^{-(x-\pi/2)^2}\cos x$. Sketch the graph of f on the interval $[0, \pi]$. Show the graph of f along with the line passing through the points $(\pi/2, f(\pi/2))$ and $(\pi/2 + h, f(\pi/2 + h))$ for small values of h.

SOLUTION: Because we will be using the `animate` and `display` commands, we first load the `plots` package and then define f.

```
> with(plots):
> f:=x->exp(-(x-Pi/2)^2)*cos(x):
```

Information about the `animate` function is obtained with `?animate`.

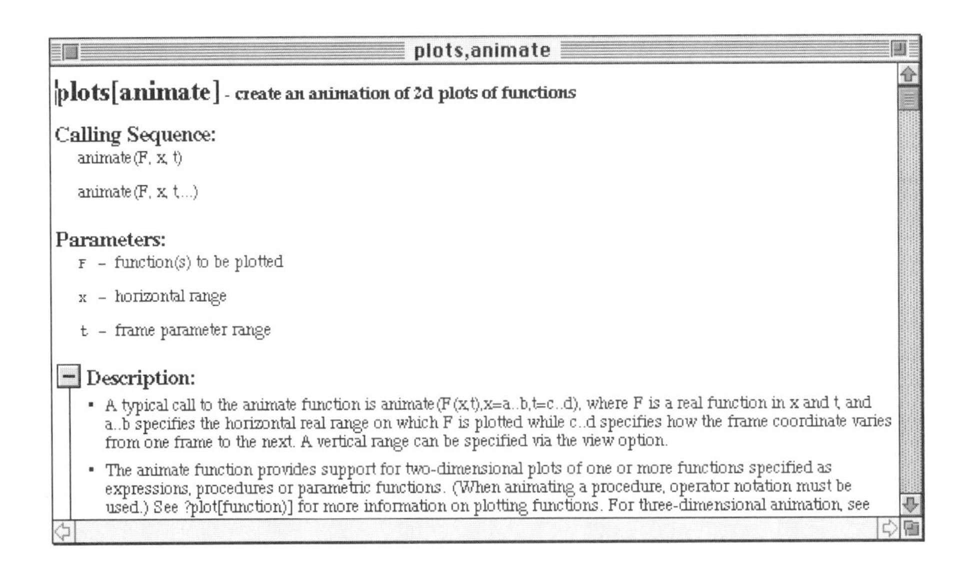

plots,animate

plots[animate] - create an animation of 2d plots of functions

Calling Sequence:

animate(F, x, t)

animate(F, x, t,...)

Parameters:

F - function(s) to be plotted

x - horizontal range

t - frame parameter range

Description:

- A typical call to the animate function is animate(F(x,t),x=a..b,t=c..d), where F is a real function in x and t, and a..b specifies the horizontal real range on which F is plotted while c..d specifies how the frame coordinate varies from one frame to the next. A vertical range can be specified via the view option.
- The animate function provides support for two-dimensional plots of one or more functions specified as expressions, procedures or parametric functions. (When animating a procedure, operator notation must be used.) See ?plot[function)] for more information on plotting functions. For three-dimensional animation, see

For this problem, $x_0 = \pi/2$ and `secline(x,h)` is the line passing through the points $(\pi/2, f(\pi/2))$ and $(\pi/2, +h, f(\pi/2+h))$.

```
> x0:=Pi/2:
> secline:=(x,h)->(f(x0+h)-f(x0))/h*(x-x0)+f(x0):
```

We then use `animate` to graph f and `secline(x,h)` for $h = 1/n, n = 1, 2, \ldots, 9$.

```
> A:=animate({f(x),secline(x,1/n)},x=0..Pi,n=1..9,
   frames=9,color=black,tickmarks=[2,4]):
```

These nine graphics are displayed as an array with `display`.

```
> display(A);
```

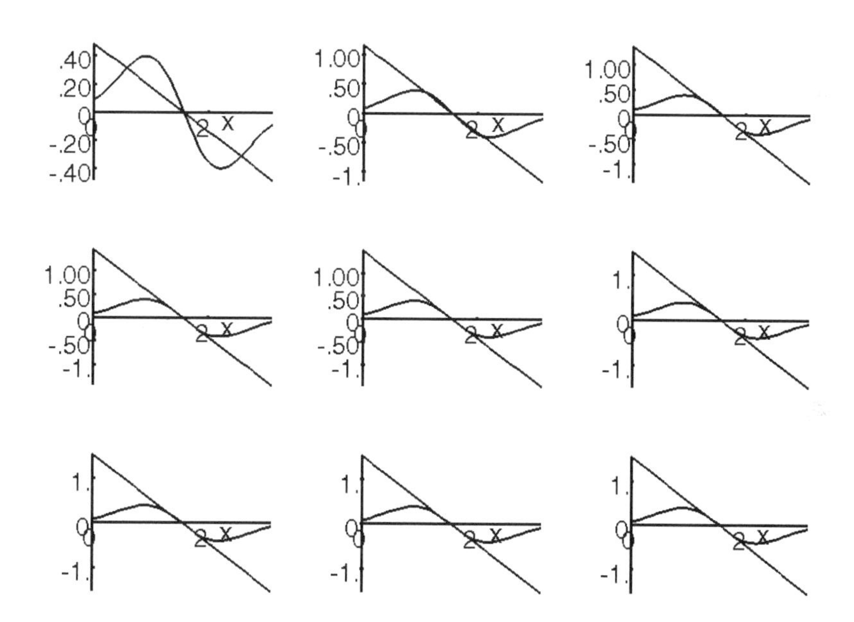

We can also use `animate` to generate two-dimensional animations. With the following command, we generate a fifty-frame animation.

```
> animate({f(x),secline(x,1/n)},x=0..Pi,n=1..10,
   frames=50,color=black,tickmarks=[2,4]);
```

Once you have selected the animation,

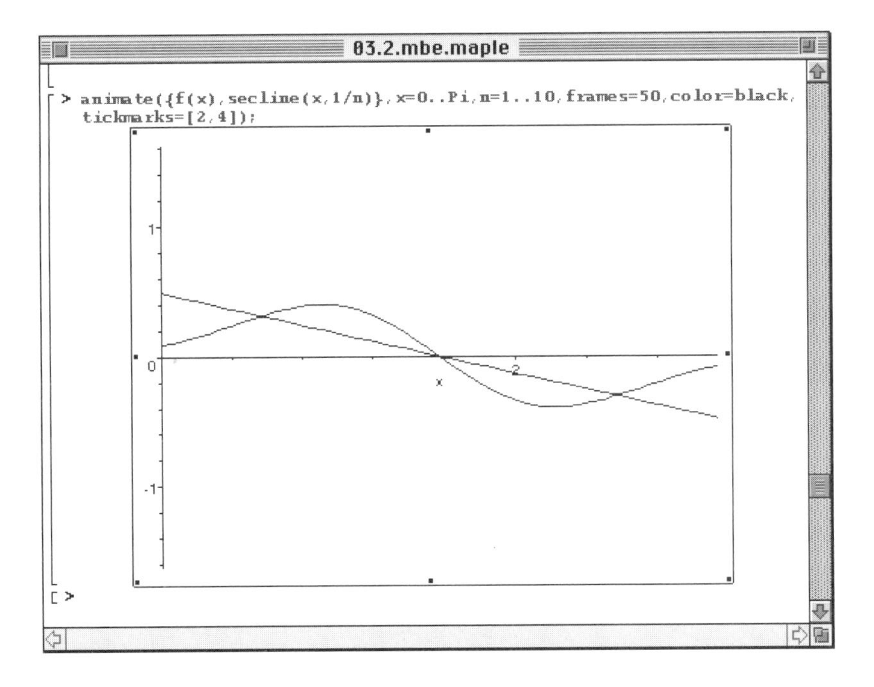

you can control the animation from the Maple menu

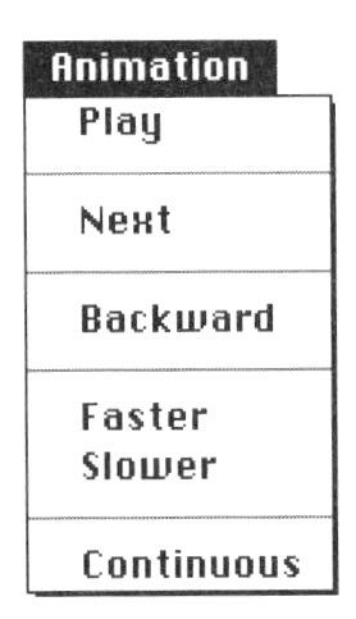

or from the control strip.

■

Maple is able to compute the derivative of a variety of functions with the commands `D` and `diff`. Note that `Diff` represents the inert form of the `diff` command. This means that `Diff(expression)` returns unevaluated; the result can be evaluated with commands like `eval` or `value`.

1. `diff(f(x),x)` computes and returns $f'(x)$.
2. `D(f)(x)` computes and returns $f'(x)$.

We use Maple help to obtain a detailed description of `diff`.

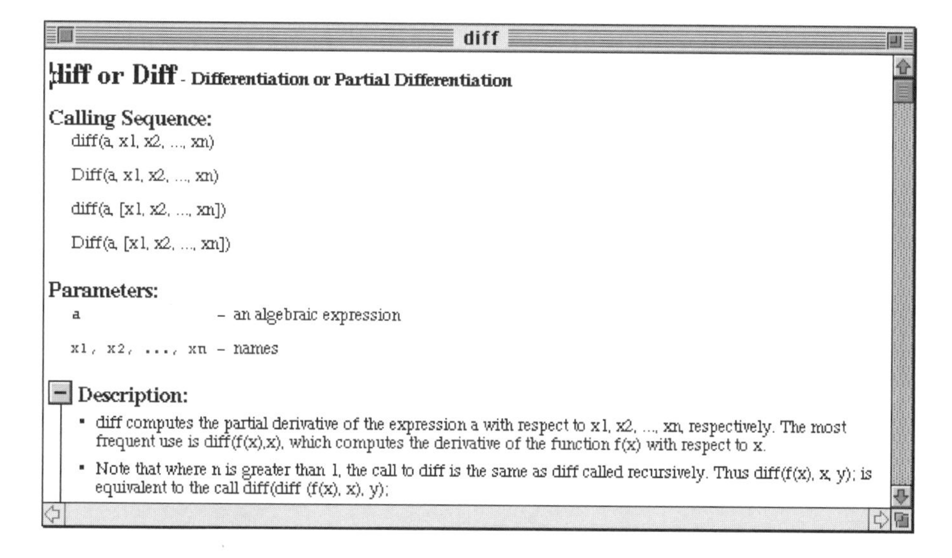

In addition to using `D` and `diff`, the button, which is located on the **Expression** palette,

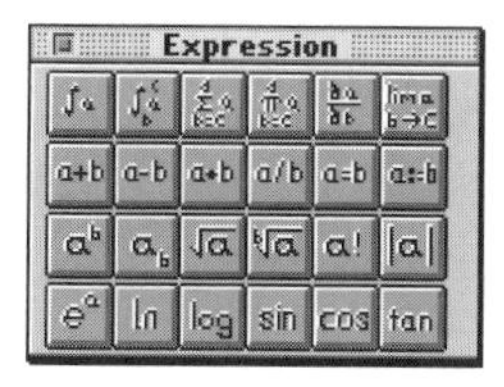

places a template at the location of the curson in the active Maple notebook.

EXAMPLE 2: Calculate the indicated derivatives:
(a) $\frac{d}{d(x)}(2x^2-7x-4)$; (b) $\frac{d}{d(x)}(\sin x)$;
(c) $\frac{d}{d(x)}((3x+4)^2(x+5)^2)$; (d) $\frac{d}{d(x)}\left(\frac{x^2+2x+1}{x^2+3x}\right)$;
(e) $f'(x)$ if $f(x)=x^3e^{-2x}$; (f) $g'(x)$ if $g(x)=x\tan^{-1}x = x\arctan x$
(g) $S'(r)$ if $S(r) = 2\pi r+\frac{75\pi}{r}$; and
(h) $\frac{d}{d(t)}\left(\frac{1}{2}t^2+9.8t\right)$.

SOLUTION: For (a)-(e), we use `diff` to compute the derivative of the indicated expression. Generally, the results from `diff` and `D` are not expressed in simplified form.

(a)

```
> diff(2*x^2-7*x-4,x);
```

$$4x-7$$

(b)

```
> diff(sin(x),x);
```

$$\cos(x)$$

(c)

```
> diff((3*x+4)^2*(x+5)^2,x);
```

$$6(3x+4)(x+5)^2+2(3x+4)^2(x+5)$$

(d)

```
> diff((x^2+2*x+1)/(x^2+3*x),x);
```

$$\frac{2x+2}{x^2+3x}-\frac{(x^2+2x+1)(2x+3)}{(x^2+3x)^2}$$

For (e), we first define f and then use `diff`.

```
> f:=x->x^3*exp(-2*x):
> diff(f(x),x);
```

$$3\,x^2\,\mathbf{e}^{(-2\,x)} - 2\,x^3\,\mathbf{e}^{(-2\,x)}$$

For (f) and (g), we illustrate the use of `D`.

```
> g:=x->x*arctan(x):
> D(g);
```

$$x \rightarrow \arctan(x) + \frac{x}{1+x^2}$$

(g)

```
> S:=r->2*Pi*r^2+75*Pi/r:
> D(S)(r);
```

$$4\,\pi\,r - 75\,\frac{\pi}{r^2}$$

(h)

```
> diff(1/2*t^2+9.8*t,t);
```

$$t + 9.8$$

■

Maple knows the familiar rules of differentiation: the product rule, quotient rule, and chain rule. After clearing all prior definitions of f and g, we compute the derivative of $f(x) \cdot g(x)$, $f(x)/g(x)$, and $(f \circ g)(x) = f(g(x))$. Note that we use `simplify` to see the familiar form of the quotient rule.

```
> f:='f':g:='g':
  diff(f(x)*g(x),x);
```

$$\left(\frac{\partial}{\partial x} \mathrm{f}(x)\right) \mathrm{g}(x) + \mathrm{f}(x) \left(\frac{\partial}{\partial x} \mathrm{g}(x)\right)$$

```
> simplify(diff(f(x)/g(x),x));
```

$$-\frac{-\left(\frac{\partial}{\partial x} \mathrm{f}(x)\right) \mathrm{g}(x) + \mathrm{f}(x) \left(\frac{\partial}{\partial x} \mathrm{g}(x)\right)}{\mathrm{g}(x)^2}$$

```
> diff(f(g(x)),x);
```

$$\mathrm{D}(f)(\mathrm{g}(x)) \left(\frac{\partial}{\partial x} \mathrm{g}(x)\right)$$

Higher-order derivatives are computed in the same way.

1. `diff(f(x),x$n)` computes $\frac{d^n}{dx^n}(f(x)) = f^{(n)}(x)$.
2. `(D@@n)(x)` computes $\frac{d^n}{dx^n}(f(x)) = f^{(n)}(x)$.

EXAMPLE 3: Compute the indicated derivatives.

(a) $\frac{d^n}{dx^n}(x^4 - 2x^3 - 36x^2 + 162x + 24)$;

(b) $\frac{d^4}{dx^4}(x^5 + 2x^4 \cos x)$;

(c) $h''(x)$ if $h(x) = (2x + 1)(3x^2 - 4x + 2)$; and

(d) $f'''(x)$ if $f(x) = \frac{\sin^{-1} x}{x^2 - 1} = \frac{\arcsin x}{x^2 - 1}$.

SOLUTION: For (a)-(c), we use `diff`.

```
> diff(x^4-2*x^3-36*x^2+162*x+24,x$2);
```

$$12 x^2 - 12 x - 72$$

```
> diff(x^5+2*x^4*cos(x),x$4);
```

$$120\,x + 48\cos(x) - 192\,x\sin(x) - 144\,x^2\cos(x) + 32\,x^3\sin(x) + 2\,x^4\cos(x)$$

```
> h:=x->(2*x+1)*(3*x^2-4*x+2):
  diff(h(x),x$2);
```

$$36\,x - 10$$

For (e), we illustrate the use of `D`.

```
> f:=x->arcsin(x)/(x^2-1):
  (D@@3)(f)(x);
```

$$3\frac{x^2}{(1-x^2)^{(5/2)}(x^2-1)} - 6\frac{x^2}{(1-x^2)^{(3/2)}(x^2-1)^2} + \frac{1}{(1-x^2)^{(3/2)}(x^2-1)}$$
$$+ 24\frac{x^2}{\sqrt{1-x^2}\,(x^2-1)^3} - 6\frac{1}{\sqrt{1-x^2}\,(x^2-1)^2} - 48\frac{\arcsin(x)\,x^3}{(x^2-1)^4} + 24\frac{\arcsin(x)\,x}{(x^2-1)^3}$$

■

Graphing Functions and Derivatives

We can also use Maple to graph a function and its derivative together. In some situations, it is instructive to graph the function and its derivative together.

EXAMPLE 4: Graph f and f' if $f(x) = x/(x^2+1)$.

SOLUTION: After clearing all prior definitions of f and defining $f(x) = x/(x^2+1)$, we compute $f'(x)$.

```
> f:='f':
  f:=x->x/(x^2+1):

  simplify(D(f)(x));
```

$$-\frac{x^2-1}{(x^2+1)^2}$$

Examining the result, we see that if $f'(x)=0$ or $x = 1$ or $x = -1$. Thus, we graph f and f' on an interval containing these values, so we see the behavior of f at these values of x. We use `plot` to graph f and f' on the interval [–5,5]. The graph of f is in black; the graph of f' is dashed.

```
> plot({f(x),D(f)(x)},x=-5..5,color=[black,black],
      linestyle=[4,1]);
```

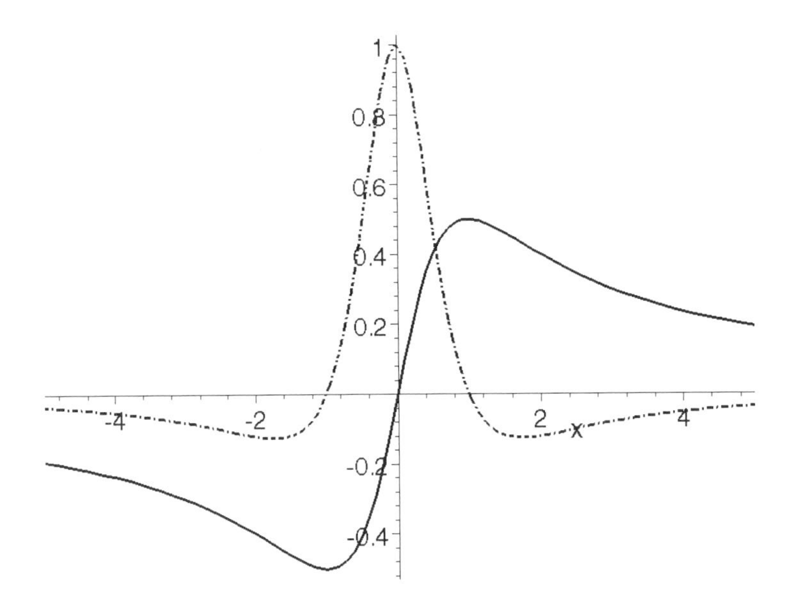

In this case, we see that an absolute maximum occurs at the point with x-coordinate $x = 1$ and an absolute minimum occurs at the point with x-coordinate $x = -1$.

■

EXAMPLE 5: Graph f and f' if $f(x) = (x-3)\sqrt[3]{(x-8)^2} = (x-3)(x-8)^{2/3}$.

SOLUTION: Proceeding in the same manner as in the previous example, we first clear all prior definitions of f define f, and then compute f'.

```
> f:='f':
  f:=x->(x-3)*(x-8)^(2/3):
  simplify(D(f)(x));
```

$$\frac{5}{3}\frac{x-6}{(x-8)^{(1/3)}}$$

From these results, we see that $f'(x) = 0$ if $x = 6$ and $f'(x)$ does not exist if $x = 8$. Thus, we graph both f and f' on an interval containing 6 and 8.

We have seen that if x is negative, Maple does not return a real number when entering `x^(1/3)` (see Example 5 in Section 2.1). However,

$$f(x) = (x-3)\sqrt[3]{(x-8)^2} = (x-3)|x-8|^{2/3}$$

and.

$$f'(x) = \frac{5(x-6)}{3\sqrt[3]{x-8}} = \begin{cases} \frac{5}{3}(x-6)(x-8)^{-1/3}, & x > 8 \\ -\frac{5}{3}(x-6)(x-8)^{-1/3}, & x < 8 \end{cases}.$$

Thus, to graph f and f', we redefine them as follows using `abs` and `piecewise`. Note that `df` corresponds to f'. The graphs of each are then generated with `plot`.

```
> f:=x->(x-3)*abs(x-8)^(2/3):
  df:=x->piecewise(x>8,5/3*(x-6)*(x-8)^(-1/3),
      x<8,-5/3*(x-6)*abs(x-8)^(-1/3)):
 plot({f(x),df(x)},x=0..16,y=-16..16,
   color=[black,gray]);
```

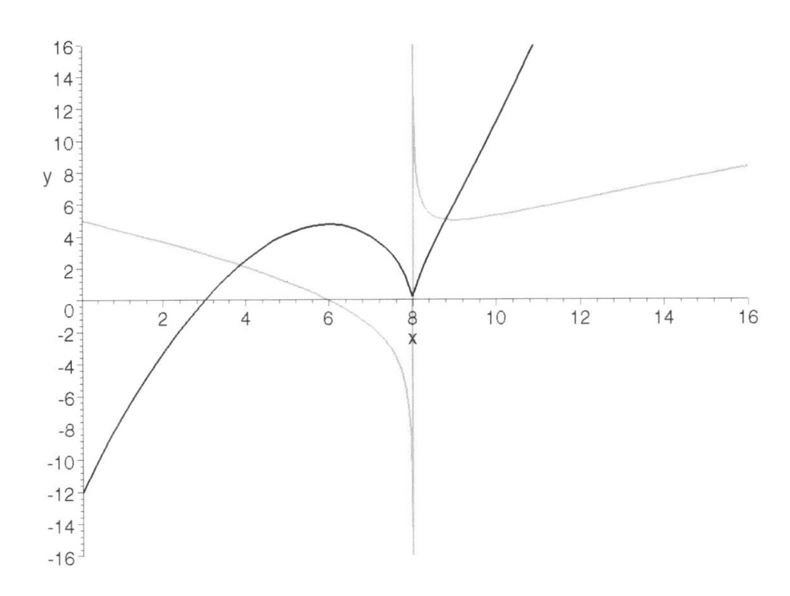

Alternatively, we can avoid redefining f and f' as piecewise-defined function if we use the `surd` function. If x is negative and n is odd, `surd(x,n)` returns the real nth root of x.

```
> f:=x->(x-3)*surd(x-8,3)^2:
  simplify(D(f)(x));
```

$$\frac{5}{3}\frac{\operatorname{surd}(x-8,3)^2\,(x-6)}{x-8}$$

Then, entering

```
> plot({f(x),D(f)(x)},x=0..16,y=-16..16,
    color=[black,gray]);
```

produces the same graph as that obtained above.

■

Tangent Lines

If f is a function for which $f'(x_0)$ exists, then $f'(x_0)$ is the slope of the line tangent to the graph of f at the point $(x_0, f(x_0))$. An equation of the line tangent to the graph of f at the point $(x_0 f(x_0))$, in point-slope form, is

$$y - f(x_0) = f'(x_0)(x - x_0)$$

while a function of x that can be graphed by Maple is given by

$$y = f'(x_0)(x - x_0) + f(x_0).$$

EXAMPLE 6: Find an equation of the line tangent to the graph of $f(x) = 2x^3 + 3x^2 - 12x + 7$ at the point $(-1, f(-1))$.

SOLUTION: After clearing all prior definitions of f, we define $f(x)$ and then compute $f'(x)$.

```
> f:='f':
  f:=x->2*x^3+3*x^2-12*x+7:
  D(f)(x);
```

$$6x^2 + 6x - 12$$

The slope of the line tangent to the graph of f at the point $(-1, f(-1))$ is $f'(-1)$.

```
> D(f)(-1);
```

$$-12$$

Finally, to find an equation of the desired tangent line, we compute the value of $f(-1)$.

```
> f(-1);
```

$$20$$

Thus, in point-slope form, an equation of the line tangent to the graph of f at the point $(-1, f(-1))$ is

$y = 20 = -12(x - (-1))$.

We now graph f along with the tangent line at the point $(-1, f(-1))$.

```
> plot({f(x),D(f)(-1)*(x+1)+f(-1)},x=-4..3,
```

```
color=[black,black],linestyle=[4,1]);
```

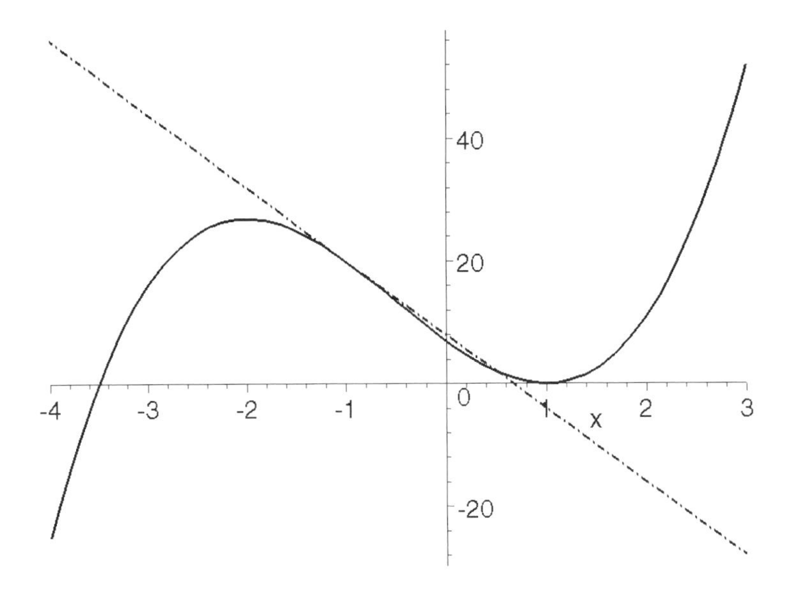

■

We can view a sequence of lines tangent to the graph of a function for a sequence of x values.

EXAMPLE 7: If $f(x) = (x-\pi/2)(3\pi/2 - x)\sin x$, graph f and the line tangent to the graph of f at 16 equally spaced points between 0 and 3π.

SOLUTION: As in Example 1, we use the `animate` command so we first load the `plots` package and then define f.

```
> with(plots):
  f:='f':f:=x->(x-Pi/2)*(3*Pi/2-x)*sin(x):
```

An equation of the line tangent to the graph of f at the point $(a, f(a))$ is $y = f'(a)(x - a) + f(a)$. We use `animate` to generate graphs for 16 equally spaced values of a between 0 and 3π.

```
> A:=animate({f(x),D(f)(a)*(x-a)+f(a)},x=0..3*Pi,
      a=0..3*Pi,view=[0..3*Pi,-9*Pi..Pi],frames=16,
      color=black,tickmarks=[4,2]):
```

The resulting array is displayed with `display`.

```
> display(A);
```

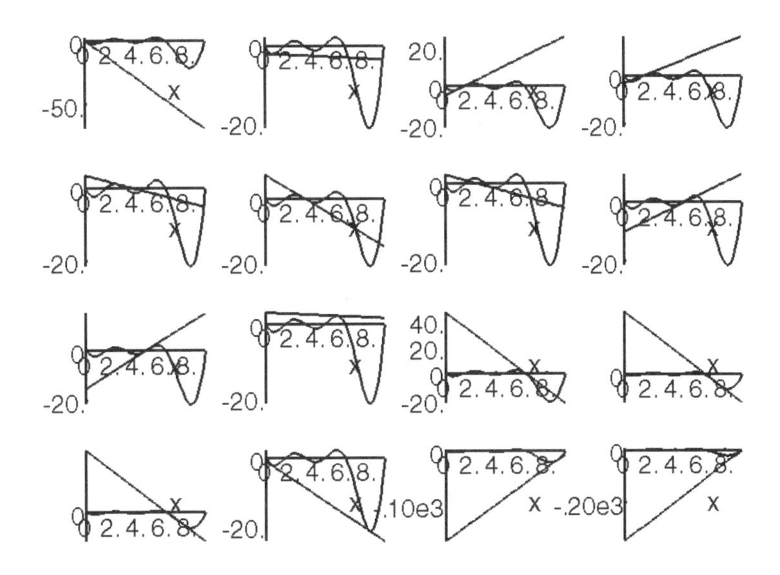

The following command generates a nice animation:

```
> animate({f(x),D(f)(a)*(x-a)+f(a)},x=0..3*Pi,a=0..3*Pi,
      view=[0..3*Pi,-9*Pi..Pi],frames=50,color=black,
      tickmarks=[4,2]);
```

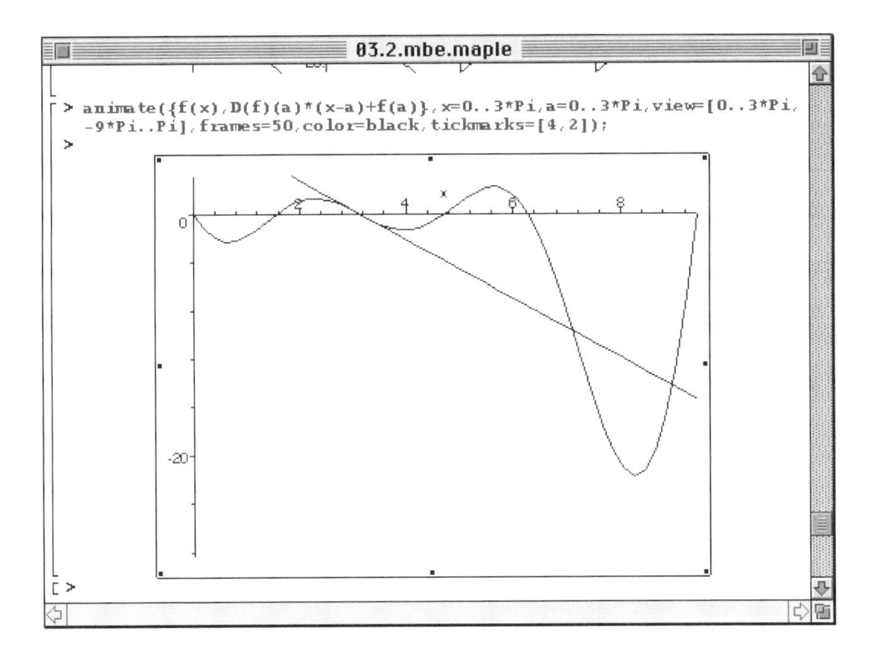

We can also use Maple to locate the values of x for which the line tangent to the graph of a particular function has certain properties. For example, the values of x for which the line tangent to the graph of f at the point $(x,f(x))$ is horizontal are the solutions of the equation $f'(x)=0$.

EXAMPLE 8: Find the values of x for which the line tangent to the graph of $h(x) = \frac{x^2-x+4}{x-1}$ is horizontal.

SOLUTION: As in the previous examples, we begin by clearing all prior definitions of h, then define h and compute h'. `simplify` is used so that h' is expressed as a single fraction.

```
> h:='h':
  h:=x->(x^2-x+4)/(x-1):
  dh:=simplify(diff(h(x),x));
```

$$dh := \frac{x^2-2x-3}{(x-1)^2}$$

The values of x for which the tangent line is horizontal are the solutions of the equation $h'(x)=0$. We can compute these numbers by either factoring the numerator of h' or using Solve.

```
> factor(dh);
```

$$\frac{(x+1)(x-3)}{(x-1)^2}$$

```
> solve(dh=0);
```

$$-1, 3$$

We conclude that the line tangent to the graph of h is horizontal when $x = -1$ and $x = 3$. These results are confirmed by examining the graph of h.

```
> plot(h(x),x=-3..5,y=-30..30,color=black);
```

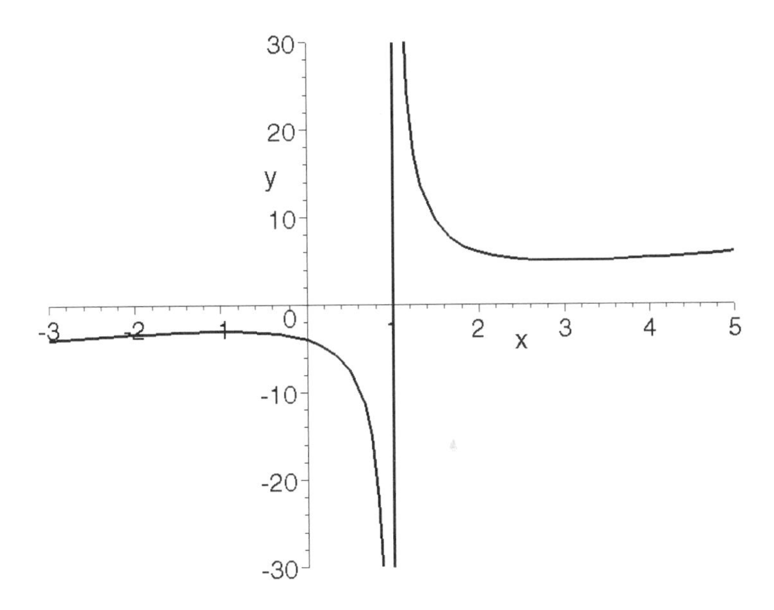

■

In some cases, the roots of an equation must be approximated numerically. With Maple, this is accomplished with the command `fsolve(eqn,var,opt)` where `eqn` represents the equation to be solved, `var` the variable, and `opt` the options which may be used with this command. These include `complex` (so that complex as well as real roots are found), `full-`

digits (so that the setting for Digits is used throughout the calculation), maxsols=n (so that only the n least roots are calculated), or interval (so that only roots on the indicated interval are determined). (This command was also discussed briefly in Chapter 2.)

EXAMPLE 9: If $w(x) = 2\sin^2 2x + \frac{5}{2}x\cos^2\frac{x}{2}$ on $[0, \pi]$, approximate the values of x for which the line tangent to the graph of w at the point $(x, w(x))$ is horizontal.

SOLUTION: Again, we are interested in determining the values of x where $w'(x) = 0$, so we begin by defining and plotting the function w.

```
> w:='w':
  w:=x->2*sin(2*x)^2+5/2*x*cos(x/2)^2:
  plot(w(x),x=0..Pi);
```

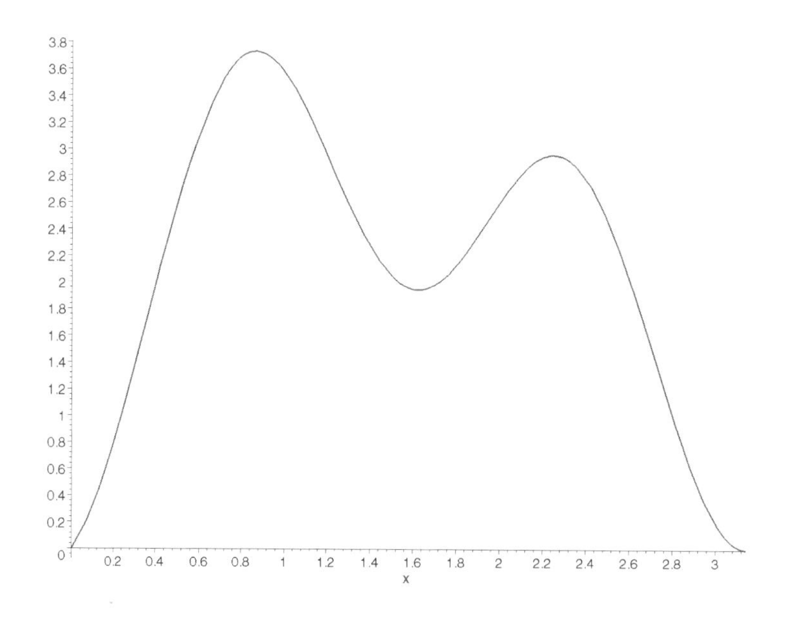

The derivative of w is then found in dw. The expression returned involves powers and products of trigonometric functions, so we employ the command combine with the trig option setting. This simplifies the expression in dw by combining terms to yield a sum of trigonometric terms.

```
> dw:=diff(w(x),x);
```

$$dw := 8\sin(2x)\cos(2x) + \frac{5}{2}\cos\left(\frac{1}{2}x\right)^2 - \frac{5}{2}x\cos\left(\frac{1}{2}x\right)\sin\left(\frac{1}{2}x\right)$$

```
> combine(dw,trig);
```

$$4\sin(4x) + \frac{5}{4}\cos(x) + \frac{5}{4} - \frac{5}{4}x\sin(x)$$

Unfortunately, the roots of this simplified expression are not easy to determine, so we plot $w'(x)$ on $[0,\pi]$ and use `fsolve` to locate the roots of $w'(x)=0$.

```
> plot(dw,x=0..Pi);
```

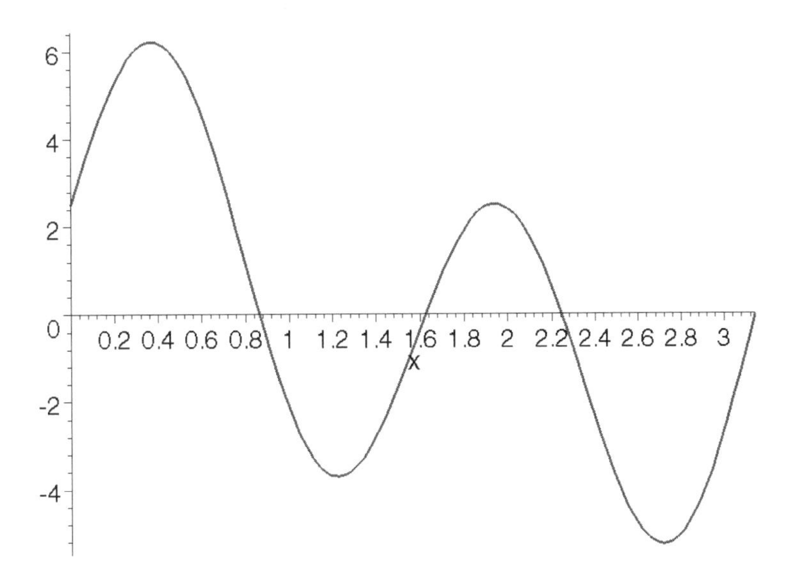

Notice that $w'(x)$ is not a polynomial, so `fsolve` returns only one solution of $w'(x)=0$, and it is not in the interval $[0, \pi]$. In order to find the solutions contained in the interval $[0, \pi]$, the `interval` option must be used. From the graph of $w'(x)$, we see that the first root is contained in the interval [0,1], the second root is contained in the interval [1.5,2], and the third is contained in [2,2.5]. Using these two intervals with `fsolve` yields the other two solutions of $w'(x)$ on $[0,\pi]$.

```
> fsolve(dw=0);
```

```
-.1647494792
```

```
> fsolve(dw=0,x,0..1);
```

.8641940613

```
> fsolve(dw=0,x,1.5..2);
```

1.623905902

```
> fsolve(dw=0,x,2..2.5);
```

2.244892845

■

Application: Locating Critical and Inflection Points

Because derivatives of functions are expressions, algebraic procedures can be performed on them. For example, Maple can be used to locate the **critical points** and **inflection points** of a function. The critical points correspond to those points on the graph of f where the tangent line is horizontal or vertical; the inflection points correspond to those points on the graph of f where the graph of f is neither concave up nor concave down.

EXAMPLE 10: Find the critical points and inflection points of $f(x)$ if $f(x) = (1 + 5x - 3x^2)(x^2 + x - 2)$.

SOLUTION: We first clear all prior definitions of f, define f, and then compute f' and f'', naming the results `df` and `ddf`, respectively.

```
> f:='f':
  f:=x->(1+5*x-3*x^2)*(x^2+x-2):
  df:=D(f)(x);
  ddf:=(D@@2)(f)(x);
```

$$df := (5 - 6x)(x^2 + x - 2) + (1 + 5x - 3x^2)(2x + 1)$$
$$ddf := -12x^2 + 4x + 14 + 2(5 - 6x)(2x + 1)$$

Next, we must solve the equations $f'(x) = 0$ and $f''(x) = 0$. We first try to factor $f'(x)$ and $f''(x)$.

```
> factor(df);
> factor(ddf);
```

$$-3\,(2\,x-3)\,(2\,x^2+2\,x-1)$$

$$-12\,(3\,x+2)\,(x-1)$$

From these results, we see that one solution of $f''(x)=0$ is $x=3/2$, while the other two solutions are the two solutions of the equation $1-2x-2x^2=0$. On the other hand, we see that the two solutions of $f''(x)=0$ are $x=1$ and $x=-2/3$. To obtain the exact solutions of the equation $f'(x)=0$, we use `solve`. The resulting list is named `critnums` and approximations of the solutions are obtained with `evalf`.

```
> critnums:=[solve(df=0)];
  evalf(critnums);
```

$$critnums := \left[\frac{3}{2}, -\frac{1}{2}+\frac{1}{2}\sqrt{3}, -\frac{1}{2}-\frac{1}{2}\sqrt{3}\right]$$

[1.500000000, .3660254040, -1.366025404]

The critical points and inflection points are then obtained by evaluating $f'(x)$ if $x=3/2$, $x=-\frac{1}{2}(1+\sqrt{3})$, and $x=-\frac{1}{2}(1-\sqrt{3})$ and if $x=1$ and $x=-2/3$, respectively.

```
> [3/2,f(3/2)];
```

$$\left[\frac{3}{2}, \frac{49}{16}\right]$$

```
> [-1/2+1/2*sqrt(3),(simplify@f)(-1/2+1/2*sqrt(3))];
```

$$\left[-\frac{1}{2}+\frac{1}{2}\sqrt{3}, \frac{27}{4}-6\sqrt{3}\right]$$

```
> [-1/2-1/2*sqrt(3),(simplify@f)(-1/2-1/2*sqrt(3))];
```

$$\left[-\frac{1}{2}-\frac{1}{2}\sqrt{3}, \frac{27}{4}+6\sqrt{3}\right]$$

```
> [-2/3,f(-2/3)];
```

$$\left[\frac{-2}{3}, \frac{220}{27}\right]$$

```
> [1,f(1)];
```

$$[1, 0]$$

Thus, the critical points are (3/2,49/16), $-\frac{1}{2}\left((1+\sqrt{3}), \frac{27}{4}+\sqrt[6]{3}\right)$, and $-\frac{1}{2}\left((1-\sqrt{3}), \frac{27}{4}-\sqrt[6]{3}\right)$, while the inflection points are (1,0) and (-2/3,220/27).

■

EXAMPLE 11: If $p(x) = \frac{1}{2}x^6 - 2x^5 - \frac{25}{2}x^4 + 60x^3 - 150x^2 - 180x - 25$, graph p along with its first and second derivatives on the interval [–6,6]. Find and classify all critical points and inflection points.

SOLUTION: First, the function p is defined so that its first and second derivatives may be found with `diff`. Next, p and p' are plotted simultaneously so that we can easily observe the values of x where $p'(x) = 0$.

```
>  p:='p':
   p:=x->1/2*x^6-2*x^5-25/2*x^4+60*x^3-
      150*x^2-180*x-25;
   diff(p(x),x);
   diff(p(x),x$2);
   plot({p(x),diff(p(x),x)},x=-6..6,
      -7500..4000);
```

$$p := x \to \frac{1}{2}x^6 - 2x^5 - \frac{25}{2}x^4 + 60x^3 - 150x^2 - 180x - 25$$

$$3x^5 - 10x^4 - 50x^3 + 180x^2 - 300x - 180$$

$$15x^4 - 40x^3 - 150x^2 + 360x - 300$$

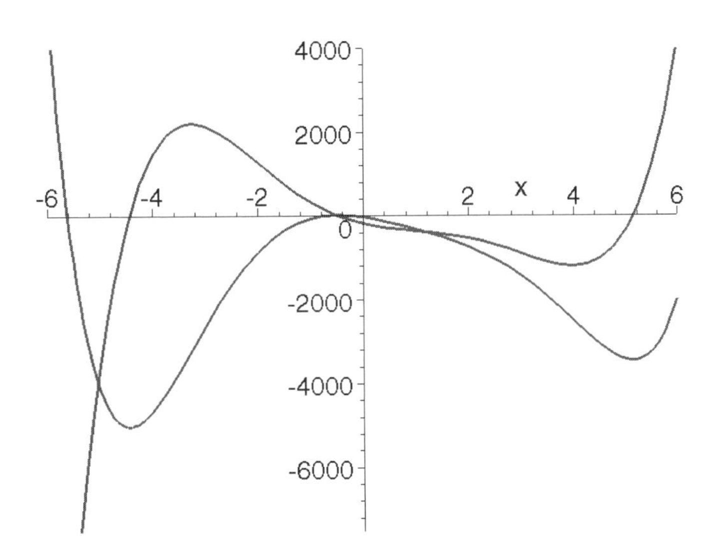

We find these values of x with `fsolve`. Note that if the `complex` option is used with `fsolve`, all roots including complex numbers are found. Otherwise, `fsolve` only returns real roots.

```
> criticalnumbers:=fsolve(diff(p(x),x)=0);
```

$$criticalnumbers := -4.443146074, -.4590960251, 5.129706562$$

```
> fsolve(diff(p(x),x)=0,x,complex);
```

$$-4.443146074, -.4590960251, 1.552934435 - 1.822768757\, I, 1.552934435 + 1.822768757\, I, 5.129706562$$

In order to find the values of $p(x)$ that correspond to the critical points found in `criticalnumbers` as well as to apply the second derivative test, an array of values with entries corresponding to x, $p(x)$, and $p''(x)$ is constructed for the critical numbers.

The commands `array` and `seq` are used to construct an array consisting of rows and columns where the first column corresponds to the list of critical numbers and each row consists of the critical number, value of p for the given critical number, and value of the second derivative of p for the given critical number. Note that `criticalnumbers[i]` yields the ith number in the output `criticalnumbers`. This information can be used to classify the critical points. The commands `array` and `seq` will be discussed in more detail in Chapters 4 and 5.

```
> array([seq(
      [criticalnumbers[i],
         subs(x=criticalnumbers[i],p(x)),
         subs(x=criticalnumbers[i],
            diff(p(x),x$2))],
         i=1..3)]);
```

$$\begin{bmatrix} -4.443146074 & -5010.782456 & 4493.748636 \\ -.4590960251 & 19.70628896 & -492.3530571 \\ 5.129706562 & -3445.427352 & 2586.621342 \end{bmatrix}$$

From this array, we see that (-4.443146074,-5010.782457) and (5.129706562, -3445.427351) are local minima and (-.4590960251,19.70628896) is a local maximum of p.

The inflection points are then located by solving $p''(x) = 0$ with `fsolve`. The values of y that correspond to the values of x given in `inflection` are then determined with `subs`. Thus, the inflection points of p are (-3.253883138,–3172.829600) and (3.981218138, –2482.684805).

```
> inflection:=fsolve(diff(p(x),x$2)=0);
```

$$inflection := -3.253883138, 3.981218138$$

```
> array([seq([inflection[i],
         subs(x=inflection[i],p(x))],
            i=1..2)]);
```

$$\begin{bmatrix} -3.253883138 & -3172.829599 \\ 3.981218138 & -2482.684803 \end{bmatrix}$$

■

Using Derivatives to Graph Functions

Maple is of great use in graphing functions. Unfortunately, if we have no idea of how the graph of a function ought to look or desire to see particular features of the graph, "randomly" choosing an interval on which to graph a particular function often yields unsatisfactory results. In these cases, information supplied by the derivative can help us locate an

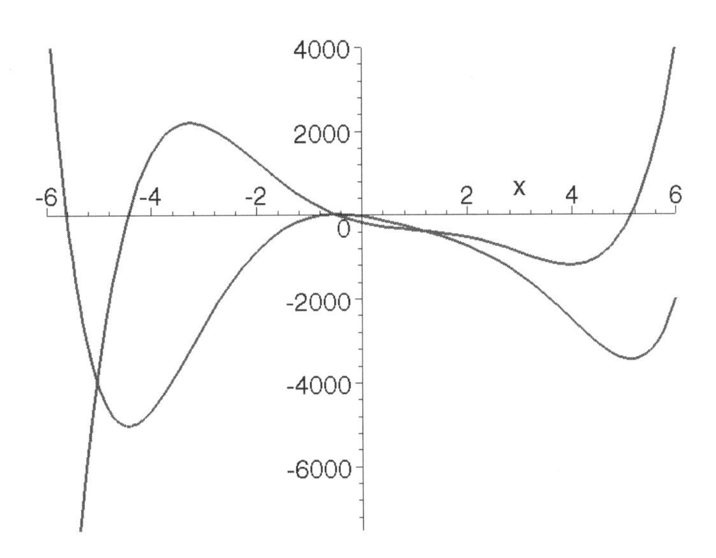

We find these values of x with fsolve. Note that if the complex option is used with fsolve, all roots including complex numbers are found. Otherwise, fsolve only returns real roots.

```
> criticalnumbers:=fsolve(diff(p(x),x)=0);
```

$$criticalnumbers := -4.443146074, -.4590960251, 5.129706562$$

```
> fsolve(diff(p(x),x)=0,x,complex);
```

$$-4.443146074, -.4590960251, 1.552934435 - 1.822768757\,I, 1.552934435 + 1.822768757\,I, 5.129706562$$

In order to find the values of $p(x)$ that correspond to the critical points found in criticalnumbers as well as to apply the second derivative test, an array of values with entries corresponding to x, $p(x)$, and $p''(x)$ is constructed for the critical numbers.

The commands array and seq are used to construct an array consisting of rows and columns where the first column corresponds to the list of critical numbers and each row consists of the critical number, value of p for the given critical number, and value of the second derivative of p for the given critical number. Note that criticalnumbers[i] yields the ith number in the output criticalnumbers. This information can be used to classify the critical points. The commands array and seq will be discussed in more detail in Chapters 4 and 5.

```
> array([seq(
      [criticalnumbers[i],
         subs(x=criticalnumbers[i],p(x)),
         subs(x=criticalnumbers[i],
            diff(p(x),x$2))],
         i=1..3)]);
```

$$\begin{bmatrix} -4.443146074 & -5010.782456 & 4493.748636 \\ -.4590960251 & 19.70628896 & -492.3530571 \\ 5.129706562 & -3445.427352 & 2586.621342 \end{bmatrix}$$

From this array, we see that (-4.443146074,-5010.782457) and (5.129706562, -3445.427351) are local minima and (-.4590960251,19.70628896) is a local maximum of p.

The inflection points are then located by solving $p''(x) = 0$ with `fsolve`. The values of y that correspond to the values of x given in `inflection` are then determined with `subs`. Thus, the inflection points of p are (-3.253883138,–3172.829600) and (3.981218138, –2482.684805).

```
> inflection:=fsolve(diff(p(x),x$2)=0);
```

$$inflection := -3.253883138, 3.981218138$$

```
> array([seq([inflection[i],
         subs(x=inflection[i],p(x))],
            i=1..2)]);
```

$$\begin{bmatrix} -3.253883138 & -3172.829599 \\ 3.981218138 & -2482.684803 \end{bmatrix}$$

■

Using Derivatives to Graph Functions

Maple is of great use in graphing functions. Unfortunately, if we have no idea of how the graph of a function ought to look or desire to see particular features of the graph, "randomly" choosing an interval on which to graph a particular function often yields unsatisfactory results. In these cases, information supplied by the derivative can help us locate an interval on which the graph of f will show the features we wish to see. In particular, the first and second derivatives of a function give us the following information:

1. The values of x for which f is **increasing** are the same as the values of x for which f' is positive.
2. The values of x for which f is **decreasing** are the same as the values of x for which f' is negative
3. The values of x for which f is **concave up** are the same as the value of x for which f'' is positive.
4. The values of x for which f is **concave down** are the same as the values of x for which f'' is negative.

EXAMPLE 12: Graph $f(x)= x^4 + 2x^3 - 72x^2 + 70x + 24.$.

SOLUTION: We proceed by clearing all prior definitions of f, defining f, and computing f' and f''.

```
> f:='f':
  f:=x->x^4+2*x^3-72*x^2+70*x+24:
  D(f)(x);
  (D@@2)(f)(x);
```

$$4x^3 + 6x^2 - 144x + 70$$
$$12x^2 + 12x - 144$$

To solve the equations $f'(x) = 0$ and $f''(x) = 0$, we use `solve`.

```
> solve(D(f)(x)=0);
```

$$-7, 5, \frac{1}{2}$$

```
> solve((D@@2)(f)(x)=0);
```

$$-4, 3$$

(Note that the solutions of these polynomial equations are rational numbers, so we could have used `Factor` to factor $f'(x)$ and $f''(x)$ and, consequently, to determine the solutions of the equations $f'(x) = 0$ and $f''(x) = 0$.) Next, we graph $f'(x)$ on an interval containing -7,1/2, and 5 and graph $f''(x)$ on an interval containing -4 and 3. The results are displayed as a graphics array.

```
> pdf:=plot(D(f)(x),x=-8..6,color=black):
  pddf:=plot((D@@2)(f)(x),x=-5..4,color=black):
  with(plots):
  A:=array([pdf,pddf]):
  display(A);
```

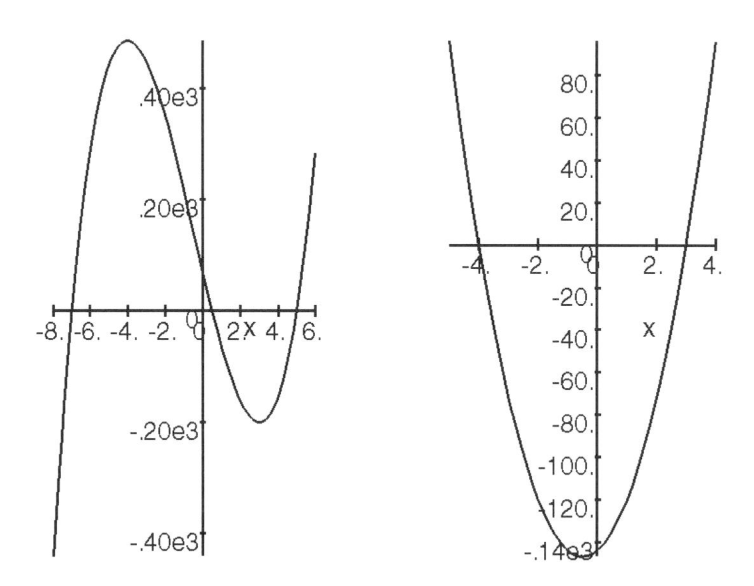

From the graphs, we see that $f'(x)$ is positive if $-7 < x < 1/2$ or $x > 5$ and $f'(x)$ is negative if $x < -7$ or $1/2 < x < 5$ while $f''(x)$ is positive if $x < -4$ or $x > 3$ and $f'(x)$ is negative if $-4 < x < 3$. Thus, f is decreasing and concave up on $(-\infty, -7)$, f is increasing and concave down on $(-7, -4)$, f is increasing and concave down on $(-4, 1/2)$, f is decreasing and concave down on $(1/2, 3)$, f is decreasing and concave up on $(3,5)$, and f is increasing and concave up on $(5, \infty)$. Last, we graph f on an interval that illustrates these features.

```
> plot(f(x),x=-9..7);
```

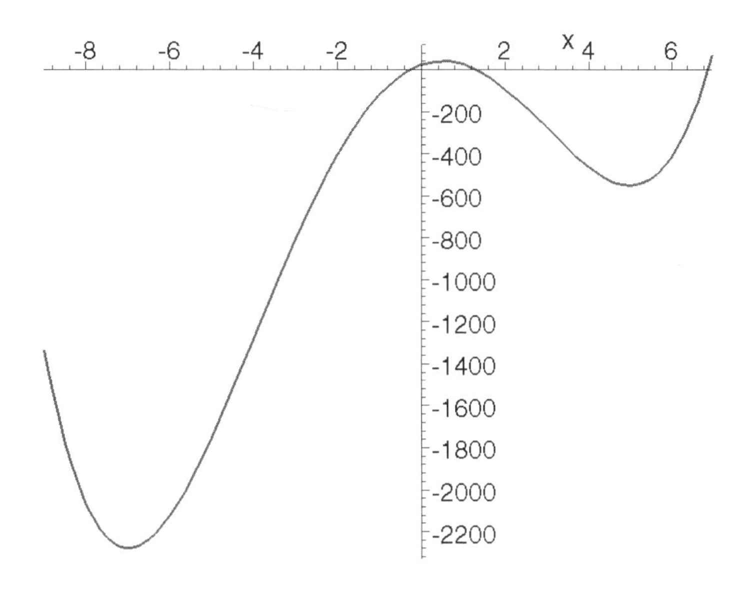

f is a polynomial of degree 4 so we know that for "large" values of x, the graph of f looks like the graph of x^4 because

$$f(x) = x^4\left(1 + \frac{2}{x} - \frac{72}{x^2} + \frac{70}{x^3} + \frac{24}{x^4}\right)$$

and for "large" values of x, $1 + \frac{2}{x} - \frac{72}{x^2} + \frac{70}{x^3} + \frac{24}{x^4}$ is close to 1. Therefore, when we graph f on a large interval, we do not see the subintervals on which f is increasing or decreasing and concave up or concave down.

```
> plot(f(x),x=-100..100);
```

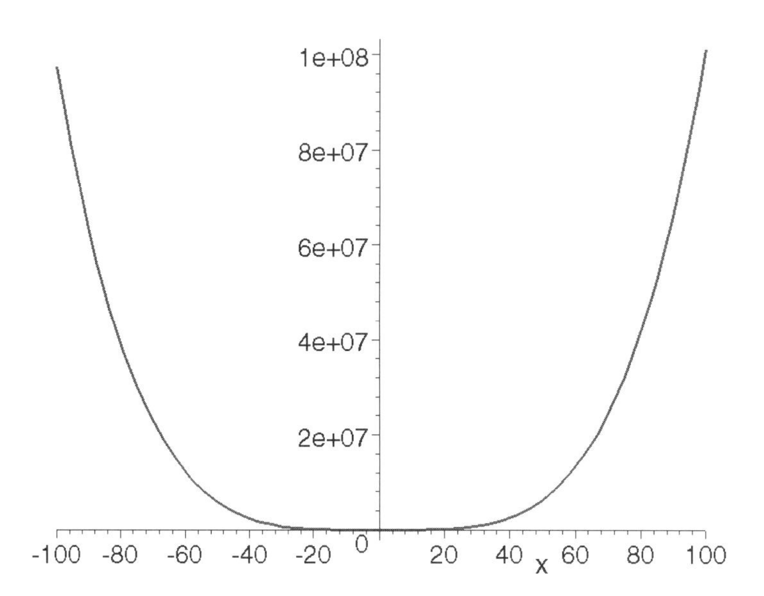

■

EXAMPLE 13: Graph $h(x) = (x-7)\sqrt[3]{x-3} = (x-7)(x-3)^{1/3}$.

SOLUTION: As we have seen before, Maple does not automatically return a real number when computing an odd root of a negative number. Therefore, we use the `surd` command when defining h:

$$\text{surd}(x-3,3) = \sqrt[3]{x-3} \quad .$$

```
> h:='h':
  h:=x->(x-7)*surd(x-3,3):
  diff(h(x),x);
  dh:=simplify(diff(h(x),x));
```

$$\text{surd}(x-3,3) + \frac{1}{3}\frac{(x-7)\,\text{surd}(x-3,3)}{x-3}$$

$$dh := \frac{4}{3}\frac{\text{surd}(x-3,3)\,(x-4)}{x-3}$$

From the simplified form of $h'(x)$, we see that the critical numbers are $x = 4$ and $x = 3$. Note that $x = 3$ is a critical number because $h(3) = 0$ exists and $h'(x)$ is not defined if $x = 3$. Thus, h has a vertical tangent at the point with x-coordinate $x = 3$. We compute the values of x for which $h''(x) = 0$ in the same manner.

```
> ddh:=simplify(diff(h(x),x$2));
```

$$ddh := \frac{4}{9} \frac{\mathrm{surd}(x-3,3)\,(x-1)}{(x-3)^2}$$

We see that $h'(x)$ if $x = 1$. Of course, $h'(x)$ does not exist if $x = 3$ so neither does $h''(x)$. Constructing a sign chart for both $h'(x)$ and $h''(x)$, or using an equivalent method, we see that $h'(x)$ is positive if $x > 4$ and is negative if $x < 4$ as long as $x \neq 3$ while $h''(x)$ is positive if $x < 1$ or $x > 3$ and negative if $1 < x < 3$. Thus, h is decreasing and concave up on $(-\infty, 1)$, decreasing and concave down on $(1,3)$, decreasing and concave up on $(3,4)$, and increasing and concave up on $(4, \infty)$.

```
> plot(h(x),x=-3..8,color=black);
```

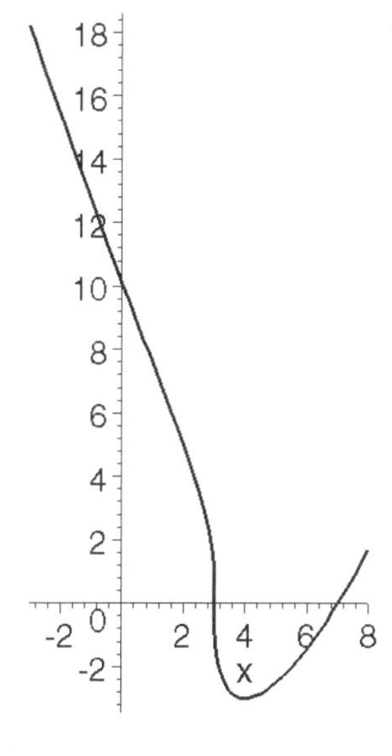

■

Application: Rolle's Theorem and The Mean-Value Theorem

Let f be a continuous function on $[a, b]$ and differentiable on (a, b).

Rolle's Theorem says that if $f(a) = f(b) = 0$, then there is at least one value of c in (a, b) satisfying $f'(c) = 0$.

The Mean-Value Theorem says that there is at least one value of c in (ab) satisfying $f'(c) = \frac{f(b) - f(a)}{b - a}$.

EXAMPLE 14: Verify that f satisfies the hypotheses of Rolle's Theorem on the interval [-3,2] if $f(x) = x3 - 7x + 6$ and find all values of c on the interval [-3,2] that satisfy the conclusion of the theorem.

SOLUTION: f is a polynomial function so f is differentiable for all real numbers and, in particular, on the interval (-3,2). We first define f and compute $f(-3)$ and $f(2)$.

```
> f:='f':
  f:=x->x^3-7*x+6:
  f(-3);
  f(2);
```

$$0$$

$$0$$

Both values are 0, so by Rolle's Theorem there is at least one value of c in the interval [-3,2] for which $f'(c) = 0$. Next, we graph f on an interval containing the interval [-3,2]. From the graph, we see that we should be able to find at least two values of c for which $f'(c) = 0$.

```
> plot(f(x),x=-4..3);
```

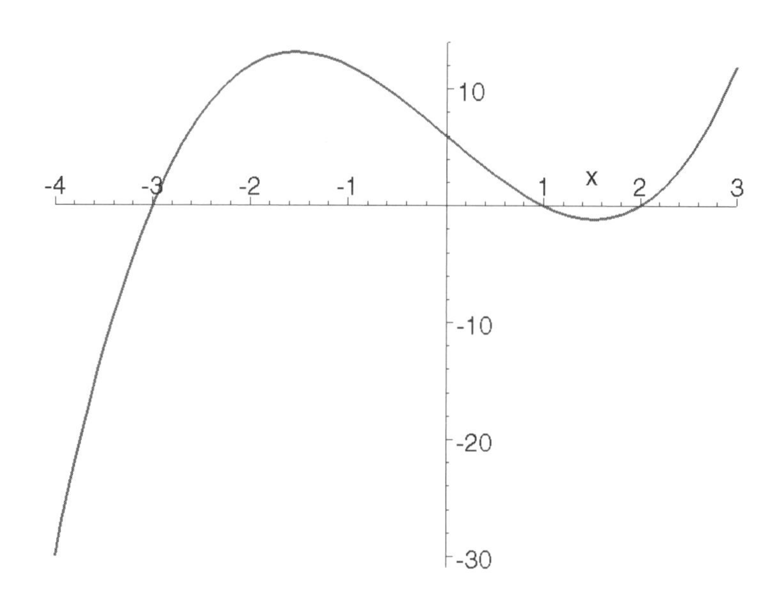

Computing $f'(x)$ and solving the equation $f'(x) = 0$ yields the desired values.

```
> D(f)(x);
```

$$3x^2 - 7$$

```
> solve(D(f)(x)=0);
```

$$\frac{1}{3}\sqrt{21}, -\frac{1}{3}\sqrt{21}$$

We conclude that the values of c for which $f'(c) = 0$ are $c = \pm\sqrt{7/3}$.

■

Generally, verifying Rolle's Theorem and the Mean-Value Theorem for particular functions is difficult, as the resulting equations that need to be solved are either very difficult or impossible to solve. In these cases, `fsolve` can be helpful in approximating solutions of equations.

EXAMPLE 15: Approximate the values of c that satisfy the conclusion of the Mean-Value theorem for $f(x) = \dfrac{\cos 3x}{x^2+1}$ on the interval $[0, \pi]$.

SOLUTION: We begin by defining and graphing f on the interval $[0, \pi]$. We name the graph of f `plotf` for later use.

```
> with(plots):
  f:='f':
  f:=x->cos(3*x)/(x^2+1):
  plotf:=plot(f(x),x=0..Pi,color=black):
  display(plotf);
```

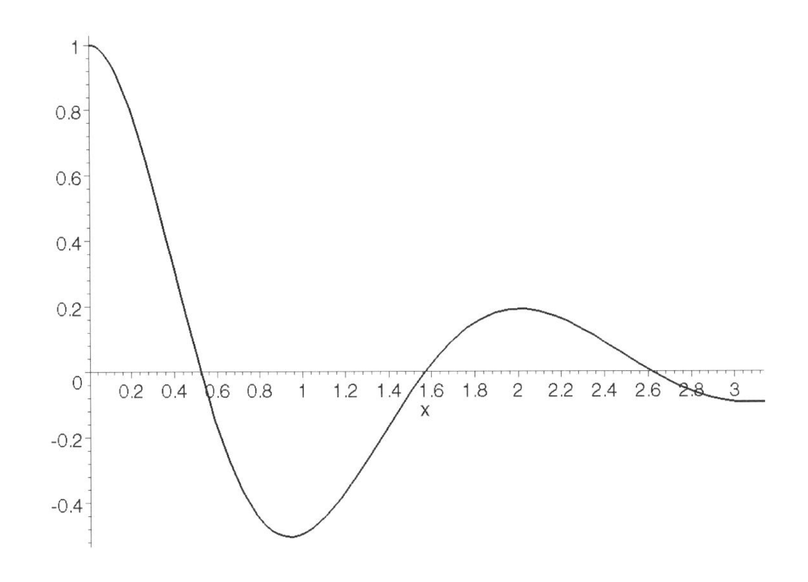

We then display the graph of f together with the graph of the line that passes through the points $(0, f(0))$ and $(\pi, f(\pi))$.

```
> p1:=plot((f(Pi)-f(0))/(Pi-0)*(x-0)+f(0),
      x=0..Pi,color=gray):
  display({plotf,p1});
```

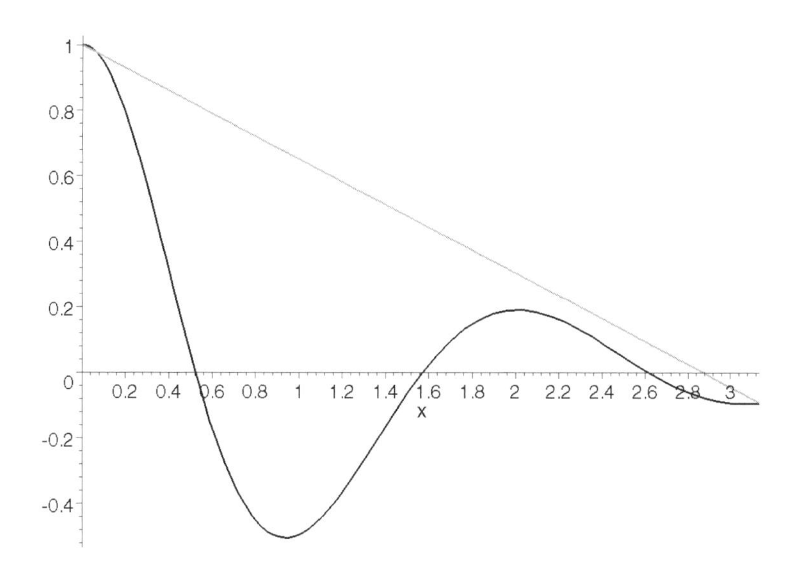

Graphically, the Mean-Value theorem tells us that we can find a number $0 < c < \pi$ so that the slope of the line tangent to the graph of f at the point $(c, f(c))$ has the same slope as the slope of the line that passes through the points $(0, f(0))$ and $(\pi, f(\pi))$. Thus, we must find the values of c in $[0, \pi]$ that satisfy the equation $f(c) = \dfrac{f(\pi) - f(0)}{\pi - 0}$. We compute $\dfrac{f(\pi) - f(0)}{\pi - 0}$ and name the number `avg`.

```
> avg:=simplify((f(Pi)-f(0))/(Pi-0));
```

$$avg := -\frac{2 + \pi^2}{(\pi^2 + 1)\,\pi}$$

```
> evalf(avg);
```

$$-.3475942900$$

`solve` cannot be used to solve the equation $f'(c) = \dfrac{f(\pi) - f(0)}{\pi - 0}$. But graphing $f'(x)$ and `avg` on the interval $[0, \pi]$ shows that there are four values of c satisfying the conclusion of the Mean-Value Theorem. We use `fsolve` to approximate these values and name the results `c1`, `c2`, `c3`, and `c4`, respectively.

```
> plot({D(f)(x),avg},x=0..Pi,color=[black,gray]);
```

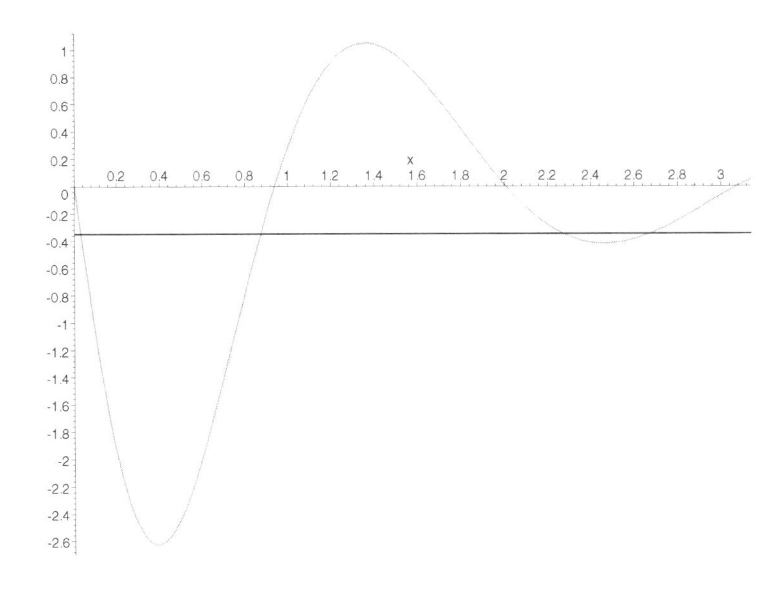


```
> c1:=fsolve(D(f)(x)=avg,x=0..0.3);
  c2:=fsolve(D(f)(x)=avg,x=0.8..1);
  c3:=fsolve(D(f)(x)=avg,x=2.2..2.4);
  c4:=fsolve(D(f)(x)=avg,x=2.6..3);
```

$$c1 := .03170213404$$
$$c2 := .8756205616$$
$$c3 := 2.268483150$$
$$c4 := 2.676831429$$

These numbers represent the values of c for which the slope of the line tangent to the graph of f at $(c, f(c))$ is the same as the slope of the line passing through $(0, f(0))$ and $(\pi, f(\pi))$. Now, we define `p2` to be a graph of the line tangent to the graph of f at the point $(0.8756, f(0.8756))$, and `p3` to be a graph of the line tangent to the graph of f at the point, $(2.6768, f(2.6768))$. The graphs `plotf`, `p1`, `p2`, and `p3` are shown together. Note that all three lines are parallel.

```
> p2:=plot(D(f)(c2)*(x-c2)+f(c2),x=0..Pi,color=gray):
  p3:=plot(D(f)(c4)*(x-c4)+f(c4),x=0..Pi,color=gray):
```

```
display({plotf,p1,p2,p3});
```

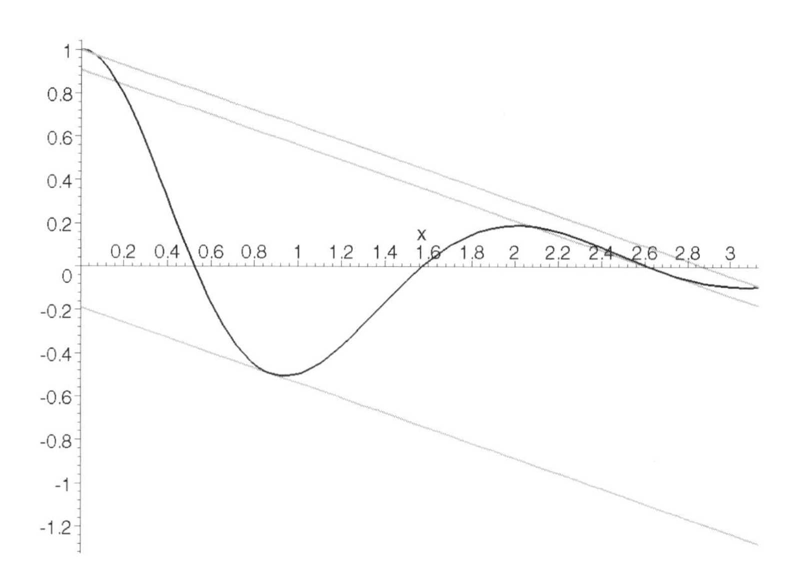

■

Application: Maxima and Minima

Maple can be used to assist in solving maximimization/minimization problems encountered in a differential calculus course.

EXAMPLE 16: A farmer has 100 feet of fencing to construct four dog kennels by first constructing a fence around a rectangular region, and then dividing that region into four smaller regions by placing fences parallel to one of the sides. What dimensions will maximize the total area?

SOLUTION: First, let y denote the length across the top and bottom of the rectangular region and let x denote the vertical length. The following figure describes this situation.

```
> with(plottools):
  rec1:=rectangle([0,0],[1,1]):
  rec2:=rectangle([1/4,0],[3/4,1]):
```

```
line1:=line([1/2,0],[1/2,1]):
display({rec1,rec2,line1},axes=NONE);
```

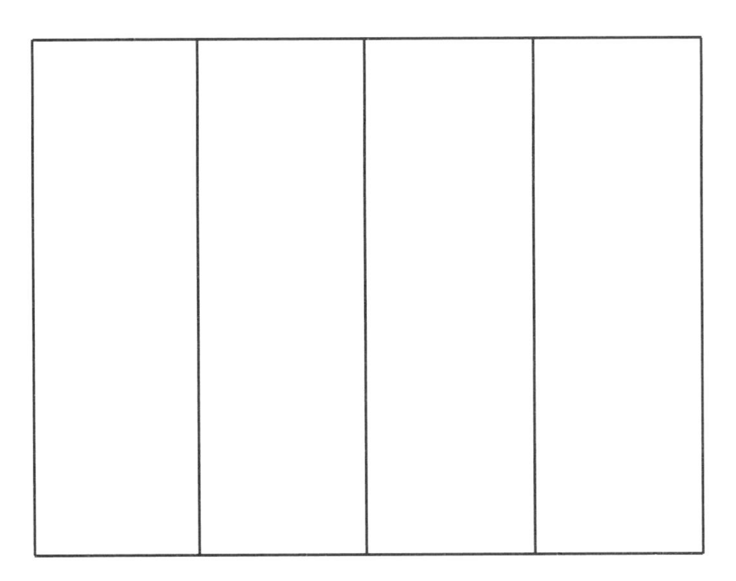

Then, because 100 feet of fencing is used, a relationship between x and y is given by the equation $2y + 5x = 100$. Solving this equation for y, we obtain $y = \frac{1}{2}(100 - 5x)$.

```
> solve(2*y+5*x=100,y);
```

$$-\frac{5}{2}x + 50$$

The area of a rectangle with height x and length y is *area* $=xy$, so the function to be maximized is

$$area(x) = x \cdot \frac{1}{2}(100 - 5x), 0 \le x \le 20.$$

After defining `area`, the value of x that maximizes the area is found by finding the critical value and observing the graph of `area(x)`.

```
> area:=x->x*1/2*(100-5*x):
```

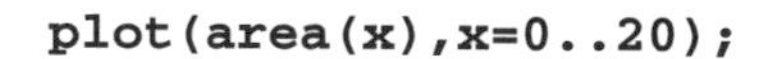

```
plot(area(x),x=0..20);
```

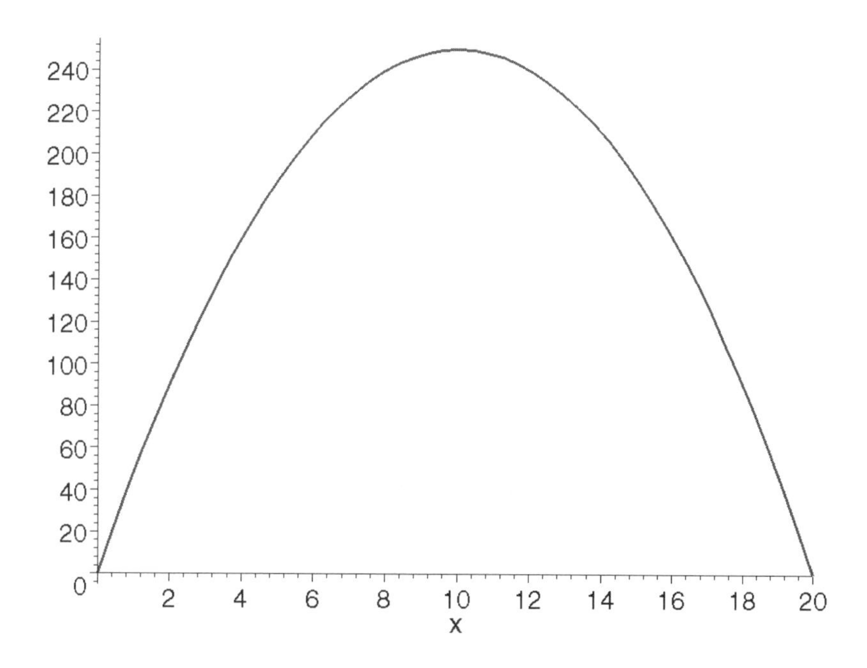

Thus, we see the value of x that maximizes area is 10. To find the other dimension, we evaluate $y = \frac{1}{2}(100 - 5x)$ if $x = 10$.

```
> solve(D(area)(x)=0);
                              10
> area(10);
                              250
> subs(x=10,1/2*(100-5*x));
                              25
```

to see that the dimensions that maximize the area are 10×25.

■

EXAMPLE 17: A woman is located on one side of a body of water 4 miles wide. Her position is directly across from a point on the other side of the body of water 16 miles from her house. If she can move across land at a rate of 10 miles per hour and move over water at a rate of 6 miles per hour, find the least amount of time for her to reach her house.

SOLUTION: The following figure illustrates the situation described in the problem.

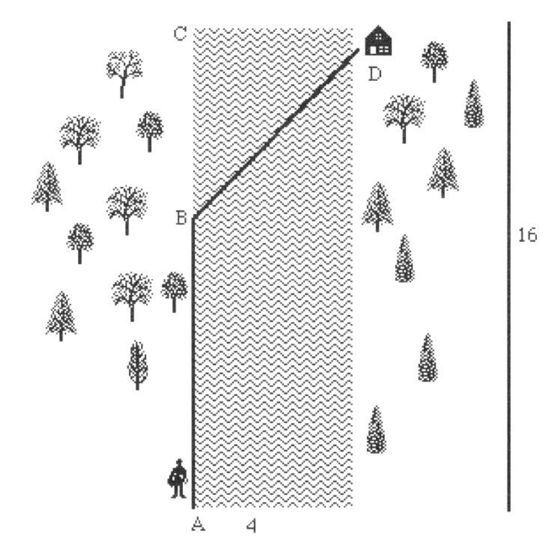

From the figure, we see that woman will travel from A to B by land and then from B to D by water. We wish to find the least time for her to complete the trip.

Let x denote the distance BC, where $0 \le x \le 16$. Then, the distance AB is given by $16 - x$ and, by the Pythagorean theorem, the distance BD is given by $\sqrt{x^2 + 4^2} = \sqrt{x^2 + 16}$. Because rate × time=distance, time = distance/rate. Thus, the time to travel from A to B is $\frac{1}{10}(16 - x)$, the time to travel from B to D is $\frac{1}{6}\sqrt{x^2 + 16}$, and the total time to complete the trip, as a function of x, is

$$time(x) = \frac{1}{10}(16 - x) + \frac{1}{6}\sqrt{x^2 + 16},\ 0 \le x \le 16.$$

We must minimize the function *time*. First, we define `TIME` and then verify that `TIME` has a minimum by graphing `TIME` on the interval [0,16]. (Note that we use `TIME` instead of `time` because `time` is a built-in Maple object.)

```
> TIME:='TIME':
  TIME:=x->(16-x)/10+sqrt(x^2+16)/6:
```

```
plot(TIME(x),x=0..16,y=2..3);
```

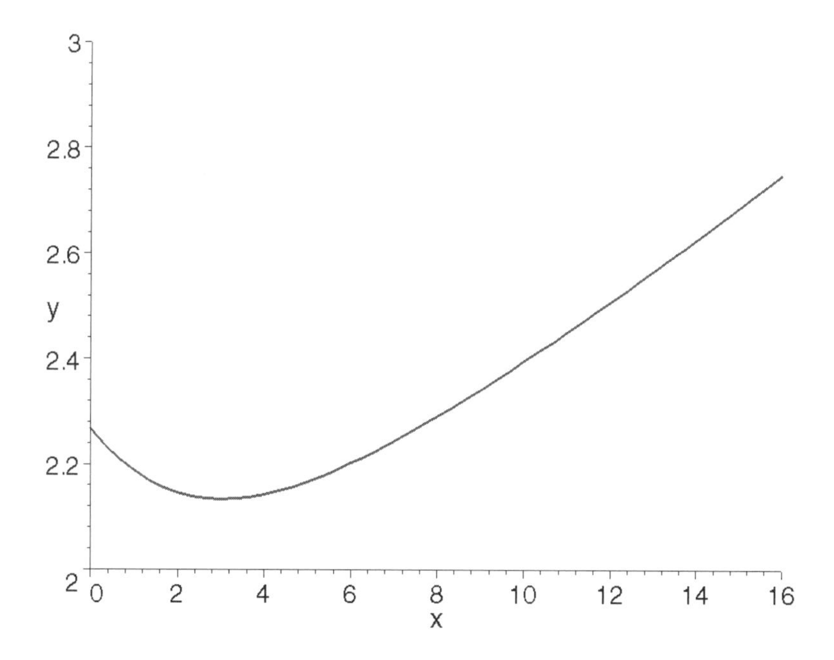

Next, we compute the derivative of `TIME` and find the values of x for which the derivative is 0 with `solve`. The resulting output is named `critnums`.

```
> simplify(D(TIME)(x),x);
```

$$-\frac{1}{10}+\frac{1}{6}\frac{x}{\sqrt{x^2+16}}$$

```
> critnums=solve(D(TIME)(x)=0);
```

$$critnums = 3$$

At this point, we can calculate the minimum time by calculating `TIME(3)`.

```
> TIME(3);
```

$$\frac{32}{15}$$

Thus, we see that the minimum time to complete the trip is 32/15 hours.

■

Our final two examples illustrate Maple's ability to symbolically manipulate algebraic expressions.

EXAMPLE 18: Let $f(x) = mx + b$ and (x_0, y_0) be any point not on the graph of f. Find the value of x for which the distance from (x_0, y_0) to $(x, f(x))$ is a minimum.

SOLUTION: The **distance** between the two points (x_1, y_1) and (x_2, y_2) is given by

$$distance((x_1, y_1), (x_2, y_2)) = \sqrt{(x_2 - x_1) + (y_2 - y_1)}.$$

In order to determine the value of x that minimizes the distance between (x_0, y_0) and $(x, f(x))$, a function that determines this distance must first be defined. The command

```
distance([x0,y0],[x1,y1])
```

determines the distance between (x_0, y_0) and $(x, f(x))$. The `distance` command is contained in the `student` package so must be loaded by entering `with(student)` or entered in the form `student[distance](...)`.

```
> with(student);
```

[D, *Diff, Doubleint, Int, Limit, Lineint, Product, Sum, Tripleint, changevar, combine, completesquare, distance, equate, extrema, integrand, intercept, intparts, isolate, leftbox, leftsum, makeproc, maximize, middlebox, middlesum, midpoint, minimize, powsubs, rightbox, rightsum, showtangent, simpson, slope, summand, trapezoid, value*]

Then the particular distance function for this problem is obtained by substituting the appropriate points (x_0, y_0) and $(x, f(x))$ into `distance` and naming the resulting output `to_min`. Note the minimizing the square of `to_min` produces the same results as minimizing `to_min`. The value of x that minimizes this function is obtained in the usual manner. (Notice how naming the distance function expression simplifies the solution of the problem.)

```
> f:='f':
  f:=x->m*x+b:
> to_min:=distance([x0,y0],[x,f(x)]);
```

$$to\_min := \sqrt{(x0 - x)^2 + (y0 - m\,x - b)^2}$$

To find the minimum, we first compute the derivative of `to_min`, name the result `dtm`, and then use `solve` to find the values of x for which the derivative is 0.

```
> dtm:=simplify(diff(to_min,x));
```

$$dtm := \frac{-x0 + x - m\,y0 + m^2\,x + m\,b}{\sqrt{x0^2 - 2\,x0\,x + x^2 + y0^2 - 2\,y0\,m\,x - 2\,y0\,b + m^2\,x^2 + 2\,m\,x\,b + b^2}}$$

```
> val:=solve(numer(dtm)=0,x);
```

$$val := -\frac{-x0 - m\,y0 + m\,b}{1 + m^2}$$

We then compute and simplify the value of $f(x)$ for the number `val` and name the result `ycoord`. Thus, (`val`,`ycoord`) is the point on the graph of f closest to (x_0, y_0). The minimum distance is then computed using `distance`.

```
> ycoord:=f(val);
```

$$ycoord := -\frac{m\,(-x0 - m\,y0 + m\,b)}{1 + m^2} + b$$

```
> simplify(distance([x0,y0],[val,ycoord]));
```

$$\sqrt{\frac{(x0\,m - y0 + b)^2}{1 + m^2}}$$

Thus, the point on the graph of $f(x)$ closest to (x_0, y_0) is

$$\left(\frac{my_0 + x_0 - bm}{m^2 + 1}, \frac{m^2y_0 + mx_0 + b}{m^2 + 1}\right)$$

and the minimum distance is .

$$\sqrt{\frac{(y_0 - mx_0 - b)^2}{m^2 + 1}} = \frac{|y_0 - mx_0 - b|}{\sqrt{m^2 + 1}}.$$

■

The next example is a familiar exercise to students in introductory differential calculus courses.

EXAMPLE 19: Find the dimensions of the cone of minimum volume that can be inscribed about a sphere of radius R.

SOLUTION: Let r and h denote the radius and height, respectively, of the right circular cone of base radius r and height h circumscribed about the sphere of radius R. A cross-section of the solid containing a diameter of the base of the cone is shown in the following figure:

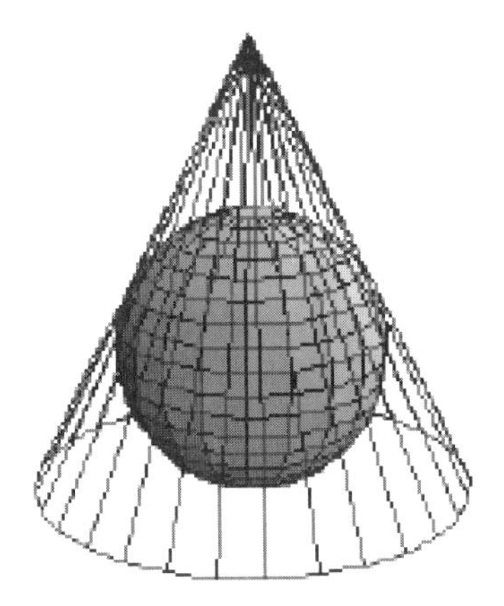

Right circular cone circumscribed about a sphere.

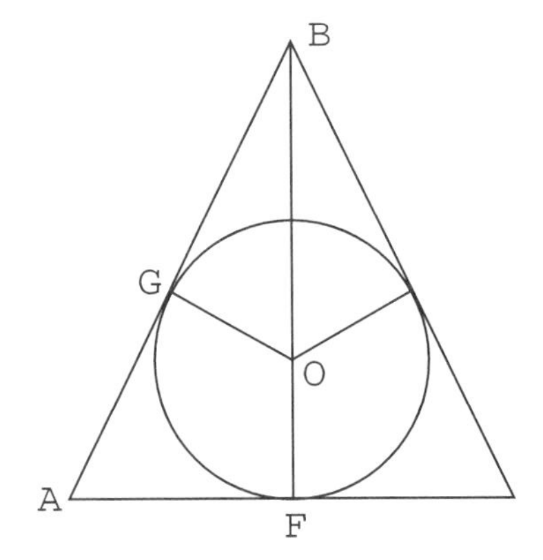

Cross-section of a right circular cone circumscribed about a sphere

From the figure, notice that triangle BOG is similar to triangle BAF. Moreover, $BO = h - R$, $OG = R$, and $AF = r$ and, by the Pythagorean Theorem, $r^2 + h^2 = BA^2$ so that $BA = \sqrt{r^2 + h^2}$. Consequently, $\frac{h - R}{R} = \frac{\sqrt{r^2 + h^2}}{r}$. We use Maple to solve this equation for h in terms of r. Note our convention of using `capr` to represent R in the equation.

```
> h:='h':r:='r':R:'R':
  solve((h-R)/R=sqrt(r^2+h^2)/r,h);
```

$$0, -2\frac{r^2 R}{R^2 - r^2}$$

The volume of the cone is given by $V = \frac{1}{3}\pi r^2 h$ and substituting $h = \dfrac{2r^2R}{r^2 - R^2}$ yields V, as a function of r,

$$V(r) = \frac{2\pi R r^4}{3(r^2 - R^2)}.$$

We define `V` to be $V(r) = \dfrac{2\pi R r^4}{3(r^2 - R^2)}$. Then, we differentiate `V` and find the values of r for which `dV` equals 0. Note that r can be neither 0 nor $-R\sqrt{2}$.

```
> V:=2*Pi*R*r^4/(3*(r^2-R^2));
```

$$V := 2\,\frac{\pi\,R\,r^4}{3\,r^2 - 3\,R^2}$$

```
> dV:=simplify(diff(V,r));
```

$$dV := -\frac{4}{3}\,\frac{\pi\,R\,r^3\,(2\,R^2 - r^2)}{(R^2 - r^2)^2}$$

```
> crit_nums:=solve(dV=0,r);
```

$$crit\_nums := \sqrt{2}\,R, -\sqrt{2}\,R, 0, 0, 0$$

The value $R\sqrt{2}$ is extracted from `crit_nums` with `crit_nums[1]`. Extracting data from lists is discussed in more detail in Chapters 4 and 5.

```
> m:=crit_nums[1];
```

$$m := \sqrt{2}\,R$$

To see that $r = R\sqrt{2}$ yields the desired minimum, we evaluate $V''(r)$ if $r = R\sqrt{2}$.

```
> ddV:=simplify(diff(V,r$2));
```

$$ddV := -\frac{4}{3}\,\frac{\pi\,R\,r^2\,(6\,R^4 - 3\,r^2\,R^2 + r^4)}{(R^2 - r^2)^3}$$

```
> subs(r=m,ddV);
```

$$\frac{32}{3}\,\pi\,R$$

The value of $V''(r)$ if $r = R\sqrt{2}$ is positive, so we conclude that $r = R\sqrt{2}$ yields the minimum volume.

```
> subs(r=m,V);
```

$$\frac{8}{3}\pi R^3$$

We conclude that the minimum volume is $V(R\sqrt{2}) = \frac{8}{3}\pi R^3$ and the cone has radius $r = R\sqrt{2}$ and height $h = \frac{2r^2R}{r^2 - R^2} = 4R$.

■

3.3 Implicit Differentiation

The commands D and diff can also be used to compute the implicit derivative of an equation.

EXAMPLE 1: Find an equation of the line tangent to the graph of

$$2x^2 - 2xy + y^2 + x + 2y + 1 = 0$$

at the points $\left(-\frac{3}{2}, -1\right)$ and $\left(-\frac{3}{2}, -4\right)$.

SOLUTION: The slope of the lines tangent to the graph of $2x^2 - 2xy + y^2 + x + 2y + 1 = 0$ at the points $(-3/2, -1)$ and $(-3/2, -4)$ is obtained by evaluating the derivative of this equation, dy/dx, at each of these points. To find the derivative, we use implicit differentiation.

After clearing all prior definitions of eq, we define Eq to be the equation $2x^2 - 2xy + y^2 + x + 2y + 1 = 0$.

```
> Eq:='Eq':
  Eq:=2*x^2-2*x*y+y^2+x+2*y+1=0;
```

$$Eq := 2x^2 - 2xy + y^2 + x + 2y + 1 = 0$$

and then we use D to differentiate Eq, naming the resulting output dEq. We interpret D(x) to be 1 and D(y) to be dy/dx.

```
> dEq:=D(Eq);
```

$$dEq := 4\,\mathrm{D}(x)\,x - 2\,\mathrm{D}(x)\,y - 2\,x\,\mathrm{D}(y) + 2\,\mathrm{D}(y)\,y + \mathrm{D}(x) + 2\,\mathrm{D}(y) = 0$$

Because we will be using the miscellaneous library function `isolate`, which lets us isolate an expression in an equation, we load the `isolate` function, replace each occurrence of `D(x)` in `dEq` by 1 and then use `isolate` to solve `DEq2` for `D(y)`.

```
>  readlib(isolate):
   dEq2:=subs(D(x)=1,dEq):
   imderiv:=isolate(dEq2,D(y));
```

$$imderiv := \mathrm{D}(y) = \frac{-4\,x + 2\,y - 1}{-2\,x + 2\,y + 2}$$

The derivative of $2x^2 - 2xy + y^2 + x + 2y + 1 = 0$ is $\frac{dy}{dx} = \frac{1 + 4x - 2y}{2(1 - x + y)}$. We use `assign` to name `D(y)` the result obtained in `imderiv`. We then use `subs` to evaluate `D(y)` at the points (–3/2, –1) and (–3/2, –4).

```
> assign(imderiv);
```

```
> m1:=subs([x=-3/2,y=-1],D(y));
  m2:=subs([x=-3/2,y=-4],D(y));
```

$$m1 := 1$$

$$m2 := 1$$

Thus, the slope of the lines tangent to the graph of $2x^2 - 2xy + y^2 + x + 2y + 1 = 0$ at the points (–3/2, –1) and (–3/2, –4) is 1.

To visualize the tangent line at these points, we graph the tangent lines simultaneously and name the result `P`. Note that `P` is not displayed because we include a colon at the end of the command. After loading the `plots` package, we use `implicitplot` to graph the equation $2x^2 - 2xy + y^2 + x + 2y + 1 = 0$ and then `display` to display the graphs `Graph_Eq` and `P` together.

```
>  P:=plot({m1*(x+3/2)-1,m2*(x+3/2)-4},
      x=-6..1,-6..1):
   with(plots):
```

```
Graph_Eq:=implicitplot(Eq,x=-6..1,y=-6..1):
display([Graph_Eq,P]);
```

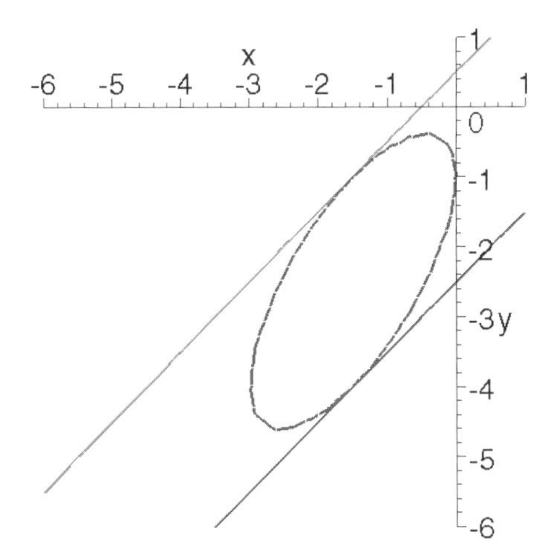

■

In the next example, we implicitly differentiate an equation in which the right-hand side of the equation is not zero.

EXAMPLE 2: Find $y' = dy/dx$ if $\cos(x + \sin y) = \sin y$.

SOLUTION: As in Example 1, we begin by clearing all prior definitions of `Eq` and defining `Eq` to be the equation $\cos(x + \sin y) = \sin y$. Notice how we declare y to be a function of x by using `y(x)` instead of `y` as in Example 1.

```
> Eq:='Eq':
  Eq:=cos(x+sin(y(x)))=sin(y(x)):
```

Next, we use diff to differentiate the equation $\cos(x + \sin y) = \sin y$ with respect to x. The symbol $\frac{\partial}{\partial x}\,y(x)$ in the output represents dy/dx.

```
> deq:=diff(Eq,x);
```

$$deq := -\sin(\,x + \sin(\,y(\,x\,)\,)\,)\left(1 + \cos(\,y(\,x\,)\,)\left(\frac{\partial}{\partial x}\,y(\,x\,)\right)\right) = \cos(\,y(\,x\,)\,)\left(\frac{\partial}{\partial x}\,y(\,x\,)\right)$$

Once again, we use `isolate` to solve for dy/dx.

```
> imderiv:=isolate(deq,diff(y(x),x));
```

$$imderiv := \frac{\partial}{\partial x}\mathrm{y}(x) = \frac{\sin(x + \sin(\mathrm{y}(x)))}{-\sin(x + \sin(\mathrm{y}(x)))\cos(\mathrm{y}(x)) - \cos(\mathrm{y}(x))}$$

Finally, we use `implicitplot` to graph the equation $\cos(x + \sin y) = \sin y$. The displayed graph corresponds is the graph of $\cos(x + \sin y) = \sin y$ on the interval $[-4\pi, 4\pi]$.

```
> implicitplot(cos(x+sin(y))=sin(y),x=-4*Pi..4*Pi,
    y=-4*Pi..4*Pi,color=black,grid=[40,40]);
```

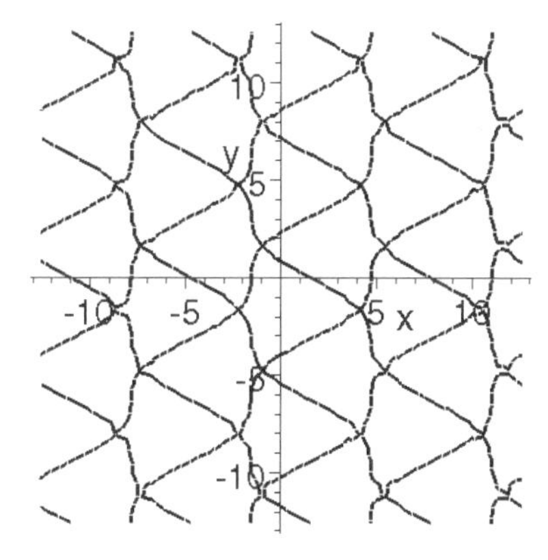

■

EXAMPLE 3: Show that the family of curves with equation $x^2 + 2xy - y^2 = C$ is orthogonal to the family of curves with equation $y^2 + 2xy - x^2 = C$.

SOLUTION: We begin by defining `Eq1` and `Eq2` to be the equations $x^2 + 2xy - y^2 = C$ and $y^2 + 2xy - x^2 = C$, respectively. Note that we use `c` to represent C to avoid conflict with the built-in symbol `C`.

```
> Eq1:=x^2+2*x*y-y^2=c:
  Eq2:=y^2+2*x*y-x^2=c:
```

Then we use `D` and `simplify` to differentiate each equation. Because c represents a constant, we interpret `D(c)` to be 0. As in Example 1, we interpet `D(x)` to be 1 and `D(y)` to be dy/dx.

```
> deq1:=simplify(D(Eq1));
  deq2:=simplify(D(Eq2));
```

$$deq1 := 2\,\mathrm{D}(x)\,x + 2\,\mathrm{D}(x)\,y + 2\,x\,\mathrm{D}(y) - 2\,\mathrm{D}(y)\,y = \mathrm{D}(c)$$
$$deq2 := 2\,\mathrm{D}(y)\,y + 2\,\mathrm{D}(x)\,y + 2\,x\,\mathrm{D}(y) - 2\,\mathrm{D}(x)\,x = \mathrm{D}(c)$$

```
> step2:=subs({D(x)=1,D(c)=0},{deq1,deq2});
```

$$step2 := \{\,2\,\mathrm{D}(y)\,y + 2\,y + 2\,x\,\mathrm{D}(y) - 2\,x = 0,\, 2\,x + 2\,y + 2\,x\,\mathrm{D}(y) - 2\,\mathrm{D}(y)\,y = 0\,\}$$

We then use `isolate` to solve each equation in `step2` for dy/dx.

```
> readlib(isolate):
  isolate(step2[1],D(y));
```

$$\mathrm{D}(y) = \frac{-2\,y + 2\,x}{2\,y + 2\,x}$$

```
> isolate(step2[2],D(y));
```

$$\mathrm{D}(y) = \frac{-2\,x - 2\,y}{-2\,y + 2\,x}$$

Because the derivatives are negative reciprocals,

$$\frac{x+y}{-x+y} = -1/\frac{x-y}{x+y},$$

we conclude that the families are orthogonal. We confirm this graphically by graphing several members of each family, and showing the results together.

```
> with(plots):
  cp1:=contourplot(x^2+2*x*y-y^2,x=-5..5,y=-5..5,
       color=black):
  cp2:=contourplot(y^2+2*x*y-x^2,x=-5..5,y=-5..5,
```

```
        color=gray):
    display({cp1,cp2});
```

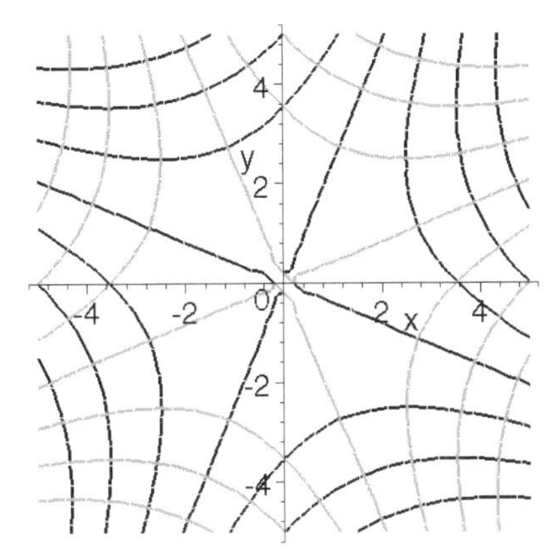

■

Application: Related Rates

Related rate problems are an important application of implicit differentiation.

EXAMPLE 4: Suppose that a lighthouse is located on an island 3 miles off the (straight) shore. If the beacon of the lighthouse revolves at a constant rate of 18° per second, how fast is the light beam moving along the shore at the two points 6 miles from the point on the shore closest to the lighthouse?

SOLUTION: A diagram of the situation described in the problem might look like the following:

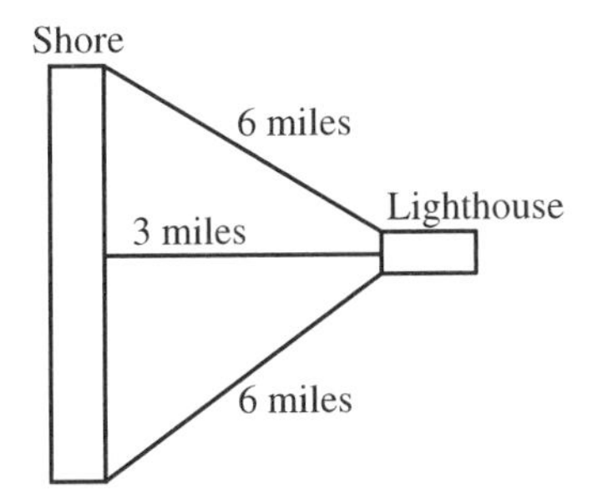

Let θ represent the angle formed by the point on the shore closest to the lighthouse, the lighthouse, and the point on the shore that intersects the light beam from the lighthouse at time t. Let x denote the distance from the point on the shore that intersects the light beam from the lighthouse at time t and the point on the shore closest to the lighthouse. Then, $\frac{1}{3}x = \tan\theta$, which we define as `eq`. We then use `diff` to differentiate `eq` with respect to t, naming the result `deq`. `deq` is equivalent to the equation $\frac{1}{3}\frac{dx}{dt} = \sec^2\theta\frac{d\theta}{dt}$ because $1 + \tan^2\theta = \sec^2\theta$.

```
> eq:=x(t)/3=tan(theta):
  deq:=D(eq);
```

$$deq := \frac{1}{3}\left(\frac{\partial}{\partial t}\mathrm{x}(t)\right) = (1 + \tan(\theta(t))^2)\left(\frac{\partial}{\partial t}\theta(t)\right)$$

At the moment when θ has value 6, equivalent to when t has value $\tan^{-1}\frac{6}{3} = \tan^{-1}2$, we must compute dx/dt, knowing that $d\theta/dt$ has the constant value 18. Thus we use the command `isolate` to isolate dx/dt in `deq` after substituting t by $\tan^{-1}2$ and $d\theta/dt$ by 18. The result means that the light beam is moving at 270 meters per second at the moment when the beam is 6 meters from the point closest on shore to the lighthouse.

```
> readlib(isolate):
  isolate(subs({
      diff(theta(t),t)=18,
      theta(t)=arctan(2)},deq),
         diff(x(t),t));
```

$$\frac{\partial}{\partial t}\mathrm{x}(t) = 270$$

■

3.4 Integral Calculus

Estimating Areas

In integral calculus courses, the definite integral is frequently motivated by an investigation of the area under the graph of a positive continuous function on a closed interval.

Let $y = f(x)$ be a positive continuous function on an interval $[a, b]$, and let n be a positive integer. If we divide $[a, b]$ into n subintervals of equal length and let $[x_{k-1}, x_k]$ denote the kth subinterval, the length of each subinterval is $(b - a)/n$ and $x_k = a + k\frac{b-a}{n}$. The area bounded by the graphs of $y = f(x)$, $x = a$, $x = b$, and the y-axis can be approximated with the sums.

$$S_{left} = \frac{b-a}{n}\sum_{k=1}^{n} f(x_{k-1}) \text{ and } S_{right} = \frac{b-a}{n}\sum_{k=1}^{n} f(x_k).$$

If f is **increasing** on $[a, b]$, S_{left} is an under approximation and S_{right} is an upper approximation. S_{left} corresponds to an approximation of the area using n inscribed rectangles; S_{right} corresponds to an approximation of the area using n circumscribed rectangles. If f is **decreasing** on $[a, b]$, S_{right} is an under approximation and S_{left} is an upper approximation. S_{right} corresponds to an approximation of the area using n inscribed rectangles; S_{left} corresponds to an approximation of the area using n circumscribed rectangles.

EXAMPLE 1: Let $f(x) = 1 + 12x - x^2$. Approximate the area bounded by the graph of $f(x)$, the y-axis, $x = 2$, and $x = 5$ using (a) 100 inscribed and (b) 100 circumscribed rectangles. (c) What is the exact value of the area?

SOLUTION: We begin by defining and graphing f.

```
> f:='f':
  f:=x->1+12*x-x^2:
  plot(f(x),x=-1..13);
```

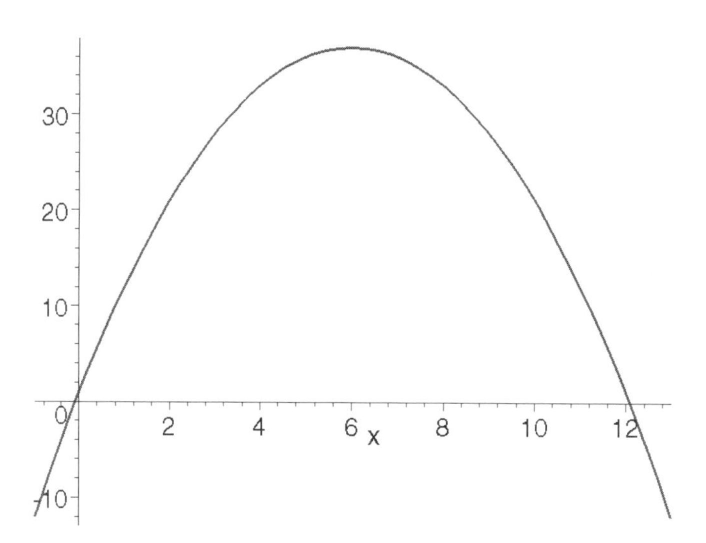

The first derivative $f'(x) = 2(6 - x)$ is positive on the interval [2, 5] so f is increasing on [2, 5]. Thus, an approximation of the area using 100 inscribed rectangles is given by $\frac{3}{100}\sum_{k=0}^{99} f\left(2 + k\frac{3}{100}\right)$ while an approximation of the area using 100 circumscribed rectangles is given by $\frac{3}{100}\sum_{k=1}^{100} f\left(2 + k\frac{3}{100}\right)$. These numbers, along with a corresponding approximation, are computed with `leftsum` and `rightsum`, respectively. Note that both `leftsum` and `rightsum` are included in the `student` package and must be loaded with the command `with(student)` before using or entered as `student[leftsum]` and `student[rightsum]`, respectively. Exact values of each of these sums are computed with `value`; approximations with `evalf`.

```
> with(student);
```

[*D*, *Diff*, *Doubleint*, *Int*, *Limit*, *Lineint*, *Product*, *Sum*, *Tripleint*, *changevar*, *combine*, *completesquare*, *distance*, *equate*, *extrema*, *integrand*, *intercept*, *intparts*, *isolate*, *leftbox*, *leftsum*, *makeproc*, *maximize*, *middlebox*, *middlesum*, *midpoint*, *minimize*, *powsubs*, *rightbox*, *rightsum*, *showtangent*, *simpson*, *slope*, *summand*, *trapezoid*, *value*]

```
left100:=leftsum(f(x),x=2..5,100);
```

$$left100 := \frac{3}{100}\left(\sum_{i=0}^{99}\left(25+\frac{9}{25}i-\left(2+\frac{3}{100}i\right)^2\right)\right)$$

```
> value(left100);
```

$$\frac{1795491}{20000}$$

```
> evalf(left100);
```

89.77455000

```
> right100:=rightsum(f(x),x=2..5,100);
```

$$right100 := \frac{3}{100}\left(\sum_{i=1}^{100}\left(25+\frac{9}{25}i-\left(2+\frac{3}{100}i\right)^2\right)\right)$$

```
> value(right100);
```

$$\frac{1804491}{20000}$$

```
> evalf(right100);
```

90.22455000

Observe that these two values appear to be close to 90. In fact, $\lim_{n \to \infty} \frac{3}{n} \sum_{k=0}^{n-1} f(x_k) = \lim_{n \to \infty} \frac{3}{n} \sum_{k=1}^{n} f(x_{k-1})$, and this number is the exact value of the area bounded by the graphs of $f(x)$, the y-axis, $x = 2$, and $x = 5$. To help us see why this is true, we use `leftbox` and `rightbox` to visualize the situation. As with `leftsum` and `rightsum`, these commands are contained in the `student` package.

```
> with(plots):
  A:=array(1..2,1..2):
  A[1,1]:=leftbox(f(x),x=2..5,4):
  A[1,2]:=leftbox(f(x),x=2..5,16):
  A[2,1]:=leftbox(f(x),x=2..5,32):
  A[2,2]:=leftbox(f(x),x=2..5,64):
  display(A);
```

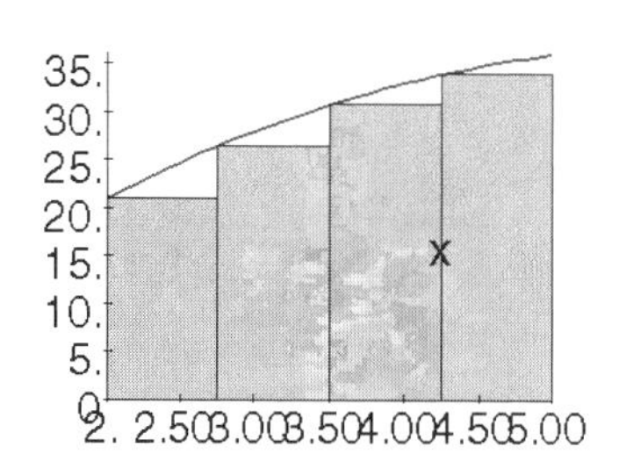

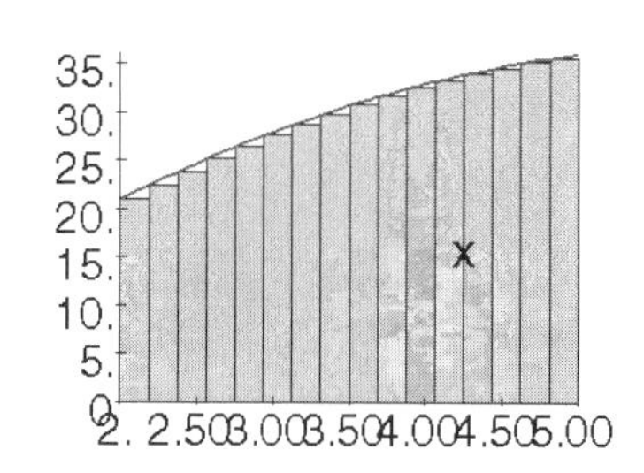

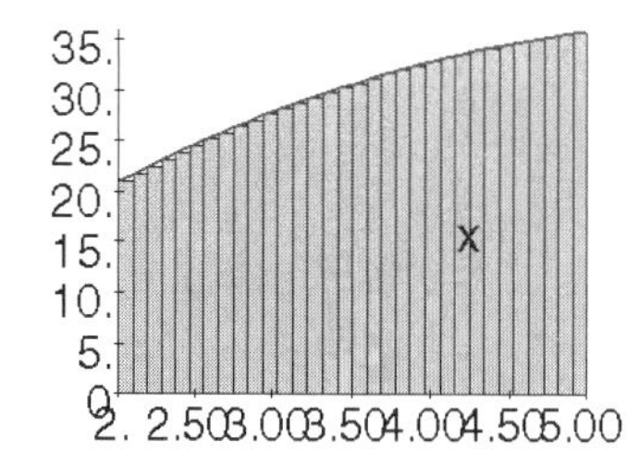

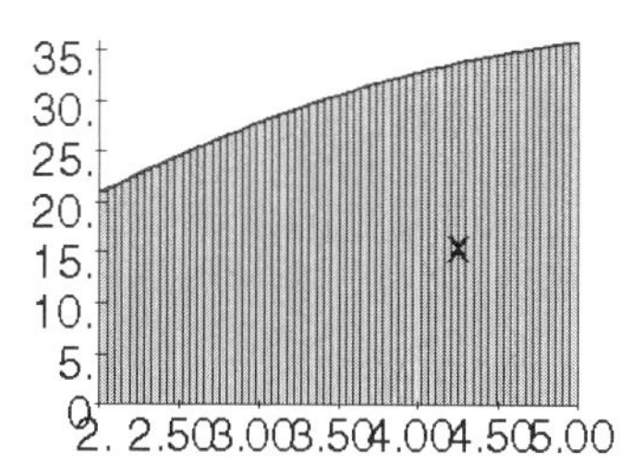


```
> A:=array(1..2,1..2):
  A[1,1]:=rightbox(f(x),x=2..5,4):
  A[1,2]:=rightbox(f(x),x=2..5,16):
  A[2,1]:=rightbox(f(x),x=2..5,32):
```

```
A[2,2]:=rightbox(f(x),x=2..5,64):
display(A);
```

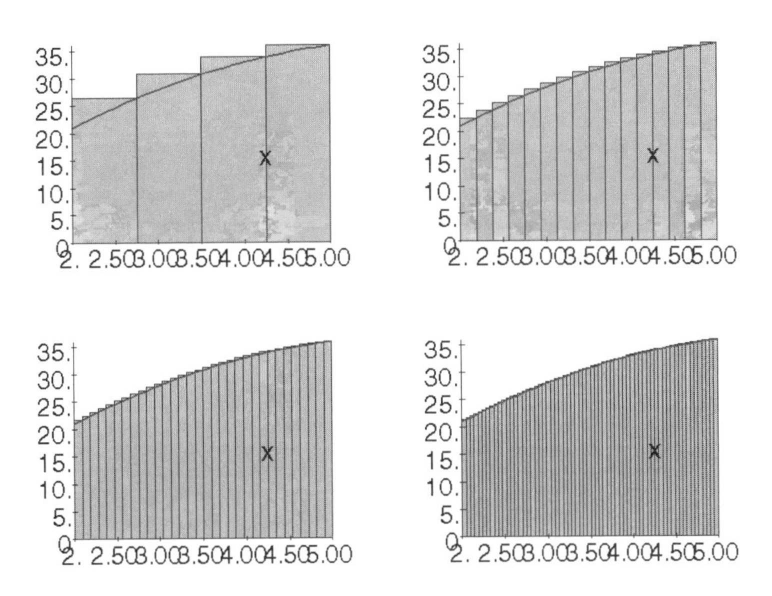

These graphs help convince us that the limit of the sum of the areas of the inscribed and circumscribed rectangles is the same. In fact, for any positive integer n, the sum of the areas of the inscribed rectangles is given by $\frac{3}{n}\sum_{k=0}^{n-1} f(2+k\frac{3}{n})$, and the sum of the areas of the circumscribed rectangles is given by $\frac{3}{n}\sum_{k=1}^{n} f(2+k\frac{3}{n})$. Closed forms of these sums are computed with `leftsum` and `rightsum` followed by using `value`. We then use `limit` to compute the limit as $n \to \infty$ of both `leftn` and `rightn`. The results, as expected, are the same.

```
> leftn:=leftsum(f(x),x=2..5,n);
```

$$leftn := 3\,\frac{\sum_{i=0}^{n-1}\left(25+36\frac{i}{n}-\left(2+3\frac{i}{n}\right)^2\right)}{n}$$

```
> closedleft:=value(leftn);
```

$$closedleft := 3\,\frac{30\,n - \frac{15}{2} - \frac{3}{2}\frac{1}{n}}{n}$$

```
> limit(closedleft,n=infinity);
```

$$90$$

```
> rightn:=rightsum(f(x),x=2..5,n);
```

$$rightn := 3\,\frac{\displaystyle\sum_{i=1}^{n}\left(25 + 36\,\frac{i}{n} - \left(2 + 3\,\frac{i}{n}\right)^2\right)}{n}$$

```
> closedright:=value(rightn);
```

$$closedright := 3\,\frac{21\,n + 12\,\frac{(n+1)^2}{n} - 12\,\frac{n+1}{n} - 3\,\frac{(n+1)^3}{n^2} + \frac{9}{2}\,\frac{(n+1)^2}{n^2} - \frac{3}{2}\,\frac{n+1}{n^2}}{n}$$

```
> limit(closedright,n=infinity);
```

$$90$$

We conclude that the exact area is 90.

■

EXAMPLE 2: Approximate the area bounded by the graphs of $y = f(x)$, the y-axis, $x = 1$, and $x = 3$ using 2, 4, 8, 16, 32, 64, 128, 256, and 512 (a) circumscribed and (b) inscribed rectangles if $f(x) = 2x^3 - 9x^2 + 30$. (c) What is the exact value of the area?

SOLUTION: As in Example 1, we begin by defining and graphing f.

```
> f:='f':
  f:=2*x^3-9*x^2+30:
  plot(f(x),x=-1..4);
```

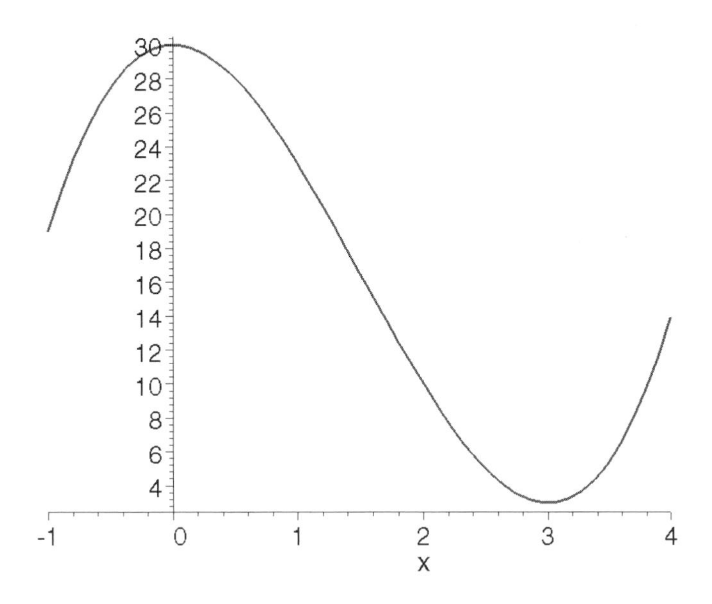

We then use `leftsum` and `rightsum` together with `seq` to approximate the area for the indicated numbers of circumscribed and inscribed rectangles. In this case, f is decreasing on the interval [1,3] so `leftsum` yields an upper approximation while `rightsum` yields a lower approximation. (Note that `array` and `seq` are discussed in more detail in Chapters 4 and 5.)

```
> array([[2^n,upper,lower],
      seq([2^n,evalf(leftsum(f(x),x=1..3,2^n)),
         evalf(rightsum(f(x),x=1..3,2^n))],
      n=1..9)]
      );
```

2^n	*upper*	*lower*
2	33.	13.
4	27.25000000	17.25000000
8	24.56250000	19.56250000
16	23.26562500	20.76562500
32	22.62890625	21.37890625
64	22.31347656	21.68847656
128	22.15649414	21.84399414
256	22.07818604	21.92193604
512	22.03907776	21.96095276

We also use the functions `leftbox` and `rightbox`, as in Example 1, to visualize various circumscribed and inscribed rectangles.

```
> A:=array(1..2,1..2):
  A[1,1]:=leftbox(f(x),x=1..3,4):
  A[1,2]:=leftbox(f(x),x=1..3,16):
  A[2,1]:=leftbox(f(x),x=1..3,32):
  A[2,2]:=leftbox(f(x),x=1..3,64):
  display(A);
```

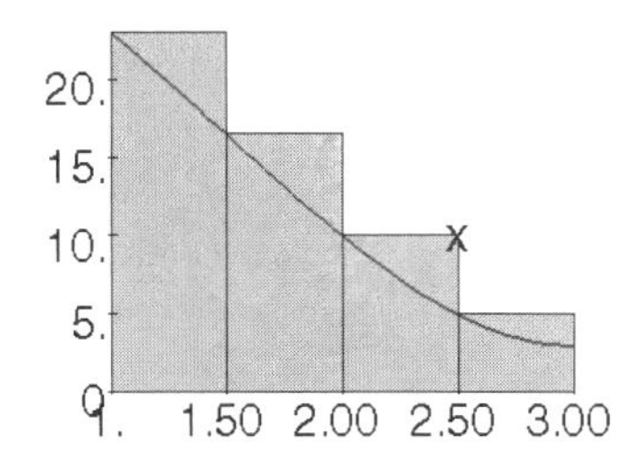

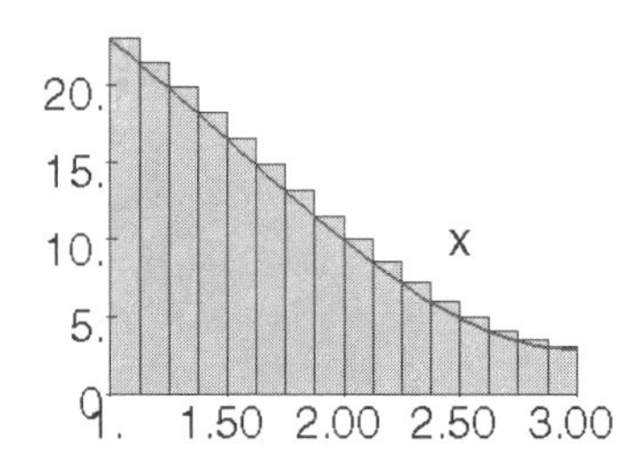

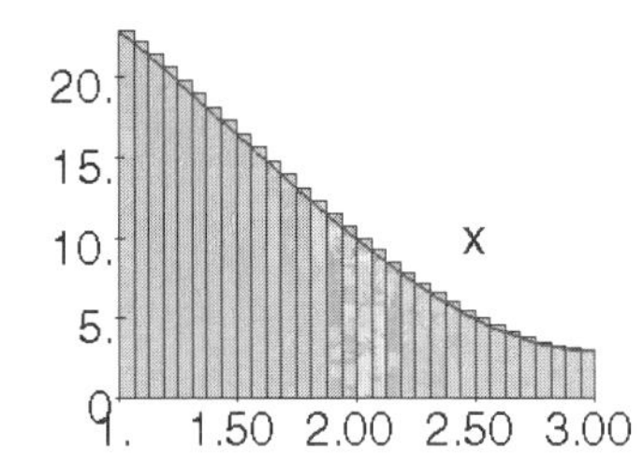

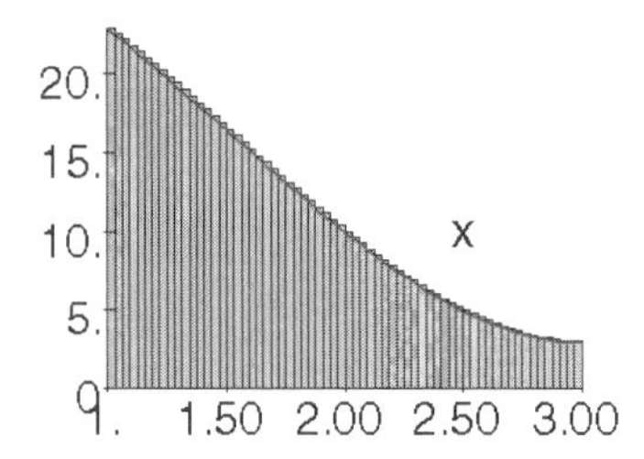


```
> A:=array(1..2,1..2):
  A[1,1]:=rightbox(f(x),x=1..3,4):
  A[1,2]:=rightbox(f(x),x=1..3,16):
  A[2,1]:=rightbox(f(x),x=1..3,32):
  A[2,2]:=rightbox(f(x),x=1..3,64):
  display(A);
```

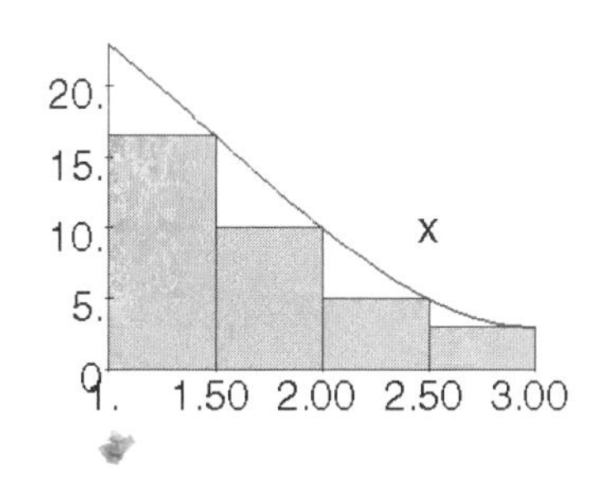

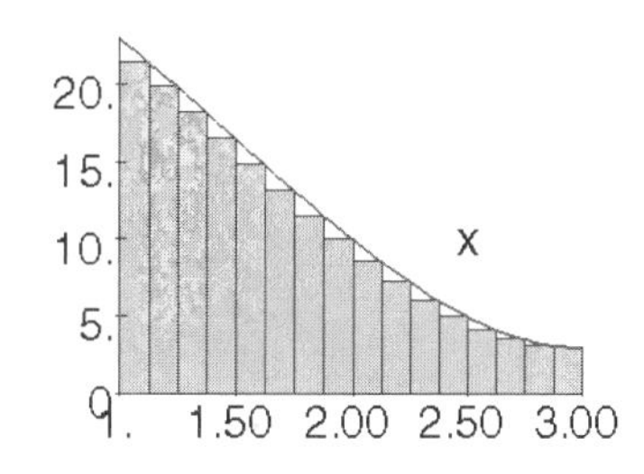

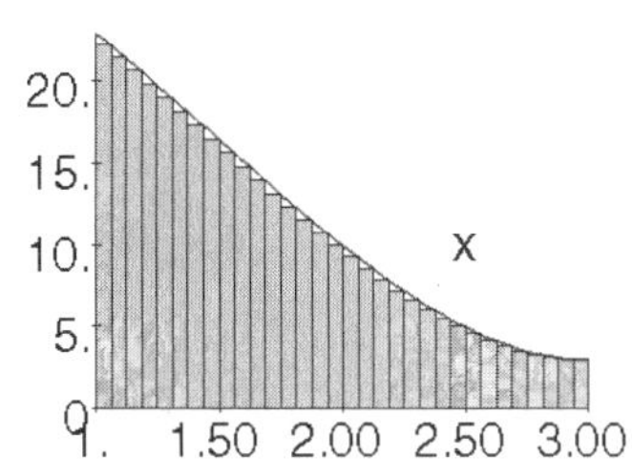

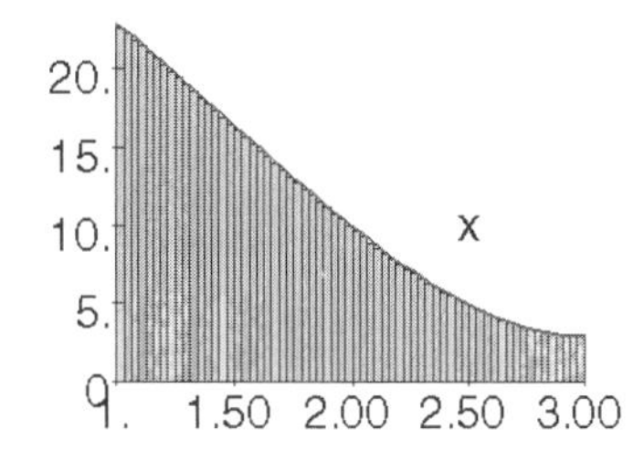

■

Computing Definite and Indefinite Integrals

Maple's `int` command can compute many definite and indefinite integrals similar to those found in typical integral calculus courses. The command

```
int(f(x),x)
```

attempts to calculate an antiderivative of $f(x)$: $\int f(x)d(x)$; the command

```
int(f(x),x=a..b)
```

attempts to compute $\int_a^b f(x)d(x)$. The inert form of the integrate command, `int`, is `Int`.

EXAMPLE 3: Compute each of the following indefinite integrals.
(a) $\int x^2(1-x^3)^5dx$; (b) $\int e^{-2x}\sin 3x dx$;
(c) $\int x^2\tan^{-1}x dx$; (d) $\int f(x)dx$ if $f(x) = \dfrac{x^2-4x}{x-2x-3}$; and
(e) $\int g(y)dy$ if $g(y) = y^3\ln^2 y$.

SOLUTION: In order to evaluate indefinite integrals, `int(expr,var)` is used. Entering

```
> int(x^2*(1-x^3)^4,x);
```

$$\frac{1}{15}x^{15}-\frac{1}{3}x^{12}+\frac{2}{3}x^{9}-\frac{2}{3}x^{6}+\frac{1}{3}x^{3}$$

computes $\int x^2(1-x^3)^5dx;$ entering

```
> int(exp(-2*x)*sin(3*x),x);
```

$$-\frac{3}{13}\mathbf{e}^{(-2x)}\cos(3x)-\frac{2}{13}\mathbf{e}^{(-2x)}\sin(3x)$$

computes $\int e^{-2x}\sin 3x dx;$ and entering

```
> int(x^2*arctan(x),x);
```

$$\frac{1}{3}x^3\arctan(x)-\frac{1}{6}x^2+\frac{1}{6}\ln(x^2+1)$$

For (d) and (e), we first define *f* and *g* and then use `int`.

```
> f:='f':g:='g':
  f:=x->(x^2-4*x)/(x^2-2*x-3):
  int(f(x),x);
```

$$x-\frac{5}{4}\ln(x+1)-\frac{3}{4}\ln(x-3)$$

```
> g:=y->y^3*ln(y)^2:
  int(g(y),y);
```

$$\frac{1}{4}y^4\ln(y)^2-\frac{1}{8}y^4\ln(y)+\frac{1}{32}y^4$$

■

Generally, Maple can compute antiderivatives of most functions encountered in an introductory integral calculus course. Integration is a difficult procedure, so it is relatively easy to make up integrals that Maple cannot calculate. Nevertheless, Maple can calculate a wide variety of integrals.

EXAMPLE 4: Calculate (a) $\int \sin x \ln x dx$;; (b) $\int \frac{1}{\sin^2 x+2} dx$;; and (c) $\int \frac{x}{\sin x + 2} dx$.

SOLUTION: Maple can compute (a), (b), and (c).

```
> int(sin(x)*ln(x),x);
```

The function `Ci(x)`, appearing in the result, represents the **cosine integral function**, *Ci*(*z*), where.

$$Ci(z) = \int_z^\infty \frac{\cos t}{t} dt = \gamma + \ln z + \int_0^z \frac{\cos t - 1}{t} dt.$$

Here, γ represents Euler's constant which is approximately 0.577216. Maple has a built-in definition of Euler's constant, `gamma`. The function `csgn` is defined by

$$csgn(z) = \begin{cases} 1, \operatorname{Re}(z) > 0 \text{ or } \operatorname{Re}(z) = 0 \text{ and } \operatorname{Im}(z) > 0 \\ -1, \operatorname{Re}(z) < 0 \text{ or } \operatorname{Re}(z) = 0 \text{ and } \operatorname{Im}(z) < 0 \end{cases}.$$

```
> int(1/(sin(x)^2+2),x)
```

$$\frac{1}{3}\frac{\arctan\left(2\frac{\tan\left(\frac{1}{2}x\right)}{\sqrt{6}+\sqrt{2}}\right)\sqrt{3}}{\sqrt{6}+\sqrt{2}} + \frac{\arctan\left(2\frac{\tan\left(\frac{1}{2}x\right)}{\sqrt{6}+\sqrt{2}}\right)}{\sqrt{6}+\sqrt{2}}$$

$$-\frac{1}{3}\frac{\arctan\left(2\frac{\tan\left(\frac{1}{2}x\right)}{\sqrt{6}-\sqrt{2}}\right)\sqrt{3}}{\sqrt{6}-\sqrt{2}} + \frac{\arctan\left(2\frac{\tan\left(\frac{1}{2}x\right)}{\sqrt{6}-\sqrt{2}}\right)}{\sqrt{6}-\sqrt{2}}$$

The result obtained for (b) is simplified considerably with `simplify`.

```
> simplify(%);
```

$$\frac{1}{3}\frac{\sqrt{3}\left(\arctan\left(\frac{\tan\left(\frac{1}{2}x\right)\sqrt{2}}{\sqrt{3}-1}\right)+\arctan\left(\frac{\tan\left(\frac{1}{2}x\right)\sqrt{2}}{\sqrt{3}+1}\right)\right)\sqrt{2}}{(\sqrt{3}+1)(\sqrt{3}-1)}$$

```
> int(x/(sin(x)+2),x);
```

$$-\frac{1}{3}\ln(-2I+I\sqrt{3})\ln\left(\frac{I(-\mathbf{e}^{(Ix)}-2I+I\sqrt{3})}{-2+\sqrt{3}}\right)\sqrt{3}$$
$$+\frac{1}{3}\ln(-2I-I\sqrt{3})\ln\left(\frac{I(\mathbf{e}^{(Ix)}+2I+I\sqrt{3})}{2+\sqrt{3}}\right)\sqrt{3}$$
$$+\frac{1}{3}\operatorname{dilog}\left(-\frac{I\,\mathbf{e}^{(Ix)}}{-2+\sqrt{3}}\right)\sqrt{3}-\frac{1}{3}\operatorname{dilog}\left(\frac{I\,\mathbf{e}^{(Ix)}}{2+\sqrt{3}}\right)\sqrt{3}$$

The `dilog` function is defined as the improper integral $dilog(x) = \int_1^x \frac{\ln t}{1-t}$.

■

Definite integrals are also computed with `Integrate`. In general, the command

```
int(expression,variable=lowerlimit..upperlimit)
```

integrates `expression` with respect to `variable` and evaluates from `lowerlimit` to `upperlimit`.

EXAMPLE 5: Calculate each definite integral.

(a) $\int_0^1 (x-x^2)dx$; (b) $\int_0^{\pi} \sin x dx$;;

(c) $\int_1^2 \sqrt{4-x^2}dx$; (d) $\int_1^2 f(x)dx$, if $f(x) = x^3e^{-4x}$; and

(e) $\int_{-\pi}^{2\pi} g(x)dx$, if $g(x) = e^{2x}\sin^2 2x$.

SOLUTION: For (a), (b), and (c), we use `int` to evaluate each definite integral. Entering

```
> int(x-x^2,x=0..1);
```

$$\frac{1}{6}$$

```
> int(sin(x),x=0..Pi);
```

$$2$$

```
> int(sqrt(4-x^2),x=1..2);
```

$$\frac{2}{3}\pi-\frac{1}{2}\sqrt{3}$$

On the other hand, for (d) and (e), we clear all prior definitions of f and g, define f and g, and then use `int` to compute the indicated definite integral.

```
> f:=x->x^3*exp(-4*x):
  int(f(x),x=1..2);
```

$$-\frac{379}{128}\mathbf{e}^{(-8)}+\frac{71}{128}\mathbf{e}^{(-4)}$$

```
> g:=x->exp(2*x)*sin(2*x)^2:
  int(g(x),x=-Pi..2*Pi);
```

$$\frac{1}{5}\mathbf{e}^{(4\pi)}-\frac{1}{5}\mathbf{e}^{(-2\pi)}$$

■

Improper integrals are computed using `int` in the same way as other definite integrals are computed.

EXAMPLE 6: Evaluate each of the following improper integrals: (a) $\int_0^4 \frac{1}{\sqrt{x}} dx$, , (b) $\int_0^\infty xe^{-x^2} dx$, (c) $\int_1^\infty \frac{1}{x\sqrt{x^2-1}} dx$, (d) $\int_0^\infty \frac{1}{x^2+x^4} dx$, and , (e) $\int_{-8}^8 \frac{1}{x^{2/3}} dx$.

SOLUTION: (a) This is an improper integral because the integrand is discontinuous on the interval $[0, 4]$ but we see that the improper integral converges to 4.

```
> int(1/sqrt(x),x=0..4);
```

$$4$$

(b) This is an improper integral because the interval of integration is infinite but we see that the improper integral converges to $1/2$.

```
> int(x*exp(-x^2),x=0..infinity);
```

$$\frac{1}{2}$$

(c) This is an improper integral because the integrand is discontinuous on the interval of integration and because the interval of integration is infinite but we see that the improper integral converges to $\pi/2$.

```
> int(1/(x*sqrt(x^2-1)),x=1..infinity);
```

$$\frac{1}{2}\pi$$

(d) As with (c), this is an improper integral because the integrand is discontinuous on the interval of integration and because the interval of integration is infinite but we see that the improper integral diverges.

```
> int(1/(x^2+x^4),x=0..infinity);
```

$$\infty$$

(e) Recall that Maple does not return a real number when we compute odd roots of negative numbers so the following result would be surprising to many students in an introductory calculus course because it contains imaginary numbers.

```
> int(1/x^(2/3),x=-8..8);
```

$$\frac{3}{2}8^{(1/3)} - \frac{3}{2}I\,8^{(1/3)}\sqrt{3}$$

Therefore, we take advantage of the `surd` command. `surd(x,3)` returns the real-valued third root of x.

```
> int(1/surd(x,3)^2,x=-8..8);
```

$$6\,8^{(1/3)}$$

■

In many cases, Maple can help illustrate the steps carried out when computing integrals using standard methods of integration like u-substitutions and integration by parts. In these situations, we take advantage of Maple's `student` package. We begin with the method of u-substitution.

EXAMPLE 7: Use the substitution $u = \cos x^3$ to compute $\int x^2 \sin x^3 \cos x^3 dx$.

SOLUTION: After defining f, we use `Int` to compute the unevaluated form of $\int x^2 \sin x^3 \cos x^3 dx$ and name the resulting output `tocompute`. Remember that `Int` represents the inert form of the `int` command; it returns results unevaluated.

SOLUTION:

```
> f:='f':
  f:=x^2*sin(x^3)*cos(cos(x^3)):
  tocompute:=Int(f,x);
```

$$tocompute := \int x^2 \sin(x^3) \cos(\cos(x^3))\,dx$$

We then use `changevar`, contained in the `student` package, to perform the substitution $u = \cos x^3$ on `tocompute`. Be sure to load the package `student` by entering `with(student)` before using the command `changevar`. The resulting output is named `aftersub`. Note that the command `changevar` is contained in the `student` package and must be loaded with the command `with(student)` or entered in the form `student[changevar]`.

```
> with(student):
  aftersub:=changevar(cos(x^3)=u,tocompute);
```

$$aftersub := \int -\frac{1}{3}\cos(u)\,du$$

`aftersub` is then evaluated with `value`, named `antideriv`, and `subs` is used to replace each occurrence of u in `antideriv` by x^3.

```
> antideriv:=value(aftersub);
```

$$antideriv := -\frac{1}{3}\sin(u)$$

```
> subs(u=cos(x^3),antideriv);
```

$$-\frac{1}{3}\sin(\cos(x^3))$$

■

Definite integrals may be considered as well.

EXAMPLE 8: Use the substitution $u = 2 + \sin x$ to compute $\int_{\pi/2}^{\pi} f(x)dx$ if $f(x) = \frac{\cos x}{(2 + \sin x)^2}$.

SOLUTION: Proceeding as in the Example 7, we first define f and `tocompute` to be the unevaluated form of the definite integral $\int_{\pi/2}^{\pi} f(x)dx$.

```
> f:='f':
  f:=x->cos(x)/(2+sin(x))^2:
  tocompute:=Int(f(x),x=Pi/2..Pi);
```

$$tocompute := \int_{1/2\,\pi}^{\pi} \frac{\cos(x)}{(2+\sin(x))^2}\,dx$$

We then use changevar to perform the substitution $u = 2 + \sin x$ on tocompute and name the resulting output aftersub. As in the previous example, be sure the student package has been loaded before the changevar command is used. Note that Maple automatically changes the limits of integration, a common error when u-substitutions are performed.

```
> aftersub:=changevar(2+sin(x)=u,tocompute);
```

$$aftersub := \int_2^3 -\frac{1}{u^2}\,du$$

value is used to obtain the exact value of the integral.

```
> value(aftersub);
```

$$\frac{-1}{6}$$

■

Maple is also useful in performing the steps needed for integration by parts.

EXAMPLE 9: The **Integration by Parts Formula** is $\int u\,dv = uv - \int v\,du$. Use Maple and integration by parts to compute $\int e^{4x}\sin 3x\,dx$.

SOLUTION: Proceeding as in the previous examples, we first define $f(x) = e^{4x}\sin 3x$ and then define tocompute to be the unevaluated form of $\int e^{4x}\sin 3x\,dx$.

```
> f:='f':
  f:=x->sin(3*x)*exp(4*x):
  tocompute:=Int(f(x),x);
```

$$tocompute := \int \sin(3\,x)\,\mathrm{e}^{(4\,x)}\,dx$$

Then we use the command intparts, contained in the student package, to compute $\int u dv = uv - \int v du$ if $u = e^{4x}$ and $dv = \sin 3x dx$. This command is entered as intparts(integral,u) in order to make the indicated substitution for u in the integral integral using integration by parts. The result, named stepone, means that

$$\int e^{4x} \sin 3x dx = -\frac{1}{3} e^{4x} \cos 3x + \int \frac{4}{3} e^{4x} \cos 3x dx.$$

```
> with(student):
  stepone:=intparts(tocompute,exp(4*x));
```

$$stepone := -\frac{1}{3} \mathbf{e}^{(4x)} \cos(3x) - \int -\frac{4}{3} \mathbf{e}^{(4x)} \cos(3x)\, dx$$

In this case, we must perform integration by parts twice. Therefore, we perform integration by parts on stepone, using the same choices for u and dv as before, and name the resulting output steptwo. The result means that

$$\int e^{4x} \sin 3x\, dx = -\frac{1}{3} e^{4x} \cos 3x + \frac{4}{9} e^{4x} \sin 3x - \int \frac{16}{9} e^{4x} \sin 3x\, dx.$$

```
> steptwo:=intparts(stepone,exp(4*x));
```

$$steptwo := -\frac{1}{3} \mathbf{e}^{(4x)} \cos(3x) + \frac{4}{9} \sin(3x)\, \mathbf{e}^{(4x)} + \int -\frac{16}{9} \sin(3x)\, \mathbf{e}^{(4x)}\, dx$$

Observe that $-\int \frac{16}{9} e^{4x} \sin 3x dx$ is extracted from steptwo with op(3,steptwo) because $-\int \frac{16}{9} e^{4x} \sin 3x dx$ is the third part of steptwo. Generally, op(n,expression) yields the *n*th part of expression; nops(expressions) yields the number of parts of expression.

```
> op(3,steptwo);
```

$$\int -\frac{16}{9}\sin(3x)\,\mathbf{e}^{(4x)}\,dx$$

Consequently,

$$\frac{25}{9}\int e^{4x}\sin 3x dx = -\frac{1}{3}e^{4x}\cos 3x + \frac{4}{9}e^{4x}\sin 3x$$

and multiplying $-\frac{1}{3}e^{4x}\cos 3x + \frac{4}{9}e^{4x}\sin 3x$ by 9/25 results in

$$\int e^{4x}\sin 3x dx = -\frac{3}{25}e^{4x}\cos 3x + \frac{3}{25}e^{4x}\sin 3x.$$

These steps are carried out with Maple next. In `lefthandside` and `righthandside`, `op(3,steptwo)` is subtracted from the left-hand side and right-hand side of the equation .

$$\int e^{4x}\sin 3x dx = -\frac{1}{3}e^{4x}\cos 3x + \frac{4}{9}e^{4x}\sin 3x - \int\frac{16}{9}e^{4x}\sin 3x dx.$$

Hence, `lefthandside` represents the expression that results on the left-hand side while `righthandside` represents the results on the right-hand side. Finally, the expression in `righthandside` is multiplied by 9/25 to yield the value of the integral.

```
> lefthandside:=combine(tocompute-op(3,steptwo));
```

$$lefthandside := \int \frac{25}{9}\sin(3x)\,\mathbf{e}^{(4x)}\,dx$$

```
> righthandside:=steptwo-op(3,steptwo);
```

$$righthandside := -\frac{1}{3}\mathbf{e}^{(4x)}\cos(3x) + \frac{4}{9}\sin(3x)\,\mathbf{e}^{(4x)}$$

```
> 9/25*righthandside;
```

$$-\frac{3}{25}e^{(4x)}\cos(3x)+\frac{4}{25}\sin(3x)e^{(4x)}$$

■

Approximating Definite Integrals

Because integration is a fundamentally difficult procedure, Maple is unable to compute a closed form of the value of many definite integrals. In these cases, numerical integration can be used to obtain an approximation of the definite integral using `evalf` together with `int` or `evalf` together with `Int`.

EXAMPLE 10: Compute both exact and approximate values of $\int_4^{10}\frac{\sqrt{4x^2-9}}{x^3}dx$.

SOLUTION: `int` is used to compute and exact value of the integral; `evalf` is used to compute an approximation of the integral.

```
> int(sqrt(4*x^2-9)/x^3,x=4..10);
```

$$-\frac{1}{200}\sqrt{391}-\frac{2}{3}\arctan\left(\frac{3}{391}\sqrt{391}\right)+\frac{1}{32}\sqrt{55}+\frac{2}{3}\arctan\left(\frac{3}{55}\sqrt{55}\right)$$

```
> evalf(int(sqrt(4*x^2-9)/x^3,x=4..10));
```

2887732708

■

In many cases, Maple can compute approximate values of definite integrals it cannot compute exactly.

EXAMPLE 11: Approximate $\int_0^{\sqrt[3]{\pi}} e^{-x^2}\cos x^3 dx$.

SOLUTION: In this case, after defining f we use `int` to attempt to compute $\int_0^{\sqrt[3]{\pi}} e^{-x^2}\cos x^3 dx$. We name the resulting output, an unevaluated integral, `tryone`.

```
> f:='f':
  f:=x->exp(-x^2)*cos(x^3):
  tryone:=int(f(x),x=0..Pi^(1/3));
```

$$tryone := \int_0^{\pi^{(1/3)}} \mathbf{e}^{(-x^2)} \cos(x^3)\, dx$$

`evalf` is then used to obtain a numerical approximation of `tryone`.

```
> evalf(tryone);
```

.7015656956

■

The `student` package contains various commands that can be used to investigate approximations of definite integrals using various numerical methods.

EXAMPLE 12: Let $f(x) = \sin(\sin(\sin(\sin x)))$. (a) Graph f on the interval $[0,\pi]$. Use (b) Simpson's rule with $n = 4$; and (c) the Trapezoidal rule with $n = 4$ to approximate $\int_0^{\pi} f(x)dx$.

SOLUTION: In this case, we use the repeated composition operator `@@` to define N$f(x) = \sin(\sin(\sin(\sin x)))$ and then graph f on $[0,\pi]$. Because $\sin x$ is composed with itself four times, we use `(sin@@4)(x)` to represent this function.

```
> f:='f':
  f:=x->(sin@@4)(x):
  plot(f(x),x=0..Pi);
```

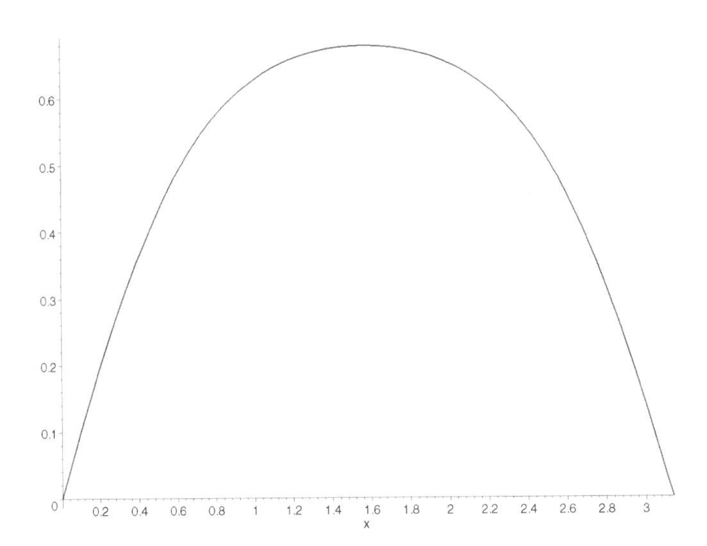

We use simpson to compute the approximation of $\int_0^{\pi} f(x)dx$. The resulting output is named simsponapprox. Note that simpson, as well as trapezoid, are contained in the student package and must be loaded with the command with(student) before they are used or entered in the form student[simpson] and student[trapezoid], respectively. The exact value of simpsonapprox is obtained with value; an approximation is computed with evalf. Simpson's rule with $n = 4$ is applied to yield an expression in unevaluated form in simpsonapprox. The exact value of simpsonapprox is then found with value, and a numerical approximation of 1.546261019 is computed with evalf.

```
> with(student):
  simpsonapprox:=simpson(f(x),x=0..Pi,4);
```

$$simpsonapprox := \frac{1}{12}\pi\left(4\left(\sum_{i=1}^{2}(\sin^{(4)})\left(\frac{1}{4}(2i-1)\pi\right)\right)+2\left(\sum_{i=1}^{1}(\sin^{(4)})\left(\frac{1}{2}i\pi\right)\right)\right)$$

```
> anapprox:=value(simpsonapprox);
```

$$anapprox := \frac{1}{12}\pi\left(8(\sin^{(3)})\left(\frac{1}{2}\sqrt{2}\right)+2(\sin^{(3)})(1)\right)$$

```
> evalf(anapprox);
```

$$1.546261019$$

In the same manner as above, `trapezoid` is used to compute an approximation of $\int_0^{\pi} f(x)dx$ using the Trapezoidal rule. In this case, the resulting output is named `trapapprox`. The exact value and an approximation of `trapapprox` are obtained with `value` and `evalf`, respectively.

```
> trapapprox:=trapezoid(f(x),x=0..Pi,4);
```

$$trapapprox := \frac{1}{4}\pi\left(\sum_{i=1}^{3}(\sin^{(4)})\left(\frac{1}{4}i\pi\right)\right)$$

```
> anotherapprox:=value(trapapprox);
```

$$anotherapprox := \frac{1}{4}\pi\left(2(\sin^{(3)})\left(\frac{1}{2}\sqrt{2}\right)+(\sin^{(3)})(1)\right)$$

```
> evalf(anotherapprox);
```

$$1.426114790$$

```
> evalf(Int(f(x),x=0..Pi));
```

$$1.534706071$$

■

Application: Area Between Curves

Finding the area of a region between two (or more) curves incorporates the commands `int` and `Int` together with `evalf` These problems use several other Maple commands (`plot`, `solve`, `fsolve`,...) that were introduced earlier in the text as well.

EXAMPLE 13: Find the area between the graphs of $y = \sin x$ and $y = \cos x$ on the interval $[0,2\pi]$.

SOLUTION: We graph $y = \sin x$ and $y = \cos x$ on the interval $[0,2\pi]$. The graph of $y = \cos x$ is gray.

```
> plot({sin(x),cos(x)},x=0..2*Pi,color=[black,gray]);
```

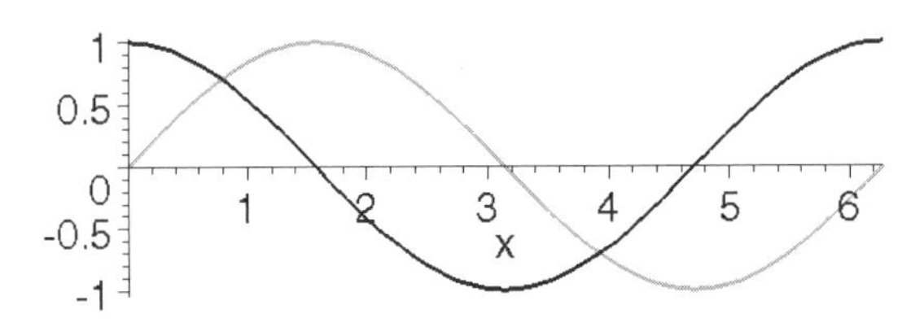

To find the upper and lower limits of integration, we must solve the equation $\sin x = \cos x$ for x. We see that `solve` is able to find one solution of this equation.

```
> solve(sin(x)=cos(x));
```

$$\frac{1}{4}\pi$$

However, we see that $\sin x = \cos x$ on the interval $[0,2\pi]$ if $x = \pi/4$ or $x = 5\pi/4$. Note that `fsolve` can be used to approximate these numbers, but the resulting area computed would not be exact.

```
> _EnvAllSolutions := true:
  solve( sin(x)=cos(x));
```

$$\frac{1}{4}\pi + \pi\_Z\sim$$

Thus, the desired area is given by

$$Area = \int_0^{\pi/4}(\cos x - \sin x)dx + \int_{\pi/4}^{5\pi/4}(\sin x - \cos x)dx + \int_{5\pi/4}^{2\pi}(\cos x - \sin x)dx.$$

```
> int(cos(x)-sin(x),x=0..Pi/4)+
      int(sin(x)-cos(x),x=Pi/4..5*Pi/4)+
         int(cos(x)-sin(x),x=5*Pi/4..2*Pi);
```

$$4\sqrt{2}$$

The area is $4\sqrt{2}$.

■

In cases when we cannot calculate the points of intersection of two graphs exactly, we can frequently use `fsolve` to estimate the points of intersection.

EXAMPLE 14: Let

$$p(x) = \frac{3}{10}x^5 - 3x^4 + 11x^3 - 18x^2 + 12x + 1$$

and

$$q(x) = -4x^3 + 28x^2 - 56x + 32.$$

Approximate the area of the region bounded by the graphs of p and q.

SOLUTION: After defining p and q, we graph them on the interval $[-1,5]$ to obtain an initial guess of the intersection points of the two graphs.

```
> p:='p':q:='q':
  p:=3*x^5/10-3*x^4+11*x^3-18*x^2+12*x+1:
  q:=-4*x^3+28*x^2-56*x+32:
  plot({p,q},x=-1..5,-15..20);
```

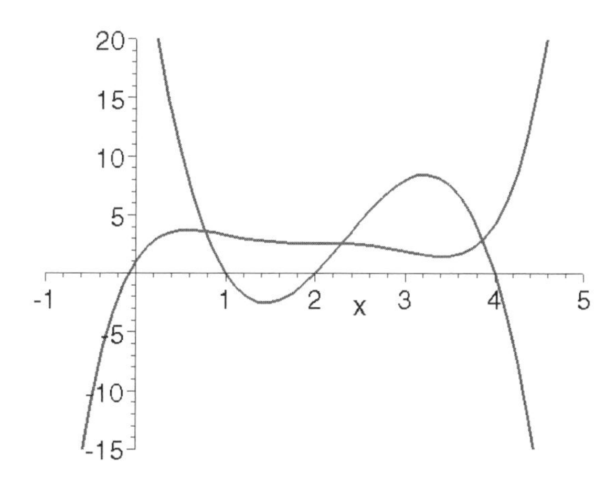

The x-coordinates of the three intersection points correspond to the solutions of the equation $p(x) = q(x)$. Although Maple can solve this equation exactly, approximate solutions are more useful for the problem and obtained with `fsolve`.

```
> intpts:=fsolve(p=q,x);
```

$$intpts := .7720583045, 2.291819211, 3.865127100$$

Using the roots to the equation $p(x) = q(x)$ and the graph, we see that $p(x) > q(x)$ for $0.772958 < x < 2.29182$ and $q(x) > p(x)$ for $2.29182 < x < 3.86513$. Hence, an approximation of the area bounded by p and q is given by the integral

$$\int_{0.772058}^{2.29182} (p(x) - q(x))dx + \int_{2.29182}^{3.86513} (q(x) - p(x))dx.$$

These two integrals are computed with `int`.

```
> intone:=evalf(Int(p-q,x=intpts[1]..intpts[2]));
  inttwo:=evalf(Int(q-p,x=intpts[2]..intpts[3]));
```

$$intone := 5.269124281$$

$$inttwo := 6.925994162$$

```
> intone+inttwo;
```

12.19511844

■

In Example 14, the exact value of the area could have been found. In cases in which the exact value of an integral cannot be computed, `evalf` and `Int` can frequently be used to approximate the exact value of an integral.

EXAMPLE 15: Let

$$f(x) = e^{-(x-2)^2 \cos \pi x} \text{ and } g(x) = 4\cos(x-2)$$

on the interval [0,4]. Approximate the area of the region bounded by the graphs of f and g.

SOLUTION: After defining f and g, we graph both on the interval [0,4].

```
> f:='f':g:='g':
  f:=x->exp(-(x-2)^2*cos(Pi*x)):
  g:=x->4*cos(x-2):
```

```
plot({f(x),g(x)},x=0..4);
```

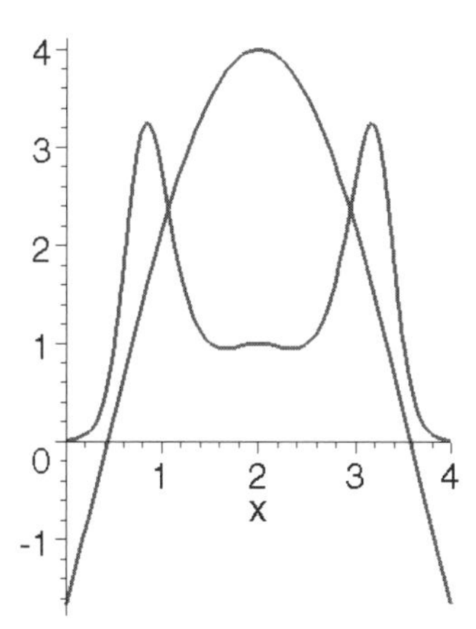

Unlike in Example 14, we cannot obtain the exact value of the intersection points. However, notice that the x-coordinate of the first intersection point is between .5 and 1.5, while the x-coordinate of the second intersection point is between 2.5 and 3.5. We then use `fsolve` with the `interval` option to approximate the x-coordinate of each intersection point, naming them `firstroot` and `secondroot`, respectively.

```
> firstroot:=fsolve(f(x)=g(x),x,.5..1.5);
  secondroot:=fsolve(f(x)=g(x),x,2.5..3.5);
```

$$firstroot := 1.062580731$$
$$secondroot := 2.937419269$$

An approximation of the area is given by evaluating the integral $\int_{1.063}^{2.937}(g(x)-f(x))dx$, which is done with `evalf` and `Int`.

```
> evalf(Int(g(x)-f(x),x=firstroot..secondroot));
```

4.174129286

■

In problems involving "circular symmetry," it is often easier to work in polar coordinates. The relationship between (x,y) in rectangular coordinates and (r,θ) in polar coordinates is

$$\begin{array}{l} x = r\cos\theta \\ y = r\sin\theta \end{array} \quad \text{and} \quad \begin{array}{l} r^2 = x^2 + y^2 \\ \tan\theta = \frac{y}{x} \end{array}.$$

If $r = f(\theta)$ is continuous and nonnegative for $\alpha \le \theta \le \beta$, then the area A of the region enclosed by the graphs of $r = f(\theta)$, $\theta = \alpha$, and $\theta = \beta$ is

$$A = \int_\alpha^\beta \frac{1}{2}(f(\theta))^2 = \int_\alpha^\beta \frac{1}{2}r^2 d\theta.$$

EXAMPLE 16: The **Lemniscate of Bernoulli** is given by

$$(x^2 + y^2)^2 = a^2(x^2 - y^2),$$

where a is a constant. (a) Graph the Lemniscate of Bernoulli if $a = 2$. (b) Find the area of the region bounded by the Lemniscate of Bernoulli.

SOLUTION: This problem is solved much more easily in polar coordinates, so we first convert the equation from rectangular to polar coordinates with `subs` and then solve for r.

```
> lofb:=(x^2+y^2)^2=a^2*(x^2-y^2):
> topolar:=subs({x=r*cos(theta),y=r*sin(theta)},lofb);
```

$$topolar := (r^2\cos(\theta)^2 + r^2\sin(\theta)^2)^2 = a^2(r^2\cos(\theta)^2 - r^2\sin(\theta)^2)$$

```
> solve(topolar,r);
```

$$0, 0, \sqrt{1 - 2\sin(\theta)^2}\,a, -\sqrt{1 - 2\sin(\theta)^2}\,a$$

These results indicate that an equation of the Lemniscate in polar coordinates is

$$r^2 = a^2(1 - \sin^2\theta).$$

The graph of the Lemniscate is then generated using `plot` together with the options `coords=polar`.

```
> plot({2*sqrt(1-2*sin(theta)^2),
      -2*sqrt(1-2*sin(theta)^2)},
```

```
theta=0..2*Pi,
   coords=polar,color=[black,black]);
```

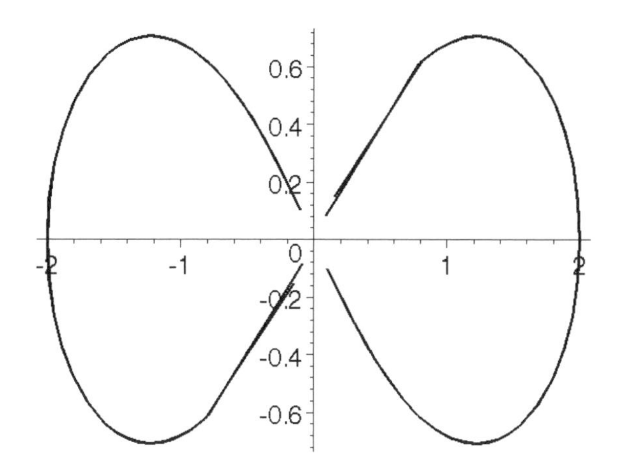

Note that the right half of the graph is obtained by graphing $r^2 = \sqrt{a^2(1-\sin^2\theta)}$ for $-\pi/4 \le \theta < \pi/4$.

```
> plot(2*sqrt(1-2*sin(theta)^2),theta=-Pi/4..Pi/4,
      coords=polar,color=black);
```

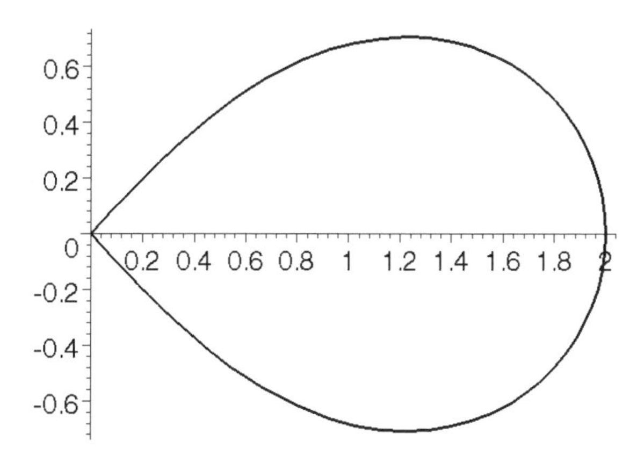

Therefore, the area of the Lemniscate is given by.

$$A = 2\int_{-\pi/4}^{\pi/4} \frac{1}{2} r^2 d\theta = \int_{-\pi/4}^{\pi/4} a^2(1 - 2\sin^2\theta)d\theta = 2a^2.$$

```
> int(2*a^2*(1-2*sin(theta)^2),theta=-Pi/4..Pi/4);
```

$$2\,a^2$$

■

Application: Arc Length

Let $f(x)$ be a function for which $f'(x)$ is continuous on an interval $[a, b]$. Then the **arc length** of the graph of f from $(a, f(a))$ to $(b, f(b))$ is given by

$$Length = \int_a^b \sqrt{1 + (f'(x))}\,dx.$$

The resulting definite integrals used for determining arc length are usually difficult to compute because they involve a radical. Maple is very helpful with approximating solutions to these types of problems.

EXAMPLE 17: Let $f(x) = \sin(x + x\sin x)$. Approximate the arc length of the graph of f from $(0, f(0))$ to $(2\pi, f(2\pi))$.

SOLUTION: We begin by defining and graphing f on the interval $[0,2\pi]$.

```
> f:='f':
  f:=x->sin(x+x*sin(x)):
  plot(f(x),x=0..2*Pi);
```

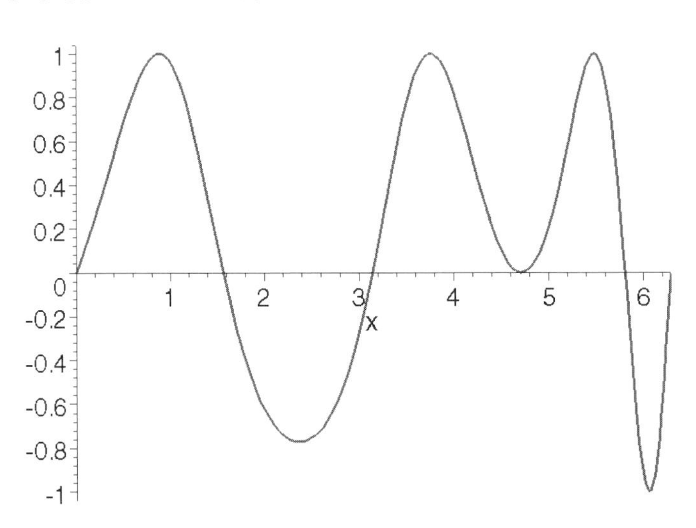

In order to evaluate the arc length formula, we first compute the derivative of $f(x)$ and then approximate $\int_0^{2\pi} \sqrt{(1+f'(x))^2}dx$ with `evalf` and `Int`.

```
> D(f)(x);
```

$$\cos(x + x\sin(x))\,(1 + \sin(x) + x\cos(x))$$

```
> evalf(Int(sqrt(1+D(f)(x)^2),x=0..2*Pi));
```

12.05642814

Thus, an approximation of the arc length is 12.0564.

■

Application: Volume of Solids of Revolution

Maple can be used to solve volume problems as well. Let $y = f(x)$ be a nonnegative continuous function on $[a,b]$. The **volume of the solid of revolution** obtained by revolving the region bounded by the graphs of $y = f(x)$, $x = a$, $x = b$, and the x-axis about the x-axis is given by

$$V_{x-axis} = \int_a^b \pi(f(x))^2 dx.$$

If $0 \le a < b$, the **volume of the solid of revolution** obtained by revolving the region bounded by the graphs of $y = f(x)$, $x = a$, and $x = b$, and the x-axis about the y-axis is given by

$$V_{y-axis} = \int_a^b 2\pi x f(x)dx$$

EXAMPLE 18: Let $g(x) = x\sin^2 x$. Find the volume of the solid obtained by revolving the region bounded by the graphs of $y = g(x)$, $x = 0$, $x = \pi$, and the x-axis about (a) the x-axis; and (b) the y-axis.

SOLUTION: In this case, after defining g, we graph g on the interval $[0,\pi]$. The volume of the solid obtained by revolving the region about the x-axis is given by $V_{x-axis} = \int_0^{\pi} \pi(g(x))^2 dx$ while the volume of the solid obtained by revolving the region about the y-axis is given by $V_{y-axis} = \int_0^{\pi} 2\pi x g(x)dx$. These integrals are computed with int and named xvol and yvol, respectively. evalf is used to approximate each volume.

```
> g:='g':
  g:=x->x*sin(x)^2:
  plot(g(x),x=0..Pi);
```

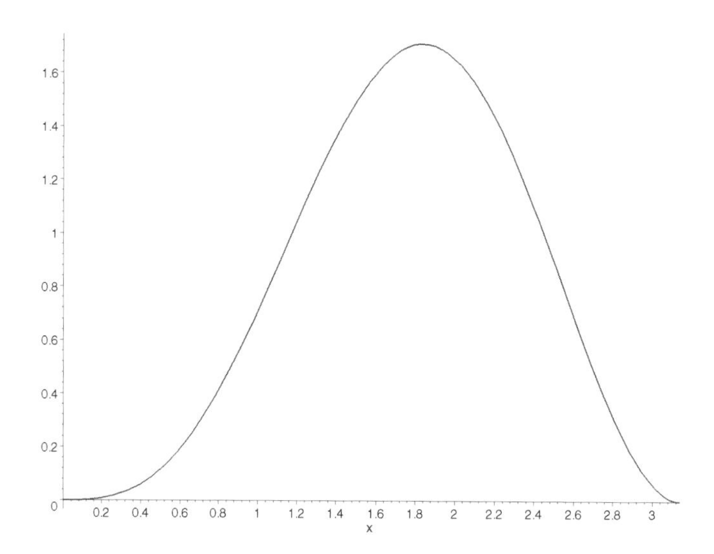


```
> xvol:=int(Pi*g(x)^2,x=0..Pi);
```

$$xvol := \frac{1}{8}\pi^4 - \frac{15}{64}\pi^2$$

```
> evalf(xvol);
```

9.862947858

```
> yvol:=int(2*Pi*x*g(x),x=0..Pi);
```

$$yvol := \frac{1}{3}\pi^4 - \frac{1}{2}\pi^2$$

```
> evalf(yvol);
```

$$27.53489482$$

We can use `plot3d` to visualize the resulting solids by parametrically graphing the equations given by $\begin{cases} x = r\cos t \\ y = r\sin t \\ z = g(r) \end{cases}$ for r between 0 and π and t between $-\pi$ and π to visualize the graph of the solid obtained by revolving the region about the y-axis and by parametrically graphing the equations given by $\begin{cases} x = r \\ y = g(r)\cos t \\ z = g(r)\sin t \end{cases}$ for r between 0 and π and t between $-\pi$ and π to visualize the graph of the solid obtained by revolving the region about the x-axis. In this case, we identify the z-axis as the y-axis. Notice that we are simply using polar coordinates for the x- and y-coordinates, and the height above the x,y-plane is given by $g(r)$ because r is replacing x in the new coordinate system.

```
> plot3d([r,g(r)*cos(t),g(r)*sin(t)],
      r=0..Pi,t=0..2*Pi,grid=[30,30],axes=FRAME);
```

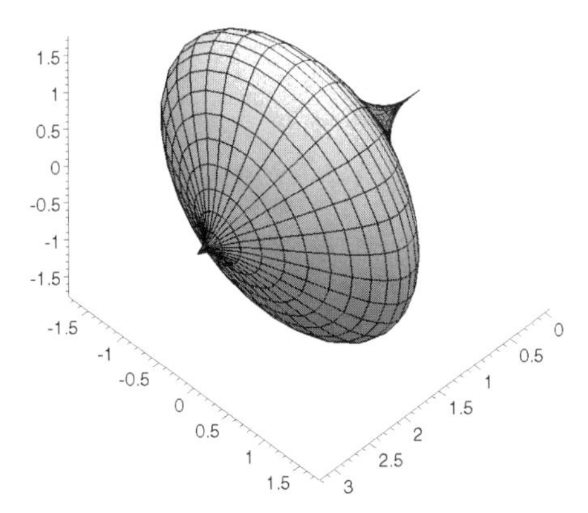


```
> plot3d([r*cos(t),r*sin(t),g(r)],
      r=0..Pi,t=0..2*Pi,grid=[30,30],axes=FRAME);
```

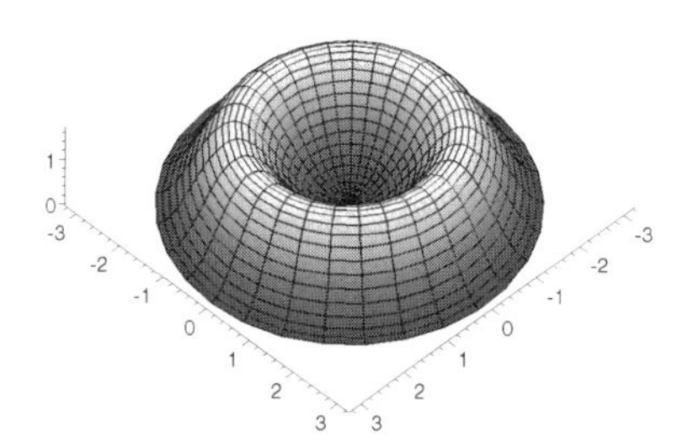

■

We now demonstrate a volume problem that requires the method of disks.

EXAMPLE 19: Let $f(x) = e^{-(x-3)\cos(4(x-3))}$. Approximate the volume of the solid obtained by revolving the region bounded by the graphs of $y = f(x)$, $x = 1$, $x = 5$, and the x-axis about the x-axis.

SOLUTION: Proceeding as in the previous example, we first define and graph f on the interval [1,5].

```
> f:='f':
  f:=x->exp(-(x-3)^2*cos(4*(x-3))):
  plot(f(x),x=1..5);
```

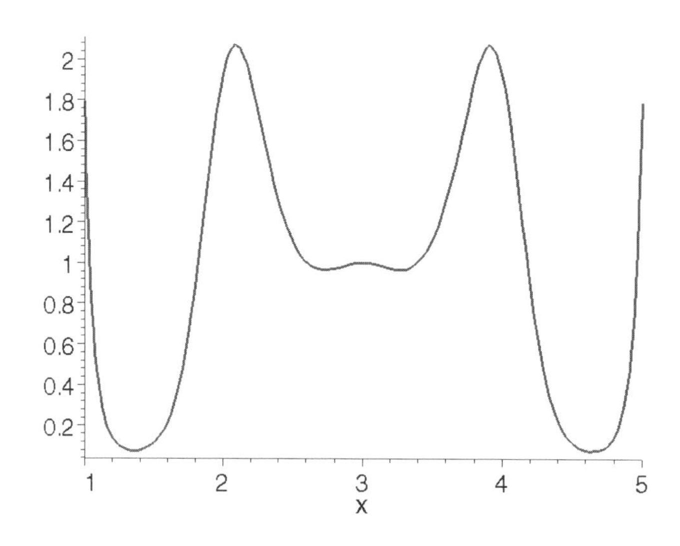

In this case, an approximation is desired, so we use `Int` and `evalf` to approximate the integral $V_{x-axis} = \int_1^5 \pi(f(x))^2 dx$.

```
> evalf(Int(Pi*f(x)^2,x=1..5));
```

16.07615212

In the same manner as before, `plot3d` can be used to visualize the resulting solid by graphing the set of equations given parametrically by $\begin{cases} x = r \\ y = f(r)\cos t \\ z = f(r)\sin t \end{cases}$ for r between 1 and 5 and t between 0 and 2π. In this case, polar coordinates are used in the y,z-plane with the distance from the x-axis given by $f(x)$. Because r replaces x in the new coordinate system, $f(x)$ becomes $f(r)$ in these equations.

```
> plot3d([r,f(r)*cos(t),f(r)*sin(t)],
      r=1..5,t=0..2*Pi,grid=[35,25],axes=FRAME);
```

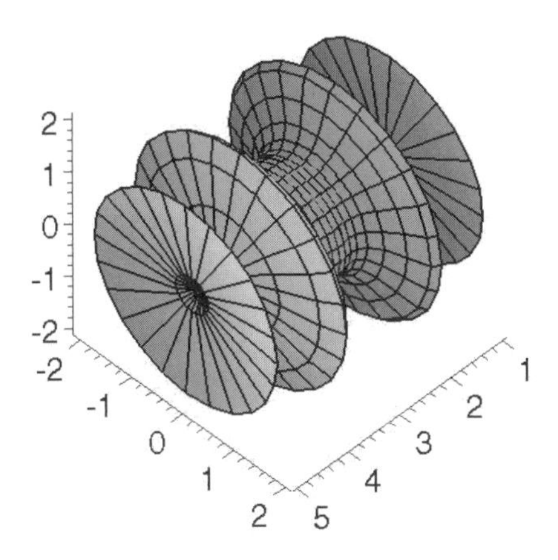

■

Application: Surface Area of a Solid of Revolution

Let $f(x)$ be a nonnegative function for which $f'(x)$ is continuous on an interval $[a,b]$. Then the **surface area** of the solid of revolution obtained by revolving the region bounded by the graphs of $y = f(x)$, $x = a$, $x = b$, and the x-axis about the x-axis is given by

$$S = \int_a^b 2\pi f(x)\sqrt{1 + (f'(x))^2}dx.$$

EXAMPLE 20: Gabriel's Horn is the solid of revolution obtained by revolving the area of the region bounded by $y = 1/x$ and the x-axis for $x \geq 1$ about the x-axis. Show that the surface area of Gabriel's Horn is infinite but that its volume is finite.

SOLUTION: After defining $f(x) = 1/x$, we use plot3d to visualize a portion of Gabriel's Horn.

```
> f:='f':
  f:=x->1/x:
  plot3d([r,f(r)*cos(t),f(r)*sin(t)],
      r=1..10,t=0..2*Pi,grid=[40,40],axes=FRAME);
```

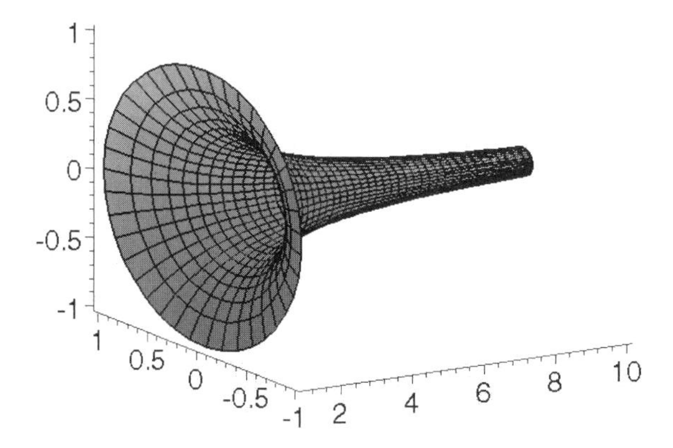

Because $f'(x) = 1/x^2$, the surface area of the solid of revolution obtained by revolving the region bounded by the graphs of $f(x) = 1/x$, $x = 1$, $x = L$, and the x-axis about the x-axis is given by

$$S = \int_1^L 2\pi \cdot \frac{1}{x}\sqrt{\left(1 + \left(-\frac{1}{x^2}\right)\right)^2}\,dx$$

```
> step1:=Int(2*Pi*f(x)*sqrt(1+D(f)(x)^2),x=1..L);
```

$$step1 := \int_1^L 2\,\frac{\pi\sqrt{1 + \frac{1}{x^4}}}{x}\,dx$$

```
> step2:=value(step1);
```

$$step2 := \pi\left(-\sqrt{\frac{L^4+1}{L^4}}\sqrt{L^4+1}+\sqrt{\frac{L^4+1}{L^4}}\ln(L^2+\sqrt{L^4+1})L^2+\sqrt{2}\sqrt{L^4+1}\right.$$
$$\left.-\ln(\sqrt{2}+1)\sqrt{L^4+1}\right)\Big/\sqrt{L^4+1}$$

so the surface area of Gabriel's Horn is given by

$$\lim_{L\to\infty}\int_1^L 2\pi\cdot\frac{1}{x}\sqrt{1+\left(-\frac{1}{x^2}\right)^2}dx=\int_1^\infty 2\pi\cdot\frac{1}{x}\sqrt{1+\left(-\frac{1}{x^2}\right)^2}dx.$$

```
> limit(step2,L=infinity);
```

$$\infty$$

Maple can compute this improper integral directly, as well.

```
> int(2*Pi*f(x)*sqrt(1+D(f)(x)^2),x=1..infinity);
```

$$\infty$$

On the other hand, the volume of Gabriel's Horn is given by the improper integral

$$\int_1^\infty \pi\cdot\left(\frac{1}{x}\right)^2dx = \lim_{L\to\infty}\int_1^L \pi\cdot\left(\frac{1}{x}\right)^2dx,$$

which converges to π.

```
> step1:=int(Pi*f(x)^2,x=1..L);
```

$$step1 := \frac{\pi(L-1)}{L}$$

```
> step2:=limit(step1,L=infinity);
```

$$step2 := \pi$$

```
> int(Pi*f(x)^2,x=1..infinity);
```

$$\pi$$

■

Application: The Mean-Value Theorem for Integrals

Another application of integrals includes the Mean-Value Theorem for Integrals. The **Mean-Value Theorem for Integrals** states that if f is continuous on $[a,b]$, then there is at least one number c between a and b satisfying $\int_a^b f(x)dx = f(c)(b-a)$.

EXAMPLE 21: Find all value(s) of c satisfying the conclusion of the Mean-Value Theorem for Integrals for the function $f(x) = x^2 - 3x + 4$ on the interval [2,6].

SOLUTION: After defining f, we compute $\int_2^6 f(x)dx$ and name the resulting output `val`. We then solve the equation $\int_2^6 f(x)dx = f(x) \cdot (6-2)$ for x and name the resulting output `exvals`. To determine which of the numbers in the list `exvals` is contained in the interval [2,6], we use `evalf` to compute an approximation of each number in `exvals`. We conclude that the only value of c satisfying the conclusion of the Mean-Value Theorem is $c = \frac{1}{6}(9 + \sqrt{273}) \approx 4.25378$.

```
> f:='f':
  f:=x->x^2-3*x+4:
  val:=int(f(x),x=2..6);
```

$$val := \frac{112}{3}$$

```
> sols:=[solve(val=f(x)*(6-2))];
```

$$sols := \left[\frac{3}{2} + \frac{1}{6}\sqrt{273}, \frac{3}{2} - \frac{1}{6}\sqrt{273}\right]$$

```
> map(evalf,sols);
```

$$[4.253785274, -1.253785274]$$

■

3.5 Series

Introduction to Series

Sequences and series are usually discussed in the third quarter or second semester of introductory calculus courses. The first topic addressed in these courses usually is the question of whether a sequence or series converges or diverges. Maple can help determine the answer to these questions in some problems either graphically or explicitly. Generally, the command `sum(f(k),k=n..m)` attempts to compute the sum $\sum_{k=n}^{m} f(k)$. (Note that m may be ∞.)

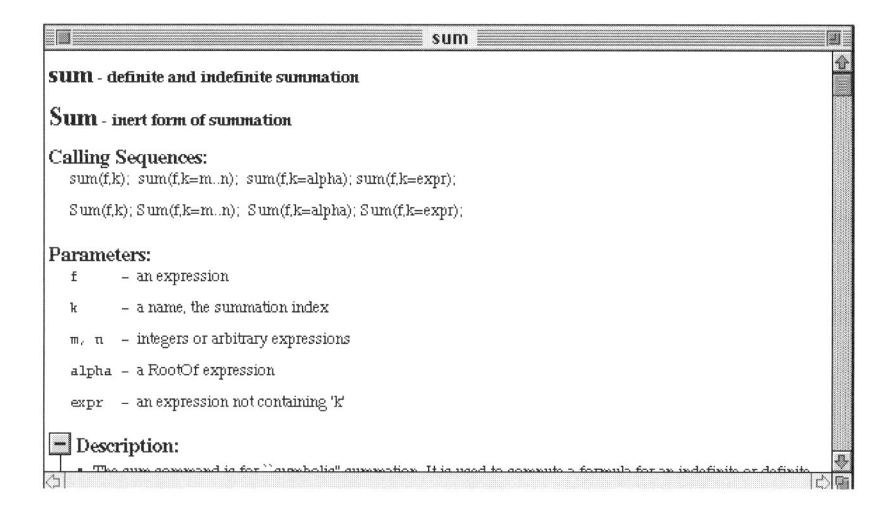
sum

sum - definite and indefinite summation

Sum - inert form of summation

Calling Sequences:

sum(f,k); sum(f,k=m..n); sum(f,k=alpha); sum(f,k=expr);

Sum(f,k); Sum(f,k=m..n); Sum(f,k=alpha); Sum(f,k=expr);

Parameters:

f - an expression

k - a name, the summation index

m, n - integers or arbitrary expressions

alpha - a RootOf expression

expr - an expression not containing 'k'

Description:

EXAMPLE 1: Find the sum of each of the following series: (a) $\sum_{n=1}^{\infty} \frac{1}{4n^2+8n+3}$; (b) $\sum_{k=1}^{\infty} x^{3k}$; and (c) $\sum_{n=1}^{\infty} \frac{3^{n/2}}{5^n}$.

SOLUTION: Sums of some infinite series are calculated with `sum`. We compute $\sum_{n=1}^{\infty} \frac{1}{4n^2+8n+3}$ by entering

```
sum(1/(4*n^2+8*n+3),n=1..infinity).
```

```
> sum(1/(4*n^2+8*n+3),n=1..infinity);
```

$$\frac{1}{6}$$

Similarly, we use sum to compute $\sum_{k=1}^{\infty} x^{3k}$. The result is valid for $|x| < 1$ because it represents the sum of a convergent geometric series.

```
> sum(x^(3*k),k=1..infinity);
```

$$-\frac{x^3}{x^3 - 1}$$

In the same manner as in (a) and (b), sum calculates $\sum_{n=1}^{\infty} \frac{3^{n/2}}{5^n}$.

```
> sum(3^(n/2)/5^n,n=1..infinity);
```

$$-\frac{\sqrt{3}}{\sqrt{3} - 5}$$

■

EXAMPLE 2: Determine whether or not the series $\sum_{k=1}^{\infty} \frac{k}{2^k}$ converges.

SOLUTION: We use the Integral test to determine whether or not the series $\sum_{k=1}^{\infty} \frac{k}{2^k}$ converges. We begin by calculating the integral $\int_1^L \frac{k}{2^k} dk$ with int and naming the resulting output step1. We then use simplify to simplify step1 and name the resulting output step2.

```
> step1:=int(k/2^k,k=1..L);
```

$$step1 := -\frac{1}{2}\frac{2+2L\ln(2)-2^L-\ln(2)\,2^L}{\ln(2)^2\,2^L}$$

```
> step2:=simplify(step1);
```

$$step2 := -\frac{1}{2}\frac{2^{(1-L)}+2^{(1-L)}L\ln(2)-1-\ln(2)}{\ln(2)^2}$$

Next, we calculate $\lim_{L\to\infty}\int_1^L \frac{k}{2^k}dk$ with `limit`.

```
> limit(step2,L=infinity);
```

$$\frac{1}{2}\frac{1+\ln(2)}{\ln(2)^2}$$

We are also able to use `int` to calculate the improper integral $\int_1^\infty \frac{k}{2^k}dk$ directly.

```
> int(k^2/2^k,k=1..infinity);
```

$$\frac{1}{2}\frac{2+2\ln(2)+\ln(2)^2}{\ln(2)^3}$$

Thus, by the Integral test, we conclude that the series $\sum_{k=1}^{\infty}\frac{k}{2^k}$ converges.

We know that the limit of the partial sums is the value of the series. Below, we use `sum` to approximate the value of the series by computing $\sum_{k=1}^{1000}\frac{k}{2^k}$.

```
> evalf(sum(k/2^k,k=1..1000));
```

2.000000000

In fact, we are able to compute the exact value of $\sum_{k=1}^{\infty}\frac{k}{2^k}$ with `sum`.

```
> sum(k/2^k,k=1..infinity);
```

2

■

EXAMPLE 3: Determine whether or not the series $\sum_{n=1}^{\infty} \frac{10^n}{n!}$ converges.

SOLUTION: After clearing all prior definitions of a, if any, we define $a_n = 10^n/n!$ and then use `seq` to calculate $a_1, a_2, \ldots a_{24}, a_{25}$, naming the resulting set of numbers `vals`. These points are then graphed with `plot`, using the options `style=POINT`. (Note that graphing lists and tables is discussed in more detail in Chapters 4 and 5.)

```
> a:'a':
  a:=n->10^n/n!:
  vals:=[seq([n,a(n)],n=1..25)];
```

$$vals := \left[[1,10],[2,50],\left[3,\frac{500}{3}\right],\left[4,\frac{1250}{3}\right],\left[5,\frac{2500}{3}\right],\left[6,\frac{12500}{9}\right],\left[7,\frac{125000}{63}\right],\right.$$
$$\left[8,\frac{156250}{63}\right],\left[9,\frac{1562500}{567}\right],\left[10,\frac{1562500}{567}\right],\left[11,\frac{15625000}{6237}\right],\left[12,\frac{39062500}{18711}\right],$$
$$\left[13,\frac{390625000}{243243}\right],\left[14,\frac{1953125000}{1702701}\right],\left[15,\frac{3906250000}{5108103}\right],\left[16,\frac{2441406250}{5108103}\right],$$
$$\left[17,\frac{24414062500}{86837751}\right],\left[18,\frac{122070312500}{781539759}\right],\left[19,\frac{1220703125000}{14849255421}\right],$$
$$\left[20,\frac{610351562500}{14849255421}\right],\left[21,\frac{6103515625000}{311834363841}\right],\left[22,\frac{30517578125000}{3430178002251}\right],$$
$$\left.\left[23,\frac{305175781250000}{78894094051773}\right],\left[24,\frac{381469726562500}{236682282155319}\right],\left[25,\frac{152587890625000}{236682282155319}\right]\right]$$

```
> plot(vals,style=POINT);
```

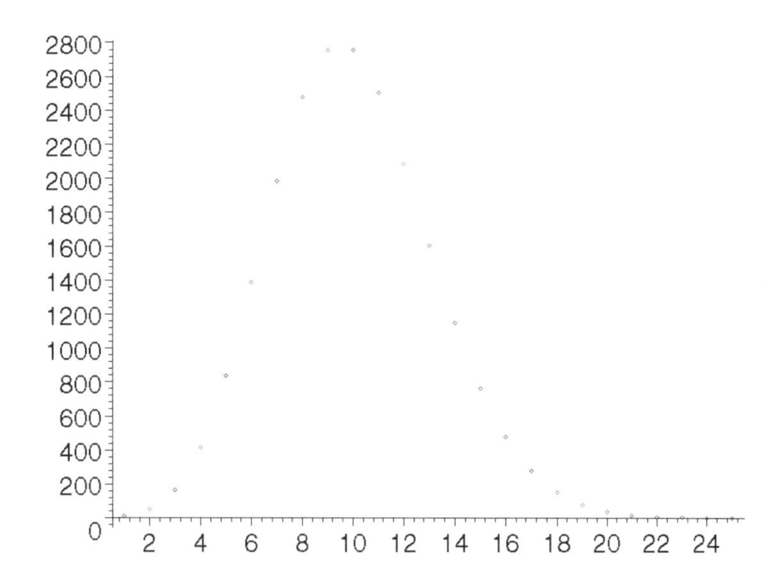

To determine whether or not the series converges, we use the Ratio test. First, we compute and simplify a_n+1/a_n.

```
>   n:='n':
    simplify(a(n+1)/a(n));
```

$$10\frac{1}{n+1}$$

The expression $\frac{a_{n+1}}{a_n} = \frac{10n!}{(n+1)!}$ is simplified to $\frac{10}{n+1}$ and we see that

$$\lim_{n\to\infty}\frac{a_{n+1}}{a_n} = \lim_{n\to\infty}\frac{10n!}{(n+1)!} = \lim_{n\to\infty}\frac{10}{n+1} = 0.$$

Thus, by the Ratio test, the series converges.

In fact, Maple can compute the exact value of the series with `sum`.

```
> sum(a(n),n=1..infinity);
```

$$e^{10}(1-e^{(-10)})$$

■

In addition to the previous examples, which are similar to those discussed in introductory calculus courses, Maple can also help determine the solution of more difficult problems.

EXAMPLE 4: Compute the 1000th partial sum of the series $\sum_{k=1}^{\infty} \frac{\sin k}{k}$.

SOLUTION: We begin by defining $a_k = \frac{\sin k}{k}$ and then using `seq` to compute a list of the values $a_{50}, a_{51}, a_{52}, \ldots a_{1049}, a_{1050}$, naming the resulting list of ordered pairs `nums`. Because a semicolon is included at the end of the command, the resulting list of 1001 ordered pairs is not displayed. We then use `plot`, together with the `style=POINT` option, to graph the list of numbers `nums`.

```
> a:='a':
  a:=k->sin(k)/k:
  nums:=[seq([k,a(k)],k=50..1050)]:
  plot(nums,style=POINT);
```

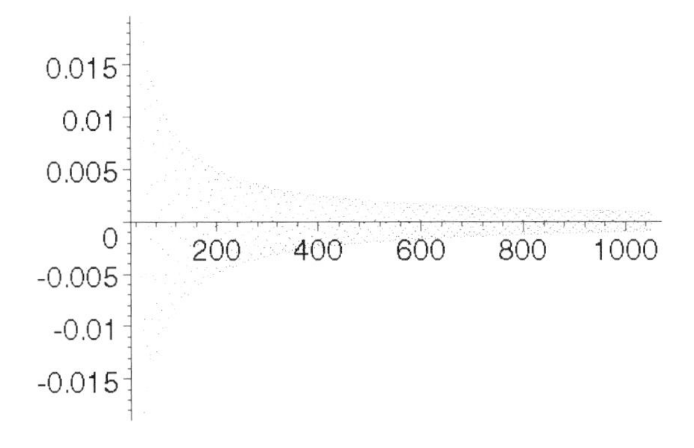

We compute $\sum_{k=1}^{1000} \frac{\sin k}{k}$ with sum. Note that we use evalf to produce an approximation of the 1000th partial sum; if evalf is not included, Maple returns the exact value.

```
> k:='k':
  evalf(sum(a(k),k=1..1000));
```

1.070694147

■

Determining the Interval of Convergence of a Power Series

After discussing sequences and series, the next topic addressed in these courses is usually power series. Given a power series $\sum_{n=0}^{\infty} a_n (x - x_0)^n$, a fundamental problem is finding the interval of convergence of the given power series.

EXAMPLE 5: Find the interval of convergence of each series: (a) $\sum_{n=0}^{\infty} \frac{1}{(-5)^n} x^{2n+1}$; and (b) $\sum_{n=0}^{\infty} \frac{4^{2n}}{n+2} (x-3)^n$.

SOLUTION: For (a), we proceed using the **Root test**. The Root test says that if $\sum a_n$ is a series with positive terms and $l = \lim_{n \to \infty} \sqrt[n]{a_n}$, then

(i) the series converges if $l < 1$

(ii) the series diverges if $l > 1$

(iii) no conclusion can be drawn if $l = 1$

We begin by defining a_n to be $\frac{1}{(-5)^n}x^{2n+1}$ and then computing and simplifying $\sqrt[n]{\frac{1}{(-5)^n}x^{2n+1}}$ with `simplify` together with the `power` and `symbolic` options. We name the resulting output `stepone`.

```
> a:='a':
  a:=n->1/(-5)^n*x^(2*n+1);
```

$$a := n \rightarrow \frac{x^{(2n+1)}}{(-5)^n}$$

```
> > stepone:=simplify(a(n)^(1/n),power,symbolic);
```

$$stepone := -\frac{1}{5}x^{\left(\frac{2n+1}{n}\right)}$$

We then compute the limit of `stepone` as n approaches infinity. This is the same as computing $\lim_{n\to\infty}\sqrt[n]{\frac{1}{(-5)^n}x^{2n+1}} = \lim_{n\to\infty}\frac{-1}{5}x^2x^{1/n}$.

```
> limit(stepone,n=infinity);
```

$$-\frac{1}{5}x^2$$

The result means that $\sum_{n=0}^{\infty}\frac{1}{(-5)^n}x^{2n+1}$ converges absolutely if $\left|\frac{-1}{5}x^2\right| < 1$. We solve this inequality next.

```
> solve(abs(-1/5*x^2)<=1,x);
```

$$\text{RealRange}(-\sqrt{5}, \sqrt{5})$$

We conclude that $\sum_{n=0}^{\infty} \frac{1}{(-5)^n} x^{2n+1}$ converges absolutely if $-\sqrt{5} < x < \sqrt{5}$. We investigate the convergence of $\sum_{n=0}^{\infty} \frac{1}{(-5)^n} x^{2n+1}$ if $x = -\sqrt{5}$ or $x = \sqrt{5}$ separately. We first substitute $x = -\sqrt{5}$ in a_n. The resulting alternating series diverges because the a_n that result do not approach zero as n approaches infinity.

```
> simplify(subs(x=-5^(1/2),a(n)));
```

$$-(-1)^n \sqrt{5}$$

Similarly, substituting $x = \sqrt{5}$ for a_n results in a divergent series.

```
> simplify(subs(x=5^(1/2),a(n)));
```

$$(-1)^{(-n)} \sqrt{5}$$

We conclude that the interval of convergence of the series $\sum_{n=0}^{\infty} \frac{1}{(-5)^n} x^{2n+1}$ is the open interval $(-\sqrt{5}, \sqrt{5})$.

For (b), we use the **Ratio test**. The Ratio test says that if $\sum a_n$ is a series with positive terms and $Nl = \lim_{n \to \infty} \frac{a_{n+1}}{a_n}$, then

(i) the series converges if $l < 1$

(ii) the series diverges if $l > 1$

(iii) no conclusion can be drawn if $l = 1$

In the same manner as in (a), we begin by defining $a_n = \frac{4^{2n}}{n+2}(x-3)^n$. We then compute and simplify a_{n+1}/a_n, naming the resulting output `stepone`.

```
> a:='a':
  a:=n->4^(2*n)/(n+2)*(x-3)^n;
```

$$a := n \to \frac{4^{(2n)} (x-3)^n}{n+2}$$

```
> stepone:=simplify(a(n+1)/a(n));
```

$$stepone := 16\frac{(x-3)(n+2)}{n+3}$$

and then compute $\lim_{n\to\infty} 16\frac{n+2}{n+3}(x-3)$, naming the resulting output `steptwo`.

```
> steptwo:=limit(stepone,n=infinity);
```

$$steptwo := 16x - 48$$

The result means that $\sum_{n=0}^{\infty} \frac{4^{2n}}{n+2}(x-3)^n$ converges absolutely if $|16x-48| < 1$. We solve this inequality and then test to see if the series converges if $x = 47/16$ or $x = 49/16$ separately.

```
> solve(abs(steptwo)<=1);
```

$$\text{RealRange}\left(\frac{47}{16}, \frac{49}{16}\right)$$

```
> simplify(subs(x=47/16,a(n)));
```

$$\frac{(-1)^n}{n+2}$$

```
> simplify(subs(x=49/16,a(n)));
```

$$\frac{1}{n+2}$$

The series $\sum_{n=0}^{\infty} \frac{(-1)^n}{n+1}$ converges by the alternating series test and $\sum_{n=0}^{\infty} \frac{1}{n+1}$ diverges by the basic comparison test, so we conclude that the interval of convergence of the series $\sum_{n=0}^{\infty} \frac{4^{2n}}{n+2}(x-3)^n$ is [47/16, 49/16)

■

Computing Power Series

The **power series** expansion of a function $f(x)$ about $x = a$ is given by the expression

$$\sum_{n=0}^{\infty} \frac{f^{(n)}(a)}{n!}(x-a)^n,$$

provided that the necessary derivatives exist.

Maple computes the power series expansion of a function $f(x)$ about the point $x = a$ up to order n with the command `series(f(x),x=a,n)`.

The symbol `O`, appearing in the output that results from the `series` command, represents the terms that are omitted from the power series for $f(x)$ expanded about the point $x = a$. The `O`-term is removed from the output of the `series` command with the `convert` command, together with the `polynom` option; the result is a polynomial function.

EXAMPLE 6: Find the first few terms of the power series for the given function about the indicated point:
(a) $\cos x$ about $x = 0$; (b) e^x about $x = 0$
(c) $\sin x$ about $x = \pi$; and (d) $\ln x$ about $x = 1$.

SOLUTION: Entering

```
> series(cos(x),x=0);
```

$$1 - \frac{1}{2}x^2 + \frac{1}{24}x^4 + O(x^6)$$

computes the terms of the power series for $\cos x$ about $x = 0$ to order 6; entering

```
> series(exp(x),x=0,7);
```

$$1 + x + \frac{1}{2}x^2 + \frac{1}{6}x^3 + \frac{1}{24}x^4 + \frac{1}{120}x^5 + \frac{1}{720}x^6 + O(x^7)$$

computes the terms of the powers series for e^x about $x = 0$ to order 7; entering

```
> series(sin(x),x=Pi,5);
```

$$-(x-\pi) + \frac{1}{6}(x-\pi)^3 + O((x-\pi)^5)$$

computes the terms of the power series for $\sin x$ about $x = \pi$ to order 5; and entering

```
> series(ln(x),x=1,8);
```

$$x - 1 - \frac{1}{2}(x-1)^2 + \frac{1}{3}(x-1)^3 - \frac{1}{4}(x-1)^4 + \frac{1}{5}(x-1)^5 - \frac{1}{6}(x-1)^6 + \frac{1}{7}(x-1)^7 + O((x-1)^8)$$

computes the terms of the power series for $\ln x$ about $x = 1$ to order 8.

■

Maple can also compute the general formula for the power series expansion of a function $y(x)$. The results of entering the following commands

```
> series(y(x),x=0);
```

$$y(0) + D(y)(0)\,x + \frac{1}{2}(D^{(2)})(y)(0)\,x^2 + \frac{1}{6}(D^{(3)})(y)(0)\,x^3 + \frac{1}{24}(D^{(4)})(y)(0)\,x^4 + \frac{1}{120}(D^{(5)})(y)(0)\,x^5 + O(x^6)$$

```
> series(y(x),x=a);
```

$$y(a) + D(y)(a)\,(x-a) + \frac{1}{2}(D^{(2)})(y)(a)\,(x-a)^2 + \frac{1}{6}(D^{(3)})(y)(a)\,(x-a)^3 + \frac{1}{24}(D^{(4)})(y)(a)\,(x-a)^4 + \frac{1}{120}(D^{(5)})(y)(a)\,(x-a)^5 + O((x-a)^6)$$

is the power series for $y(x)$ about $x = 0$ and $x = a$ to order 6, respectively.

We can also use `series` to compute Taylor and Maclaurin polynomials. If $y = f(x)$ is a function with n derivatives at $x = a$, then the **nth-degree Taylor polynomial of f** at $x = a$ is

$$p_n(x) = \sum_{k=0}^{n} \frac{f^{(k)}(a)}{k!}(x-a)^k.$$

The ***n*th-degree Maclaurin polynomial of *f*** is the nth-degree Taylor polynomial of f at $x = 0$.

However, the result of entering a `series` command is not a function that can be evaluated if x is a particular number. We remove the remainder term of the power series `series(f(x),x=a,n)` with the command `convert(series(f(x),x=a,n),polynom)`, and we can then evaluate the resulting polynomial. Hence, with the `convert` command, a polynomial is obtained. This polynomial serves as an approximation to the function $f(x)$.

Maclaurin and Taylor polynomials are constructed from a series object by converting the series object to a polynomial object that is achieved by removing from the series the O-term that represents the higher-order omitted terms of the series.

EXAMPLE 7: Find the fifth-degree Maclaurin polynomial for $f(x) = \tan^{-1} x$. Compare the graphs of $f(x) = \tan^{-1} x$ and the polynomial.

SOLUTION: We define `ser1` to be the power series expansion for the function $f(x) = \tan^{-1} x$ about $x = 0$ to order 6. To illustrate that the resulting output is not a function, we attempt to evaluate `ser1` if $x = 1$. Note the error messages that occur.

```
> ser1:=series(arctan(x),x=0,6);
```

$$ser1 := x - \frac{1}{3}x^3 + \frac{1}{5}x^5 + O(x^6)$$

```
> subs(x=1,ser1);
```

```
Error, invalid substitution in series
```

However, we can use `convert` together with the `polynom` option to remove the O-term that represents the omitted higher-order terms of the series. We name the resulting output `poly`. Note that `poly` is an expression that can be evaluated for particular numbers. In fact, `poly` represents the fifth-degree Maclaurin polynomial for $f(x) = \tan^{-1} x$.

```
> poly:=convert(ser1,polynom);
```

$$poly := x - \frac{1}{3}x^3 + \frac{1}{5}x^5$$

Finally, we use `plot` to compare the graphs of `poly` and $f(x) = \tan^{-1} x$. The graph of $f(x) = \tan^{-1} x$ is gray, and the graph of `poly` is in black. Note that `poly` appears to approximate f well on an interval containing 0.

```
> plot({arctan(x),poly},x=-2..2,y=-2..2,
  color=[gray,black]);
```

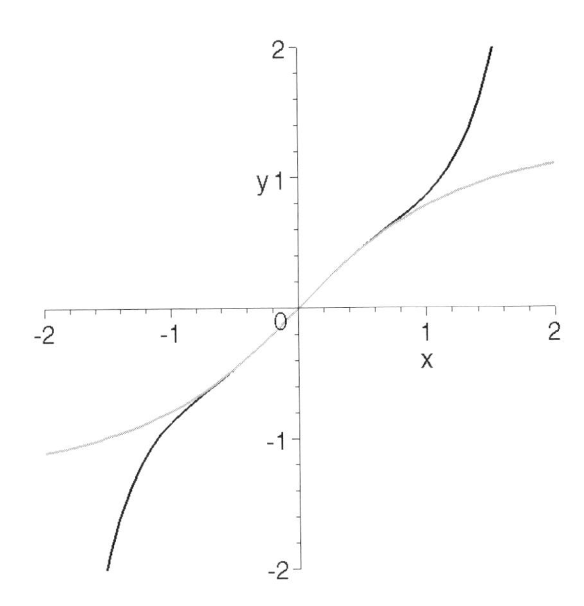

■

EXAMPLE 8: Find the eighth-degree Taylor polynomial for $f(x) = e^{-(x-1)^2(x+1)^2}$ about $x = 1$. Compare the graphs of the polynomial and $f(x)$.

SOLUTION: After clearing all prior definitions of $f(x)$, we define and graph $f(x)$, naming the result `plotf`.

```
> with(plots):
  f:='f':
  f:=x->exp(-(x-1)^2*(x+1)^2):
  plotf:=plot(f(x),x=-2..2):
  display(plotf);
```

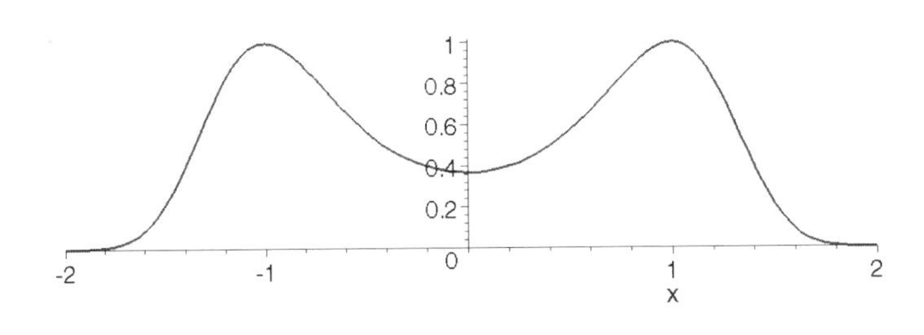

Next, we define `ser` to be the power series for $f(x)$ about $x = 1$ to order 9. We then remove the `O`-term from `ser` with `convert` and name the resulting polynomial `poly`, which represents the eighth-degree Taylor polynomial for $f(x)$ about $x = 1$.

```
> ser:=series(f(x),x=1,9);
```

$$ser := 1 - 4(x-1)^2 - 4(x-1)^3 + 7(x-1)^4 + 16(x-1)^5 + \frac{4}{3}(x-1)^6 - 28(x-1)^7 - \frac{173}{6}(x-1)^8 + O((x-1)^9)$$

```
> poly:=convert(ser,polynom);
```

$$poly := 1 - 4(x-1)^2 - 4(x-1)^3 + 7(x-1)^4 + 16(x-1)^5 + \frac{4}{3}(x-1)^6 - 28(x-1)^7 - \frac{173}{6}(x-1)^8$$

Next, we graph `poly` and name the resulting graph `plotpoly`. `plotf` and `plotpoly` are displayed, together with `display`.

```
> plotpoly:=plot(poly,x=0..2):
  display({plotf,plotpoly},view=[-2..2,-2..2]);
```

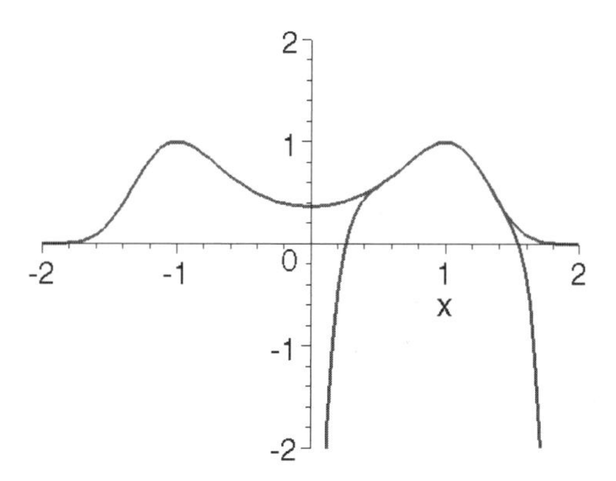

■

If $f(x) = \sum_{k=0}^{\infty} \frac{f^{(k)}(a)}{k!}(x-a)^k$, Maple's animation capabilities can help us visualize the partial sums of $\sum_{k=0}^{\infty} \frac{f^{(k)}(a)}{k!}(x-a)^k$ converging to $f(x)$.

EXAMPLE 9: Let $p_n(x)$ denote the Maclaurin polynomial of degree n for $f(x) = \cos x$. Graph $p_n(x)$ together with $f(x)$ for $n = 0, 2, \ldots, 38$.

SOLUTION: The `display` command together with the option `insequence=true` animates a set of graphics. After defining n to be 20 and $f(x) = \cos x$,

```
> with(plots):
  :=x-> cos(x):
  n := 20:
```

we define `A` to be the set of graphs of $p_n(x)$ for $n = 0, 2, \ldots, 38$ on the interval $[-3\pi/2, 3\pi/2]$. In the `seq` command, `convert(series(f(x),x=0,2*i-1), polynom)` corresponds to $p_{2i-2}(x)$.

```
> A := display(
        seq(plot(convert(series(f(x),x=0,2*i-1), polynom),
           x=-3*Pi/2..3*Pi/2,y=-2..2,color=black),
        i=1..n),insequence=true):
```

We then generate twenty graphs of $f(x) = \cos x$ on the interval $[-3\pi/2, 3\pi/2]$ with `animate` and name the result `B`.

```
> B := animate(f(x),x=-3*Pi/2..3*Pi/2,
        y=-2..2,color=gray,frames=n):
```

`A` and `B` are then displayed together with `display`. This is one frame from the resulting animation:

```
> display(A,B,view=[-3*Pi/2..3*Pi/2,-2..2]);
```

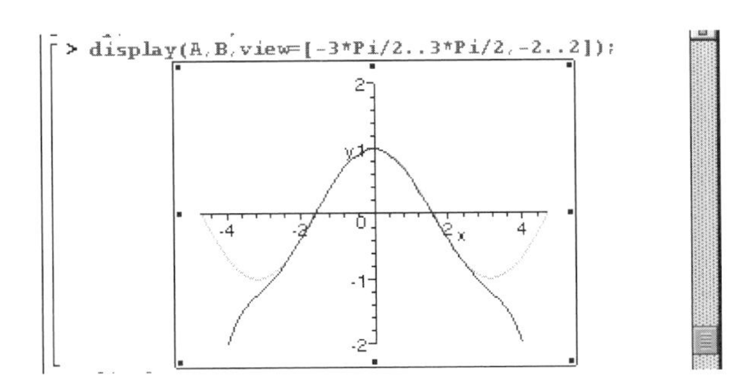

The previous result is displayed as a graphics array with `display` as well.

```
> display(%);
```

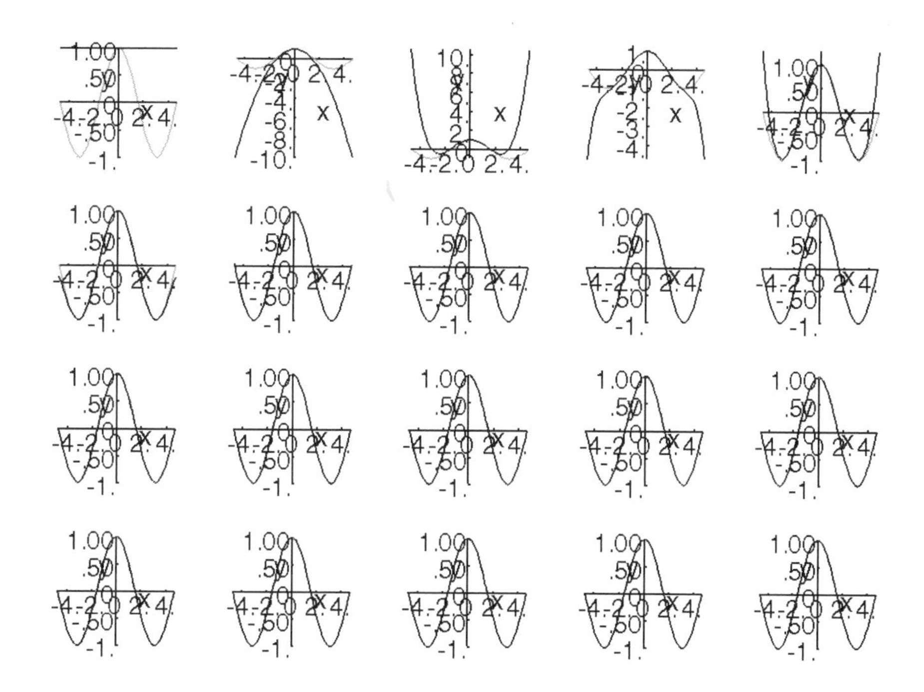

■

Application: Approximating the Remainder

Let f have (at least) $n+1$ derivatives in an interval containing a. **Taylor's Theorem** says that if x is any number in the interval, then

$$f(x) = \underbrace{\sum_{k=0}^{n} \frac{f^{(k)}(a)}{k!}(x-a)^k}_{\substack{n\text{th degree Taylor} \\ \text{polynomial of } f \\ \text{at } x = a.}} + \underbrace{\frac{f^{(n+1)}(z)}{(n+1)!}(x-a)^{n+1}}_{n\text{th degree remainder}}$$

where z is between a and x. We may use Taylor's Theorem to estimate the error involved when we use a Taylor polynomial to approximate a given function.

EXAMPLE 10: Find an upper bound on the error that results when you use the fourth-degree Maclaurin polynomial for $f(x)$ to approximate $f(x)$ on the interval $[0,1/2]$ if $f(x) = \frac{x}{x^2+1}$. What is an upper bound on the error that results when you use the tenth-degree Maclaurin polynomial to approximate $f(x)$ on the interval $[0,1/2]$?

SOLUTION: We proceed by clearing all prior definitions of *f*, defining *f*, and then graphing *f*.

```
> f:='f':
  f:=x->x/(x^2+1):
  plot(f(x),x=-4..4);
```

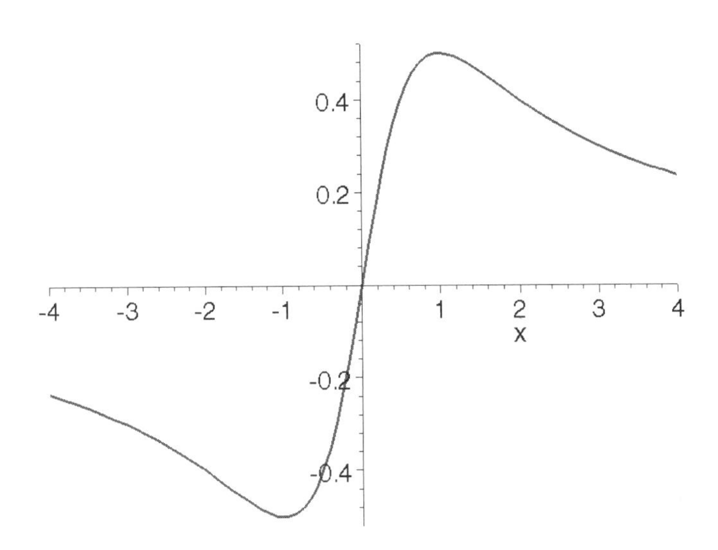

We will be computing several Maclaurin polynomials for *f*, so we define `mp` to compute the *n*th Maclaurin polynomial of *f*. We then use `mp` to compute the seventh-degree Maclaurin polynomial of *f*.

```
> mp:=proc(n)
      convert(series(f(x),x=0,n+1),polynom) end:
> mp(4);
```

$$x - x^3$$

Because we will examine the $(n+1)$st derivative when we estimate an upper bound on the error, we next define `d` to compute the $(n+1)$st derivative of f and then replace the x's by z's. We then compute `d(4)`.

```
> d:=proc(n)
      simplify(subs(x=z,diff(f(x),x$(n+1)))) end:
> d(4);
```

$$-120\frac{z^6-15\,z^4+15\,z^2-1}{(z^2+1)^6}$$

To estimate the maximum value of `d(4)` on the interval [0,1/2], we graph `d(4)`. We see that the maximum value of `d(4)` on the interval [0,1/2] is 120, which occurs if $z=0$.

```
> plot(d(4),z=0..2);
```

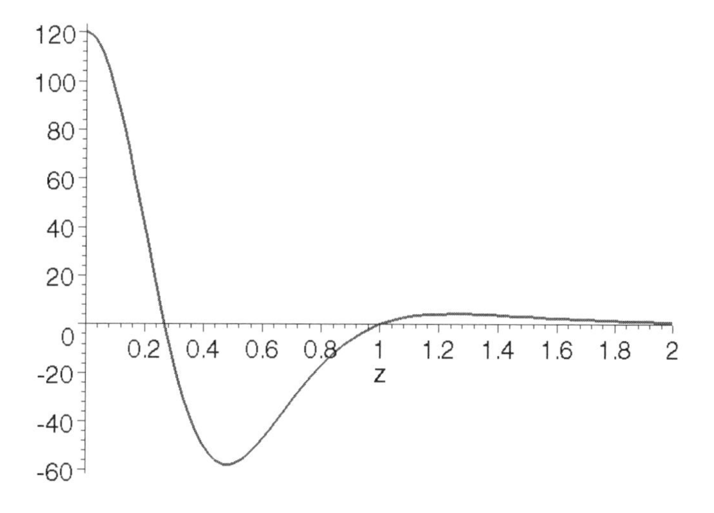

Next, we define `r` to compute the nth remainder and compute `r(4)`.

```
> r:=n->d(n)*x^(n+1)/(n+1)!:
  r(4);
```

$$-\frac{(z^6-15\,z^4+15\,z^2-1)\,x^5}{(z^2+1)^6}$$

Because we know that the maximum value of d(4) on [1,1/2] is 120, it follows that the maximum possible value of r(4) for any value of x in the interval [0,1/2] is

$$\frac{120}{120}\left(\frac{1}{2}\right)^5 = \frac{1}{32} \approx 0,03125.$$

We use plot to graph both f and the fourth-degree Maclaurin polynomial on the interval [0,1/2].

```
> plot({f(x),mp(4)},x=0..1/2);
```

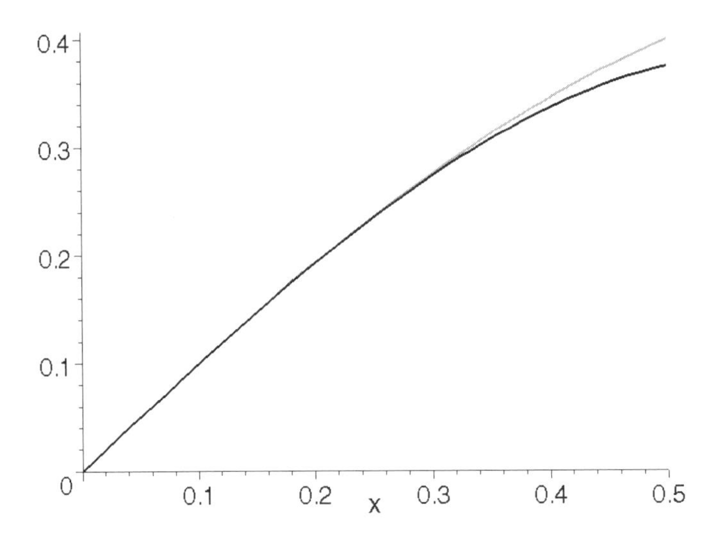

We proceed in the same manner in estimating an upper bound on the error when using the tenth-degree Maclaurin polynomial. First, we compute the eleventh derivative of f, and then we graph this function on the interval [0,1/2].

```
> d(10);
```

$$-39916800 \frac{-66 z^2 + z^{12} + 495 z^8 - 66 z^{10} - 924 z^6 + 495 z^4 + 1}{(z^2+1)^{12}}$$

```
> plot(d(10),z=0..1);
```

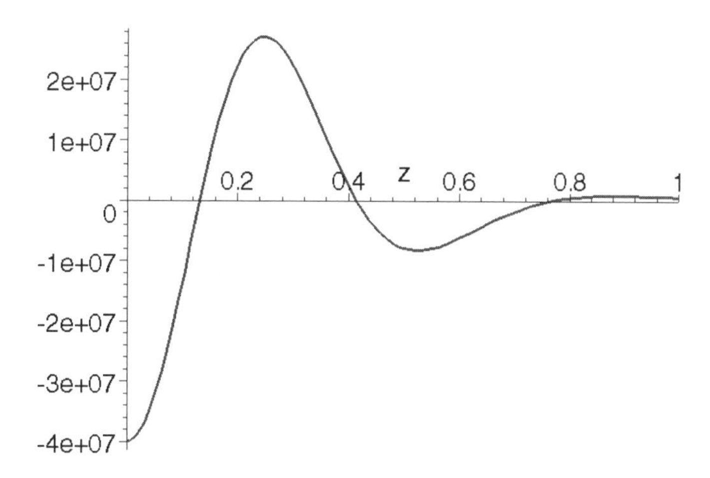

Next, we compute the tenth remainder.

```
> r(10);
```

$$-\frac{(-66\,z^2+z^{12}+495\,z^8-66\,z^{10}-924\,z^6+495\,z^4+1)\,x^{11}}{(z^2+1)^{12}}$$

Thus, the maximum possible error is $\frac{4\times10^7}{11!}\left(\frac{1}{2}\right)^{11} \approx 0.00048299.$

In the graph, we see that the graphs of f and the tenth Maclaurin polynomial are virtually identical on the interval [0,1/2].

```
> plot({f(x),mp(10)},x=0..1/2,color=[black,gray]);
```

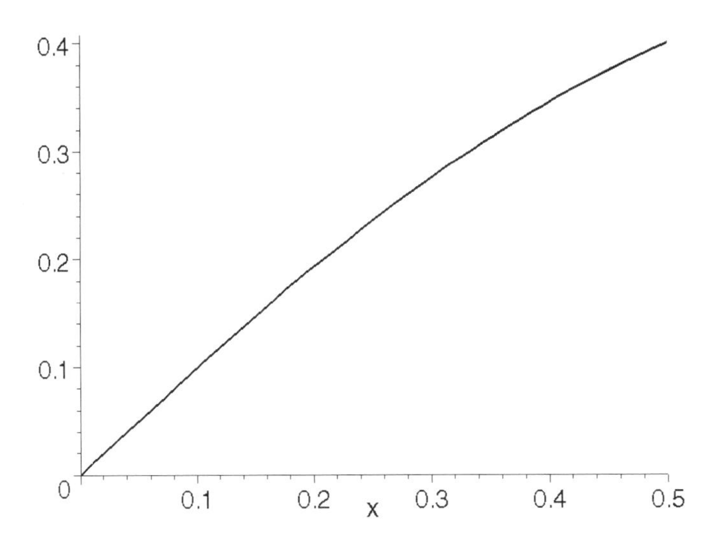

■

Other Series

In calculus, we show that if $f(x) = \sum_{n=0}^{\infty} a_n x^n$ is a power series with radius of convergence $r > 0$, then f is differentiable and integrable on its interval of convergence. However, if f is not a power series, this result is not true in general. For example, the function $f(x) = \sum_{n=0}^{\infty} \frac{\cos(3^n x)}{2^n}$ is continuous for all values of x but nowhere differentiable. We can use Maple to help us see why this function is not differentiable. Let $f_k(x) = \sum_{n=0}^{k} \frac{\cos(3^n x)}{2^n}$. We use Maple to recursively define f_k.

```
> f:='f':
  f:=proc(k) option remember;
  f(k-1)+cos(3^k*x)/2^k end:
  f(0):=cos(x):
```

We define f using the form

```
f:=proc(k) option remember; ...
```

so that Maple "remembers" the values it computes. Thus, to compute `f(5)`, Maple uses the previously computed values, namely `f(4)`, to compute `f(5)`.

Note that we can produce the same results by defining *f* with the command

```
f:=proc(k) sum(cos(3^n*x)/2^n,n=0..k) end:
```

The disadvantage of defining *f* in this manner is that Maple does not "remember" the previously computed values and thus takes longer to compute `f(k)` for larger values of *k*.

Next, we use `seq` to generate `f(3)`, `f(6)`, `f(9)`, and `f(12)`.

```
> ints:=seq(3*i,i=1..4);
```

$$ints := 3, 6, 9, 12$$

```
> seq(f(k),k=ints);
```

$$\cos(x)+\frac{1}{2}\cos(3x)+\frac{1}{4}\cos(9x)+\frac{1}{8}\cos(27x),\ \cos(x)+\frac{1}{2}\cos(3x)+\frac{1}{4}\cos(9x)$$
$$+\frac{1}{8}\cos(27x)+\frac{1}{16}\cos(81x)+\frac{1}{32}\cos(243x)+\frac{1}{64}\cos(729x),\ \cos(x)$$
$$+\frac{1}{2}\cos(3x)+\frac{1}{4}\cos(9x)+\frac{1}{8}\cos(27x)+\frac{1}{16}\cos(81x)+\frac{1}{32}\cos(243x)$$
$$+\frac{1}{64}\cos(729x)+\frac{1}{128}\cos(2187x)+\frac{1}{256}\cos(6561x)+\frac{1}{512}\cos(19683x),$$
$$\cos(x)+\frac{1}{2}\cos(3x)+\frac{1}{4}\cos(9x)+\frac{1}{8}\cos(27x)+\frac{1}{16}\cos(81x)$$
$$+\frac{1}{32}\cos(243x)+\frac{1}{64}\cos(729x)+\frac{1}{128}\cos(2187x)+\frac{1}{256}\cos(6561x)$$
$$+\frac{1}{512}\cos(19683x)+\frac{1}{1024}\cos(59049x)+\frac{1}{2048}\cos(177147x)$$
$$+\frac{1}{4096}\cos(531441x)$$

We now graph each of these functions and show the results as a graphics array with `display`.

```
> with(plots):
  A:=array(1..2,1..2):
```

```
A[1,1]:=plot(f(3),x=0..3*Pi,color=black):
A[1,2]:=plot(f(6),x=0..3*Pi,color=black):
A[2,1]:=plot(f(9),x=0..3*Pi,color=black):
A[2,2]:=plot(f(12),x=0..3*Pi,color=black):
display(A);
```

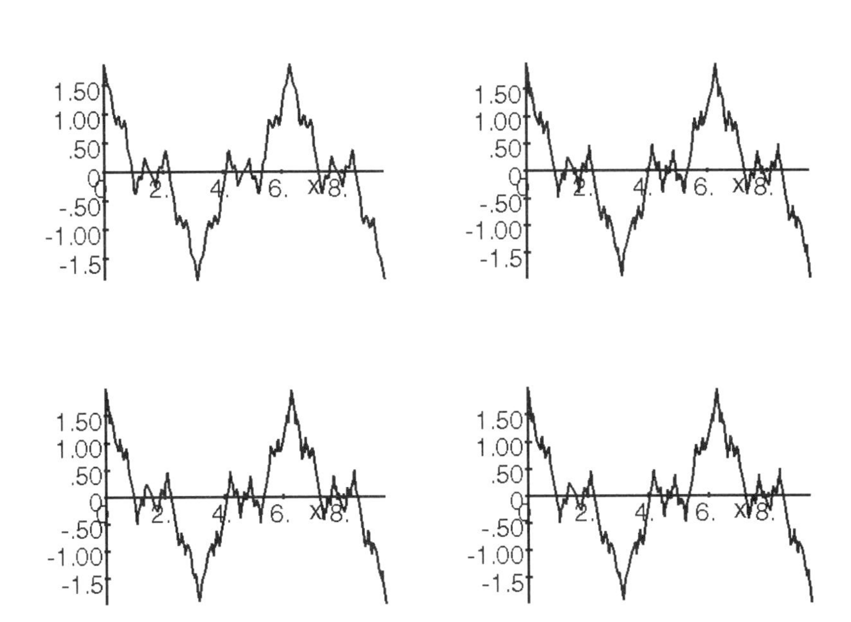

From these graphs, we see that for large values of k, the graph of $f_k(x)$, although actually smooth, appears "jagged," and thus we might suspect that $f(x)$ is indeed continuous everywhere but differentiable nowhere.

3.6 Multi-Variable Calculus

Maple is useful is investigating functions involving more than one variable. In particular, the analysis of functions that depend on two (or more) variables is enhanced with the help of Maple's graphics capabilities.

Limits of Functions of Two Variables

One of the first topics discussed in multi-variable calculus courses is limits of functions of two variables. Maple's graphics and numerical capabilities are helpful in investigating these problems.

EXAMPLE 1: Show that the limit $\lim_{(x,y)\to(0,0)} \frac{xy}{x^2+y^2}$ does not exist.

SOLUTION: We begin by clearing all prior definitions of f, if any, and defining $f(x,y) = \frac{xy}{x^2+y^2}$. Next, we use `plot3d` to graph f on the rectangle $[-2, 2] \times [-2, 2]$ and `contourplot` to graph several level curves on the same rectangle. Be sure to load the `plots` package before using the `plot3d` and `contourplot` commands.

```
> f:='f':
  with(plots):
  f:=(x,y)->x*y/(x^2+y^2):
  plot3d(f(x,y),x=-2..2,y=-2..2,
     grid=[40,40],axes=BOXED);
  contourplot(f(x,y),x=-2..2,y=-2..2,
     grid=[40,40],axes=FRAMED);
```

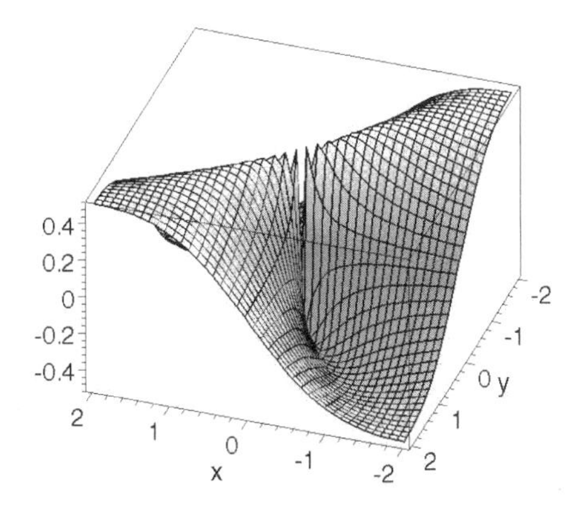

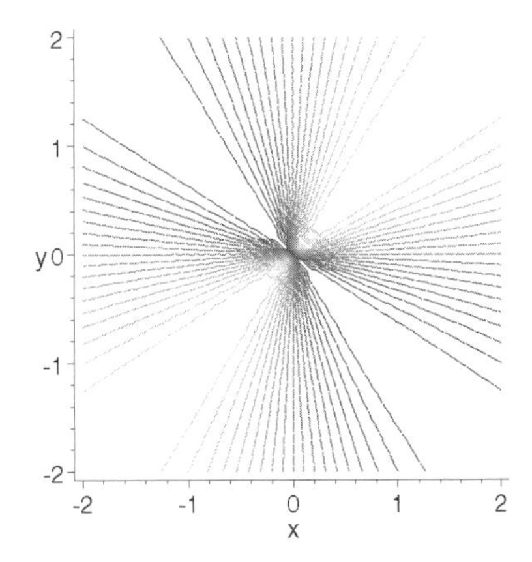

From these graphs, we see that f behaves strangely near (0,0). In fact, from the graph of the level curves, we see that near (0,0), f attains many different values. We can obtain further evidence that the limit does not exist by computing the value of f for various points chosen randomly near (0,0). We use `seq` and `rand` to generate thirteen ordered pairs (x, y) near (0,0) and the corresponding values of $f(x, y)$. Because `rand` is included in the calculation, your results will almost certainly be different from those shown here. Generally, `rand()` returns a twelve-digit nonnegative random integer. (Note that tables and lists are discussed in more detail in Chapters 4 and 5.) The first column corresponds to the x-coordinate, the second column to the y-coordinate, and the third column to the value of $f(x, y)$.

```
> r:='r':
  r:=proc(n)
     local a,b;
     a:=evalf(rand()*12^(-12-n));
     b:=evalf(rand()*12^(-12-n));
     [a,b,f(a,b)]
     end:
> seq(r(n),n=0..12);
```

$[.04793796027, .03601470117, .4802281104], [.003211728002, .004432581881, .4751279037], [.0004349643177, .0005816209138, .4796122261], [.2081013642\ 10^{-5}, .00004692497627, .04426062785], [.3268561734\ 10^{-5}, .4032685324\ 10^{-5}, .4891656677], [.1171056756\ 10^{-6}, .1397610813\ 10^{-6}, .4922809732], [.2994288814\ 10^{-7}, .1471250566\ 10^{-8}, .04901688565], [.2767953294\ 10^{-9}, .3006444334\ 10^{-8}, .09129349917], [.2120426056\ 10^{-9}, .1183556144\ 10^{-9}, .4255788133], [.1399912437\ 10^{-10}, .2001135850\ 10^{-10}, .4696972302], [.1722731318\ 10^{-11}, .2653442115\ 10^{-12}, .1504559168], [.2348632968\ 10^{-13}, .6481655897\ 10^{-13}, .3202964558], [.6609425775\ 10^{-14}, .3429074467\ 10^{-14}, .4087835854]$

From the third column, we see that f does not appear to approach any particular value for points chosen randomly near (0,0). In fact, along the line $y = x$ we see that $f(x, y) = f(x, x) = 1/2$, while along the line $y = -x$, $f(x, y) = f(x, -x) = -1/2$. Thus, f does not have a limit as $(x, y) \to (0, 0)$.

```
> simplify(f(x,x));
  simplify(f(x,-x));
```

$$\frac{1}{2}$$
$$\frac{-1}{2}$$

■

Partial Derivatives

Partial derivatives are easily calculated with Maple with either D or diff. We illustrate the use of each command in the following examples.

EXAMPLE 2: Calculate $\frac{\partial f}{\partial x}$, $\frac{\partial f}{\partial y}$, $\frac{\partial^2 f}{\partial x \partial y}$ $\frac{\partial^2 f}{\partial y \partial x}$, and the value of $\frac{\partial^2 f}{\partial y \partial x}$ if $x = \pi/2$ and $y = 1$ if $f(x, y) = \sin xy$.

SOLUTION: After defining *f*, we graph *f* on the rectangle $[\pi, \pi] \times [-\pi, \pi]$. The option grid=[40,40] is included in the plot3d command to help assure that the resulting displayed graph is smooth.

```
> f:='f':
  f:=(x,y)->sin(x*y):
  plot3d(f(x,y),x=-Pi..Pi,y=-Pi..Pi,
       grid=[40,40],axes=BOXED);
```

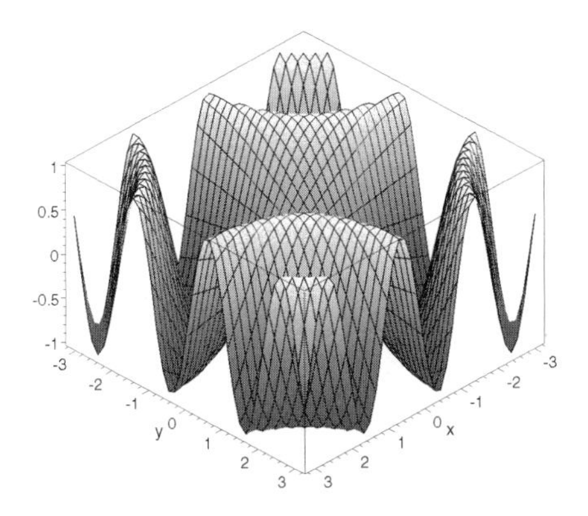

We use `diff` and then `D` to compute $\frac{\partial f}{\partial x}$.

```
> diff(f(x,y),x);
```

$$\cos(x\,y)\,y$$

`D[1]` instructs Maple to differentiate with the respect to the first argument, which in this case is x.

```
> D[1](f)(x,y);
```

$$\cos(x\,y)\,y$$

Similarly, we use `diff` and `D` to compute $\frac{\partial f}{\partial y}$.

```
> diff(f(x,y),y);
```

$$\cos(x\,y)\,x$$

`D[2]` instructs Maple to differentiate with respect to the second argument, which in this case is y.

```
> D[2](f)(x,y);
```

$$\cos(x\,y)\,x$$

We use `D` to differentiate with respect to the second argument, y, and then differentiate with respect to the first argument, x.

```
> D[1,2](f)(x,y);
```

$$-\sin(x\,y)\,y\,x+\cos(x\,y)$$

```
> diff(f(x,y),y,x);
```

$$-\sin(x\,y)\,y\,x+\cos(x\,y)$$

Similarly, we use `D` to differentiate with respect to the first argument, x, and then differentiate with respect to the second argument, y.

```
> D[2,1](f)(x,y);
```

$$-\sin(x\,y)\,y\,x+\cos(x\,y)$$

```
> diff(f(x,y),x,y);
```

$$-\sin(x\,y)\,y\,x+\cos(x\,y)$$

We use D to compute the value of $\frac{\partial^2 f}{\partial y \partial x}$ if $x = \pi/2$ and $y = 1$.

```
> D[2,1](f)(Pi/2,1);
```

$$-\frac{1}{2}\pi$$

■

The command diff(function,variable$n) computes the nth derivative of function with respect to variable.

EXAMPLE 3: Calculate $\frac{\partial^2 h}{\partial x^2}$ and $\frac{\partial^2 h}{\partial y^2}$ if $h(x, y) = \sin^2 x \cos y^2$.

SOLUTION: Proceeding as in Example 2, we first define and graph h and then use diff to calculate the indicated partial derivatives.

```
> h:='h':
  h:=(x,y)->sin(x)^2*cos(y^2):
  plot3d(h(x,y),x=0..2*Pi,
    y=-Pi..Pi,grid=[45,45],axes=BOXED);
```

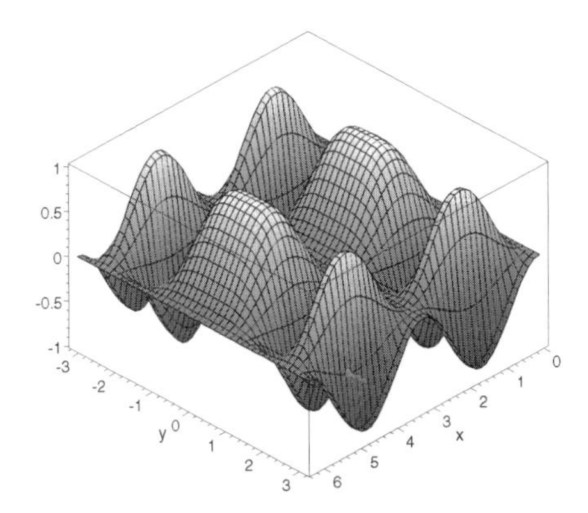

We use diff to compute $\frac{\partial^2 h}{\partial x^2}$ and $\frac{\partial^2 h}{\partial y^2}$

```
> hxx:=diff(h(x,y),x$2);
```

$$hxx := 2\cos(x)^2\cos(y^2) - 2\sin(x)^2\cos(y^2)$$

```
> hyy:=diff(h(x,y),y$2);
```

$$hyy := -4\sin(x)^2\cos(y^2)\,y^2 - 2\sin(x)^2\sin(y^2)$$

Finally, we illustrate how to compute a derivative like $\frac{\partial^4 h}{\partial y^2 \partial x^2}$

```
> hxxyy:=diff(h(x,y),x$2,y$2);
```

$$hxxyy := -8\cos(x)^2\cos(y^2)\,y^2 - 4\cos(x)^2\sin(y^2) + 8\sin(x)^2\cos(y^2)\,y^2 + 4\sin(x)^2\sin(y^2)$$

■

Application: Classifying Critical Points

Maple can be used to assist in determining certain properties of functions of more than one variable. Just as we found critical points of single variable functions in an earlier example, we can perform a similar task in the case of multi-variable functions. Let *f* be a real-valued function of two variables with continuous second-order partial derivatives. A **critical point** of *f* is a point (x_0, y_0) in the interior of the domain of *f* for which $f_x(x_0, y_0) = \frac{\partial f}{\partial x}(x_0, y_0) = 0$ and $f_y(x_0, y_0) = \frac{\partial f}{\partial y}(x_0, y_0) = 0$. Critical points are classified by the Second Derivatives test. Let $f_x(x, y) = \frac{\partial f}{\partial x}(x, y)$, $f_y(x, y) = \frac{\partial f}{\partial y}(x, y)$, $f_{xx}(x, y) = \frac{\partial^2 f}{\partial x^2}(x, y)$, $f_{yy}(x, y) = \frac{\partial^2 f}{\partial y^2}(x, y)$, and $f_{yy}(x, y) = \frac{\partial^2 f}{\partial x \partial y}(x, y)$.

Second Derivatives Test for Extrema

Let (x_0, y_0) be a critical point of a function f of two variables, and let

$$D(f,(x_0,y_0)) = \begin{vmatrix} f_{xx}(x_0,y_0) & f_{xy}(x_0,y_0) \\ f_{xy}(x_0,y_0) & f_{yy}(x_0,y_0) \end{vmatrix} = (f_{xx}(x_0,y_0)f_{yy}(x_0,y_0) - f_{xy}(x_0,y_0))^2.$$

1. If $D(f,(x_0,y_0)) > 0$ and $f_{xx}(x_0,y_0) > 0$, then *f* has a relative minimum at (x_0,y_0).
2. If $D(f,(x_0,y_0)) > 0$ and $f_{xx}(x_0,y_0) < 0$, then *f* has a relative maximum at (x_0,y_0).
3. If $D(f,(x_0,y_0)) < 0!$, then *f* has a saddle at (x_0,y_0).
4. If $Df_x(x_0,y_0) = 0$, no conclusion can be drawn and (x_0,y_0) is called a **degenerate critical point**.

EXAMPLE 4: Locate and classify all the critical points of the function

$$f(x,y) = 120x^3 - 30x^4 + 18x^5 + 5x^6 + 30xy^2.$$

SOLUTION: After clearing all prior definitions of *f*, we define *f*.

```
> f:='f':
  f:=(x,y)->-120*x^3-30*x^4+18*x^5+5*x^6+30*x*y^2:
```

The critical points of f correspond to the solutions of the system of equations $\begin{cases} f_x(x,y) = 0 \\ f_y(x,y) = 0 \end{cases}$. In order to find the critical points of *f*, the partial derivatives $f_x(x,y)$ and $f_y(x,y)$ are calculated and set equal to zero. We then locate the critical points by solving the system of equations $\begin{cases} f_x(x,y) = 0 \\ f_y(x,y) = 0 \end{cases}$ with `solve` and naming the resulting list of numbers `critpts`.

```
> dfx:=diff(f(x,y),x);
```

$$dfx := -360\,x^2 - 120\,x^3 + 90\,x^4 + 30\,x^5 + 30\,y^2$$

```
> dfy:=diff(f(x,y),y);
```

$$dfy := 60\,x\,y$$

```
> critpts:=solve({dfx=0,dfy=0});
```

$$critpts := \{y = 0, x = 0\}, \{y = 0, x = 0\}, \{y = 0, x = 2\}, \{y = 0, x = -3\}, \{y = 0, x = -2\}$$

Next, we define dfxx, dfyy, and dfxy to be $f_{xx}(x,y)$, $f_{yy}(x,y)$, and $f_{xy}(x,y) = f_{yx}(x,y)$ respectively:

```
>  dfxx:=diff(f(x,y),x$2);
   dfyy:=diff(f(x,y),y$2);
   dfxy:=diff(f(x,y),x,y);
```

$$dfxx := -720\,x - 360\,x^2 + 360\,x^3 + 150\,x^4$$

$$dfyy := 60\,x$$

$$dfxy := 60\,y$$

and disc to be $(f_{xx}(x,y))(f_{yy}(x,y)) - (f_{xy}(x,y))^2$.

```
> disc:=simplify(dfxx*dfyy-dfxy^2);
```

$$disc := -43200\,x^2 - 21600\,x^3 + 21600\,x^4 + 9000\,x^5 - 3600\,y^2$$

In order to classify the critical points, we define data to be the ordered quadruple [x,y,dfxx,disc] and evaluate data for each set of ordered pairs in critpts. Entering

```
> data:=[x,y,dfxx,disc]:
```

defines data. Entering

```
> classification:=
      array([seq(subs(critpts[i],data),i=1..5)]);
```

$$classification := \begin{bmatrix} 0 & 0 & 0 & 0 \\ 0 & 0 & 0 & 0 \\ 2 & 0 & 2400 & 288000 \\ -3 & 0 & 1350 & -243000 \\ -2 & 0 & -480 & 57600 \end{bmatrix}$$

creates an array by substituting the values of x and y in `critpts` into `data`.

By the Second Derivatives test we conclude that (0,0) is a degenerate critical point, (2,0) is a relative minimum, (–3,0) is a saddle, and (–2,0) is a relative maximum.

■

EXAMPLE 5: Classify the extrema of $f(x, y) = 3x^5 - 3xy^2 + y^3 + 3y^3$.

SOLUTION: We begin by defining $f(x, y) = 3x^5 - 3xy^2 + y^3 + 3y^3$.

```
> f:='f':
  f:=(x,y)->3*x^5-3*x*y^2+y^3+3*y^2:fx:=D[1](f)(x,y);
  fy:=D[2](f)(x,y);
```

$$fx := 15\,x^4 - 3\,y^2$$
$$fy := -6\,x\,y + 3\,y^2 + 6\,y$$

The critical points of f are found by solving the system $\{f_x(x, y) = 0, f_y(x, y) = 0\}$

```
> critpts:=[solve({fx=0,fy=0})];
```

$$critpts := [\,\{y = 0, x = 0\}, \{y = 0, x = 0\}, \{y = 0, x = 0\}, \{y = 0, x = 0\},$$
$$\{y = 2\operatorname{RootOf}(5\_Z^4 - 4\_Z^2 + 8\_Z - 4) - 2, x = \operatorname{RootOf}(5\_Z^4 - 4\_Z^2 + 8\_Z - 4)\}]$$

and finding the numerical approximation of the solutions with `allvalues`, `map` and `evalf` in `critpts`.

```
> critpts:=evalf(map(allvalues,critpts));
```

$$critpts := [\,\{y = 0, x = 0\}, \{y = 0, x = 0\}, \{y = 0, x = 0\}, \{y = 0, x = 0\},$$
$$\{y = -1.105572809 + 1.666645963\,I, x = .4472135956 + .8333229814\,I\},$$
$$\{x = .4472135956 - .8333229814\,I, y = -1.105572809 - 1.666645963\,I\},$$
$$\{y = -.802129706, x = .5989351464\}, \{y = -4.986724676, x = -1.493362338\}]$$

Next, we find the second-order derivatives

```
> fxx:=diff(f(x,y),x$2);
  fyy:=diff(f(x,y),y$2);
```

```
fxy:=diff(f(x,y),x,y);
```

$$fxx := 60\,x^3$$
$$fyy := -6\,x + 6\,y + 6$$
$$fxy := -6\,y$$

in order to compute the discriminant.

```
> disc:=fxx*fyy-fxy^2;
```

$$disc := 60\,x^3\,(-6\,x + 6\,y + 6) - 36\,y^2$$

Finally, we apply the Second Derivatives test. We evaluate the discriminant at each critical point in `critpts`. If this value is positive, we must determine whether $f_{xx} > 0$ or $f_{xx} < 0$. We can find this information individually for each critical point, or we can generate all of the values at one time in a table with `seq`. To do this, we state the desired quantities in `data` and use Maple to give this information at each point as a list.

```
> data:=[x,y,dfxx,disc]:
  classification:=
      array([seq(subs(critpts[i],data),i=1..8)]);
```

classification :=

$[0, 0, 0, 0]$

$[0, 0, 0, 0]$

$[0, 0, 0, 0]$

$[0, 0, 0, 0]$

$[.4472135956 + .8333229814\,I, -1.105572809 + 1.666645963\,I, -493.8650106 - 1007.201084\,I, 247.2074538 - 104.3392839\,I]$

$[.4472135956 - .8333229814\,I, -1.105572809 - 1.666645963\,I, -493.8650106 + 1007.201084\,I, 247.2074538 + 104.3392839\,I]$

$[.5989351464, -.802129706, -463.7246158, -54.18388488]$

$[-1.493362338, -4.986724676, -180.5453641, 2094.168980]$

We see that f has a local maximum value at the point (–1.49336, –4.98672), and f has a saddle point at (–0.598935, –0.80213). Because the discriminant is 0 at (0, 0), the Second Derivatives test fails at (0, 0). However, we can use Maple to investigate the behavior of f near (0, 0). Many times, in a three-dimensional plot of a function, we cannot see the places where the local extrema are occuring (even when we make sure that we include the critical points in our x and y ranges).

```
> plot3d(f(x,y),x=-2..1,y=-5..1/2,grid=[40,40],
      axes=FRAME);
```

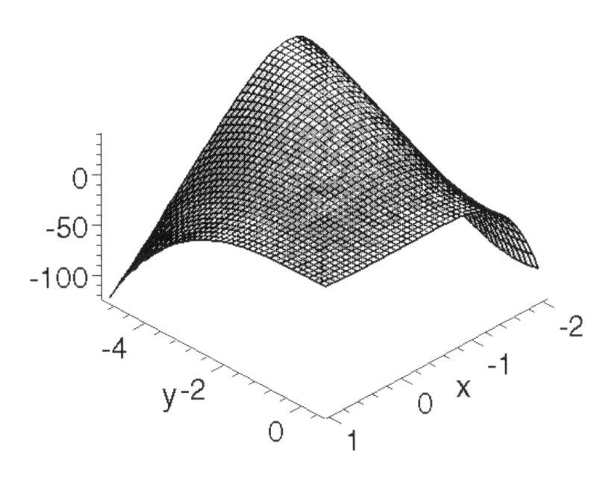

In these cases, `contourplot` may be more useful.

```
> with(plots):
  contourplot(f(x,y),x=-2..1,y=-8..2,
      contours=30,grid=[60,60],color=black);
```

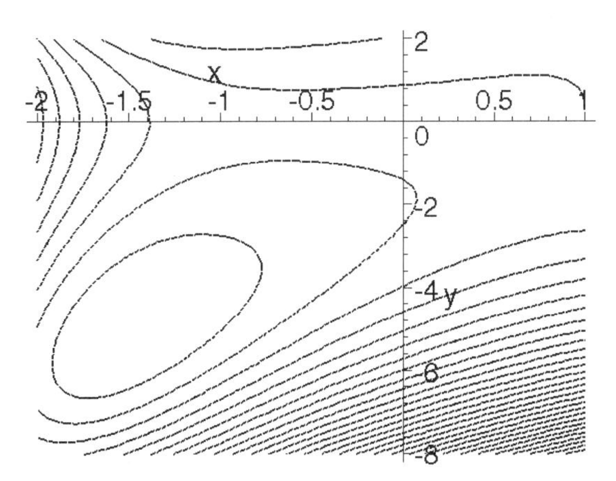

Closing in on the individual critical points gives us an even better idea of the behavior of f near each point. This procedure is particularly useful when the value of the discriminant is 0.

```
> contourplot(f(x,y),x=-2..1,y=-8..0,
      contours=30,grid=[60,60],color=black);
```

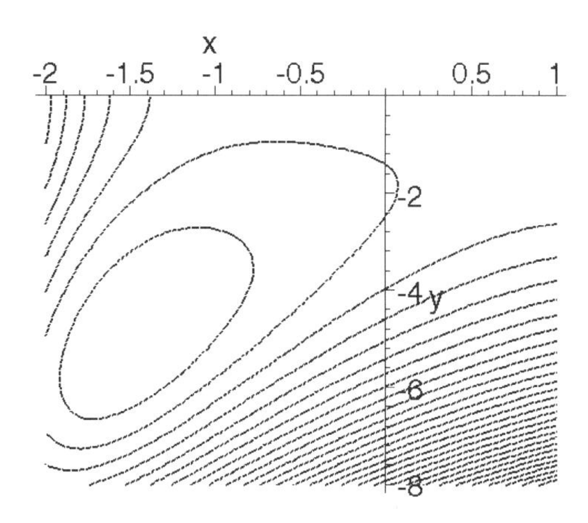


```
> contourplot(f(x,y),x=-0.5..1,y=-1..1,
      contours=30,grid=[60,60],color=black);
```

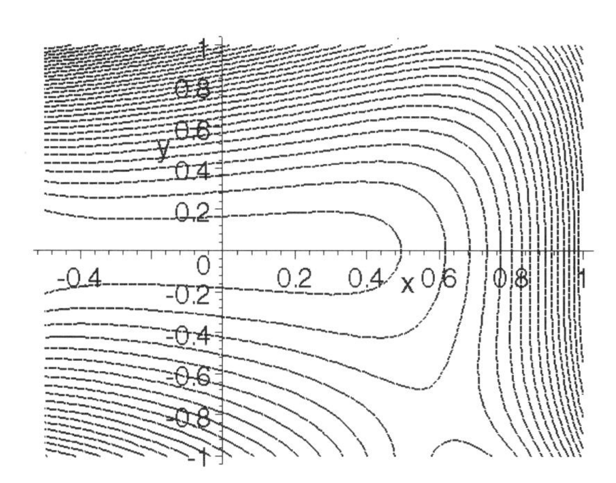

Closing in even more on (0,0), we see that the function demonstrates the behavior of either a local minimum or a local maximum value at (0,0). The three-dimensional graph of f near (0,0) shows that f has a local minimum value at (0,0).

```
> contourplot(f(x,y),x=-0.5..0.5,y=-0.5..0.5,
      contours=30,grid=[60,60],color=black);!
```

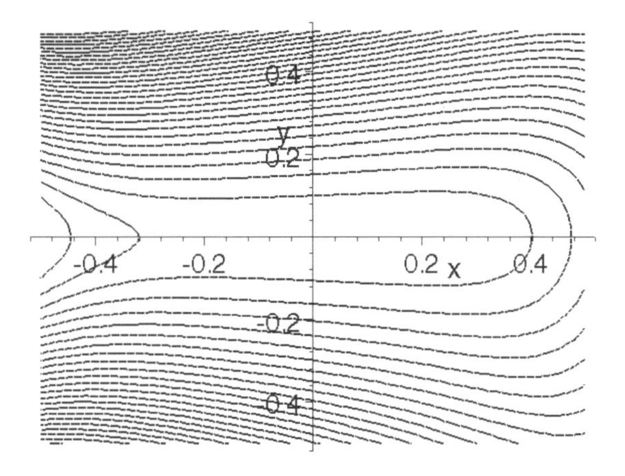


```
> plot3d(f(x,y),x=-0.25..0.25,y=-0.25..0.25,
      grid=[40,40],axes=NORMAL);
```

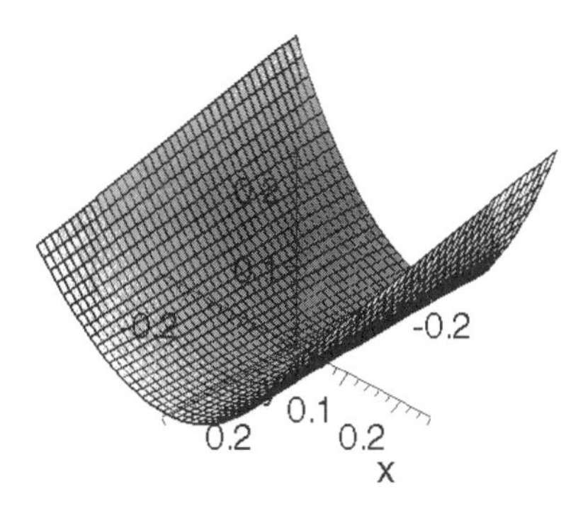

■

Application: Tangent Planes

Maple can be used to determine the equation of the plane tangent to a three-dimensional surface at a particular point as well as to graph this plane simultaneously with the surface. Let f be a real-valued function of two variables. If both $f_x(x_0,y_0)$ and $f_y(x_0,y_0)$ exist, then an equation of the plane tangent to the graph of f at the point $(x_0, y_0, f_x(x_0,y_0))$ is given by

$$f_x(x_0,y_0)(x-x_0)+f_y(x_0,y_0)(y-y_0)-(z-f(x_0,y_0))=0.$$

Solving for z yields the function (of two variables)

$$z = f_x(x_0, y_0)(x - x_0) + f_y(x_0, y_0)(y - y_0) + f(x_0, y_0).$$

EXAMPLE 6: Find an equation of the plane tangent to the graph of $k(x, y) = 4/(x^2 + y^2 + 1)$ at (1/2,1,16/9).

SOLUTION: We begin by defining and graphing k on the rectangle $[-2, 2] \times [-2, 2]$. The resulting graphics object is named `plotk`...

```
> k:='k':
  k:=(x,y)->4/(x^2+y^2+1):
  plotk:=plot3d(k(x,y),x=-2..2,y=-2..2,
      view=[-2..2,-2..2,0..4],grid=[30,30],axes=BOXED):
  with(plots):
  display(plotk);
```

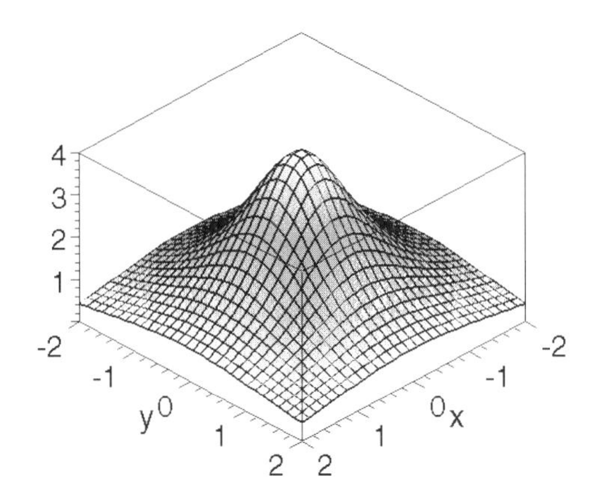

To find an equation of the tangent plane, we calculate the partial derivatives $\partial k/\partial x$ and $\partial k/\partial y$ and evaluate each if $x = 1/2$ and $y = 1$, naming the resulting output `kx` and `ky`, respectively.

```
kx:=D[1](k)(1/2,1);
```

$$kx := \frac{-64}{81}$$

```
> ky:=D[2](k)(1/2,1);
```

$$ky := \frac{-128}{81}$$

Hence, the tangent plane is defined by the function $z = kx(x+1) + ky(y-2) + 16/9$. . We define and graph z on the rectangle $[-2, 2] \times [-2, 2]$. Last, we use `display3d` to show `plotk` and `plotz` together.

```
> z:=kx*(x+1)+ky*(y-2)+k(1/2,1):
  plotz:=plot3d(z,x=-2..2,y=-2..2):
  display3d({plotk,plotz},view=[-2..2,-2..2,0..4]);
```

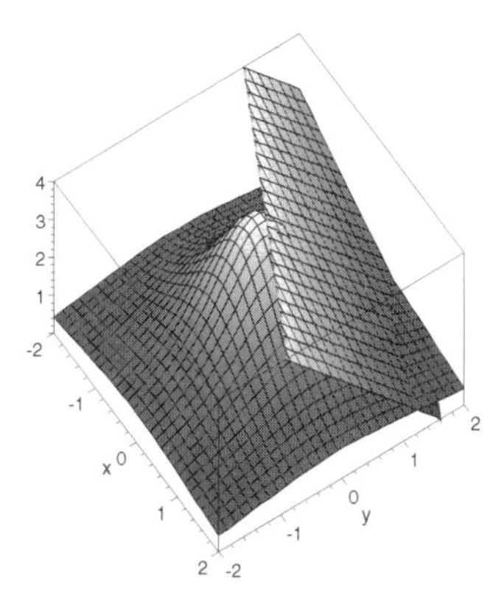

■

Application: The Method of Lagrange Multipliers

Certain types of optimization problems can be solved using the method of Lagrange multipliers, which is based on **Lagrange's theorem**: Let $f(x, y)$ and $g(x, y)$ be real-valued functions with continuous partial derivatives and let f have an extreme value at a point (x_0, y_0) on the smooth constraint curve $g(x, y) = c$. If $g_x(x_0, y_0) \neq 0$ and $g_y(x_0, y_0) \neq 0$, then there is a real number λ satisfying $f_x(x_0, y_0) = \lambda g_x(x_0, y_0)$ and $f_y(x_0, y_0) = \lambda g_y(x_0, y_0)$.

The points (x_0, y_0) at which the extreme values occur correspond to the points where the level curves of $f(x, y)$ are tangent to the graph of $g(x, y) = c$.

EXAMPLE 7: Find the maximum and minimum values of $f(x, y) = x^2 + 4y^3$, subject to the constraint $x^2 + 4y^2 = 1$.

SOLUTION: We use `spacecurve` to graph $f(x, y) = x^2 + 4y^3$ on the ellipse $x^2 + 4y^2 = 1$. A parametrization of $x^2 + 4y^2 = 1$ is given by .

$$\begin{cases} x(t) = \cos t \\ y(t) = \frac{1}{2}\sin t \end{cases}, 0 \le t \le 2\pi.$$

To graph this equation in space, we set the z-coordinate equal to 0. Be sure to load the `plots` package before using the `spacecurve` and `display3d` commands.

```
> with(plots):
  f:=(x,y)->x^2+4*y^3:
  SC1:=spacecurve([cos(t),sin(t)/2,0],
      t=0..2*Pi,axes=BOXED,color=BLACK):
  display3d(SC1);
```

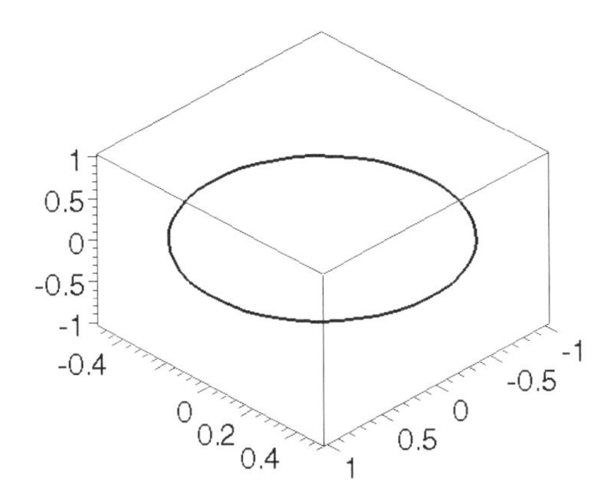

To graph $f(x, y) = x^2 + 4y^3$ on the ellipse $x^2 + 4y^2 = 1$, we use `spacecurve` to graph the set of points

$$\begin{cases} (x(t) = \cos t) \\ y(t) = \frac{1}{2}\sin t \\ \left(z(t) = f\left(\cos t, \frac{1}{2}\sin t\right)\right) \end{cases}.$$

To see the graph of f on the ellipse, we use `display3d` to display `SC1` and `SC2` together.

```
> SC2:=spacecurve([
  cos(t),sin(t)/2,f(cos(t),sin(t)/2)],
      t=0..2*Pi,color=BLACK):
  display3d([SC1,SC2],axes=NORMAL);
```

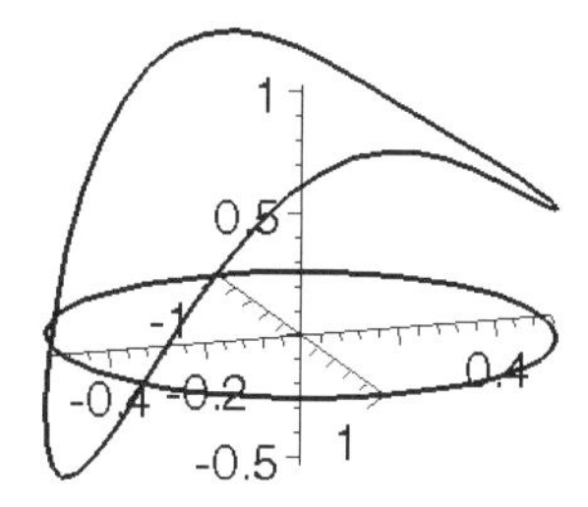

From the graphs, we see that $f(x, y) = x^2 + 4y^3$ attains a minimum and maximum on the ellipse $x^2 + 4y^2 = 1$. The minimum and maximum values occur at the points where the level curves of *f* are tangent to the graph of $x^2 + 4y^2 = 1$. We can see these points using `contourplot`. The graph of the equation $x^2 + 4y^2 = 1$ is the graph of the level curve of $g(x, y) = x^2 + 4y^2$, corresponding to 1. We use `contourplot` to graph this level curve and name the resulting graphics object `cp1`.

```
> g:=(x,y)->x^2+4*y^2:
  cp1:=contourplot(g(x,y),x=-2..2,y=-2..2,
      contours=[1],color=gray,grid=[60,60]):
> display(cp1);
```

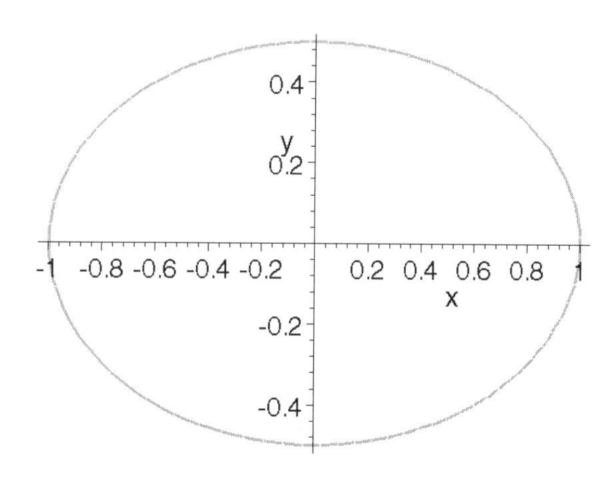

Next, we use contourplot to graph several level curves of f and name the resulting graphics object cp2. The graphs cp1 and cp2 are shown together with display.

```
> cp2:=contourplot(f(x,y),x=-2..2,y=-2..2,
      contours=30,color=black,grid=[60,60]):
> display({cp1,cp2});
```

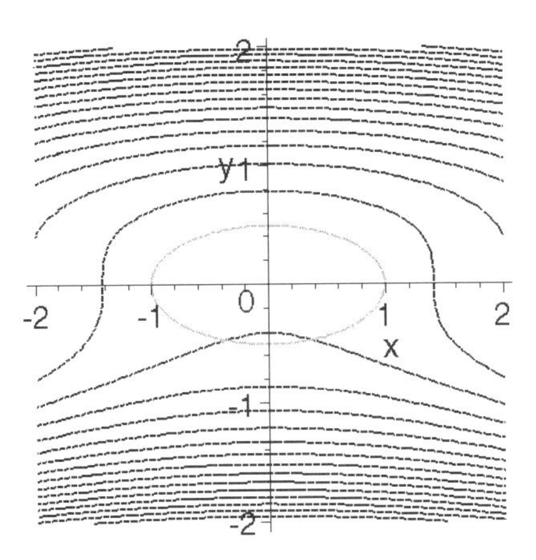

In order to find the points at which the minimum and maximum values are computed, the first-order derivatives (with respect to x and y) of f and g are computed so that Lagrange's Theorem can be applied. We then define eq1, eq2, and eq3 to be the equations representing $\partial f/\partial x = \lambda \partial g/\partial x$, $\partial f/\partial y = \lambda \partial g/\partial y$, and $g(x, y) = 0$, respectively.

```
> dfx:=D[1](f)(x,y):
  dfy:=D[2](f)(x,y):
  dgx:=D[1](g)(x,y):
  dgy:=D[2](g)(x,y):
> eq1:=dfx=lambda*dgx:
  eq2:=dfy=lambda*dgy:
  eq3:=g(x,y)=1:
```

The values of x, y, and λ that satisfy the system of three equations in Lagrange's Theorem are determined using solve, and the resulting output is named extpoints. The solutions of this system are ordered triples (x, y, λ). The values of x and y in each ordered triple represent the point at which f may have a maximum or minimum value.

```
> extpoints:=solve({eq1,eq2,eq3},{x,y,lambda});
```

$$extpoints := \{x=0, y=\frac{1}{2}, \lambda=\frac{3}{4}\}, \{x=0, y=\frac{-1}{2}, \lambda=\frac{-3}{4}\}, \{y=0, x=1, \lambda=1\},$$
$$\{y=0, x=-1, \lambda=1\}, \{y=\frac{2}{3}, x=\frac{1}{3}\,\mathrm{RootOf}(\_Z^2+7), \lambda=1\}$$

Thus, the maximum and minimum values of f are found by substituting these points back into the function $f(x,y)$ and comparing the resulting values of f. We may evaluate each point directly as we have done next to compute $f(0,-1/2)$.

```
> f(0,-1/2);
```

$$\frac{-1}{2}$$

However, we may also compute all four values with a single command. We compute $f(x,y)$ for each value in the table `extpoints`. We use `seq` and `array` to display the result in row-and-column form.

```
> array([seq(subs(extpoints[i],[x,y,f(x,y)]),i=1..4)]);
```

$$\begin{bmatrix} 0 & \frac{1}{2} & \frac{1}{2} \\ 0 & \frac{-1}{2} & \frac{-1}{2} \\ 1 & 0 & 1 \\ -1 & 0 & 1 \end{bmatrix}$$

Thus, we see that maximum values of 1 occur at (–1,0) and (1,0); a minimum value of –1/2 occurs at (0,–1/2). (The imaginary results, containing the symbol `I`, which represents $i = \sqrt{-1}$, are ignored.)

If we do not wish to use Maple to perform the steps necessary to solve the problem using Lagrange multipliers, we can instead use the command `extrema`, which uses the method of Lagrange multipliers to solve certain optimization problems. We begin by defining `eq` to be the equation $x^2 + 4y^2 = 1$, and then we use `extrema` to compute the minimum and maximum values of k, subject to the constraint equation.

```
> readlib(extrema):
  eq:=x^2+4*y^2=1:
  extrema(f(x,y),{eq},{x,y},'s');
```

$$\{1, \frac{-1}{2}\}$$

In this case, we interpret the result to mean that the minimum and maximum values must occur at the points given in s.

```
> s;
```

$$\{\{x=0, y=\frac{1}{2}\}, \{x=0, y=\frac{-1}{2}\}, \{y=0, x=1\}, \{y=0, x=-1\},$$
$$\{y=\frac{2}{3}, x=\frac{1}{3}\,\text{RootOf}(\_Z^2+7)\}\}$$

To see that the maximum and minimum values of f on the ellipse are indeed 1 and –1/2, we substitute the values obtained in s into k. We first note that the ith ordered pair in s is extracted from s with s[i]. For example, the first ordered pair in s is returned by entering s[1], as shown in the following command. (Note that extracting elements from lists is discussed in more detail in Chapters 4 and 5.)

```
> s[1];
```

$$\{x=0, y=\frac{1}{2}\}$$

We then use seq and array to compute the ordered triple $(x, y, f(x, y))$ for each of the ordered pairs (x, y) in s.

```
> array([seq(subs(s[i],[x,y,f(x,y)]),i=1..4)]);
```

$$\begin{bmatrix} 0 & \frac{1}{2} & \frac{1}{2} \\ 0 & \frac{-1}{2} & \frac{-1}{2} \\ 1 & 0 & 1 \\ -1 & 0 & 1 \end{bmatrix}$$

Once again, we see that maximum values of 1 occur at (−1, 0) and (1, 0); a minimum value of −1/2 occurs at (0, −1/2). (The imaginary results, containing the symbol `I`, which represents $i = \sqrt{-1}$, are ignored.)

■

Double Integrals

The command `int`, used to compute single integrals, is used to compute iterated integrals. The command that computes the iterated integral

$$\int_a^b \int_c^d f(x, y)\,dy\,dx$$

is

```
int(int(f(x,y),y=c..d),x=a..b),
```

and the definite integral $\int_a^b \int_c^d f(x, y)\,dy\,dx$ is numerically evaluated with the command

```
evalf(Int(Int(f(x,y),y=c..d),x=a..b)).
```

Note that the `student` package contains the command `Doubleint` that can be used to form double integrals as well.

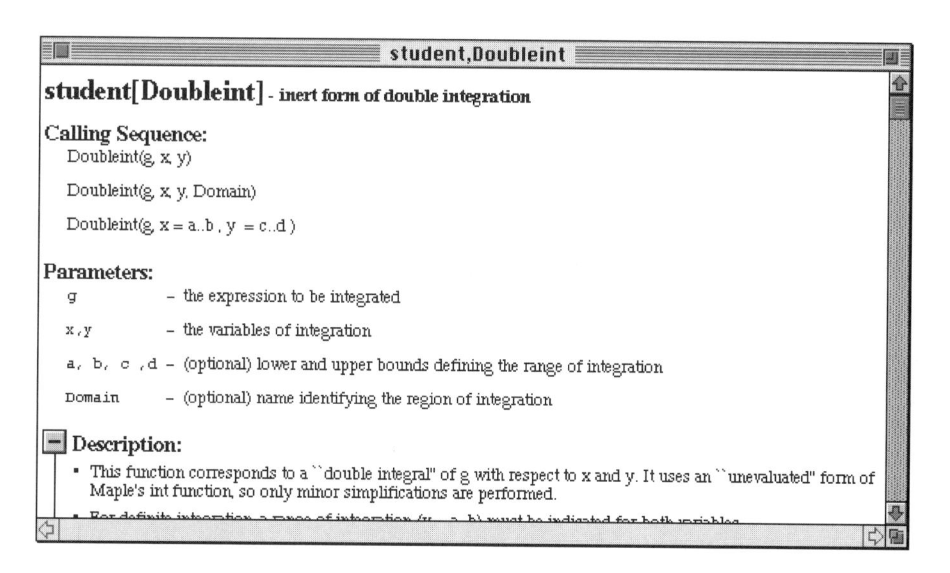

EXAMPLE 8: Evaluate each of the integrals: (a) $\iint xy^2 dydx$; (b) $\int_1^2 \int_{1-y}^{\sqrt{y}} xy^2 dxdy$; and (c) $\int_{\pi/6}^{\pi/4} \int_0^{\cos y} e^2 \sin y \, dx \, dy$.

SOLUTION: Entering

```
> int(int(x*y^2,x),y);
```

$$\frac{1}{6} x^2 y^3$$

computes $\iint xy^2 dydx$; entering

```
> int(x*y^2,x=1-y..sqrt(y));
```

$$\frac{1}{2} y^2 (y - (1-y)^2)$$

computes $\int_{x=1-y}^{x=\sqrt{y}} xy^2 dx$; entering

```
> int(int(x*y^2,x=1-y..sqrt(y)),y=1..2);
```

$$\frac{163}{120}$$

computes $\int_1^2 \int_{1-y}^{\sqrt{y}} xy^2 dxdy$; and entering

```
> int(int(exp(x)*sin(y),x=0..cos(y)),y=Pi/6..Pi/4);
```

$$-\mathbf{e}^{(1/2\sqrt{2})} + \frac{1}{2}\sqrt{2} + \mathbf{e}^{(1/2\sqrt{3})} - \frac{1}{2}\sqrt{3}$$

computes $\int_{\pi/6}^{\pi/4} \int_0^{\cos y} e^2 \sin y \, dx \, dy$.

■

After loading the `student` package, we can use the `Doubleint` command to form double integrals. The results are evaluated with `value` or approximated with `evalf`.

Thus, entering

```
> with(student):
  step1:=Doubleint(exp(x)*sin(y),x=0..cos(y),
```

```
        y=Pi/6..Pi/4);
```

$$-\mathbf{e}^{(1/2\sqrt{2})}+\frac{1}{2}\sqrt{2}+\mathbf{e}^{(1/2\sqrt{3})}-\frac{1}{2}\sqrt{3}$$

forms the iterated integral $\int_{\pi/6}^{\pi/4}\int_{0}^{\cos y} e^2 \sin y \, dx\, dy$.. On the other hand, entering

```
> value(step1);
```

$$-\mathbf{e}^{(1/2\sqrt{2})}+\frac{1}{2}\sqrt{2}+\mathbf{e}^{(1/2\sqrt{3})}-\frac{1}{2}\sqrt{3}$$

computes the exact value of this integral, while entering

```
> evalf(step1);
```

$$.1904090710$$

approximates the value of the integral.

In cases in which Maple cannot produce an exact value of an integral, evalf and int can frequently be combined to numerically approximate an integral.

EXAMPLE 9: Approximate $\int_0^1\int_0^1 \sin(e^{xy})dxdy$.

SOLUTION: In this case, we first attempt to use int to compute the value of $\int_0^1\int_0^1 \sin(e^{xy})dxdy$ and name the resulting output cant_evaluate. We then approximate cant_evaluate with evalf.

```
> cant_evaluate:=int(int(sin(exp(x*y)),x=0..1),
        y=0..1);
```

$$cant\_evaluate := \int_0^1 -\frac{-\mathrm{Si}(\mathbf{e}^y)+\mathrm{Si}(1)}{y}\,dy$$

```
> evalf(cant_evaluate);
```

$$.9174020997$$

■

The inert integration command, Int, can be used in a similar manner to approximate the value of a definite iterated integral.

EXAMPLE 10: Approximate the value of $\int_0^{\sqrt{\pi}}\int_0^{\sqrt{\pi}} \cos(x^2-y^2)\,dx\,dy$.

SOLUTION: In this case, we use the inert integration command Int. First, we define step1 to be the iterated integral $\int_0^{\sqrt{\pi}}\int_0^{\sqrt{\pi}} \cos(x^2-y^2)\,dx\,dy$. After defining step1, we approximate step1 with evalf.

```
> value:=Int(Int(cos(x^2-y^2),
      y=0..sqrt(Pi)),x=0..sqrt(Pi));
```

$$step1 := \int_0^{\sqrt{\pi}}\int_0^{\sqrt{\pi}} \cos(x^2-y^2)\,dy\,dx$$

```
> evalf(step1);
```

1.240116434

■

Application: Volume

A typical application of iterated integrals is determining the volume of a region in three-dimensional space.

EXAMPLE 11: Find the volume of the region between the graphs of $q(x,y) = e^{-x^2}\cos(x^2+y^2)$ and $w(x,y) = 3-x^2-y^2$ on the domain $[-1,1]\times[-1,1]$.

SOLUTION: We begin by defining q and w and then graphing the two functions on the rectangle $[-1,1]\times[-1,1]$.

```
> q:=(x,y)->cos(x^2+y^2)*exp(-x^2):
  w:=(x,y)->3-x^2-y^2:
  plot3d({q(x,y),w(x,y)},x=-1..1,y=-1..1,
     grid=[30,30],axes=BOXED);
```

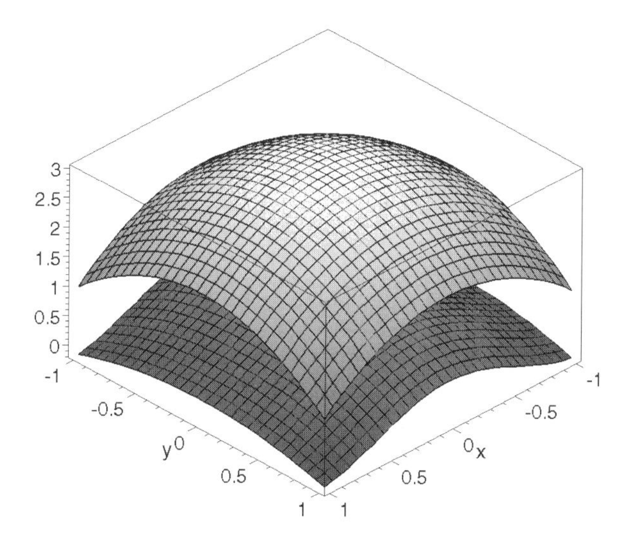

In this case, we see that q and w do not intersect on the rectangle. However, to compute the volume of the region, we must know which function is larger on the interval. We proceed by graphing q and w on the interval $[-3, 3] \times [-3, 3]$ and computing $q(0, 0)$ and $w(0, 0)$. Because $w(0, 0)$ is greater than $q(0, 0)$, we conclude that w is larger than q on the rectangle $[-1, 1] \times [-1, 1]$.

```
> plot3d({q(x,y),w(x,y)},
    x=-3..3,y=-3..3,grid=[35,35],axes=BOXED);
  q(0,0),w(0,0);
```

```
1, 3
```

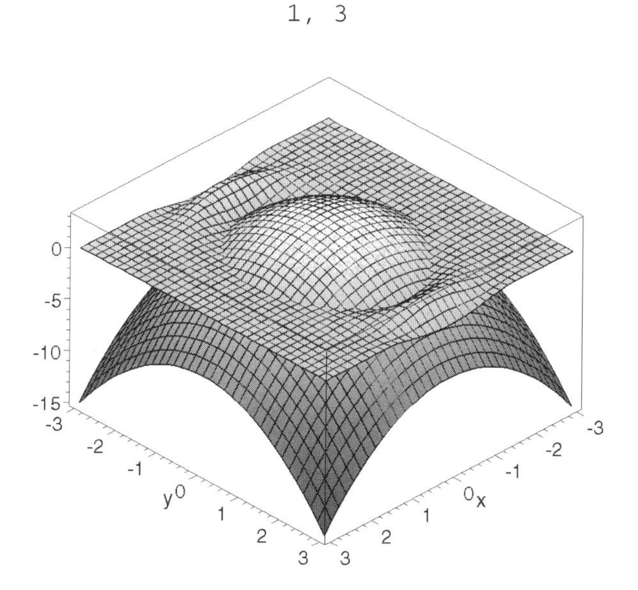

Thus, the region is bounded above by w and below by q so that the volume is given by the double integral $\iint_{[-1,1]\times[-1,1]} (w(x,y)-q(x,y))dA$, which is equivalent to the iterated integral $\int_{-1}^{1}\int_{-1}^{1}(w(x,y)-q(x,y))dydx$. We define `volume` to be $\int_{-1}^{1}\int_{-1}^{1}(w(x,y)-q(x,y))dydx$, which we then approximate with `evalf`.

```
> volume:=Int(Int(w(x,y)-q(x,y),y=-1..1),x=-1..1);
```

$$volume := \int_{-1}^{1}\int_{-1}^{1} 3-x^2-y^2-\cos(x^2+y^2)\,\mathbf{e}^{(-x^2)}\,dy\,dx$$

```
> evalf(volume);
```

7.027069074

■

In the next example, the surfaces intersect so we must determine the region of integration. We also make use of the `plots` package.

EXAMPLE 12: Find the volume of the solid bounded by the graphs of $f(x,y) = 1-x-y$ and $g(x,y) = 2-x^2-y^2$.

SOLUTION: Proceeding as in the previous example, we first define f and g. In this case, we plot f, named `Plotf`, and g, named `Plotg`, separately on the rectangle [2,2] × [–2,2], using a different number of sample points for each graph so that the graphs are easier to distinguish on the final plot. In order to display two (or more) three-dimensional graphics objects, we must use the command `display3d`, which is contained in the package `plots`. Consequently, we load the package `plots` before entering the `display3d` command.

```
> f:='f':g:='g':
  with(plots):
  f:=(x,y)->1-x-y:
  g:=(x,y)->2-x^2-y^2:
  Plotf:=plot3d(f(x,y),x=-2..2,y=-2..2,
```

```
     grid=[30,30],axes=BOXED):
Plotg:=plot3d(g(x,y),x=-2..2,y=-2..2,
     grid=[35,35],axes=BOXED):
display3d({Plotf,Plotg});
```

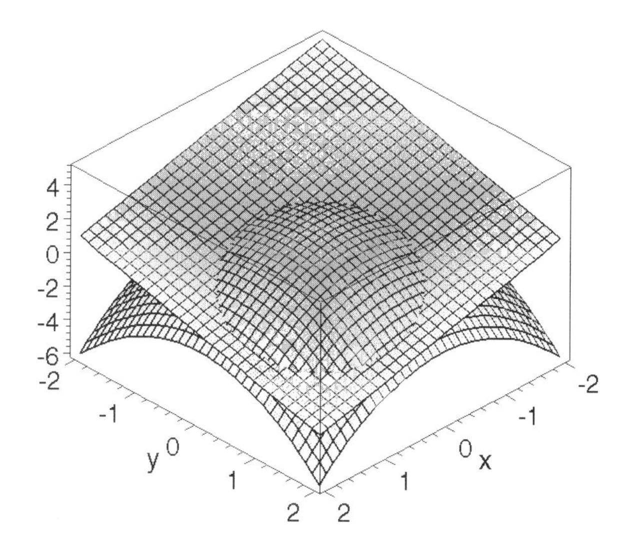

From the graph, we see that the region is bounded above by the paraboloid and below by the plane. In this case, we note that the graphs of $f(x, y) = 1 - x - y$ and $g(x, y) = 2 - x^2 - y^2$ intersect when $f(x, y) = g(x, y)$, which is equivalent to the equation $1 - x - y = 2 - x^2 - y^2$. Simplifying and completing the square yields

$$x^2 - x + y^2 - y = (x - 1/2)^2 + (y - 1/2)^2 = (\sqrt{3/2})^2$$

so that the graph of the intersection is a circle with center (1/2,1/2) and radius $\sqrt{3/2}$. Instead of completing the square by hand, we can also take advantage of the command `completesquare` contained in the `student` package. Thus, entering

```
> with(student):
  step1:=completesquare(x^2-x+y^2-y-1=0,x);
```

$$step1 := \left(x - \frac{1}{2}\right)^2 - \frac{5}{4} + y^2 - y = 0$$

completes the square with respect to x, and entering

```
> step2:=completesquare(step1,y);
```

$$step2 := \left(y - \frac{1}{2}\right)^2 - \frac{3}{2} + \left(x - \frac{1}{2}\right)^2 = 0$$

completes the square with respect to y. The result in `step2` is equivalent to the one we obtained previously.

Let R denote the interior and boundary of the circle:

$$R = \{(x, y) : (x - 1/2)^2 + (y - 1/2)^2 \leq 3/2\}$$

Then the volume of the region bounded by the graphs of f and g is given by the double integral $\iint_R (g(x, y) - f(x, y))dA$, because g is the larger of the two functions over R, as shown in the last graph. To rewrite this double integral as an iterated integral, we first use `solve` to solve the equation $1 - x - y = 2 - x^2 - y^2$ for y and name the resulting output `int_pts`. These numbers represent the upper and lower limit of integration with respect to y and are extracted from `int_pts` with `int_pts[1]` and `int_pts[2]`, respectively. Note that an alternate method of solution can be performed in the same way by solving the equation for x. Note that the result of entering `int_pts[1]` and `int_pts[2]` corresponds to the equations $y = \frac{1}{2}(1 + \sqrt{5 + 4x - 4x^2})$ and $y = \frac{1}{2}(1 - \sqrt{5 + 4x - 4x^2})$.

```
> int_pts:=solve(f(x,y)=g(x,y),y);
```

$$int\_pts := \frac{1}{2} + \frac{1}{2}\sqrt{5 - 4x^2 + 4x}, \frac{1}{2} - \frac{1}{2}\sqrt{5 - 4x^2 + 4x}$$

```
> int_pts[1],int_pts[2];
```

$$\frac{1}{2} + \frac{1}{2}\sqrt{5 - 4x^2 + 4x}, \frac{1}{2} - \frac{1}{2}\sqrt{5 - 4x^2 + 4x}$$

Then, to find the upper and lower limits of integration with respect to x, we solve the equation $5 + 4x - 4x^2 = 0$. We name the resulting solutions `xvals`. Note that entering `xvals[2]` would yield $\frac{1}{2}(1 - \sqrt{6})$, while `xvals[1]` would yield $\frac{1}{2}(1 + \sqrt{6})$.

```
> xvals:=solve(5+4*x-4*x^2=0);
```

$$xvals := \frac{1}{2}+\frac{1}{2}\sqrt{6}, \frac{1}{2}-\frac{1}{2}\sqrt{6}$$

Thus, the volume is given by the iterated integral,

$$\iint_R (g(x,y)-f(x,y))dA = \int_{(1-\sqrt{6})/2}^{(1+\sqrt{6})/2}\int_{(1-\sqrt{5+4x-4x^2})/2}^{(1+\sqrt{5+4x-4x^2})/2}(g(x,y))-f(x,y)dydx$$

which is evaluated in `volume`. We also obtain an approximation of `volume` with `evalf`.

```
> volume:=int(int(g(x,y)-f(x,y),
      y=int_pts[2]..int_pts[1]),
         x=xvals[2]..xvals[1]);
```

$$volume := \frac{9}{8}\pi$$

```
> evalf(volume);
```

3.534291736

■

Triple Integrals

Triple iterated integrals are calculated in the same manner as are double iterated integrals.

EXAMPLE 13: Evaluate $\int_0^3\int_1^x\int_0^{z+x} e^{2x}(2y-z)dydzdx$.

SOLUTION: Entering

```
> int(int(int(exp(2*x)*(2*y-z),
      y=z-x..z+x),z=1..x),x=0..3);
```

$$\frac{145}{16}e^6+\frac{11}{16}$$

computes $\int_0^3\int_1^x\int_0^{z+x} e^{2x}(2y-z)dydzdx$.

■

We illustrate how triple integrals can be used to find the volume of a solid when spherical coordinates are used.

EXAMPLE 14: Find the volume of the torus with equation in spherical coordinates $\rho = 5\sin\phi$.

SOLUTION: In general, the volume of the solid region D is given by $\iiint_D 1\,dV$. We proceed by graphing the torus. The equation of the torus is given in spherical coordinates, so we use the command `sphereplot` to graph the torus. The command `sphereplot` is not a built-in command but is contained in the `plots` package. Thus, we first load the package `plots` and then use `sphereplot` to graph (ρ, ϕ, θ).

```
> with(plots):
  sphereplot(5*sin(phi),theta=0..2*Pi,phi=0..2*Pi,
     axes=BOXED,grid=[40,40]);
```

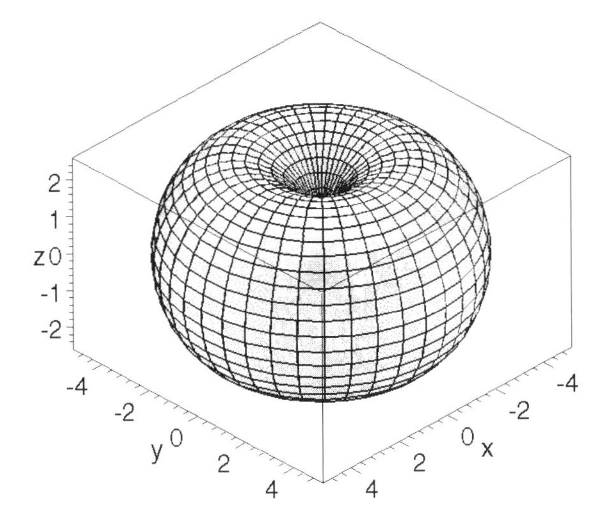

From the graph, we see that the volume of the torus is the volume of the set points with spherical coordinates (ρ, ϕ, θ) satisfying $0 \le \theta \le 2\pi$, $0 \le \rho \le 5\sin\phi$, and $0 \le \rho \le 5\sin\phi$. Thus, the volume is given by the triple integral $\iiint_D 1\,dV = \int_0^{2\pi}\int_0^{\pi}\int_0^{5\sin\theta} \rho^2 \sin\phi\, d\rho\, d\phi\, d\theta$.

```
> int(int(int(r^2*sin(phi),r=0..5*sin(phi)),
  phi=0..2*Pi),theta=0..Pi);
```

$$\frac{125}{4}\pi^2$$

Thus, the volume of the torus is $\frac{125}{4}\pi^2$.

■

Higher-Order Integrals

Higher-order iterated integrals are computed in the same manner as are double and triple iterated integrals.

EXAMPLE 15: Evaluate $\int_0^1\int_0^x\int_0^{x+y}\int_0^{x+y+z} x\,y\,z\,dw\,dz\,dy\,dx$.

SOLUTION: Entering

```
> int(int(int(int(x*y*z*w,w=0..x+y+z),
  z=0..x+y),y=0..x),x=0..1);
```

$$\frac{731}{1920}$$

computes $\int_0^1\int_0^x\int_0^{x+y}\int_0^{x+y+z} x\,y\,z\,dw\,dz\,dy\,dx$.

■

CHAPTER 4

Introduction to Sets, Lists, and Tables

4.1 Operations on Lists and Tables

Defining Lists

A **list** is a Maple object of the form `[element[1],element[2], ...,element[n]]`, where `element[i]` represents the ith element of the list. Elements of a list are separated by commas. Notice that lists are always contained in square brackets (`[...]`), and each element of a list can be almost any Maple object—even other lists. Since lists are Maple objects, they can be named. For easy reference, we will usually name lists.

Lists may be defined in a variety of ways. They may be completely typed in, or they may be created with the `seq` command. For a function *f* with domain nonnegative integers and a positive integer *n*, the command `seq(f(i),i=1..n)` creates the list `[f(1),f(2),...,f(n)]`.

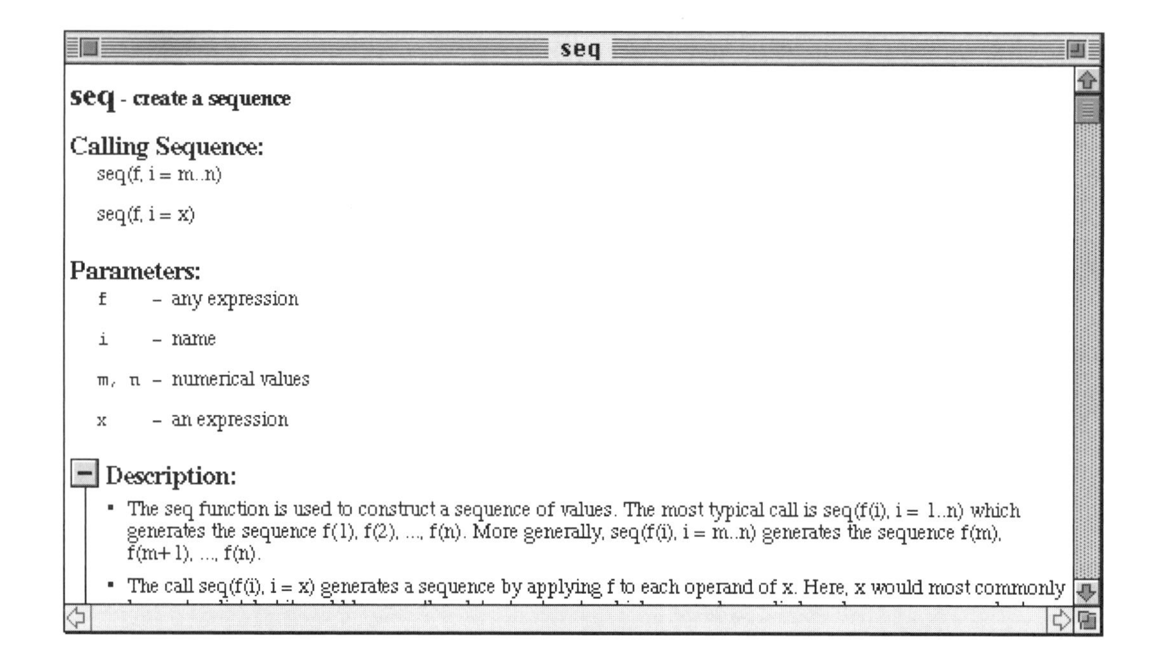

EXAMPLE 1: Use Maple to define `ListOne` to be the list of numbers consisting of 1, 3/2, 2, 5/2, 3, 7/2, and 4.

SOLUTION: In Maple, lists are contained in square brackets (`[...]`), and elements of lists are separated by commas.

```
> ListOne:=[1,3/2,2,5/2,3,7/2,4];
```

$$ListOne := \left[1, \frac{3}{2}, 2, \frac{5}{2}, 3, \frac{7}{2}, 4\right]$$

The ith element of `ListOne` is extracted from `ListOne` with `ListOne[i]`. For example, entering

```
> ListOne[3];
```

$$2$$

returns the third element of `ListOne`.

■

EXAMPLE 2: Use `seq` to define `ListTwo` to be the list consisting of 2, 5, 10, 17, and 26.

SOLUTION: In this case, we recognize that the elements of this list satisfy the relationship $f(i) = i^2 + 1$ for $i = 1, \ldots, 5$. This function is entered in the `seq` command to yield the correct list of numbers in `ListTwo`.

```
> ListTwo:=[seq(i^2+1,i=1..5)];
```

$$ListTwo := [2, 5, 10, 17, 26]$$

■

The elements of a list may also be lists.

EXAMPLE 3: Use `seq` to define the list consisting of the order pairs $(i, i^3 - i)$ for $i = 1, \ldots, 5$.

SOLUTION: Elements of the form `[i,i^3]` are included in the `seq` command, so that elements of the correct form are generated.

```
> ListThree:=[seq([i,i^3-i],i=1..5)];
```

$$ListThree := [[1, 0], [2, 6], [3, 24], [4, 60], [5, 120]]$$

■

Maple will display a list, like any other output, on successive lines that may sometimes be difficult to read or interpret. The command `array(list)` displays list in row-and-column form as illustrated in the following example.

EXAMPLE 4: Define `ListFour` to be the list consisting of the ordered pairs $(i, \sqrt{i})$ for $i = 1, \ldots, 5$. Compute a numerical approximation of each number in `ListFour`, and display `ListFour` in column form.

SOLUTION: We begin by defining `ListFour` with `seq`.

```
> ListFour:=[seq([i,sqrt(i)],i=1..5)];
```

$$ListFour := [[1, 1], [2, \sqrt{2}], [3, \sqrt{3}], [4, 2], [5, \sqrt{5}]]$$

Note that when a built-in function is applied to a list, the function is applied to each element of the list. Hence, `evalf` is used to compute a numerical approximation of each number in `ListFour`.

```
> evalf(ListFour);
```

[[1.,1.],[2.,1.414213562],[3.,1.732050808],[4.,2.],[5.,2.236067978]]

`array` is used to display `ListFour` in row-and-column form.

```
> array(ListFour);
```

$$\begin{bmatrix} 1 & 1 \\ 2 & \sqrt{2} \\ 3 & \sqrt{3} \\ 4 & 2 \\ 5 & \sqrt{5} \end{bmatrix}$$

■

EXAMPLE 5: The **Fibonacci numbers** are defined by the recursive relationship

$$f(0) = 1, f(1) = 1, \text{ and } (f(n) = f(n-1) + f(n-2)$$

Create a list, `fiblist`, consisting of the first ten Fibonacci numbers.

SOLUTION: We begin by defining *f*. Note that we define *f* using the form

```
f:=proc(n) option remember;...
```

so that Maple remembers the functional values it computes and thus avoids recomputing functional values previously computed. This would be particularly advantageous if we were to compute the value of *f* for large values of *n*. We then use `seq` to compute a list consisting of the values of *f* for *n*=0, 1, ... , 9. The resulting list is a list of the first ten Fibonacci numbers.

```
> n:='n':f:='f':
  f:=proc(n) option remember; f(n-1)+f(n-2) end:
  f(0):=1:
  f(1):=1:
  fiblist:=seq(f(n),n=1..10);
```

$$fiblist := 1, 2, 3, 5, 8, 13, 21, 34, 55, 89$$

The ninth Fibonacci number is extracted from `fiblist` with `fiblist[9]`.

```
> fiblist[9];
```

55

The purpose of this example is to illustrate how to define functions so that Maple remembers the functional values computed and to illustrate the `seq` command. In fact, Maple contains a `fibonacci` function in the combinat package. We load the `combinat` package and then use `fibonacci` to compute the hundredth Fibonacci number.

```
> with(combinat);
```

[Chi, *bell*, binomial, *cartprod*, *character*, *choose*, *composition*, *conjpart*, *decodepart*, *encodepart*, *fibonacci*, *firstpart*, *graycode*, *inttovec*, *lastpart*, *multinomial*, *nextpart*, *numbcomb*, *numbcomp*, *numbpart*, *numbperm*, *partition*, *permute*, *powerset*, *prevpart*, *randcomb*, *randpart*, *randperm*, *stirling1*, *stirling2*, *subsets*, *vectoint*]

```
> fibonacci(100);
```

354224848179261915075

■

In addition, we can use `seq` to generate lists consisting of the same or similar objects.

EXAMPLE 6: (a) Generate a list consisting of five copies of the letter a. (b) Generate a table consisting of ten random integers between –10 and 10.

SOLUTION: Entering

```
> a:='a':
  a$5;
```

a, a, a, a, a

generates a table consisting of five copies of the letter a. For (b), we use the command `rand` to generate the desired table. First note that entering

```
> (-1)^(rand() mod 2)*(rand() mod 11);
```

7

generates a random integer between –10 and 10. Thus, entering

```
> seq((-1)^(rand() mod 2)*(rand() mod 11),
i=1...10);
```

-2,-9,5,-6,-3,0,6,-3,-2,10

generates a list of ten random integers between −10 and 10. Because we have used the `rand` command above, your output will almost certainly be different from that obtained above.

■

As indicated in the previous examples, elements of lists may be numbers, ordered pairs, functions, or even other lists. For example, Maple has built-in definitions of many commonly used special functions. Consequently, lists of special functions can be created quickly.

EXAMPLE 7: The **Hermite polynomials**, $H_n(x)$, satisfy the differential equation $y''-2xy'+2ny=0$. The Maple command `H(n,x)` yields the Hermite polynomial $H_n(x)$. `H` is contained in the package `orthopoly`. Therefore, the command must be entered in the form `orthopoly[H]`, or the package `orthopoly` must be loaded before `H` is used.

(a) Create a table of the first five Hermite polynomials; (b) evaluate each Hermite polynomial if $x=1$, and then compute the value of each Hermite polynomial for $j=0, 2/10, 4/10, \ldots, 2$; (c) compute the derivative of each Hermite polynomial in the table; and (d) graph the five Hermite polynomials on the interval [-2,2].

SOLUTION: We proceed by loading the package `orthopoly` and then defining `hermite_table` to be a table consisting of the first five Hermite polynomials.

```
> with(orthopoly);
```

$$[G,H,L,P,T,U]$$

```
> hermite_table:=[seq(H(n,x),n=1..5)];
```

$$hermite\_table:=[2x, 4x^2-2, 8x^3-12x, 16x^4-48x^2+12, 32x^5-160x^3+120x]$$

We then use `subs` to evaluate each member of `hermite_table` when x is replaced by 1.

```
> subs(x=1,hermite_table);
```

[2,2,-4,-20,-8]

For (b), we first define `vals` to be the list of numbers 0, 2/10, 4/10, ... , 2.

```
> vals:=seq(2/10*j,j=0..10);
```

$$vals := 0, \frac{1}{5}, \frac{2}{5}, \frac{3}{5}, \frac{4}{5}, 1, \frac{6}{5}, \frac{7}{5}, \frac{8}{5}, \frac{9}{5}, 2$$

We then compute a table of the values of `hermite_table` if x is replaced by each element of `vals`.

```
> seq(subs(x=i,hermite_table),i=vals);
```

$$[0, -2, 0, 12, 0], \left[\frac{2}{5}, \frac{-46}{25}, \frac{-292}{125}, \frac{6316}{625}, \frac{71032}{3125}\right],$$

$$\left[\frac{4}{5}, \frac{-34}{25}, \frac{-536}{125}, \frac{2956}{625}, \frac{119024}{3125}\right],$$

$$\left[\frac{6}{5}, \frac{-14}{25}, \frac{-684}{125}, \frac{-2004}{625}, \frac{124776}{3125}\right],$$

$$\left[\frac{8}{5}, \frac{14}{25}, \frac{-688}{125}, \frac{-7604}{625}, \frac{76768}{3125}\right], [2, 2, -4, -20, -8],$$

$$\left[\frac{12}{5}, \frac{94}{25}, \frac{-72}{125}, \frac{-14964}{625}, \frac{-165168}{3125}\right],$$

$$\left[\frac{14}{5}, \frac{146}{25}, \frac{644}{125}, \frac{-12884}{625}, \frac{-309176}{3125}\right],$$

$$\left[\frac{16}{5}, \frac{206}{25}, \frac{1696}{125}, \frac{-3764}{625}, \frac{-399424}{3125}\right],$$

$$\left[\frac{18}{5}, \frac{274}{25}, \frac{3132}{125}, \frac{-15276}{625}, \frac{-351432}{3125}\right], [4, 14, 40, 76, -16]$$

In the same manner as when a built-in function is applied to a list of numbers, a built-in function applied to a list of functions results in each member of the list's being evaluated for the given function. Therefore, we use `diff` to compute the derivative of each term of `hermite_table`. Note that `int` could be used in the same manner to integrate each term of a table or list.

```
> diff(hermite_table,x);
```

$$[2, 8x, 24x^2-12, 64x^3-96x, 160x^4-480x^2+120]$$

Last, we use `plot` to plot each function in the list `hermite_table` on the interval [–2,2]. In this case, we specify that the displayed y-values consist of the interval [–50,50].

```
> plot(hermite_table,x=-2..2,-50..50);
```

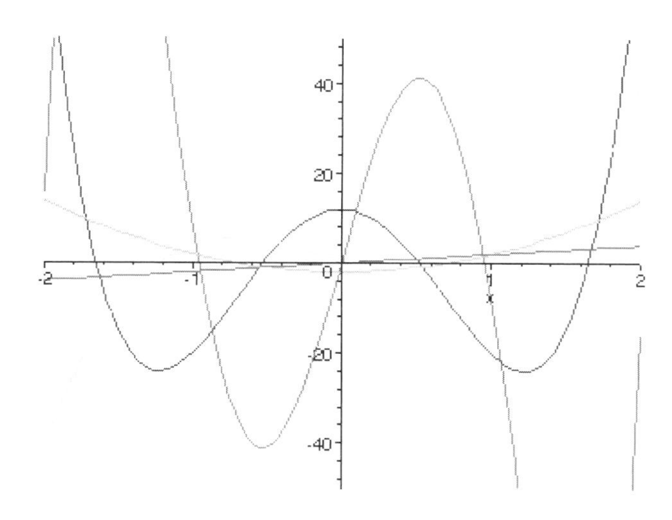

■

4.2 Operations on Lists

Extracting Elements of Lists

Elements of lists and tables are extracted with square brackets (`[...]`) or with the `op` function. Generally, both `list[i]` and `op(i,list)` return the ith element of `list`; `list[i..j]` and `op(i..j,list)` return the *i*th through *j*th elements of list. If the elements of a list are lists, we can generalize these commands. For example, if list is a list, then list[i][j] extracts the *j*th element of the *i*th element of list.

EXAMPLE 1: (a) Construct a list of the first fifteen prime numbers. (b) What is the third prime number? What is the thirteenth prime number? Construct a list of the third through thirteenth prime numbers.

SOLUTION: The built-in function `ithprime(i)` yields the ith prime number. We use `seq` and `ithprime` to compute a list of the first fifteen prime numbers and name the resulting list `prime_list`.

```
> prime_list:=[seq(ithprime(i),i=1..15)];
```

prime_list:=[2 ,3 ,5 ,7 ,11 ,13, 17, 19, 23, 29, 31, 37, 41, 43, 47]

We see that prime_list contains fifteen elements with `nops`, which counts the number of elements in a list.

```
> nops(prime_list);
```

15

Square brackets (`[...]`) are used to extract elements of lists: if `list` is a list, then `list[i]` yields the ith member of `list`. Therefore, the third and thirteenth elements of `prime_list` are extracted with the following two commands:

```
> prime_list[3];
```

5

```
> prime_list[13];
```

41

The same resutls are obtained with `op`.

```
> op(3,prime_list);
```

5

```
> op(13,prime_list);
```

41

Last, we use both square brackets and `op` to extract the third through thirteenth elements of `prime_list`.

```
> prime_list[3..13];
```

$$[5, 7, 11, 13, 17, 19, 23, 29, 31, 37, 41]$$

```
> op(3..13,prime_list);
```

$$5, 7, 11, 13, 17, 19, 23, 29, 31, 37, 41$$

■

EXAMPLE 2: Compute a 101-digit approximation of π. How many times does each digit occur in the approximation? What is the sixty-sixth digit to the right of the decimal place?

SOLUTION: We begin by setting the global variable `Digits`, which has default value 10, to 101. This instructs Maple to carry 101 digits in its operations on floating point values. We then use `evalf` to compute a 101-digit approximation of π. Note that 100 digits are to the right of the decimal place.

```
> Digits:=101:
  pi_approx:=evalf(Pi,100);
```

pi_approx:=3.14159265358979323846264338327950288419\
716939937510582097494459230781640628620899862803482\
5342117068

We now convert this approximation to a list of digits to the right of the decimal place with trunc and frac:

trunc(x) returns the largest integer smaller than x if x is positive and the smallest integer larger than x if x is negative.

frac(x) returns x modulo 1 (the decimal part of x)

For example, frac(pi_approx*10^2) returns the decimal 0.159265...; frac(pi_approx*10^2)*10 returns 1.59265...; and trunc(frac(pi_approx*10^n)*10) returns 1. Thus, entering

```
> digitsright:=[seq(trunc(frac(pi_approx*10^n)*10),
     n=0..99)];
```

digitsright:=[1, 4, 1, 5, 9, 2, 6, 5, 3, 5, 8, 9, 7, 9, 3, 2, 3, 8, 4, 6, 2, 6, 4, 3,
3, 8, 3, 2, 7, 9, 5, 0, 2, 8, 8, 4, 1, 9, 7, 1, 6, 9, 3, 9, 9, 3, 7, 5, 1, 0, 5, 8, 2, 0, 9,
7, 4, 9, 4, 4, 5, 9, 2, 3, 0, 7, 8, 1, 6, 4, 0, 6, 2, 8, 6, 2, 0, 8, 9, 9, 8, 6, 2, 8, 0, 3,
4, 8, 2, 5, 3, 4, 2, 1, 1, 7, 0, 6, 8, 0]

returns a list of the digits to the right of the decimal place in pi_approx. We use nops to see that this list has 100 elements, as desired.

```
> nops(digitsright);
```

100

The sixty-sixth digit to the right of the decimal place in the approximation is the sixty-sixth element of digitsright.

```
> digitsright[66];
```

7

To determine the number of times each digit occurs, we use the tally command that is contained in the transform subpackage of the stats package.

```
> stats[transform, tally](digitsright);
```

[Weight(4, 10),Weight(5, 8),Weight(6, 9),Weight(7, 7)

Weight(8, 13),Weight(9, 13),Weight(0, 9),Weight(1, 8)

Weight(2, 12),Weight(3, 11)]

This result means that 4 occurs ten times, 5 occurs eight times, 6 occurs nine times, and so on.

■

Evaluation of Lists by Functions

Another helpful command is `map(f,list)`, which creates a list consisting of elements obtained by evaluating `f` for each element of `list`, provided that each member of `list` is an element of the domain of `f`. To avoid errors, be sure to check that each element of `list` is in the domain of `f` prior to executing the command `map(f,list)`.

EXAMPLE 3: Create a table, named `oddints`, consisting of the first twenty-five odd integers. Square each number in `oddints`.

SOLUTION: We begin by using `seq` to create a list of the first twenty-five odd integers, and we name the resulting list `oddints`.

```
> oddints:=[seq(2*i-1,i=1..25)];
```

oddhints := [1,3,5,7,9,11,13,15,17,19,21,23,25,27,29,31,33,3
5,37,39,41,43,45,47,49]

Next, we define $f(x) = x^2$ and use `map` to compute $f(x)$ for each x in `oddints`.

```
> f:=x->x^2:
  map(f,oddints);
```

[1, 9, 25, 49, 81, 121, 169, 225, 289, 361, 441, 529, 625, 729, 841, 961,
1089, 1225, 1369, 1521, 1681, 1849, 2025, 2209, 2401]

■

We can use `map` on any list, including lists of functions.

EXAMPLE 4: The **Legendre polynomials**, $P_n(x)$, are solutions of the ordinary differential equation $(1-x^2)y'' - 2xy' + n(n+1)y = 0$. The function `P(n,x)` represents the Legendre polynomial $P_n(x)$. The function `P` is contained in the `orthopoly` package. Verify that $P_n(x)$ satisfies $(1-x^2)y'' - 2xy' + n(n+1)y = 0$ for n=1, 2, ... , 5.

SOLUTION: After loading the `orthopoly` package, we use `seq` to define `lps` to be the list consisting of the ordered pairs $(nP_n(x))$ for n=1, 2, ... , 5.

```
> with(orthopoly):
  x:='x':n:='n':
  lps:=[seq([n,P(n,x)],n=1..5)];
```

$$lps = [1, x], \left[2, \frac{3}{2}x^2 - \frac{1}{2}\right], \left[3, \frac{5}{2}x^3 - \frac{3}{2}x\right], \left[4, \frac{35}{8}x^4 - \frac{15}{4}x^2 + \frac{3}{8}\right], \left[5, \frac{63}{8}x^5 - \frac{35}{4}x^3 + \frac{15}{8}x\right]]$$

Next, we define the function f, which, given an ordered pair (n,y), computes and simplifies $(1-x^2)y''-2xy'+n(n+1)y=0$. The argument of `f`, `fnc`, must be an ordered pair: `n:=fnc[1]` and `y:=fnc[2]` define `n` and `y` to be the first and second parts of `fnc`, respectively.

```
> f:=proc(fnc)
        local n,y;
        n:=fnc[1];
        y:=fnc[2];
        simplify((1-x^2)*diff(y,x$2)-2*x*diff(y,x)+
           n*(n+1)*y);
        end:
```

Last, we use `map` to compute the value of `f` for each of the ordered pairs in `lps`. As expected, the resulting output is a list of five zeros.

```
> map(f,lps);
```

[0,0,0,0,0]

■

EXAMPLE 5: Compute a table of the values of the trigonometric functions $\sin x$ and $\cos x$, for the principal angles.

SOLUTION: We first construct a set of the principal angles. This is accomplished by defining `setone` to be the set of numbers $n\pi/4$ for n=0,1,...8 and then defining `settwo` be the set of numbers $n\pi/6$ for n=0,1,...12. We then use union to define `setthree` to be the union of `setone` and `settwo`. `setthree` could also have been obtained with the command `setone union settwo`. Note that sets do not contain duplicate elements and are sorted in a "standard" order, unlike lists. (Note also that lists are contained in brackets `[...]` and sets are contained in braces `{...}`.)

```
> setone:={seq(n*Pi/4,n=0..8)};
  settwo:={seq(n*Pi/6,n=0..12)};
  setthree:=`union`(setone,settwo);
```

$$setone := \{0, \pi, \frac{3}{2}\pi, \frac{5}{4}\pi, 2\pi, \frac{3}{4}\pi, \frac{1}{2}\pi, \frac{1}{4}\pi, \frac{7}{4}\pi\}$$

$$settwo := \{0, \pi, \frac{3}{2}\pi, 2\pi, \frac{11}{6}\pi, \frac{5}{6}\pi, \frac{4}{3}\pi, \frac{1}{2}\pi, \frac{5}{3}\pi, \frac{7}{6}\pi, \frac{2}{3}\pi, \frac{1}{6}\pi, \frac{1}{3}\pi\}$$

$$setthree := \{0, \pi, \frac{3}{2}\pi, \frac{5}{4}\pi, 2\pi, \frac{11}{6}\pi, \frac{5}{6}\pi, \frac{3}{4}\pi, \frac{4}{3}\pi, \frac{1}{2}\pi, \frac{1}{4}\pi, \frac{7}{4}\pi, \frac{5}{3}\pi, \frac{7}{6}\pi, \frac{2}{3}\pi, \frac{1}{6}\pi, \frac{1}{3}\pi\}$$

Since our result will have duplicate elements (for example, $\sin\pi/6=\cos\pi/3$), which we do not want omitted, we first convert `setthree` from type `set` to type `list` and name the resulting list `prin_vals`.

```
> prin_vals:=convert(setthree,list);
```

$$prin\_vals := \left[0, \pi, \frac{3}{2}\pi, \frac{5}{4}\pi, 2\pi, \frac{11}{6}\pi, \frac{5}{6}\pi, \frac{3}{4}\pi, \frac{4}{3}\pi, \frac{1}{2}\pi, \frac{1}{4}\pi, \frac{7}{4}\pi, \frac{5}{3}\pi, \frac{7}{6}\pi, \frac{2}{3}\pi, \frac{1}{6}\pi, \frac{1}{3}\pi\right]$$

This is then sorted into the standard order with `sort`.

```
> g:=(x,y)->is(x<y):
  prin_vals:=sort(prin_vals,g);
```

$$prin\_vals := \left[0, \frac{1}{6}\pi, \frac{1}{4}\pi, \frac{1}{3}\pi, \frac{1}{2}\pi, \frac{2}{3}\pi, \frac{3}{4}\pi, \frac{5}{6}\pi, \pi, \frac{7}{6}\pi, \frac{5}{4}\pi, \frac{4}{3}\pi, \frac{3}{2}\pi, \frac{5}{3}\pi, \frac{7}{4}\pi, \frac{11}{6}\pi, 2\pi\right]$$

We then define f to be the function that when given x returns the ordered triple $(x, \sin x, \cos x)$. We compute the ordered triple in $f(x)$ for each number x in the list `prin_vals` with the `map` command. The list that results is called `s_and_c`.

```
> f:=x->[x,sin(x),cos(x)]:
  s_and_c:=array(map(f,prin_vals));
```

$$s\_and\_c := \begin{bmatrix} 0 & 0 & 1 \\ \frac{1}{6}\pi & \frac{1}{2} & \frac{1}{2}\sqrt{3} \\ \frac{1}{4}\pi & \frac{1}{2}\sqrt{2} & \frac{1}{2}\sqrt{2} \\ \frac{1}{3}\pi & \frac{1}{2}\sqrt{3} & \frac{1}{2} \\ \frac{1}{2}\pi & 1 & 0 \\ \frac{2}{3}\pi & \frac{1}{2}\sqrt{3} & \frac{-1}{2} \\ \frac{3}{4}\pi & \frac{1}{2}\sqrt{2} & -\frac{1}{2}\sqrt{2} \\ \frac{5}{6}\pi & \frac{1}{2} & -\frac{1}{2}\sqrt{3} \\ \pi & 0 & -1 \\ \frac{7}{6}\pi & \frac{-1}{2} & -\frac{1}{2}\sqrt{3} \\ \frac{5}{4}\pi & -\frac{1}{2}\sqrt{2} & -\frac{1}{2}\sqrt{2} \\ \frac{4}{3}\pi & -\frac{1}{2}\sqrt{3} & \frac{-1}{2} \\ \frac{3}{2}\pi & -1 & 0 \\ \frac{5}{3}\pi & -\frac{1}{2}\sqrt{3} & \frac{1}{2} \\ \frac{7}{4}\pi & -\frac{1}{2}\sqrt{2} & \frac{1}{2}\sqrt{2} \\ \frac{11}{6}\pi & \frac{-1}{2} & \frac{1}{2}\sqrt{3} \\ 2\pi & 0 & 1 \end{bmatrix}$$

■

Graphing Lists of Points

Lists of numbers and ordered pairs are graphed with the `plot` command using the option `style=POINT`.

EXAMPLE 6: Graph the set of points consisting of the ordered pairs (1,2), (3,4), (5,6), (7,8), and (9,10).

SOLUTION: Both of the commands

```
plot([1,2,3,4,5,6,7,8,9,10],style=POINT)
```

and

```
plot([[1,2],[3,4],[5,6],[7,8],[9,10]],style=POINT)
```

graph the set of points consisting of the ordered pairs (1,2), (3,4), (5,6), (7,8), and (9,10).

```
>.plot([[1,2],[3,4],[5,6],[7,8],[9,10]],
   style=POINT);
```

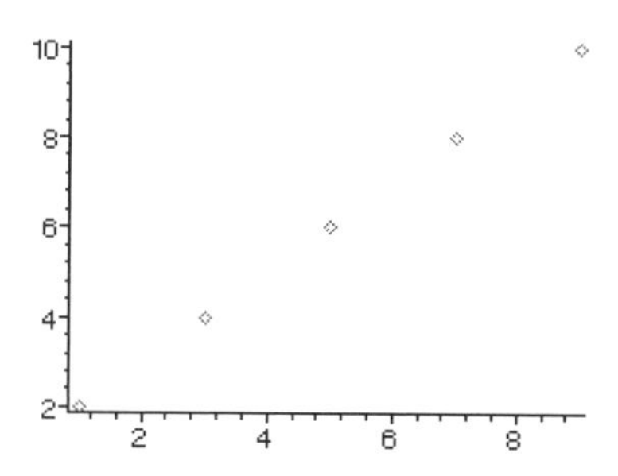

Note that you can change the appearance of the points in the plot using the option `symbol=x`, where `x` is one of `CROSS`, `DIAMOND`, `POINT`, `CIRCLE`, or `BOX`. Alternatively, you can select the graphic generated and then go to the Maple menu, select **`Style`**, followed by **`Symbol`**, followed by the desired symbol.

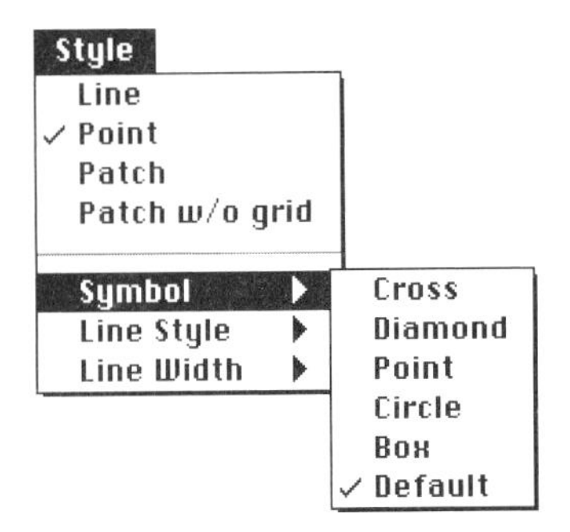

■

The commands `seq`, `table`, and `array` can all be used to construct sets of points, which can then be graphed with `plot`.

EXAMPLE 7: Graph the set of points (n,sin n) for n=0, 1, 2, ... , 1000.

SOLUTION: In this problem, we first use `seq` to compute the list of ordered pairs (n,sin n) for n=0, 1, 2, ..., 1000 and name the resulting set of ordered pairs `ordered_pairs`. Note that the output is suppressed because the command is followed by a colon. We then graph `ordered_pairs` using `plot`.

```
> n:='n':
  ordered_pairs:=[seq([n,sin(n)],n=1..1000)]:
  plot(ordered_pairs,style=POINT,symbol=POINT,

       color=BLACK);
```

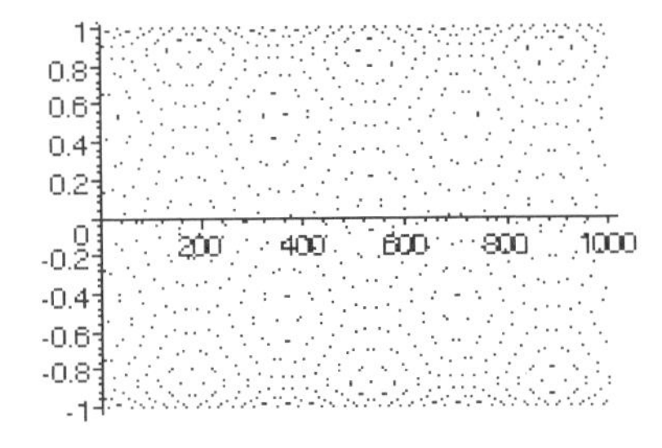

■

Lists of points can be graphed, and then the line segments connecting consecutive points can be displayed in the resulting graph if the option `style=POINT` is not included in the `plot` command.

EXAMPLE 8: In the following table, the percentage of the U.S. labor force that belonged to unions during certain years is displayed. Graph the data represented in the table.

Year	Union Membership as a Percentage of the Labor Force
1930	11.6
1935	13.2
1940	26.9
1945	35.5
1950	31.5
1955	33.2
1960	31.4
1965	28.4
1970	27.3
1975	25.5
1980	21.9
1985	18.0
1990	16.1

Source: *The World Almanac and Book of Facts*, 1993

SOLUTION: We begin by entering the data represented in the table as `dataunion`: the x-coordinate of each point corresponds to the year, where x is the number of years past 1900, and the y-coordinate of each point corresponds to the percentage of the U.S. labor force that belonged to unions in the given year. Finally, we use `plot` to graph the set of points represented in `dataunion`.

```
> dataunion:=[[30,11.6],[35,13.2],[40,26.9],
      [45,35.5],[50,31.5],
```

```
        [55,33.2],[60,31.4],[65,28.4],[70,27.3],
        [75,25.5],[80,21.9],
        [85,18.0],[90,16.1]]:
> plot(dataunion);
  plot(dataunion,style=POINT);
```

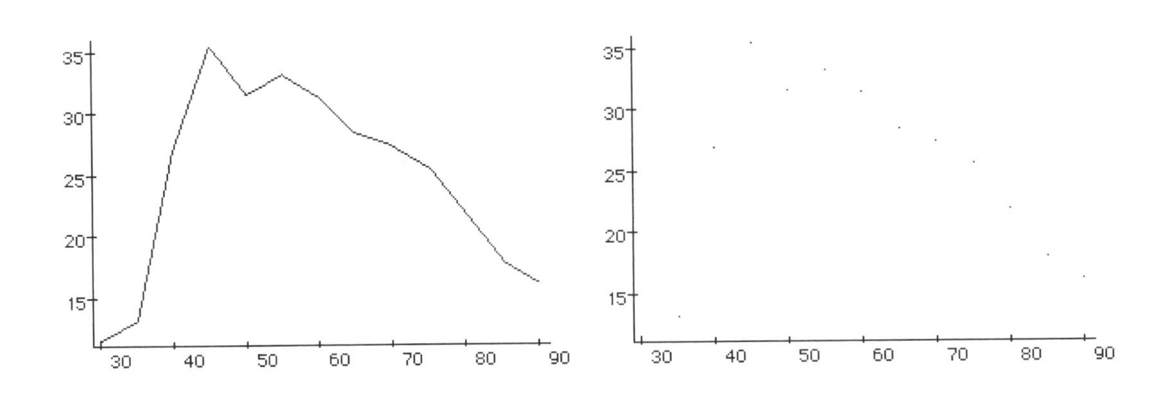

■

4.3 Mathematics of Finance

Lists and tables are quite useful in economic applications that deal with interest rates, annuities, and amortization. Maple is of great use in these types of problems because it can show the results of problems in tabular form. Also, if a change is made in the problem, Maple can easily recompute the results.

Application: Compound Interest

A common problem in economics is the determination of the amount of interest earned from an investment. If P dollars are invested for t years at an annual interest rate of r% compounded m times per year, the **compound amount**, $A(t)$, at time t is given by

$$A(t) = P\left(1 + \frac{r}{m}\right)^{mt}.$$

If P dollars are invested for t years at an annual interest rate of r% compounded continuously, the **compound amount**, $A(t)$, at time t is given by $A(t) = Pe^{rnt}$.

EXAMPLE 1: Suppose $12,500 is invested at an annual rate of 7% compounded daily. How much money has accumulated, and how much interest has been earned at the end of each five-year period for $t = 0, 5, 10, 15, 20, 25, 30$?

SOLUTION: We proceed by defining `ac(t)` to be $12500\left(1 + \frac{07}{365}\right)^{365t}$, giving the total value of the investment at the end of t years, and `Interest(t)` to be `ac(t)-12500`, giving the cumulative amount of interest earned at the end of t years.

```
> ac:='ac':interest:='interest':
  ac:=t->12500*(1+.07/365)^(365*t):
  Interest:=t->ac(t)-12500:
```

We then define `Time` to be the sequence consisting of 0, 5, 10, ... , 30, and we compute an array of the ordered pairs `(t,ac(t))` for each of the values in `Time`. This corresponds to the amount of money accumulated at the end of each time period.

```
> Time:=seq(5*n,n=0..6);
  array([seq([t,ac(t)],t=Time)]);
```

$$Time := 0, 5, 10, 15, 20, 25, 30$$

$$\begin{bmatrix} 0 & 12500 \\ 5 & 17737.75488 \\ 10 & 25170.23585 \\ 15 & 35717.07790 \\ 20 & 50683.26183 \\ 25 & 71920.58199 \\ 30 & 102056.7723 \end{bmatrix}$$

In the same manner, we compute the amount of money and interest earned at the end of each time period. Hence, the first column in the following array represents time, the second column the amount of money accumulated, and the third column the amount of interest earned.

```
> array([seq([t,ac(t),Interest(t)],t=Time)]);
```

$$\begin{bmatrix} 0 & 12500 & 0 \\ 5 & 17737.75488 & 5237.75488 \\ 10 & 25170.23585 & 12670.23585 \\ 15 & 35717.07790 & 23217.07790 \\ 20 & 50683.26183 & 38183.26183 \\ 25 & 71920.58199 & 59420.58199 \\ 30 & 102056.7723 & 89556.7723 \end{bmatrix}$$

■

The problem can be redefined for arbitrary values of t, P, r, and n as follows: We define results by declaring time, P, r, and n to be variables for the process `results`. Note that `time` represents the ordered triple, as discussed above. The first element of time, corresponding to `t0`, is extracted from `Time` with `Time[1]`; the second element of time, corresponding to `t1`, is extracted from `time` with `Time[2]`; and the third element of `time`, corresponding to the step size, is extracted from `Time` with `Time[3]`; (2) declaring `k`, `vals` and `t` local to the procedure `results`; (3) defining `vals` to be the sequence consisting of `t0`, `t0+m`, `t0+2m`, `...` , `t1`; and (4) computing the array corresponding to the compound amount of investing `P` dollars at an annual rate of `r`% compounded `n` times per year for the increments of time from `t0` to `t1` in steps of `m`.

```
> ac:='ac':Interest:='Interest':
  results:='results':
  ac:=(t,P,r,n)->P*(1+r/n)^(n*t):
  Interest:=(t,P,r,n)->ac(t,P,r,n)-P:
  results:=proc(Time,P,r,n)
      local k,vals,t;
      vals:=seq(
      Time[1]+k*Time[3],
         k=0..(Time[2]-Time[1])/Time[3]);

   array(
   [seq([t,ac(t,P,r,n),Interest(t,P,r,n)],
      t=vals)])
   end:
```

EXAMPLE 2: Suppose \$10,000 is invested at an interest rate of 12% compounded daily. Create a table consisting of the total value of the investment and the interest earned at the end of 0, 5, 10, 15, 20, and 25 years. What is the total value and interest earned on an investment of \$15,000 invested at an interest rate of 15% compounded daily at the end of 0, 10, 20, and 30 years?

SOLUTION: We use `results` to compute the accrued amount and interest earned when \$10,000 is invested at an annual rate of 12% compounded daily for 0, 5, 10, 15, 20, and 25 years. Hence, the information in the columns of the array represents the time, the accrued amount, and the interest earned, respectively.

```
> results([0,25,5],10000,.12,365);
```

$$\begin{bmatrix} 0 & 10000 & 0 \\ 5 & 18219.38723 & 8219.38723 \\ 10 & 33194.60710 & 23194.60710 \\ 15 & 60478.54008 & 50478.54008 \\ 20 & 110188.1941 & 100188.1941 \\ 25 & 200756.1376 & 190756.1376 \end{bmatrix}$$

We perform similar calculations corresponding to investing \$15,000 at an annual rate of 15% for 0, 10, 20, and 30 years and investing \$1000 at an annual rate of 9 1/2% for 10, 12, 14, ... , 18, and 20 years.

```
> results([0,30,10],15000,.15,365);
```

$$\begin{bmatrix} 0 & 15000 & 0 \\ 10 & 67204.64831 & 52204.64831 \\ 20 & 301097.6502 & 286097.6502 \\ 30 & .1349010779\ 10^7 & .1334010779\ 10^7 \end{bmatrix}$$

```
> results([10,20,2],1000,.09,365);
```

$$\begin{bmatrix} 10 & 2459.327183 & 1459.327183 \\ 12 & 2944.283140 & 1944.283140 \\ 14 & 3524.867805 & 2524.867805 \\ 16 & 4219.938251 & 3219.938251 \\ 18 & 5052.069985 & 4052.069985 \\ 20 & 6048.290192 & 5048.290192 \end{bmatrix}$$

■

Application: Future Value

If R dollars are deposited at the end of each period for n periods in an annuity that earns interest at a rate of $j\%$ per period, the **future value** of the annuity is given by:

$$S_{future} = R\frac{(1+j)^n - 1}{j}$$

EXAMPLE 3: Define a function `future` that calculates the future value of an annuity. Compute the future value of an annuity, where \$250 is deposited at the end of each month for 60 months at a rate of 7% per year. Make a table of the future values of the annuity, where \$150 is deposited at the end of each month for $12t$ months at a rate of 8% per year for $t = 1, 5, 9, 13, \ldots, 21, 25$.

SOLUTION: After defining `future`, we use `future` to calculate that the future value of an annuity where \$250 is deposited at the end of each month for 60 months at a rate of 7% per year is \$17898.22.

```
> future:=(r,j,n)->r*(((1+j)^n-1)/j):

> future(250,.07/12,5*12);
                              17898.22411
```

For the second problem, we first define `tocompute` to be the sequence of numbers 1, 5, 9, 13, ... , 21, 25 and then use `array`, `seq`, and `future` to compute the future values of the annuity, where \$150 is deposited at the end of each month for $12t$ months at a rate of 8% per year for $t = 1, 5, 9, 13, \ldots, 21, 25$. Hence, the first column in the table corresponds to the time (in years), and the second column corresponds to the future value of the annuity.

```
> tocompute:=seq(1+4*i,i=0..(25-1)/4);
```

$$tocompute := 1, 5, 9, 13, 17, 21, 25$$

```
> array([seq([t,future(150,.08/12,12*t)],
      t=tocompute)]);
```

$$\begin{bmatrix} 1 & 1867.488997 \\ 5 & 11021.52910 \\ 9 & 23614.43196 \\ 13 & 40938.06180 \\ 17 & 64769.59246 \\ 21 & 97553.82186 \\ 25 & 142653.9756 \end{bmatrix}$$

■

Application: Annuity Due

Another type of annuity is as follows: If R dollars are deposited at the beginning of each period for n periods with an interest rate of $j\%$ per period, the **annuity due** is given by:

$$S_{due} = R\left(\frac{(1+j)^{n+1}-1}{j} - 1\right)$$

EXAMPLE 4: Define a function due that computes the annuity due. Use due to (a) compute the annuity due of $500 deposited at the beginning of each month at an annual rate of 12% compounded monthly for 3 years; and (b) calculate the annuity due of $100k deposited at the beginning of each month at an annual rate of 9% compounded monthly for 10 years for $k = 1, 2, 3, \ldots, 10$.

SOLUTION: In the same manner as in Example 3, we first define due and then use due to compute the annuity due of $500 deposited at the beginning of each month at an annual rate of 12% compounded monthly for 3 years.

```
> due:=(r,j,n)->r*(((1+j)^(n+1)-1)/j)-r:

> due(500,.12/12,3*12);
                    21753.82360
```

We then use array, seq, and due to calculate the annuity due of $100k deposited at the beginning of each month at an annual rate of 9% compounded monthly for 10 years for k=1, 2, 3, ... ,10. Notice that the first column corresponds to the amount deposited each month at an annual rate of 9% compounded monthly, and the second column corresponds to the value of the annuity.

```
> array([seq([100*k,due(100*k,.09/12,10*12)],
      k=1..10)]);
```

$$\begin{bmatrix} 100 & 19496.56341 \\ 200 & 38993.12683 \\ 300 & 58489.69024 \\ 400 & 77986.25365 \\ 500 & 97482.81707 \\ 600 & 116979.3805 \\ 700 & 136475.9439 \\ 800 & 155972.5073 \\ 900 & 175469.0707 \\ 1000 & 194965.6341 \end{bmatrix}$$

■

EXAMPLE 5: Compare the annuity due on \$100k monthly investement at an annual rate of 8% compounded monthly for t=5, 10, 15, 20 and k=1, 2, 3, 4, 5.

SOLUTION: In this case, we define `times` to be the list of numbers 5, 10, 15, and 20. We then use `array`, `seq`, and `due` to calculate `due(100*k,.08/12,t*12)`, corresponding to the annuity due of \$100k deposited monthly at an annual rate of 8% compounded monthly for t years, for k=1, 2, 3, 4, and and t=5, 10, 15, and 20. Notice that the rows correspond to the annuity due on \$100, \$200, \$300, \$400, and \$500 monthly investment for 5, 10, 15, and 20 years, respectively.

```
> times:=seq(5*i,i=1..4):
  array([seq([seq(due(100*k,.08/12,t*12),t=times)],
      k=1..5)]);
```

$$\begin{bmatrix} 7396.670645 & 18416.56889 & 34834.51730 & 59294.72777 \\ 14793.34129 & 36833.13778 & 69669.03460 & 118589.4555 \\ 22190.01193 & 55249.70667 & 104503.5519 & 177884.1832 \\ 29586.68258 & 73666.27556 & 139338.0692 & 237178.9110 \\ 36983.35322 & 92082.84445 & 174172.5865 & 296473.6388 \end{bmatrix}$$

■

Application: Present Value

Yet another type of problem deals with determining the amount of money that must be invested in order to insure a particular return on the investment over a certain period of time. The **present value**, P, of an annuity of n payments of R dollars each at the end of consecutive interest periods with interest compounded at a rate of j% per period is given by:

$$P = R\frac{1-(1+j)^{-n}}{j}$$

EXAMPLE 6: Define a function `present` to compute the present value of an annuity. (a) Find the amount of money that would have to be invested at 7 1/2% compounded annually to provide an ordinary annuity income of \$45,000 per year for 35 years; and (b) find the amount of money that would have to be invested at 8% compounded annually to provide an ordinary annuity income of \$20000+\$5000k per year for 35 years for k=0, 1, 2, 3, 4, and 5 years.

SOLUTION: In the same manner as in the previous examples, we first define the function `present`, which calculates the present value of an annuity. We then use `present` to calculate the amount of money that would have to be invested at 7 1/2% compounded annually to provide an ordinary annuity income of $45,000 per year for 35 years.

```
> r:='r':j:='j':n:='n':
  present:=(r,j,n)->r*((1-(1+j)^(-n))/j):
> present(45000,.075,40);
```

566748.3899

Also, we use array and seq to find the amount of money that would have to be invested at 8% compounded annually to provide an ordinary annuity income of \$20000+\$5000k per year for 35 years for k=0, 1, 2, 3, 4, and 5. Notice that the first column corresponds to the annuity income, and the second column corresponds to the present value of the annuity.

```
> array([seq(
        [20000+5000*k,present(20000+5000*k,.08,35)],
                                k=0..5)]);
```

$$\begin{bmatrix} 20000 & 233091.3644 \\ 25000 & 291364.2054 \\ 30000 & 349637.0465 \\ 35000 & 407909.8876 \\ 40000 & 466182.7286 \\ 45000 & 524455.5698 \end{bmatrix}$$

■

Application: Deferred Annuities

Deferred annuities can also be considered. The present value of a **deferred annuity** of R dollars per period for n periods deferred for k periods with interest rate j per period is given by:

$$P_{def} = R\left[\frac{1-(1+j)^{-(n+k)}}{j} - \frac{1-(1+j)^{-k}}{j}\right].$$

EXAMPLE 7: Define a function `def(r,n,k,j)` that computes the value of a deferred annuity, where *r* equals the amount of the deferred annuity, *n* equals the number of years in which the annuity is received, *k* equals the number of years in which the lump-sum investment is made, and *j* equals the rate of interest. Use `def` to compute the lump sum that would have to be invested for 30 years at a rate of 15% compounded annually to provide an ordinary annuity income of $35,000 per year for 35 years. How much money would have to be invested at the ages of 25, 35, 45, 55, and 65 at a rate of 8 1/2% compounded annually to provide an ordinary annuity income of $30,000 per year for 40 years beginning at age 65?

SOLUTION: As in the previous examples, we first define `def` and then use `def` to compute the lump sum that would have to be invested for 30 years at a rate of 15% compounded annually to provide an ordinary annuity income of $35,000 per year for 35 years.

```
> def:=(r,n,k,j)->
      r*((1-(1+j)^(-(n+k)))/j-(1-(1+j)^(-k))/j):

> def(35000,35,30,.15);
```

3497.584370

To answer the second question, we note that the number of years the annuity is deferred is equal to 65 (the age at retirement) minus the age at which the money is initially invested. We proceed by defining `k_vals` to be the list of numbers 25, 35, 45, 55, and 65 and then use `array`, `seq`, and `def` to compute the amount of money that would have to be invested at the ages of 25, 35, 45, 55, and 65 at a rate of 8 1/2% compounded annually to provide an ordinary annuity income of $30,000 per year for 40 years beginning at age 65. Note that the first column corresponds to the current age of the individual, the second column corresponds the number of years from retirement, and the third column corresponds to the present value of the annuity.

```
> k_vals:=seq(25+10*k,k=0..4):
  array([seq([k,65-k,def(30000,40,65-k,.085)],
   k=k_vals)]);!
```

$$\begin{bmatrix} 25 & 40 & 12988.76520 \\ 35 & 30 & 29367.38340 \\ 45 & 20 & 66399.16809 \\ 55 & 10 & 150127.4196 \\ 65 & 0 & 339435.6102 \end{bmatrix}$$

■

Application: Amortization

In addition to calculating annuity information, Maple also contains several commands that are loaded with the command `readlib(finance)` that are useful in computing financial information. A loan is **amortized** if both the principal and interest are paid by a sequence of equal periodic payments. A loan of P dollars at interest rate j per period may be amortized in n equal periodic payments of R dollars made at the end of each period, where

$$R = \frac{pj}{1-(1+j)^{-n}}.$$

The function `amort(p,j,n)`, defined next, determines the monthly payment needed to amortize a loan of `p` dollars with an interest rate of `j` compounded monthly over `n` months. A second function, `totintpaid(p,j,n)`, calculates the total amount of interest paid to amortize a loan of `p` dollars with an interest rate of `j`% compounded monthly over `n` months.

```
> amort:=(p,j,n)->p*j/(1-(1+j)^(-n)):
  totintpaid:=(p,j,n)->n*amort(p,j,n)-p:
```

EXAMPLE 8: What is the monthly payment required to amortize a loan of \$75,000 at an annual rate of 9 1/2% compounded monthly over 20 years?

SOLUTION: The first calculation uses `amort` to determine the necessary monthly payment to amortize the loan. The second calculation determines the total amount paid on a loan of \$75,000 at a rate of 9.5% compounded monthly over twenty years, while the third shows how much of this amount was paid toward the interest.

```
> amort(75000,0.095/12,20*12);
                         699.0983810

> 240*amort(75000,0.095/12,20*12);
                         167783.6114

> totintpaid(75000,0.095/12,20*12);
                         92783.6114
```

■

Amortization schedules for loans can be constructed with the command `amortization`, which is contained in the `finance` package. Detailed information regarding the `finance` package is obtained with `?finance`.

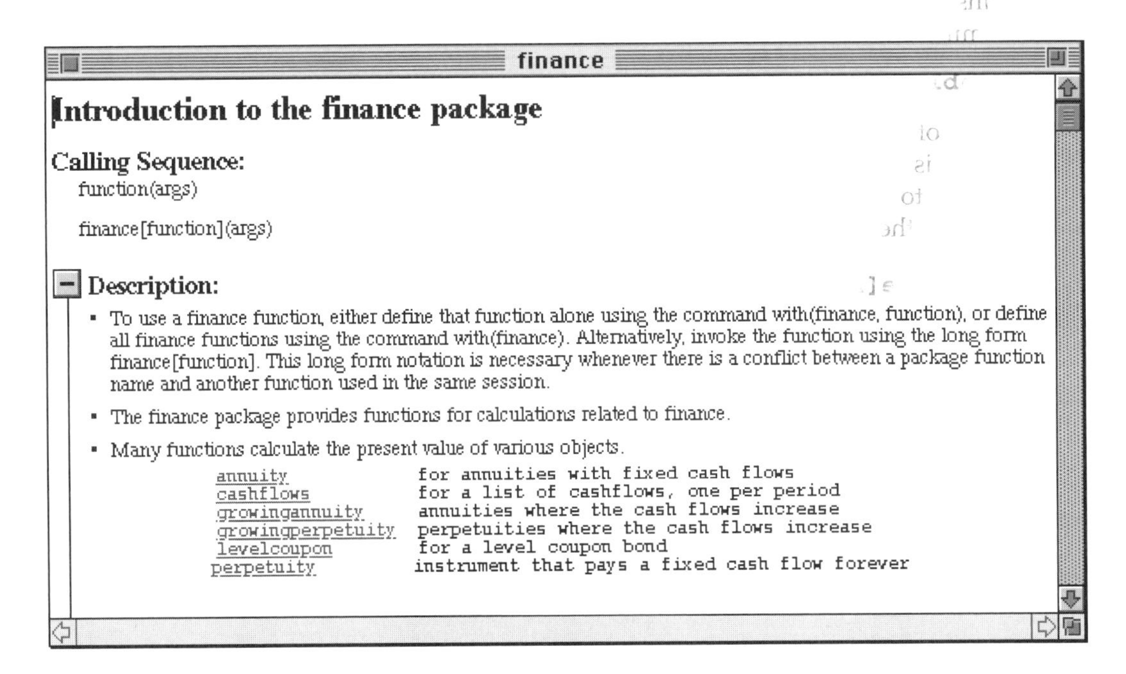

finance

Introduction to the finance package

Calling Sequence:

function(args)

finance[function](args)

Description:

- To use a finance function, either define that function alone using the command with(finance, function), or define all finance functions using the command with(finance). Alternatively, invoke the function using the long form finance[function]. This long form notation is necessary whenever there is a conflict between a package function name and another function used in the same session.
- The finance package provides functions for calculations related to finance.
- Many functions calculate the present value of various objects.

```
annuity             for annuities with fixed cash flows
cashflows           for a list of cashflows, one per period
growingannuity      annuities where the cash flows increase
growingperpetuity   perpetuities where the cash flows increase
levelcoupon         for a level coupon bond
perpetuity          instrument that pays a fixed cash flow forever
```

EXAMPLE 9: Create an amortization schedule for a loan of $75,000 with an annual interest rate of 9 1/2% compounded monthly and monthly payments of $699.10. What amount of the second loan payment is for interest on the principal? What is the balance of the principal after the second payment?

SOLUTION: First, we load the `finance` package.

```
> with(finance);
```

$[amortization, annuity, blackscholes, cashflows, effectiverate, futurevalue, growingannuity, growingperpetuity, levelcoupon, perpetuity, presentvalue, yieldtomaturity]$

We then use amortization to construct the amortization table for a loan of $75,000 at an annual rate of interest of 9 1/2% compounded monthly amortized at rate of $699.10. The resulting table is named a_table. The resulting table, which contains 240 entries, is not displayed because a colon (:) is included at the end of the command.

```
a_table:=amortization(75000,.095/12,699.10):
```

Elements of tables are extracted from tables with square brackets ([...]). Hence, if table is a table, the *i*th element of table is extracted with table[i]. In particular, to see the second element of a_table, corresponding to the second payment of the loan, we enter a_table[2]:

```
> a_table[2];
```

[2, 699.10,592.92, 106.18, 74788.47]

Note that each element of a_table is also a table of five elements, where the first element corresponds to the payment number, the second element corresponds to the payment amount, the third element corresponds to the amount of the payment for payment of interest, the fourth element corresponds to the amount of the payment applied to the principal, and the fifth element corresponds to the balance of the principal. Thus, we see that $592.92 of the second payment is for payment of interest.

```
> a_table[2][3];
```

592.92

■

In many cases, the amount paid toward the principal of the loan and the total amount that remains to be paid after a certain payment need to be computed.

EXAMPLE 10: Use the amortization table, a_table, from Example 9 to compute the total amount of interest paid during the life of the loan. What is the principal balance at the end of the fifth, tenth, fifteenth, and twentieth payments?

SOLUTION: Each array contains three operands. The number of operands of an object object is obtained with nops(object).

```
> nops(eval(a_table));
```

3

Note that the second operand, which is obtained below with op(2,eval(a_table)), yields the number of entries in a_table. Thus, to compute the total interest paid during the life of the loan, we must add the third entry of the *i*th entry of a_table, corresponding to the interest paid after the *i*th payment, for i=1, 2, ... , 240. Therefore, we are merely adding the interest paid after each payment.

```
> op(2,eval(a_table));
```

$$0..240$$

```
> sum(a_table[i][3],i=1..240);
```

92782.86

In this case, we use a for loop instead of seq. Hence, the principal balance at the end of the fifth, tenth, fifteenth, and twentieth payments are given in the last column in each row.

```
> for j from 0 by 5 to 20 do a_table[j] od;
```

$$[0, 0, 0, 0, 75000.00]$$
$$[5, 699.10, 590.37, 108.73, 74464.85]$$
$$[10, 699.10, 586.00, 113.10, 73908.17]$$
$$[20, 699.10, 576.72, 122.38, 72726.74]$$

■

Alternatively, this can be accomplished with the functions unpaidbalance and curprinpaid defined, using the functions amort and present, which were previously defined.

```
> unpaidbalance:=(p,j,n,m)->present(amort(p,j,n),j,n-m):
  unpaidbalance(p,j,n,m);
```

$$\frac{p(1-(1+j)^{(-n+m)})}{1-(1+j)^{(-n)}}$$

```
> curprinpaid:=(p,j,n,m)->p-unpaidbalance(p,j,n,m):
  curprinpaid(p,j,n,m);
```

$$\frac{p(1-(1+j)^{(-n+m)})}{1-(1+j)^{(-n)}}$$

EXAMPLE 11: What is the unpaid balance of the principal at the end of the fifth year of a loan of $60,000 with an annual interest rate of 8% scheduled to be amortized with monthly payments over a period of 10 years? What is the total interest paid immediately after the sixtieth payment?

SOLUTION: We use the functions `unpaidbalance` and `curprinpaid`, defined above, to calculate that of the original $60,000 loan, $24,097.90 has been paid at the end of five years; $35,902.10 is still owed on the loan.

```
> unpaidbalance(60000,0.08/12,120,60);
```

$$35902.12153$$

```
> curprinpaid(60000,0.08/12,120,60);
```

$$24097.87847$$

■

Maple can also be used to determine the total amount of interest paid on a loan using the following function:

```
> curintpaid:=(p,j,n,m)->m*amort(p,j,n)-
      curprinpaid(p,j,n,m):
  curintpaid(p,j,n,m);
```

$$\frac{mpj}{1-(1+j)^{(-n)}} - p + \frac{p(1-(1+j)^{(-n+m)})}{1-(1+j)^{(-n)}},$$

where `curintpaid(p,j,n,m)` computes the interest paid on a loan of $`p` amortized at a rate of `j` per period over `n` periods immediately after the `m`th payment.

EXAMPLE 12: What is the total interest paid on a loan of $60,000 with an interest rate of 8% compounded monthly amortized over a period of 10 years (120 months) immediately after the sixtieth payment?

SOLUTION: Using `curintpaid`, we see that the total interest paid is $19,580.10.

```
> curintpaid(60000,0.08/12,120,60);
```

$$19580.05407$$

■

Using the functions defined above, amortization tables can be created that show a breakdown of the payments made on a loan.

EXAMPLE 13: What is the monthly payment necessary to amortize a loan of $45,000 with interest rate of 7% compounded monthly over a period of 15 years (180 months)? What is the total principal and interest paid after 0, 3, 6, 9, 12, and 15 years?

SOLUTION: We first use amort to calculate the monthly payment necessary to amortize the loan.

```
> amort(45000,0.07/12,15*12);
                    404.4727349
```

Next, we use `array`, `seq`, `curprinpaid`, and `curintpaid` to determine the interest and principal paid at the end of 0, 3, 6, 9, 12, and 15 years.

```
> tvals:=seq(3*t,t=0..5):
  array([seq([t,curprinpaid(45000,0.07/12,15*12,12*t),
      curintpaid(45000,0.07/12,15*12,12*t)],t=tvals)]);
```

$$\begin{bmatrix} 0 & 0 & 0 \\ 3 & 5668.98524 & 8892.03322 \\ 6 & 12658.42214 & 16463.61477 \\ 9 & 21275.87760 & 22407.17777 \\ 12 & 31900.55882 & 26343.51501 \\ 15 & 45000 & 27805.09228 \end{bmatrix}$$

Note that the first column represents the number of years, the second column represents the principal paid, and the third column represents the interest paid. Thus, at the end of 12 years, $31,900.60 of the principal has been paid and $26,343.50 has been paid in interest.

■

Because `curintpaid(p,j,n,y)` computes the interest paid on a loan of $`p` amortized at a rate of `j` per period over `n` periods immediately after the `y`th payment, and `curintpaid(p,j,n,y-12)` computes the interest paid on a loan of $`p` amortized at a rate of `j` per period over `n` periods immediately after the (`y-12`)th payment,

```
curintpaid(p,j,n,y)-curintpaid(p,j,n,y-12)
```

yields the amount of interest paid on a loan of $p amortized at a rate of j per period over n periods between the (y-12)th and yth payment. Consequently, the interest paid and the amount of principal paid over a year can also be computed.

EXAMPLE 14: Suppose that a loan of $45,000 with interest rate of 7% compounded monthly is amortized over a period of 15 years (180 months). What is the principal and interest paid during each of the first five years of the loan?

SOLUTION: We begin by defining the functions annualintpaid and annualprinpaid that calculate the interest and principal paid during the yth year on a loan of $p amortized at a rate of j per period over n periods.

```
> annualintpaid:=(p,j,n,y)->curintpaid(p,j,n,y)-
      curintpaid(p,j,n,y-12):
  annualprinpaid:=(p,j,n,y)->curprinpaid(p,j,n,y)-
      curprinpaid(p,j,n,y-12):
```

We then use these functions along with seq and array to calculate the principal and interest paid during the first five years of the loan. Note that the first column represents the number of years the loan has been held, the second column represents the interest paid on the loan during the year, and the third column represents the amount of the principal that has been paid.

```
> array([seq([t,annualintpaid(45000,0.07/12,15*12,12*t),
      annualprinpaid(45000,0.07/12,15*12,12*t)],t=1..5
```

$$\begin{bmatrix} 1 & 3094.263699 & 1759.40912 \\ 2 & 2967.075879 & 1886.59694 \\ 3 & 2830.693642 & 2022.97918 \\ 4 & 2684.45231 & 2169.22051 \\ 5 & 2527.63920 & 2326.03361 \end{bmatrix}$$

For example, we see that during the third year of the loan, $2830.69 was paid in interest, and $2022.98 was paid on the principal.

■

Application: Financial Planning

We can use many of the functions defined above and in the `finance` package to help make decisions about financial planning.

EXAMPLE 15: Suppose a retiree has \$1,200,000. If she can invest this sum at 7%, compounded annually, what level payment can she withdraw annually for a period of 40 years?

SOLUTION: The answer to the question is the same as the monthly payment necessary to amortize a loan of \$1,200,000 at a rate of 7% compounded annually over a period of 40 years. Thus, we use amort to see that she can withdraw \$90,011 annually for 40 years.

```
> amort(1200000,0.07,40);
                              90010.96665
```

■

EXAMPLE 16: Suppose an investor begins investing at a rate of d dollars per year at an annual rate of j%. Each year the investor increases the amount invested by i%. How much has the investor accumulated after m years?

SOLUTION: The following table illustrates the amount invested each year and the value of the annual investment after m years.

Year	Rate of Increase	Annual Interest	Amount Invested	Value after m Years
0		j%	d	$(1+j\%)^m d$
1	i%	j%	$(1+i\%)d$	$(1+i\%)(1+j\%)^{m-1}d$
2	i%	j%	$(1+i\%)^2 d$	$(1+i\%)^2(1+j\%)^{m-2}d$
3	i%	j%	$(1+i\%)^3 d$	$(1+i\%)^3(1+j\%)^{m-3}d$
k	i%	j%	$(1+i\%)^k d$	$(1+i\%)^k(1+j\%)^{m-k}d$
m	i%	j%	$(1+i\%)^m d$	$(1+i\%)^m d$

It follows that the total value of the amount invested for the first k years after m years is given by:

Year	Total Investment
0	$(1+j\%)^m d$
1	$(1+j\%)^m d+(1+i\%)(1+j\%)^{m-1} d$
2	$(1+j\%)^m d+(1+i\%)(1+j\%)^{m-1} d+(1+i\%)^2(1+j\%)^{m-2} d$
3	$\sum_{n=0}^{3} (1+i\%)^n (1+j\%)^{m-n} d$
k	$\sum_{n=0}^{k} (1+i\%)^n (1+j\%)^{m-n} d$
m	$\sum_{n=0}^{m} (1+i\%)^n (1+j\%)^{m-n} d$

The command `sum` can be used to find a closed form of the sums $\sum_{n=0}^{k} (1+i\%)^n (1+j\%)^{m-n} d$ and $\sum_{n=0}^{m} (1+i\%)^n (1+j\%)^{m-n} d$. Next, we use `sum` to find the sum $\sum_{n=0}^{k} (1+i\%)^n (1+j\%)^{m-n} d$ and name the result `closedone`.

```
> closedone:=sum((1+i)^n*(1+j)^(m-n)*d,n=0..k);
```

$$closedone := -\frac{(1+j)^m d\left(\frac{1+i}{1+j}\right)^{(k+1)}(1+j)}{-i+j} + \frac{(1+j)^m d(1+j)}{-i+j}$$

In the same manner, `sum` is used to find a closed form of $\sum_{n=0}^{m} (1+i\%)(1+j\%)^{m-n} d$, naming the result `closedtwo`.

```
> closedtwo:=sum((1+i)^n*(1+j)^(m-n)*d,n=0..m);
```

$$closedtwo := -\frac{(1+j)^m d\left(\frac{1+i}{1+j}\right)^{(m+1)}(1+j)}{-i+j} + \frac{(1+j)^m d(1+j)}{-i+j}$$

These results are used to define the functions `investment(d,i,j,k,m)` and `investmenttot(d,i,j,m)`, which return the value of the investment after `k` and m years, respectively. In each case, notice that output cells can be edited like any other input or text cell. Consequently, we use editing features to copy and paste the result when we define these functions.

```
> investment:=(d,i,j,k,m)->
      -(1+j)^m*d*((1+i)/(1+j))^(k+1)*(1+j)/
         (-i+j)+(1+j)^m*d*(1+j)/(-i+j):
  investmenttot:=(d,i,j,m)->
      -(1+j)^m*d*((1+i)/(1+j))^(m+1)*(1+j)/
         (-i+j)+(1+j)^m*d*(1+j)/(-i+j):
```

Finally, investment and investmenttot are used to illustrate various financial scenarios. In the first example, investment is used to compute the value after 25 years of investing $6500 the first year and then increasing the amount invested 5% per year for 5, 10, 15, 20, and 25 years assuming a 15% rate of interest on the amount invested. In the second example, investmenttot is used to compute the value after 25 years of investing $6500 the first year and then increasing the amount invested 5% per year for 25 years, assuming various rates of interest.

```
> tvals:=seq(5*t,t=1..5):
      array([seq([t,investment(6500,0.05,0.15,t,25)],
         t=tvals)]);
```

$$\begin{bmatrix} 5 & .1035064556\ 10^7 \\ 10 & .1556077818\ 10^7 \\ 15 & .1886680271\ 10^7 \\ 20 & .2096459926\ 10^7 \\ 25 & .2229572983\ 10^7 \end{bmatrix}$$

```
> ivals:=seq(0.08+0.02*i,i=0..6):
      array([seq([i,investmenttot(6500,0.05,i,25)],
         i=ivals)]);
```

$$\begin{bmatrix} .08 & 832147.4473 \\ .10 & .1087125500\ 10^7 \\ .12 & .1437837092\ 10^7 \\ .14 & .1921899153\ 10^7 \\ .16 & .2591635686\ 10^7 \\ .18 & .3519665383\ 10^7 \\ .20 & .4806524115\ 10^7 \end{bmatrix}$$

■

EXAMPLE 17: Ada is 50 years old and has $500,000 that she can invest at a rate of 7% annually. Furthermore, she wishes to receive a payment of $50,000 the first year. Future annual payments should include cost-of-living adjustments at a rate of 3% annually. Is $500,000 enough to guarantee this amount of annual income if she lives to be 80 years old?

SOLUTION: In this case, we take advantage of the command `growingannuity`, which is contained in the `finance` package. The command `growingannuity(p,i,j,n)` returns the present value of an annuity of `n` periods invested at a rate of `i` per period with initial payment `p`. The payments increase at a rate `j` per period. Thus, entering

```
> with(finance):
  growingannuity(50000,.07,.03,30);
```

```
851421.9086
```

shows that to receive the desired income, she must invest $851,422. On the other hand, using `fsolve` we see that

```
> fsolve(growingannuity(x,0.07,0.03,30)=500000,x);
```

```
29362.64589
```

if she invests her $500,000 at 7% annually, she can receive an initial payment of $29,363 with subsequent 3% annual increases for 30 years. If, on the other hand, she wishes to receive $50,000 her first year and guarantee annual increases of 3% annually forever, `growingperpetuity`, which is also contained in the `finance` package,

```
> growingperpetuity(50000,0.07,0.03);
```

$$.1250000000\ 10^7$$

shows us that she must initially invest $1,250,000. On the other hand, her $500,000 investment is enough to guarantee a first-year income of $20,000, with subsequent annual increases of 3% per year forever.

```
> fsolve(growingperpetuity(x,0.07,0.03)=500000);
```

```
20000.
```

■

We can also investigate certain other problems. For example, using `annuity`, which is also contained in the `finance` package, we see that a 30-year mortgage of $80,000 at 8 1/8% requires an annual payment of $7,190 or approximately $600 per month.

```
fsolve(annuity(x,0.08125,30)=80000);
```

```
7190.169059
```

On the other hand, using growingannuity, we see that if the amount of the payments is increased by 3% each year, the 30-year mortgage is amortized in 17 years!

```
> fsolve(growingannuity(7200,0.08125,0.03,k)=80000);
```

```
17.35372050
```

4.4 Other Applications

We now discuss several other interesting applications that require the manipulation of lists.

Application: Approximating Lists by Functions

Given a set of data points, we frequently want to approximate the data with a particular function. The command

`fit[leastsquare [[x,y]], function, unknown parameters]([xcoords], [ycoords])` fits the list of data points `[xcoords]`, `[ycoords]` using the function `function`, $y=f(x)$, containing parameters `unknown paramaters` to be determined by the method of least-squares. The unknown parameters must appear linearly. If `function` and `unknown parameters` are not specified, a linear fit is found. Note that `fit` is contained in the `stats` package, so it is loaded by entering `with(stats)` before being used.

Recall from Section 4.2 that when we graph lists of points with `plot`, the lists are in the form

$$(x_1,y_1),(x_2,y_2),\ldots,(x_n,y_n).$$

However, when we use fit to find an approximating function, the lists are of the form

$$(x_1,x_2,\ldots,x_n),(y_1,y_2,\ldots,y_n),$$

as indicated above. The following example illustrates how to use `seq` to transform a list from the form $(x_1,y_1),(x_2,y_2),\ldots,(x_n,y_n)$ to the form $(x_1,x_2,\ldots,x_n),(y_1,y_2,\ldots,y_n)$.

EXAMPLE 1: Define `datalist` to be the list of points consisting of (1,1.14479), (2,1.5767), (3,2.68572), (4,2.5199), (5,3.58019), (6,3.84176), (7,4.09957), (8,5.09166), (9,5.98085), (10,6.49449), and (11,6.12113). (a) Find a linear approximation of the points in `datalist`. (b) Find a fourth-degree polynomial approximation of the points in `datalist`.

SOLUTION: We begin by loading the `stats` package and defining `datalist`.

```
> with(stats);
```

[anova, describe, fit, importdata, random, statevalf, statplots, transform]

```
> datalist:=[[1,1.14479],[2,1.5767],[3,2.68572],
     [4,2.5199],[5,3.58019],[6,3.84176],[7,4.09957],
     [8,5.09166],[9,5.98085],[10,6.49449],[11,6.12113]]:
```

Next, we transform `datalist` from a list of the form $(x_1,y_1),(x_2,y_2),\ldots,(x_n,y_n)$ to a list of the form $(x_1,x_2,\ldots,x_n),(y_1,y_2,\ldots,y_n)$, with `seq`. `nops` returns the number of elements in a list.

```
> datalist2:=[[seq(datalist[i,1],i=1..nops(datalist))],
     [seq(datalist[i,2],i=1..nops(datalist))]];
```

$$datalist2 := [[1, 2, 3, 4, 5, 6, 7, 8, 9, 10, 11], [1.14479, 1.5767, 2.68572, 2.5199, 3.58019, 3.84176, 4.09957, 5.09166, 5.98085, 6.49449, 6.12113]]$$

We then use `fit` to find the linear least-squares function that approximates the data.

```
> y:='y':
  fit1:=fit[leastsquare[[x,y]]](datalist2);
```

$$fit1 := y = .6432790909 + .5463740909\, x$$

Note that the same results would have been obtained with the command

```
fit[leastsquare[[x,y],y=a*x+b,{a,b}]](datalist2).
```

We then use `assign` to name `y` the result obtained in `fit1`. The approximating function obtained via the least-squares method with `fit` is plotted along with the data points. Notice that many of the data points are not very close to the approximating function.

```
> assign(fit1):
  with(plots):
  p1:=plot(y,x=0..12):
  p2:=plot(datalist,style=POINT):
  display({p1,p2});
```

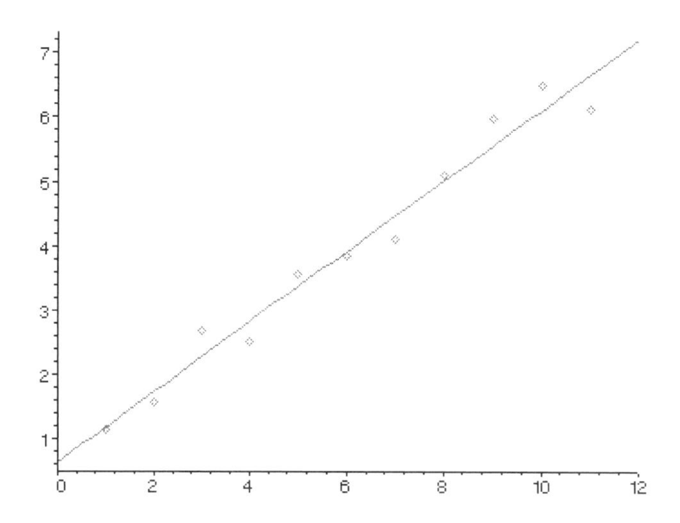

A better approximation is obtained using a polynomial of higher degree (4).

```
> y:='y':
  fit2:=fit[leastsquare[[x,y],y=a*x^4+b*x^3+c*x^2+d*x+e,
       {a,b,c,d,e}]](datalist2);
```

$$fit2 := y = -.003109847999\,x^4 + .07092011267\,x^3 - .5322814690\,x^2 + 2.027437282\,x - .5413294697$$

To check its accuracy, this second approximation is graphed together with the data points.

```
> assign(fit2):
  p3:=plot(y,x=0..12):
  display({p2,p3});
```

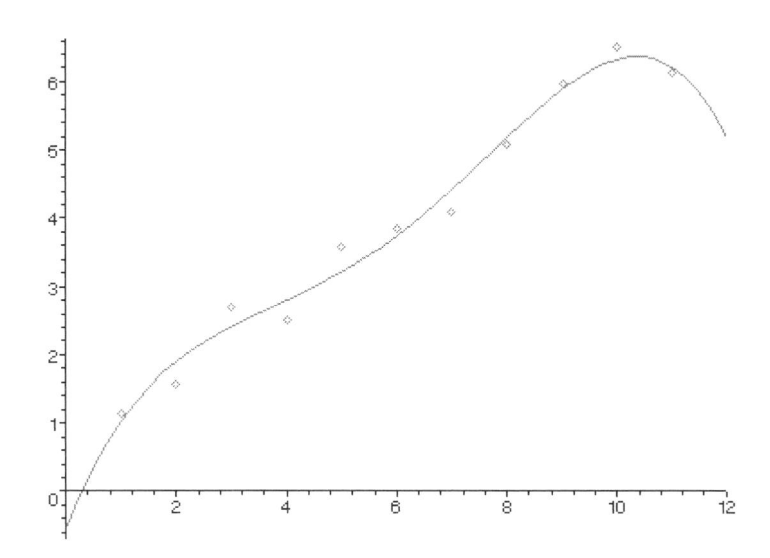

■

`interp(xcoords,ycoords,x)` fits the list of data points `data` with an n–1 degree polynomial in the variable `x`.

EXAMPLE 2: The following table shows the average percentage of petroleum products imported to the United States for certain years. (a) Graph the points corresponding to the data in the table, and connect the consecutive points with line segments. (b) Use `interp` to find an interpolating polynomial that approximates the data in the table. (c) Find a fourth-degree polynomial approximation of the data in the table. (d) Find a trigonometric approximation of the data in the table.

Year	Percent
1973	35
1974	35
1975	36
1976	41
1977	47
1978	42
1979	43

Year	Percent
1983	28
1984	30
1985	27
1986	33
1987	35
1988	38
1989	42

Continued

Year	Percent
1980	37
1981	34
1982	28

Year	Percent
1990	42
1991	40

SOLUTION: We begin by defining `dataset` to be the set of ordered pairs represented in the table: The x-coordinate of each point represents the number of years past 1900, and the y-coordinate represents the percentage of petroleum products imported to the United States.

```
> dataset:=[[73,35],[74,35],[75,36],[76,41],[77,47],
      [78,42],[79,43],[80,37],[81,34],[82,28],[83,28],
      [84,30],[85,27],[86,33],[87,35],[88,38],[89,42],
      [90,42],[91,40]]:
```

Next, we transform dataset from a list of the form $(x_1,y_1),(x_2,y_2),\ldots,(x_n,y_n)$ to a list of the form $(x_1,x_2,\ldots,x_n),(y_1,y_2,\ldots,y_n)$ with `seq`.

```
> dataset2:=[[seq(dataset[i,1],i=1..nops(dataset))],
      [seq(dataset[i,2],i=1..nops(dataset))]];
```

Then, `interp` is used to find a polynomial approximation of the data in the table.

```
> y:='y':
  fit1:=interp(dataset2[1],dataset2[2],x);
```

$$
\begin{aligned}
fit1 := &-\frac{8901154279393747643198828831}{120120}x+\frac{43232709577448294198481 71}{603542016000}x^{8}\\
&+\frac{37044510282008835960660324027 97}{482431950}x^{2}-\frac{75404011043895143 23}{804722688000}x^{11}\\
&+\frac{481141615128786053}{7242504192000}x^{12}+\frac{373836119}{31384184832000}x^{16}-\frac{711649}{41845579776000}x^{17}\\
&-\frac{41241822239551}{110702592000}x^{13}+\frac{101442473918507}{62768369664000}x^{14}-\frac{6083042893}{1162377216000}x^{15}\\
&+\frac{73387}{6402373705728000}x^{18}+3375889156787146497746 71\\
&-\frac{906481906162650160371686834132 3}{18162144000}x^{3}\\
&+\frac{745480258740212832390851369598 1}{326918592000}x^{4}\\
&-\frac{203387957734798634635203175776 1}{2615348736000}x^{5}\\
&+\frac{48291295759359883750572754739 9}{23538138624000}x^{6}-\frac{27342153315368424111375 07}{6386688000}x^{7}\\
&-\frac{31480890189278780466 31}{32514048000}x^{9}+\frac{93032613055846958008 7}{877879296000}x^{10}
\end{aligned}
$$

We then graph `fit1` along with the data in the table for the years corresponding to 1973 to 1991. Notice that the interpolating polynomial oscillates wildly.

```
> with(plots):
  p1:=plot(dataset,style=POINT,color=BLACK):
  p2:=plot(fit1,x=73..91,y=0..60):
```

```
display({p2,p1});
```

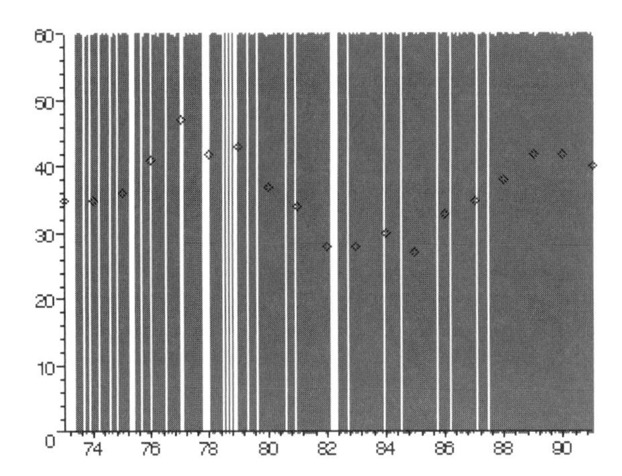

In fact, it may be difficult to believe that the interpolating polynomial agrees with the data exactly, so we use `seq` and `subs` to substitute the x-coordinates into the polynomial to confirm that it does agree exactly.

```
> seq(subs(x=t,fit1),t=dataset2[1]);
```

$$35, 35, 36, 41, 47, 42, 43, 37, 34, 28, 28, 30, 27, 33, 35, 38, 42, 42, 40$$

To find a polynomial that approximates the data but does not oscillate wildly, we use `fit`. Again, we graph the fit and display the graph of the fit and the data simultaneously. In this case, the fit does not identically agree with the data, but it also does not oscillate wildly.

```
> with(stats):
  y:='y':
  fit2:=fit[leastsquare[[x,y],y=a*x^4+b*x^3+c*x^2+d*x+e,
      {a,b,c,d,e}]](dataset2);
```

$$fit2 := y = -\frac{2659}{653752}x^4 + \frac{4002953}{2941884}x^3 - \frac{333893095}{1961256}x^2 + \frac{27775296511}{2941884}x - \frac{122751569}{627}$$

```
> assign(fit2):
  p3:=plot(y,x=73..91):
```

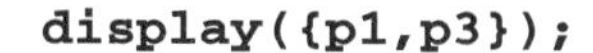

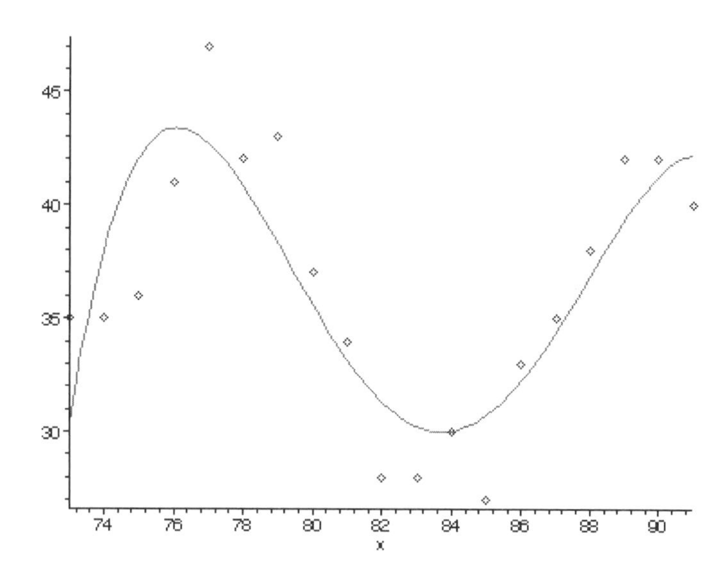

In addition to curve-fitting with polynomials, Maple can also fit the data with trigonometric functions. In this case, we use `fit` to find an approximation of the data of the form $y=a+b\sin x+c\sin(x/2)+d\cos x+e\cos(x/2)$. As in the previous two cases, we graph the fit and display the graph of the fit and the data simultaneously.

```
> y:='y':
  fit3:=fit[leastsquare[[x,y],y=a+b*sin(x)+c*sin(x/2)+
      d*cos(x)+e*cos(x/2),{a,b,c,d,e}]](evalf(dataset2));
```

$$fit3 := y = 35.36378125 + .1147371434\sin(x) + 6.159409185\sin\left(\frac{1}{2}x\right) - .8594797302\cos(x) + 4.267714122\cos\left(\frac{1}{2}x\right)$$

```
> assign(fit3):
  p4:=plot(y,x=73..91):
  display({p1,p4});
```

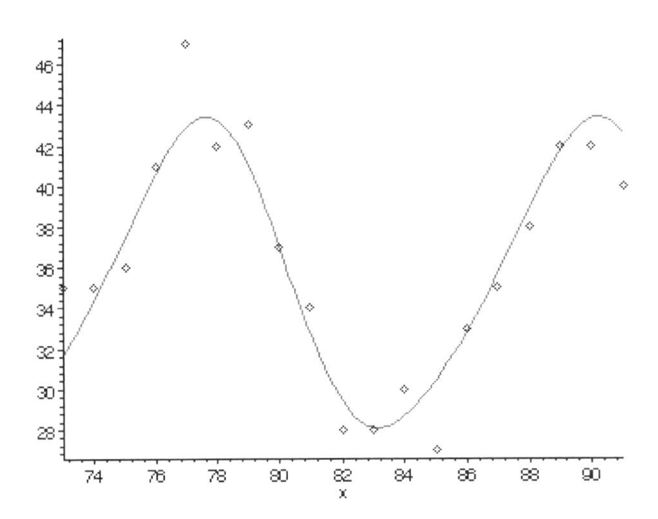

■

Application: Introduction to Fourier Series

Many problems in applied mathematics are solved through the use of Fourier series. Maple assists in the computation of these series in several ways. First, we restate the following standard definitions. The **Fourier series of a periodic function *f*(*x*) with period 2L** is the trigonometric series

$$a_0 + \sum_{n=1}^{\infty} \left[a_n \cos\left(\frac{n\pi x}{L}\right) + b_n \sin\left(\frac{n\pi x}{L}\right) \right],$$

where $a_0 + \frac{1}{2L}\int_{-L}^{L} f(x)dx$, $a_n = \frac{1}{L}\int_{-L}^{L} f(x)\cos\left(\frac{n\pi x}{L}\right)dx$, and $b_n = \frac{1}{L}\int_{-L}^{L} f(x)\sin\left(\frac{n\pi x}{L}\right)dx$. The ***k*th term of the Fourier series** $a_0 + \sum_{n=1}^{\infty} \left[a_n \cos\left(\frac{n\pi x}{L}\right) + b_n \sin\left(\frac{n\pi x}{L}\right) \right]$ is

$$a_k \cos\left(\frac{k\pi x}{L}\right) + b_k \sin\left(\frac{k\pi x}{L}\right).$$

The ***k*th partial sum of the Fourier series** $a_0 + \sum_{n=1}^{\infty} \left[a_n \cos\left(\frac{n\pi x}{L}\right) + b_n \sin\left(\frac{n\pi x}{L}\right) \right]$ is

$$a_0 + \sum_{n=1}^{k} \left[a_n \cos\left(\frac{n\pi x}{L}\right) + b_n \sin\left(\frac{n\pi x}{L}\right) \right].$$

It is a well-known theorem that if $f(x)$ is a periodic function with period $2L$ and $f'(x)$ is continuous on $[-L,L]$ except at finitely many points, then at each point x the Fourier series corresponding to f converges and

$$a_0 + \sum_{n=1}^{\infty}\left[a_n \cos\left(\frac{n\pi x}{L}\right) + b_n \sin\left(\frac{n\pi x}{L}\right)\right] = \frac{\lim_{y \to x+} f(y) + \lim_{y \to x-} f(y)}{2}.$$

In fact, if the series $\sum_{n=1}^{\infty} (|a_n| + |b_n|)$ converges, then the Fourier series

$$a_0 + \sum_{n=1}^{\infty}\left[a_n \cos\left(\frac{n\pi x}{L}\right) + b_n \sin\left(\frac{n\pi x}{L}\right)\right]$$

converges uniformly on $\Re$.

EXAMPLE 3: Let $f(x) = \begin{cases} 1, 0 \le x < 1 \\ -x, -1 \le x < 0 \\ f(x-2), x \ge 1 \end{cases}$. Compute and graph the first few partial sums of the Fourier series for f.

SOLUTION: We begin by clearing all prior definitions of f. We then define the piecewise-defined function using a procedure. Notice that in this procedure, $f(x)$ is assigned the value of 1 if $0 \le x < 1$, $-x$ if $-1 \le x < 0$, and $f(x-2)$ if $x \ge 1$. Note that `elif`, which is used to avoid repeated use of `fi`, means "else if." f is then plotted on the interval [–1,5], using 150 points. The function f and the variable x are contained in single quotation marks ('), so that a delayed evaluation takes place. Of course, since f is defined as a procedure, operator notation may also be used to graph f.

```
> f:='f':
  f:=proc(x) if x>=0 and x<1 then 1
      elif x<0 and x>=-1 then -x
      elif x>=1 then f(x-2) fi end:
> plot('f(x)','x'=-1..5,numpoints=150);
```

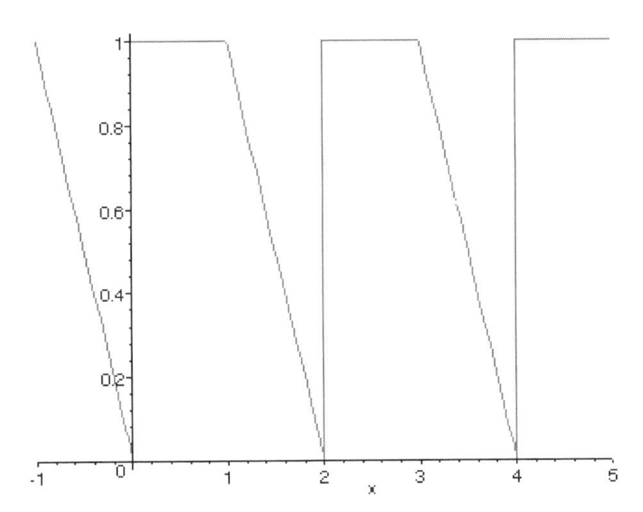

To assist in future calculations, we next define the two pieces of f in `f1` and `f2`.

```
> f1:=x->-x:
  f2:=x->1:
```

The table `a` is then defined to store the values of a_0, a_1, a_2,... . The integral to determine the coefficient a_0 is defined in two parts by integrating `f1` and `f2` over the intervals [–1,0] and [0,1], respectively. `evalf` is used to yield a numerical approximation of these values.

```
> a:='a':
  a:=table():
  a[0]:=evalf(1/2*(int(f1(x),x=-1..0)+
            int(f2(x),x=0..1))):
```

We use a `for` loop to determine a_1 through a_{12}. As with a_0, integration is performed over the two intervals [–1,0] and [0,1]. Notice that the variable x is contained in single quotation marks so that the integrals are not evaluated until they are given a value of i.

```
> j:='j':
  for j from 1 to 12 do
      a[j]:=evalf(Int(f1(x)*cos(j*Pi*x),'x'=-1..0)+
              f2(x)*Int(cos(j*Pi*x),'x'=0..1)) od:
```

A similar loop is then used to compute the coefficients b_1 through b_{12}.

```
> b:='b':
  b:=table():
```

```
j:='j':
for j from 1 to 12 do
    b[j]:=evalf(Int(f1(x)*sin(j*Pi*x),'x'=-1..0)+
          f2(x)*Int(sin(j*Pi*x),'x'=0..1)) od:
```

We now display the coefficients computed above. The elements in the second column of the array represent the a_i's, and the third column represents the b_i's .

```
> array([seq([i,a[i],b[i]],i=1..12)]);
```

$$\begin{bmatrix} 1 & -.2026423673 & .3183098862 \\ 2 & .6579099405\ 10^{-16} & .1591549431 \\ 3 & -.02251581859 & .1061032954 \\ 4 & -.3289549703\ 10^{-15} & .07957747155 \\ 5 & -.008105694691 & .06366197726 \\ 6 & -.1315819881\ 10^{-15} & .05305164770 \\ 7 & -.004135558516 & .04547284089 \\ 8 & .1151342396\ 10^{-15} & .03978873577 \\ 9 & -.002501757621 & .03536776513 \\ 10 & .8223874256\ 10^{-17} & .03183098862 \\ 11 & -.001674730308 & .02893726238 \\ 12 & .9868649108\ 10^{-16} & .02652582385 \end{bmatrix}$$

The kth term of the Fourier series, $a_k\cos(k\pi x)+b_k\sin(k\pi x)$, is defined in `kterm`. Hence, the nth partial sum of the series is given by

$$a_0 + \sum_{k=1}^{n} a_k\cos(k\pi x) + b_k\sin(k\pi x) = \mathtt{a[0]} + \sum_{k=1}^{n} \mathtt{fs[k,x]},$$

which is defined in `fapprox`. We illustrate the use of `fapprox` by finding `fapprox(2)`.

```
> i:='i':k:='k':
  kterm:=k->a[k]*cos(k*Pi*x)+b[k]*sin(k*Pi*x):
  fapprox:=n->a[0]+sum(kterm(k),i=1..n):
> fapprox(2);
```

$$.7500000000 - .2026423673\cos(\pi x) + .3183098862\sin(\pi x) \\ + .6579099405\ 10^{-16}\cos(2\pi x) + .1591549431\sin(2\pi x)$$

To see how the Fourier series approximates the periodic function, we plot the function together with the Fourier approximation `fapprox(n)` for $n = 2, 4, 6, 8, 10, 12$, using `animate`. Again, we make use of the delayed evaluation of f in the `animate` command.

```
> with(plots):
  A:=animate({'f(x)',fapprox(n)},x=-1..5,n=2..12,
      frames=6,color=black,tickmarks=[2,4]):
  display(A);
```

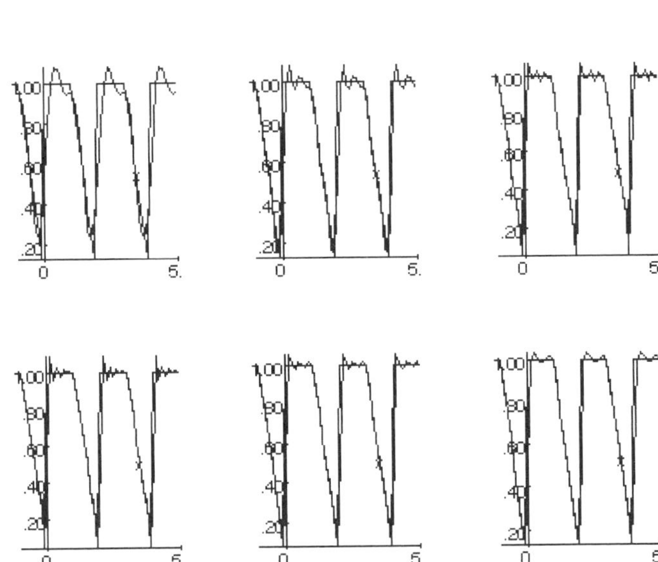

■

Application: The One-Dimensional Heat Equation

A typical problem in applied mathematics that involves the use of Fourier series is that of the one-dimensional heat equation. The boundary value problem that describes the temperature in a uniform rod with insulated surface is given by:

1. $k\frac{\partial u}{\partial x^2} = \frac{\partial u}{\partial t}, 0 < x < a, t > 0$

2. $u(0, t) = T_0, t > 0$

3. $u(a, t) = T_a, t > 0$

4. $u(x, 0) = f(x), 0 < x < a$

In this case, the rod has "fixed end temperatures" at $x=0$ and $x=a$. $f(x)$ is the initial temperature distribution. The solution to the problem is

$$u(x,t) = \underbrace{T_0 + \frac{1}{a}x(T_a - T_0)}_{v(x)} + \sum_{n=1}^{\infty} b_n \sin(\lambda_n x) e^{-\lambda_n^2 kt},$$

where $\lambda_n = n\pi/a$ and $b_n = \frac{2}{a}\int (f(x) - v(x))\sin\left(\frac{n\pi a}{a}\right)dx$, and is obtained through separation of variables techniques. The coefficient b_n in the solution, $u(x,t)$, is the Fourier series coefficient b_n of the function $f(x)-v(x)$, where $v(x)$ is the steady-state temperature.

EXAMPLE 4: Solve

$$\begin{cases} \frac{\partial u}{\partial x^2} = \frac{\partial u}{\partial t}, 0 < x < a, t > 0 \\ u(0, t) = 10, u(1, t) = 10, t > 0 \\ u(x, 0) = 10 + 20\sin^2 \pi x \end{cases}.$$

SOLUTION: In this case, $a=1$ and $k=1$. The fixed end temperatures are $T_0=T_a=10$, and the initial heat distribution is $f(x)=10+20\sin^2\pi x$. The steady-state temperature is $v(x)=10$. The function $f(x)$ is defined and plotted. Also, the steady-state temperature, $v(x)$, and the eigenvalue are defined.

```
> f:='f':
  f:=x->10+20*sin(Pi*x)^2:
  plot(f(x),x=0..1,y=0..30,color=BLACK);
```

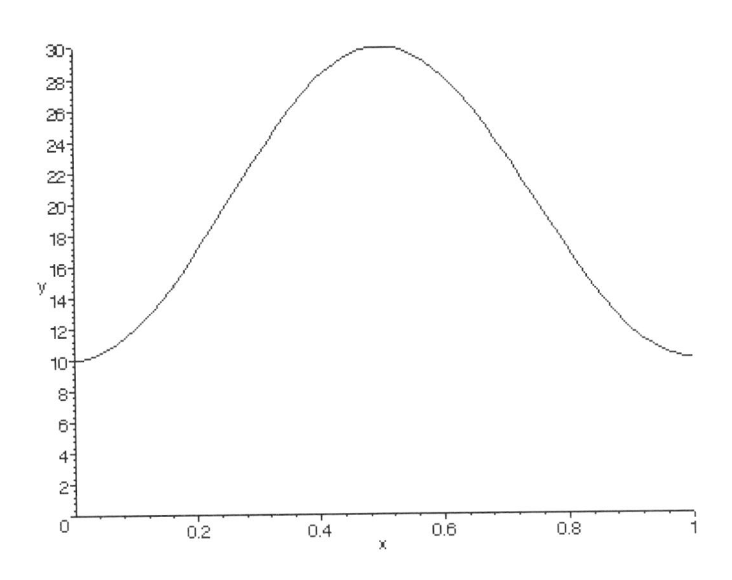


```
> v:=x->10:
  lambda:=n->n*Pi:
```

The procedure b is then defined to compute the solution series coefficients b_n. Notice that because `option remember` is used in the definition, the coefficients are stored as they are calculated so that they will not have to be recomputed when they are used at a later time in the definition of the solution. The coefficient b_1 is then determined.

```
> b:=proc(n) option remember;
      evalf(2*Int((f(x)-v(x))*sin(n*Pi*x),x=0..1))
      end:

> b(1);
```

16.97652726

We define the series solution to this problem in a manner similar to the Fourier series in Example 3. However, in this case, we define the function recursively. Let $S_m = b_m \sin(\lambda_m x) e^{-\lambda_m^2 t}$. Then, the desired solution $u(x,t)$ is given by

$$u(x,t) = v(x) + \sum_{m=1}^{\infty} S_m.$$

Let $u(x, t, n) = v(x) + \sum_{m=1}^{n} S_m$. Notice that $u(x,t,n) = u(x,t,n-1) + S_n$. Consequently, approximations of the solution to the heat equation are obtained recursively, taking advantage of Maple's ability to compute recursively. The solution is first defined for n=1 by `u(1)`. Subsequent partial sums, `u(n)`, are obtained by adding the nth term of the series, $S_n = b_n \sin(\lambda_n x) e^{-\lambda_n^2 t}$, to `u(n-1)`.

```
> u:='u':
  u:=proc(n) option remember;
      u(n-1)+b(n)*sin(lambda(n)*x)*
         exp(-lambda(n)^2*t)
      end:
  u(1):=v(x)+b(1)*sin(lambda(1)*x)*
         exp(-lambda(1)^2*t):
> u(2);
```

$$10 + 16.97652726 \sin(\pi x)\, \mathbf{e}^{(-\pi 2 t)}$$
$$-.8353273230\ 10^{-15} \sin(2\pi x) \mathbf{e}^{-4\pi^2 t}$$

Note that in order to compute u_8, Maple must first compute u_3, u_4, ..., u_7 (u_2 is already known from the previous calculation). However, once u_8 has been calculated, it will not have to be recomputed, due to the fact that `option remember` is used to define `u`.

```
> u(8);
```

$$10 + 16.97652726 \sin(\pi x)\, \mathbf{e}^{(-\pi^2 t)} - .8353273230\ 10^{-15} \sin(2\pi x)\, \mathbf{e}^{(-4\pi^2 t)}$$
$$- 3.395305452 \sin(3\pi x)\, \mathbf{e}^{(-9\pi^2 t)} + .1530231911\ 10^{-14} \sin(4\pi x)\, \mathbf{e}^{(-16\pi^2 t)}$$
$$- .4850436360 \sin(5\pi x)\, \mathbf{e}^{(-25\pi^2 t)} - .3640831544\ 10^{-14} \sin(6\pi x)\, \mathbf{e}^{(-36\pi^2 t)}$$
$$- .1616812120 \sin(7\pi x)\, \mathbf{e}^{(-49\pi^2 t)} + .9559705176\ 10^{-15} \sin(8\pi x)\, \mathbf{e}^{(-64\pi^2 t)}$$

Recall that the series solution given in `u` that is a function of x and t represents the temperature distribution in a rod. We can plot the solution over time by substituting values for t to view the behavior of the temperature along the rod. We do this next by plotting the solution for t=0 to t=1, using increments of 1/12. The nine graphs that result from the `animation` command are shown as a graphics array with `display`. Notice that the solution approaches the steady-state temperature $v(x) = 10$ as t increases.

```
> j:='j':
  with(plots):
  A:=animate(subs(t=j,u(8)),x=0..1,j=0..2,
       frames=12,tickmarks=[2,2],color=BLACK,
          view=[0..1,0..30]):
  display(A);
```

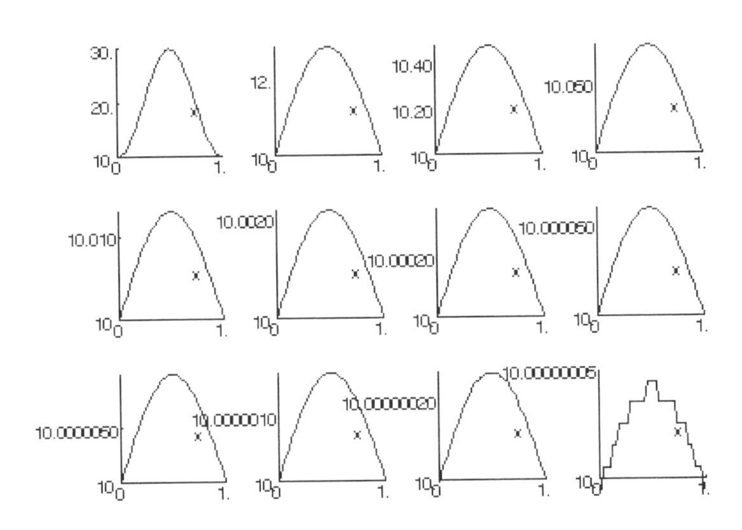

We show one frame from the animation that results from the following command:

```
> animate(subs(t=j,u(8)),x=0..1,j=0..2,frames=30,
   tickmarks=[2,2],color=BLACK);
```

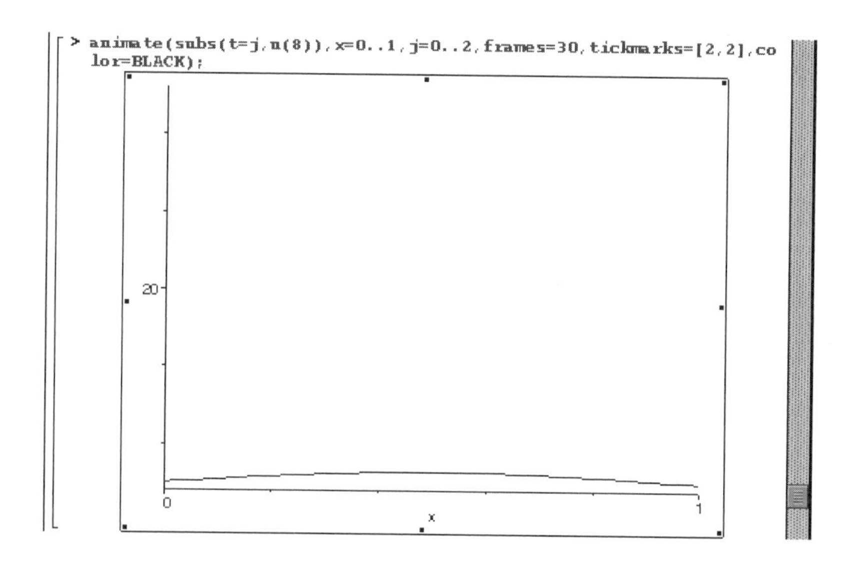
> animate(subs(t=j,u(8)),x=0..1,j=0..2,frames=30,tickmarks=[2,2],co
lor=BLACK);
20
0
1
x

■

CHAPTER 5

Nested Lists: Matrices and Vectors

5.1 Nested Lists: Introduction to Matrices, Vectors, and Matrix Operations

Defining Matrices and Vectors

Matrix algebra can be performed with Maple V in a variety of ways. Before the operations involved in matrix algebra are introduced, the method by which a matrix is entered must be discussed. In Maple, a **matrix** is simply a list of lists in which each list represents a row of the matrix. Therefore, the $m \times n$ matrix

$$\mathbf{A} = (a_{ij}) = \begin{pmatrix} a_{11} & a_{12} & a_{13} & \cdots & a_{1n} \\ a_{21} & a_{22} & a_{23} & \cdots & a_{2n} \\ a_{31} & a_{32} & a_{33} & \cdots & a_{3n} \\ \vdots & \vdots & \vdots & & \vdots \\ a_{m1} & a_{m2} & a_{m3} & \cdots & a_{mn} \end{pmatrix}$$

is entered in the following manner:

```
A:=array(1..n,1..m),
```

if each entry is not assigned a value immediately, or with either

```
A:=array(1..n,1..m,list_of_rows) or A:=array(list_of_rows),
```

where `list_of_rows` is a list of lists in which the ith list in `list_of_rows` corresponds to the entries in the ith row of the matrix **A**, if the entries of **A** are assigned a value immediately.

EXAMPLE 1: Define the matrices

$$\mathbf{A} = (a_{ij}) = \begin{pmatrix} a_{11} & a_{12} & a_{13} \\ a_{21} & a_{22} & a_{23} \\ a_{31} & a_{32} & a_{33} \end{pmatrix} \text{ and } \begin{pmatrix} b_{11} & b_{12} & b_{13} & b_{14} \\ b_{21} & b_{22} & b_{23} & b_{24} \end{pmatrix}.$$

SOLUTION: After first clearing prior definitions of `A`, if any, we then use `array` to define `A` to be a three-by-three array. `print` is used to view `A` in a standard row-and-column form.

```
> A:='A':
  A:=array(1..3,1..3);
```

$$A := \text{array}(1..3, 1..3, [\ ])$$

```
> print(A);
```

$$\begin{bmatrix} A_{1,1} & A_{1,2} & A_{1,3} \\ A_{2,1} & A_{2,2} & A_{2,3} \\ A_{3,1} & A_{3,2} & A_{3,3} \end{bmatrix}$$

In the same manner, we define `B` to be the two-by-four matrix $\begin{pmatrix} b_{11} & b_{12} & b_{13} & b_{14} \\ b_{21} & b_{22} & b_{23} & b_{24} \end{pmatrix}$.

```
> B:=array(1..2,1..4):
  print(B);
```

$$\begin{bmatrix} B_{1,1} & B_{1,2} & B_{1,3} & B_{1,4} \\ B_{2,1} & B_{2,2} & B_{2,3} & B_{2,4} \end{bmatrix}$$

■

Of course, matrices can be defined by entering the definition directly.

EXAMPLE 2: Define **A** to be the matrix $\begin{pmatrix} 10 & -6 & -9 \\ 6 & -5 & -7 \\ -10 & 9 & 12 \end{pmatrix}$ and **C** to be the 3×4 matrix (c_{ij}), where c_{ij}, the entry in the ith row and jth column of **C**, is the numerical value of $\cos(j_2 - i_2)\sin(i_2 - j_2)$.

SOLUTION: We define **A** directly.

```
> A:='A':
  A:=array([[10,-6,-9],[6,-5,-7],[-10,9,12]]);
```

$$A := \begin{bmatrix} 10 & -6 & -9 \\ 6 & -5 & -7 \\ -10 & 9 & 12 \end{bmatrix}$$

Alternatively, we can take advantage of the **Matrix Palette**.

Clicking on the button inserts a 3×3 template at the location of the cursor.

```
[ > A:=matrix([[%?, %?, %?], [%?, %?, %?], [%?, %?, %?]])
```

We then replace each `%?` with the desired value and enter the result.

```
> A:=matrix([[10,-6,-9],[6,-5,-7],[-10,9,12]]);
```

$$A := \begin{bmatrix} 10 & -6 & -9 \\ 6 & -5 & -7 \\ -10 & 9 & 12 \end{bmatrix}$$

To define **C**, we first clear all prior definitions of `C`, `i`, and `j` and then first define `C` to be a three-by-four array. The entries of `C` are then assigned using a `for` loop. The matrix `C` is then displayed with `print`.

```
> C:='C':i:='i':j:='j':
  C:=array(1..3,1..4):
  for i to 3 do
  for j to 4 do
      C[i,j]:=evalf(cos(j^2-i^2)*sin(i^2-j^2))
  od od:
  print(C);
```

$$\begin{bmatrix} 0 & .1397077491 & .1439516583 & .4940158121 \\ -.1397077491 & 0 & .2720105555 & .4527891810 \\ -.1439516583 & -.2720105555 & 0 & -.4953036778 \end{bmatrix}$$

Note that square brackets (`[...]`) are used to extract elements of matrices. Hence, the entry in the second row and first column of **A** is extracted with `A[2,1]`; the entry in the third row and second column of **C** is extracted with `C[3,2]`.

```
> A[2,1],C[3,2];
```

$$6, -2720105555$$

■

EXAMPLE 3: Define the matrix $\begin{pmatrix} 1 & 0 & 0 \\ 0 & 1 & 0 \\ 0 & 0 & 1 \end{pmatrix}$.

SOLUTION: Generally, the command

```
A:=array(identity,1..n,1..n);
```

defines **A** to be the $n \times n$ identity matrix, the matrix with 1's down the diagonal and 0's elsewhere. Thus, entering

```
> A:=array(identity,1..3,1..3);
  print(A);
```

$$\begin{bmatrix} 1 & 0 & 0 \\ 0 & 1 & 0 \\ 0 & 0 & 1 \end{bmatrix}$$

returns the 3×3 identity matrix.

■

The `linalg` package contains a variety functions to manipulate matrices and vectors. A list of the commands available in the package is obtained by entering `?linalg`.

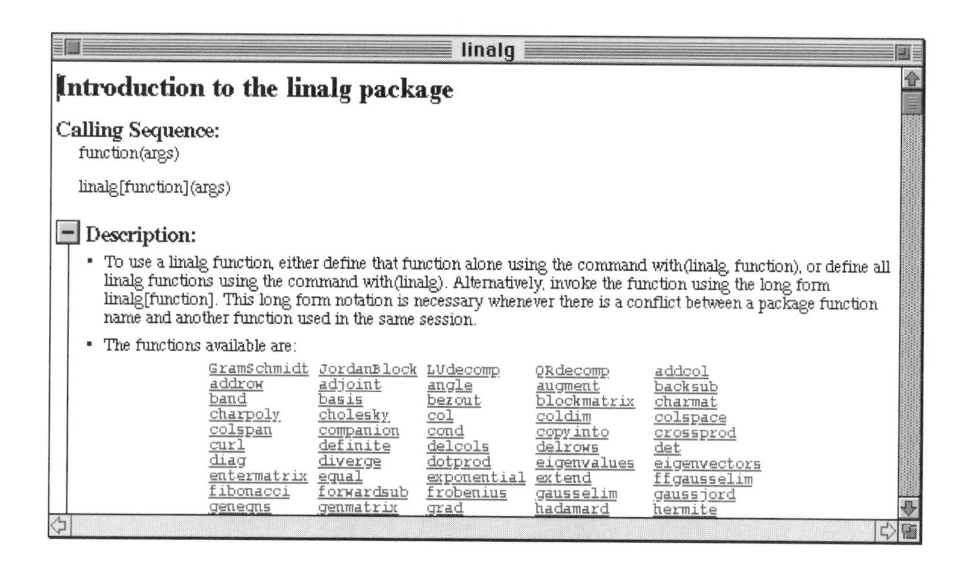

linalg

Introduction to the linalg package

Calling Sequence:

function(args)

linalg[function](args)

Description:

- To use a linalg function, either define that function alone using the command with(linalg, function), or define all linalg functions using the command with(linalg). Alternatively, invoke the function using the long form linalg[function]. This long form notation is necessary whenever there is a conflict between a package function name and another function used in the same session.
- The functions available are:

GramSchmidt	JordanBlock	LUdecomp	QRdecomp	addcol
addrow	adjoint	angle	augment	backsub
band	basis	bezout	blockmatrix	charmat
charpoly	cholesky	col	coldim	colspace
colspan	companion	cond	copyinto	crossprod
curl	definite	delcols	delrows	det
diag	diverge	dotprod	eigenvalues	eigenvectors
entermatrix	equal	exponential	extend	ffgausselim
fibonacci	forwardsub	frobenius	gausselim	gaussjord
geneqns	genmatrix	grad	hadamard	hermite

EXAMPLE 4: Generate a 2×3 matrix where the entries are randomly chosen integers between –4 and 4.

SOLUTION: In this case, we take advantage of the `randmatrix` command, which is contained in the `linalg` package. After we load the linalg package,

```
>with(linalg);
```

[*BlockDiagonal, GramSchmidt, JordanBlock, LUdecomp, QRdecomp,Wronskian, addcol, addrow, adj, adjoint, angle, augment, backsub, band,basis, bezout, blockmatrix, charmat, charpoly, cholesky, col, coldim, colspace,colspan, companion, concat, cond, copyinto, crossprod, curl, definite, delcols,delrows, det, diag, diverge, dotprod, eigenvals, eigenvalues, eigenvectors,eigenvects, entermatrix, equal, exponential, extend, ffgausselim, fibonacci,forwardsub, frobenius, gausselim, gaussjord, geneqns, genmatrix, grad,hadamard, hermite, hessian, hilbert, htranspose, ihermite, indexfunc,innerprod, intbasis, inverse, ismith, issimilar, iszero, jacobian, jordan, kernel,laplacian, leastsqrs, linsolve, matadd, matrix, minor, minpoly, mulcol, mulrow,multiply, norm, normalize, nullspace, orthog, permanent, pivot, potential,randmatrix, randvector, rank, ratform, row, rowdim, rowspace, rowspan, rref,scalarmul, singularvals, smith, stackmatrix, submatrix, subvector, sumbasis,swapcol, swaprow, sylvester, toeplits, trace, transpose, vandermonde,vecpotent, vectdim, vector, wronskian*]

entering

```
> randmatrix(2,3);
```

$$\begin{bmatrix} 45 & -8 & -93 \\ 92 & 43 & -62 \end{bmatrix}$$

generates a 2×3 matrix where the entries are randomly chosen integers between –99 and 99. To generate a 2×3 matrix where the entries are randomly chosen integers between –4 and 4, we define the function `mod5`. Given no argument, `mod5()` returns a randomly chosen integer between –4 and 4.

```
> mod5:=proc() (-1)^rand()*(rand() mod 5) end:
```

For example, entering

```
> seq(mod5(),i=1..10);
```

```
-4, 0, 2, -2, 0, -3, -3, -3, -2, -4
```

generates a list of ten randomly chosen integers between –4 and 4. Thus, entering

```
> randmatrix(2, 2, entries = mod5);
```

$$\begin{bmatrix} 1 & 3 \\ 4 & -1 \end{bmatrix}$$

2×3 matrix where the entries are randomly chosen integers between –4 and 4.

■

In Maple, a **vector** is a one-dimensional array, and thus, it is entered in the same manner as matrices with `array` or with the command `vector`, which is contained in the `linalg` package.

EXAMPLE 5: Define $\mathbf{w} = \begin{pmatrix} -4 \\ -5 \\ 2 \end{pmatrix}$, $v = (v_1 \ v_2 \ v_3 \ v_4)$, and $\mathbf{0} = (0 \ 0 \ 0 \ 0 \ 0)$.

SOLUTION: We use array to define **w** and **v**.

```
> w:=array([[-4],[-5],[2]]);
```

$$w = \begin{bmatrix} -4 \\ -5 \\ 2 \end{bmatrix}$$

```
> v:=array(1..4):
  print(v);
```

$$[v_1, v_2, v_3, v_4]$$

On the other hand, we use `vector` to form $\mathbf{0} = (0 \ 0 \ 0 \ 0 \ 0)$.

```
> with(linalg):
  zerovec:=vector(5,0);
```

$$zerovec = [0, 0, 0, 0, 0]$$

Extracting Elements of Matrices

The entry in the ith row and jth column of the $m \times n$ matrix

$$\mathbf{A} = (a_{ij}) = \begin{pmatrix} a_{11} & a_{12} & a_{13} & \cdots & a_{1n} \\ a_{21} & a_{22} & a_{23} & \cdots & a_{2n} \\ a_{31} & a_{32} & a_{33} & \cdots & a_{3n} \\ \vdots & \vdots & \vdots & & \vdots \\ a_{m1} & a_{m2} & a_{m3} & \cdots & a_{mn} \end{pmatrix}$$

is extracted from **A** with `A[i,j]`. In addition, rows and columns are extracted from **A** using the `row` and `col` commands contained in the `linalg` package.

EXAMPLE 6: Define $\mathbf{A}=\begin{pmatrix} 10 & -6 & -9 \\ 6 & -5 & -7 \\ -10 & 9 & 12 \end{pmatrix}$. (a) Extract the element in the first row and third column of **A**. (b) Extract the third row of **A**. (c) Extract the second column of **A**.

SOLUTION: First, define **A**.

```
> A:=array([[10,-6,-9],[6,-5,-7],[-10,9,12]]);
```

$$A := \begin{bmatrix} 10 & -6 & -9 \\ 6 & -5 & -7 \\ -10 & 9 & 12 \end{bmatrix}$$

The element in the first row and third column of **A** is extracted with `A[1,3]`.

```
> A[1,3];
```

$$-9$$

To use the `row` and `col` commands, we first load the `linalg` package. Then, entering

```
> with(linalg):
  row(A,3);
```

$$[-10, 9, 12]$$

returns the 3rd row and **A**, and entering

```
> col(A,2);
```

$$[-6, -5, 9]$$

returns the second column of **A**. Multiple rows and columns can be selected as well. For example,

```
> col(A,1..2);
```

$$[10, 6, -10], [-6, -5, 9]$$

returns the first and second columns of **A**, while

```
> row(A,2..3);
```

$$[6, -5, -7], [-10, 9, 12]$$

returns the second and third rows of **A**.

■

Basic Operations from Linear Algebra

Maple V can perform many of the calculations encountered in elementary linear algebra courses. At this point, we briefly discuss some of the basic computations that Maple V can perform on matrices; additional topics are discussed in Chapter 6. To evaluate basic calculations involving matrices—for example, addition, multiplication, and exponentiation—we use `evalm`. For example, if **A** and **B** are matrices, `evalm(A+B)` adds **A** and **B**; `evalm(A*B)` computes the product of **A** and **B**, provided that the product is defined; and `evalm(A^n)` raises **A** to the nth power, if **A** is a square matrix.

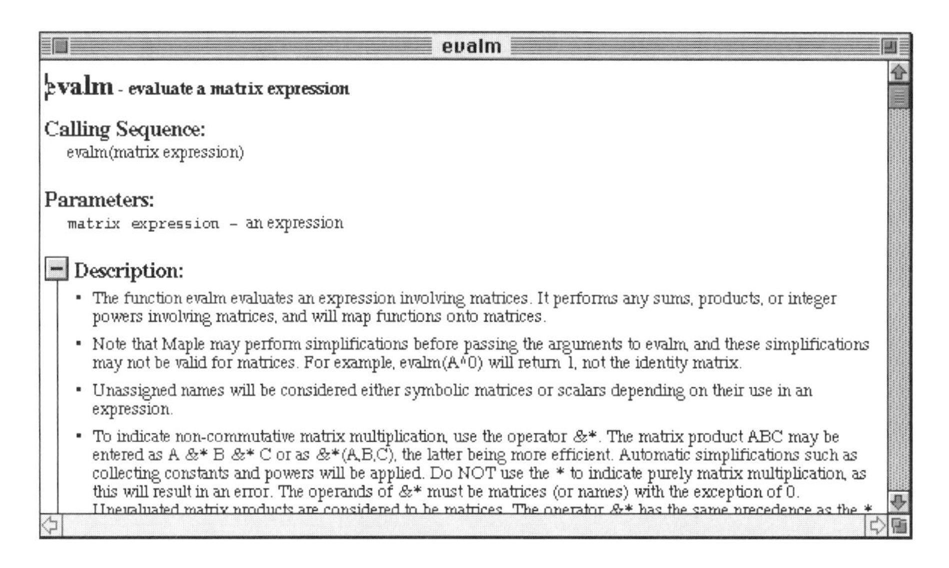
evalm

evalm - evaluate a matrix expression

Calling Sequence:
evalm(matrix expression)

Parameters:
matrix expression - an expression

Description:

- The function evalm evaluates an expression involving matrices. It performs any sums, products, or integer powers involving matrices, and will map functions onto matrices.
- Note that Maple may perform simplifications before passing the arguments to evalm, and these simplifications may not be valid for matrices. For example, evalm(A^0) will return 1, not the identity matrix.
- Unassigned names will be considered either symbolic matrices or scalars depending on their use in an expression.
- To indicate non-commutative matrix multiplication, use the operator &*. The matrix product ABC may be entered as A &* B &* C or as &*(A,B,C), the latter being more efficient. Automatic simplifications such as collecting constants and powers will be applied. Do NOT use the * to indicate purely matrix multiplication, as this will result in an error. The operands of &* must be matrices (or names) with the exception of 0. Unevaluated matrix products are considered to be matrices. The operator &* has the same precedence as the *

Nearly all of the built-in functions that perform more complex operations on matrices are contained in the `linalg` package. Hence, in order for the commands contained in the linear algebra package to be used, either the package must first be loaded with the command `with(linalg)`, or the commands must be entered in the form `linalg[command name](argument)`. Some of the commands contained in the linear algebra package to perform elementary operations on vectors and matrices include:

`det`, which computes the determinant of square matrices

`scalarmul`, which multiplies each entry of a matrix or vector by a given number

`multiply`, which multiplies matrices, provided that matrix multiplication is defined

`inv`, which computes the inverse of a square matrix, provided that the inverse is defined

`transpose`, which computes the transpose of a matrix.

EXAMPLE 7: If A $= \begin{pmatrix} 3 & -4 & 5 \\ 8 & 0 & -3 \\ 5 & 2 & 1 \end{pmatrix}$ and $\begin{pmatrix} 10 & -6 & -9 \\ 6 & -5 & -7 \\ -10 & 9 & 12 \end{pmatrix}$, compute (a) $\mathbf{A}+\mathbf{B}$; (b) $4\mathbf{A}$; (c) $\mathbf{B}-4\mathbf{A}$; (d) $(\mathbf{A}\cdot\mathbf{B})^{-1}$; (e) $((\mathbf{A}-2\mathbf{B})\cdot\mathbf{B})^{t}$; and (f) $\det(\mathbf{A}) = |\mathbf{A}|$.

SOLUTION: In this case, we first define **A** and **B** and then load the package `linalg`

```
> A:='A':B:='B':
  A:=array([[3,-4,5],[8,0,-3],[5,2,1]]):
  B:=array([[10,-6,-9],[6,-5,-7],[-10,9,12]]):
  with(linalg):
```

Entering

```
> evalm(A+B);
```

$$\begin{bmatrix} 13 & -10 & -4 \\ 14 & -5 & -10 \\ -5 & 11 & 13 \end{bmatrix}$$

adds **A** and **B**. Similarly,

```
> evalm(4*A);
```

$$\begin{bmatrix} 12 & -16 & 20 \\ 32 & 0 & -12 \\ 20 & 8 & 4 \end{bmatrix}$$

multiplies each entry of **A** by 4. Entering

```
> add(B,scalarmul(A,-4));
```

$$\begin{bmatrix} -2 & 10 & -29 \\ -26 & -5 & 5 \\ -30 & 1 & 8 \end{bmatrix}$$

computes $\mathbf{B} - 4\mathbf{A}$. Generally, matrix multiplication is not commutative. In Maple, the symbol `&*` represents noncommutative multiplication. Thus, to compute $(\mathbf{A} \cdot \mathbf{B})^{-1}$ we can use `evalm`, `&*`, and `inverse`, which is contained in the `linalg` package

```
> inverse(evalm(A&*B));
```

$$\begin{bmatrix} \frac{59}{380} & \frac{53}{190} & \frac{-167}{380} \\ \frac{-223}{570} & \frac{-92}{95} & \frac{979}{570} \\ \frac{49}{114} & \frac{18}{19} & \frac{-187}{114} \end{bmatrix}$$

or we can use `inverse` together with `multiply`, which is also contained in the `linalg` package.

```
> inverse(multiply(A,B));
```

$$\begin{bmatrix} \frac{59}{380} & \frac{53}{190} & \frac{-167}{380} \\ \frac{-223}{570} & \frac{-92}{95} & \frac{979}{570} \\ \frac{49}{114} & \frac{18}{19} & \frac{-187}{114} \end{bmatrix}$$

We use `transpose`, which is contained in the `linalg` package, together with `evalm` to compute $((\mathbf{A}-2\mathbf{B})\cdot\mathbf{B})^t$.

```
> transpose(evalm((A-2*B)&*B));!
```

$$\begin{bmatrix} -352 & -90 & 384 \\ 269 & 73 & -277 \\ 373 & 98 & -389 \end{bmatrix}$$

Last, we use `det`, which is contained in the `linalg` package to compute $\det(\mathbf{A}) = |\mathbf{A}|$.

```
> det(A);
```

    190

■

Maple V can also perform operations on nonsquare matrices, as long as the operation is defined.

EXAMPLE 8: Compute **AB** and **BA** if $\mathbf{A} = \begin{pmatrix} -1 & -5 & -5 & -4 \\ -3 & 5 & 3 & -2 \\ -4 & 4 & 2 & -3 \end{pmatrix}$ and $\mathbf{B} = \begin{pmatrix} 1 & -2 \\ -4 & 3 \\ 4 & -4 \\ -5 & -3 \end{pmatrix}$.

SOLUTION: Because **A** is a 3×4 and **B** is a 4×2 matrix, **AB** is defined and is a 3×2 matrix. We define **A** and **B** and then use `evalm` together with `&*` to compute **AB**.

```
> A:='A':B:='B':
  A:=array([[-1,-5,-5,-4],[-3,5,3,-2],[-4,4,2,-3]]):
  B:=array([[1,-2],[-4,3],[4,-4],[-5,-3]]):
> evalm(A&*B);
```

$$\begin{bmatrix} 19 & 19 \\ -1 & 15 \\ 3 & 21 \end{bmatrix}$$

However, the matrix product **BA** is not defined, and Maple produces an error message when we attempt to compute **BA**.

```
> evalm(B&*A);
Error, (in linalg[multiply]) nonmatching dimensions for
                    vector/matrix product
```

■

Computations with vectors are performed in the same way.

EXAMPLE 9: Let $\mathbf{v} = \begin{pmatrix} 0 \\ 5 \\ 1 \\ 2 \end{pmatrix}$ and $\mathbf{w} = \begin{pmatrix} 3 \\ 0 \\ 4 \\ -2 \end{pmatrix}$. (a) Calculate $\mathbf{v} - 2\mathbf{w}$ and $\mathbf{v} \bullet \mathbf{w}$. (b) Find a unit vector with same direction as **v** and a unit vector with the same

direction as **w**.

SOLUTION: We begin by defining **v** and **w** and then compute $\mathbf{v} - 2\mathbf{w}$.

```
> v:='v':w:='w':
  v:=array([0,5,1,2]):
  w:=array([3,0,4,-2]):
```

```
evalm(v-2*w);
```

$$[-6, 5, -7, 6]$$

The dot product $\mathbf{v} \bullet \mathbf{w}$ is computed with `dotprod`, which is contained in the `linalg` package.

```
> dotprod(v,w);
```

$$0$$

The **norm** of a vector $\mathbf{v} = \begin{pmatrix} v_1 \\ v_2 \\ \vdots \\ v_n \end{pmatrix}$ is $\|\mathbf{v}\| = \sqrt{v_1^2 + v_2^2 + \ldots + v_n^2} = \sqrt{\mathbf{v} \bullet \mathbf{v}}$, and it is computed with `norm(v,frobenius)`. Note that `norm` is contained in the `linalg` package. If k is a scalar, the direction of $k\mathbf{v}$ is the same as the direction of $\mathbf{v}$. Thus, if $\mathbf{v}$ is a nonzero vector, the vector $\frac{\mathbf{v}}{\|\mathbf{v}\|}$ has the same direction as $\mathbf{v}$ and because $\left\|\frac{\mathbf{v}}{\|\mathbf{v}\|}\right\| = \frac{1}{\|\mathbf{v}\|}\|\mathbf{v}\| = 1$, $\frac{\mathbf{v}}{\|\mathbf{v}\|}$ is a unit vector. We then compute $\frac{\mathbf{v}}{\|\mathbf{v}\|}$, calling the result `uv`, and $\frac{\mathbf{w}}{\|\mathbf{w}\|}$. The results correspond to unit vectors with the same direction as $\mathbf{v}$ and $\mathbf{w}$, respectively.

```
> uv:=evalm(v/norm(v,frobenius));>
```

$$uv := \left[0, \frac{1}{6}\sqrt{30}, \frac{1}{30}\sqrt{30}, \frac{1}{15}\sqrt{30}\right]$$

```
uw:=evalm(w/norm(w,frobenius));
```

$$uw := \left[\frac{3}{29}\sqrt{29}, 0, \frac{4}{29}\sqrt{29}, -\frac{2}{29}\sqrt{29}\right]$$

Alternatively, we can take advantage of the `normalize` command, which is contained in the `linalg` package. If $\mathbf{v} \neq \mathbf{0}$, `normalize(v)` returns $\frac{\mathbf{v}}{\|\mathbf{v}\|}$. Thus, entering

```
> uw:=normalize(w);
```

$$uw := \left[\frac{3}{29}\sqrt{29}, 0, \frac{4}{29}\sqrt{29}, -\frac{2}{29}\sqrt{29}\right]$$

returns the same result as that obtained previously.

■

5.2 Linear Systems of Equations

Maple V can solve a variety of systems of equations using a variety of methods.

Calculating Solutions of Linear Systems of Equations

To solve the system of linear equations $\mathbf{Ax} = \mathbf{b}$, where $\mathbf{A}$ is the coefficient matrix, $\mathbf{b}$ is the known vector, and $\mathbf{x}$ is the unknown vector, we proceed in the usual manner: if $\mathbf{A}^{-1}$ exists, then $\mathbf{A}^{-1}\,\mathbf{Ax} = \mathbf{A}^{-1}\mathbf{b}$ so $\mathbf{x} = \mathbf{A}^{-1}\mathbf{b}$.

EXAMPLE 1: Solve the matrix equation $\begin{pmatrix} 3 & 0 & 2 \\ -3 & 2 & 2 \\ 2 & -3 & 3 \end{pmatrix}^{-1} \begin{pmatrix} x \\ y \\ z \end{pmatrix} = \begin{pmatrix} 3 \\ -1 \\ 4 \end{pmatrix}$.

SOLUTION: The solution is given by .

$$\begin{pmatrix} x \\ y \\ z \end{pmatrix} = \begin{pmatrix} 3 & 0 & 2 \\ -3 & 2 & 2 \\ 2 & -3 & 3 \end{pmatrix}^{-1} \begin{pmatrix} 3 \\ -1 \\ 4 \end{pmatrix}.$$

We proceed by defining `A` and `b` and then using `inverse` to calculate `inverse(matrixa)&*b`, naming the resulting output `xvec`.

```
> A:='A':
  with(linalg):
  A:=array([[3,0,2],[-3,2,2],[2,-3,3]]):
  b:=vector([3,-1,4]):
  inverse(A);
```

$$\begin{bmatrix} \frac{6}{23} & \frac{-3}{23} & \frac{-2}{23} \\ \frac{13}{46} & \frac{5}{46} & \frac{-6}{23} \\ \frac{5}{46} & \frac{9}{46} & \frac{3}{23} \end{bmatrix}$$

```
> xvec:=evalm(inverse(A)&*b);
```

$$xvec := \left[\frac{13}{23}, \frac{-7}{23}, \frac{15}{23}\right]$$

We verify that the result is the desired solution by calculating A&*xvec. Because the result of this procedure is

$$\begin{pmatrix} 3 \\ -1 \\ 4 \end{pmatrix},$$

we conclude that the solution to the system is

$$\begin{pmatrix} x \\ y \\ z \end{pmatrix} = \begin{pmatrix} 13/23 \\ -7/23 \\ 15/23 \end{pmatrix}.$$

```
> evalm(A&*xvec);
```

$$[3, -1, 4]$$

Last, we note that this matrix equation is equivalent to the system of equations

$$\begin{cases} 3x + 2z = 3 \\ -3x + 2y + 2z = -1 \\ 2x - 3y + 3z = 4 \end{cases}$$

that we are able to solve with `solve`. We are able to form this system from the matrix equation with `geneqns`, which is also contained in the `linalg` package, as follows:

```
> sys:=geneqns(A,[x,y,z],b);
```

$$sys := \{2x - 3y + 3z = 4, -3x + 2y + 2z = -1, 3x + 2z = 3\}$$

```
> solve(sys);
```

$$\{x = \frac{13}{23}, z = \frac{15}{23}, y = \frac{-7}{23}\}$$

■

Maple offers several commands for solving systems of linear equations, like `solve`, that do not depend on the computation of the inverse of **A**.

EXAMPLE 2: Solve the system of three equations

$$\begin{cases} x - 2y + z = -4 \\ 3x + 2y - z = 8 \\ -x + 3y + 5z = 0 \end{cases}$$ for x, y, and z.

SOLUTION: In this case, we use `solve` to solve the system directly.

```
> sys:={x-2*y+z=-4,3*x+2*y-z=8,-x+3*y+5*z=0}:
  solve({x-2*y+z=-4,3*x+2*y-z=8,-x+3*y+5*z=0});
```

$$\{y = 2, z = -1, x = 1\}$$

An alternative approach takes advantage of the command `linsolve`, contained in the package `linalg`. The command `linsolve(A,b)` solves the equation **Ax** = **b** for **x**. In this case, the matrix equation corresponding to the system $\begin{cases} x-2y+z=-4 \\ 3x+2y-z=8 \\ -x+3y+5z=0 \end{cases}$ is $\begin{pmatrix} 1 & -2 & 1 \\ 3 & 2 & -1 \\ -1 & 3 & 5 \end{pmatrix}\begin{pmatrix} x \\ y \\ z \end{pmatrix} = \begin{pmatrix} -4 \\ 8 \\ 0 \end{pmatrix}$, which we form using `genmatrix`. Entering

```
> A:=genmatrix(sys, [x,y,z], 'b');
```

$$A := \begin{bmatrix} -1 & 3 & 5 \\ 1 & -2 & 1 \\ 3 & 2 & -1 \end{bmatrix}$$

and defines **b** to be the vector $\begin{pmatrix} -4 \\ 8 \\ 0 \end{pmatrix}$.

```
> print(b);
```

$$[0, -4, 8]$$

We then use `linsolve` to solve the equation $\begin{pmatrix} 1 & -2 & 1 \\ 3 & 2 & -1 \\ -1 & 3 & 5 \end{pmatrix}\begin{pmatrix} x \\ y \\ z \end{pmatrix} = \begin{pmatrix} -4 \\ 8 \\ 0 \end{pmatrix}$. Of course, the results are the same as those obtained above.

```
> linsolve(A,b);
```

$$[1, 2, -1]$$

■

As with other operations, results can be saved and manipulated later in the Maple session.

EXAMPLE 3: Solve the system $\begin{cases} 2x-4y+z=-1 \\ 3x+y-2z=3 \\ -5x+y-2z=4 \end{cases}$. Verify that the result returned satisfies the system.

SOLUTION: As in the previous example, we first use `solve` to solve the system. We name the resulting output `sol`.

```
> sys:={2*x-4*y+z=-1,3*x+y-2*z=3,-5*x+y-2*z=4}:
  sol:=solve(,{x,y,z});
```

$$sol := \{x = \frac{-1}{8}, z = \frac{-51}{28}, y = \frac{-15}{56}\}$$

The values obtained with `solve` are substituted into the system of equations with `subs` to verify that the values obtained are the solutions of the system. An alternative approach would be to use `assign` to assign x, y, and z the values obtained with `solve`.

```
> subs(sol,sys);
```

$$\{3 = 3, -1 = -1, 4 = 4\}$$

We also use `linsolve` to solve the system. The matrix equation corresponding to this system is $\begin{pmatrix} 2 & -4 & 1 \\ 3 & 1 & -2 \\ -5 & 1 & -2 \end{pmatrix}\begin{pmatrix} x \\ y \\ z \end{pmatrix} = \begin{pmatrix} -1 \\ 3 \\ 4 \end{pmatrix}$.

```
> A:=genmatrix(sys,[x,y,z],'b'):
  x:=linsolve(A,b);
```

$$x := \left[\frac{-1}{8}, \frac{-15}{56}, \frac{-51}{28}\right]$$

In this case, we name the solution vector `x` and then use `evalm` to verify that `x` is the solution vector to the matrix equation.

```
> evalm(A&*x);
```

[-1, 3, 4]

■

EXAMPLE 4: Solve the system of equations
$$\begin{cases} 4x_1 + 5x_2 - 5x_3 - 8x_4 - 2x_5 = 5 \\ 7x_1 + 2x_2 - 10x_3 - x_4 - 6x_5 = -4 \\ 6x_1 + 2x_2 + 10x_3 - 10x_4 + 7x_5 = -7. \\ -8x_1 - x_2 - 4x_3 + 3x_5 = 5 \\ 8x_1 - 7x_2 - 3x_3 + 10x_4 + 5x_5 = 7 \end{cases}$$

SOLUTION: The system is equivalent to the matrix equation

$$\begin{pmatrix} 4 & 5 & -5 & -8 & -2 \\ 7 & 2 & -10 & -1 & -6 \\ 6 & 2 & 10 & -10 & 7 \\ -8 & -1 & -4 & 0 & 3 \\ 8 & -7 & -3 & 10 & 5 \end{pmatrix} \begin{pmatrix} x_1 \\ x_2 \\ x_3 \\ x_4 \\ x_5 \end{pmatrix} = \begin{pmatrix} 5 \\ -4 \\ -7 \\ 5 \\ 7 \end{pmatrix}.$$

After defining **A** and **b**, we use `linsolve`.

```
> A:='A':b:='b':
  with(linalg):
  A:=array([[4,5,-5,-8,-2],[7,2,-10,-1,-6],
       [6,2,10,-10,7],[-8,-1,-4,0,3],
       [8,-7,-3,10,5]]):
  b:=vector([5,-4,-7,5,7]):
  linsolve(A,b);
```

$$\left[\frac{1245}{6626}, \frac{113174}{9939}, \frac{-7457}{9939}, \frac{38523}{6626}, \frac{49327}{9939}\right]$$

■

Gauss-Jordan Elimination

Given the matrix equation $\mathbf{Ax} = \mathbf{b}$, where $\mathbf{A} = \begin{bmatrix} a_{11} & a_{12} & \cdots & a_{1n} \\ a_{21} & a_{22} & \cdots & a_{2n} \\ \vdots & \vdots & \ddots & \vdots \\ a_{m1} & a_{m2} & \cdots & a_{mn} \end{bmatrix}$, $\mathbf{x} = \begin{pmatrix} x_1 \\ x_2 \\ \vdots \\ x_n \end{pmatrix}$, and $\mathbf{b} = \begin{pmatrix} b_1 \\ b_2 \\ \vdots \\ b_m \end{pmatrix}$, the $m \times n$ matrix $\mathbf{A}$ is called the **coefficient matrix** for the matrix equation $\mathbf{Ax} = \mathbf{b}$, and the $m \times (n + 1)$ matrix $(\mathbf{A}|\mathbf{b}) = \begin{pmatrix} a_{11} & a_{12} & \cdots & a_{1n} & b_1 \\ a_{21} & a_{22} & \cdots & a_{2n} & b_2 \\ \vdots & \vdots & \ddots & \vdots & \vdots \\ a_{m1} & a_{m2} & \cdots & a_{mn} & b_m \end{pmatrix}$ is called the **augmented matrix** for the matrix equation.

`augment(matrixlist)` yields the matrix that results when the rows of the matrices in `matrixlist` are joined. `augment` is contained in the linear algebra package `linalg`, so it must be loaded first with the command `with(linalg)` or entered in the form `linalg[augment](matrixlist)`.

EXAMPLE 5: Construct the augmented matrix for the matrix equation

$$\begin{pmatrix} 3 & 0 & 2 \\ -3 & 2 & 2 \\ 2 & -3 & 3 \end{pmatrix} \begin{pmatrix} x_1 \\ x_2 \\ x_3 \end{pmatrix} = \begin{pmatrix} 3 \\ -1 \\ 4 \end{pmatrix}.$$

SOLUTION: The matrix of coefficients is defined as the 3×3 matrix `A`, the right-hand side values in `b`, and the augmented matrix is computed with `augment` and named `M`.

```
> with(linalg):
  A:=matrix(3,3,[3,0,2,-3,2,2,2,-3,-3]):
  b:=vector([3,-1,4]):
```

```
M:=augment(A,b);
```

$$M := \begin{bmatrix} 3 & 0 & 2 & 3 \\ -3 & 2 & 2 & -1 \\ 2 & -3 & -3 & 4 \end{bmatrix}$$

■

gaussjord(mat) performs Gauss-Jordan elimination on the matrix mat so that the matrix that results is of reduced row echelon form. Note that gaussjord(mat) produces the same result as rref(mat). Another useful command is backsub(rremat), which performs back substitution on the matrix rremat. All three of these commands, gaussjord, rref, and backsub, are contained in the linear algebra package linalg.

EXAMPLE 6: Solve the system $\begin{cases} -2x + y - 2z = 4 \\ 2x - 4y - 2z = -4 \\ x - 4y - 2z = 3 \end{cases}$ using Gauss-Jordan elimination.

SOLUTION: First, the matrix of coefficients and the vector of right-hand-side values are entered in A and b, respectively. The augmented matrix that corresponds to the system of equations is formed in C with augment.

```
> A:=array([[-2,1,-2],[2,-4,-2],[1,-4,-2]]):
  b:=vector([4,-4,3]):
  C:=augment(A,b);
```

$$C := \begin{bmatrix} -2 & 1 & -2 & 4 \\ 2 & -4 & -2 & -4 \\ 1 & -4 & -2 & 3 \end{bmatrix}$$

Gauss-Jordan elimination is performed with gaussjord in solmatrix. The result of this command is a matrix in reduced row echelon form, so backsub is used to obtain the solution, which is x = -7, y = -4, and z = 3.

```
> solmatrix:=gaussjord(C);
```

$$solmatrix := \begin{bmatrix} 1 & 0 & 0 & -7 \\ 0 & 1 & 0 & -4 \\ 0 & 0 & 1 & 3 \end{bmatrix}$$

```
> sols:=backsub(solmatrix);
```

$$sols := [-7, -4, 3]$$

From this result, we see that the solution is $\begin{pmatrix} x \\ y \\ z \end{pmatrix} = \begin{pmatrix} -7 \\ -4 \\ 3 \end{pmatrix}$. We verify this by replacing each occurrence of x, y, and z on the left-hand side of each equation by -7, -4, and 3, respectively, and noting that the components of the result are equal to the right-hand side of each equation.

```
> subs({x=-7,y=-4,z=3},
  [-2*x+y-2*z,2*x-4*y-2*z,x-4*y-2*z]);
```

$$[4,-4,3]$$

5.3 Maxima and Minima Using Linear Programming

The Standard Form of a Linear Programming Problem

We call the linear programming problem of the form the **standard form**: "Minimize $Z = \underbrace{c_1x_1 + c_2x_2 + \ldots + c_nx_n}_{\text{function}}$, subject to the restrictions

$$\begin{cases} a_{11}x_1 + a_{12}x_2 + \dots + a_{1n}x_n \geq b_1 \\ a_{21}x_1 + a_{22}x_2 + \dots + a_{2n}x_n \geq b_2 \\ \vdots \\ a_{m1}x_1 + a_{m2}x_2 + \dots + a_{mn}x_n \geq b_m \end{cases}, \text{ and } x_1 \geq 0,\ x_2 \geq 0,\ \dots,\ x_n \geq 0.\text{''}$$

The Maple V command `minimize`, contained in the `simplex` package, solves the standard form of the linear programming problem. Similarly, the command `maximize`, also contained in the `simplex` package, solves the linear programming problem: Maximize $Z = \underbrace{c_1x_1 + c_2x_2 + \dots + c_nx_n}_{\texttt{function}}$ subject to the restrictions.

$$\begin{cases} a_{11}x_1 + a_{12}x_2 + \dots + a_{1n}x_n \geq b_1 \\ a_{21}x_1 + a_{22}x_2 + \dots + a_{2n}x_n \geq b_2 \\ \vdots \\ a_{m1}x_1 + a_{m2}x_2 + \dots + a_{mn}x_n \geq b_m \end{cases}, \text{ and } x_1 \geq 0,\ x_2 \geq 0,\ \dots,\ x_n \geq 0.$$

As when you use commands contained in any other package, be sure to load the `simplex` package before you use the commands `minimize` and `maximize`.

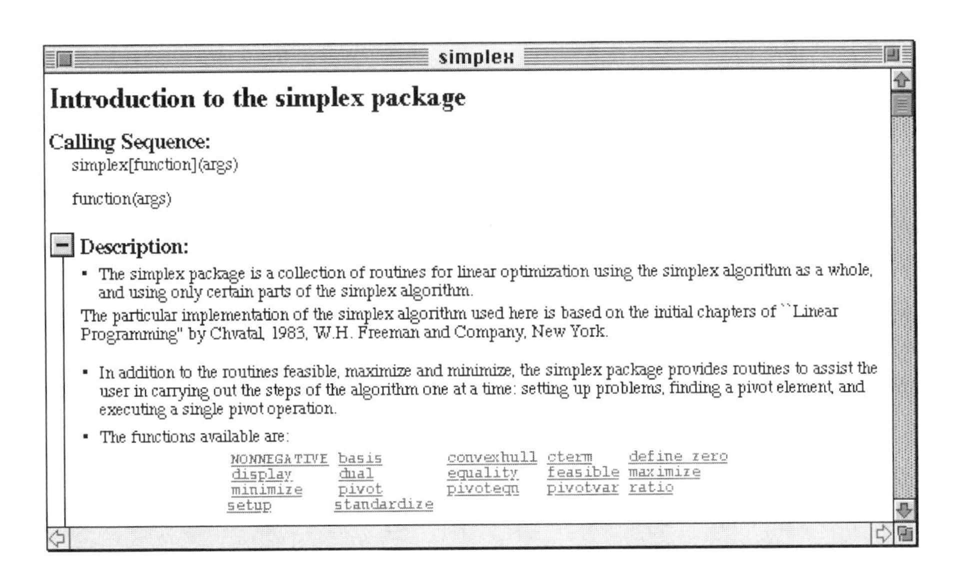

EXAMPLE 1: Maximize $Z(x_1, x_2, x_3) = 4x_1 - 3x_2 + 2x_3$ subject to the constraints $3x_1 - 5x_2 + 2x_3 \le 60$, $x_1 - x_2 + 2x_3 \le 10$, $x_1 + x_2 - x_3 \le 20$, and x_1, x_2, x_3 all nonnegative.

SOLUTION: In this case, we begin by clearing all prior definitions of the variables we will use, if any, and then loading the simplex package. After loading the simplex package, we define Z to be $4x_1 - 3x_2 + 2x_3$ and constraints to be the set of inequalities consisting of $3x_1 - 5x_2 + 2x_3 \le 60$, $x_1 - x_2 + 2x_3 \le 10$, and $x_1 + x_2 - x_3 \le 20$. Note that Maple represents the symbol "≤" with "<="; similarly, the symbol "≥" is denoted with ">=".

```
> Z:='Z':x1:='x1':x2:='x2':x3:='x3':
  with(simplex):
  Z:=4*x1-3*x2+2*x3:
  constraints:={3*x1-5*x2+x3<=60,
       x1-x2+2*x3<=10,x1+x2-x3<=20}:
```

The solution to the problem in which the nonnegativity constraint is *not* considered is determined with maximize and named sols_one, so that the maximum value of Z can be determined with subs(sols_one,Z).

```
> sols_one:=maximize(Z,constraints);
```

$$sols\_one := \{x3 = -5, x2 = \frac{-5}{2}, x1 = \frac{35}{2}\}$$

```
> subs(sols_one,Z);
```

$$\frac{135}{2}$$

The nonnegative constraints are indicated with NONNEGATIVE in the maximize command or entered explicitly along with the other constraints. We make use of the NONNEGATIVE setting to find the solution in sols_two. The maximum value of Z in this case is then found to be 45.

```
> sols_two:=maximize(Z,constraints,NONNEGATIVE);
```

$$sols\_two := \{x3 = 0, x2 = 5, x1 = 15\}$$

```
> assign(sols_two);
  Z;
```

$$45$$

■

We demonstrate the use of `minimize` in the following example:

EXAMPLE 2: Minimize $Z(x, y, z) = 4x - 3y + 2z$ subject to the constraints $3x - 5y + 2z \leq 60$, $x - y + 2z \leq 10$, $x + y - z \leq 20$, and x, y, and z all nonnegative.

SOLUTION: We begin by loading the `simplex` package. The point at which the minimum value of Z is found with `minimize` is named `vals`. The value of $Z(x, y, z)$ at this point is then found to be -90 with `subs`, which substitutes the values of x, y, and z that were determined with `minimize` into the function $Z(x, y, z)$.

```
> with(simplex):
  vals:=minimize(4*x-3*y+2*z,{3*x-5*y+2*z<=60,
      x-y+2*z<=10,x+y-z<=20},NONNEGATIVE);
```

$$vals := \{x = 0, z = 30, y = 50\}$$

```
> subs(vals,4*x-3*y+2*z);
```

$$-90$$

■

The Dual Problem

Given the standard form of the linear programming problem: Minimize $Z = \sum_{j=1}^{n} c_j x_j$ subject to the constraints $\sum_{j=1}^{n} a_{ij} x_j \geq b_i$ for i=1, 2, ... , m and $x_j \geq 0$ for $j = 1, 2, \ldots, n$, the **dual problem** is: Maximize $Y = \sum_{i=1}^{m} b_i y_i$ subject to the constraints $\sum_{i=1}^{m} a_{ij} y_i \leq c_j$ for $j = 1, 2, \ldots, n$ and $y_i \geq 0$ for $i = 1, 2, \ldots, m$. Similarly, for the problem: Maximize $Z = \sum_{j=1}^{n} c_j x_j$ subject to the constraints $\sum_{j=1}^{n} a_{ij}x_j \ge b_i$ for i=1, 2, ... , m and $x_j \ge 0$ for j = 1, 2, ... , n, the dual problem is: Minimize $Y = \sum_{i=1}^{m} b_i y_i$ subject to the constraints $\sum_{i=1}^{m} a_{ij}y_i \le c_j$ for j = 1, 2, ... , n and $y_i \ge 0$ for i = 1, 2, ... , m.

The `simplex` package contains the command `dual(func,conlist,var)`, which produces the dual of the linear programming problem in standard form with objective function `func` and constraints `conlist`. The dual problem that results is given in trems of the dual variables `var1`, `var2`, ..., `varn`.

EXAMPLE 3: Maximize $Z = 6x + 8y$ subject to the constraints $5x + 2y \le 20$, $x + 2y \le 10$, $x \ge 0$ and $y \ge 0$. State the dual problem and find its solution.

SOLUTION: First, we solve the problem in its original form by using steps similar to those used in the previous example. The point at which the maximum value of the objective function occurs is determined and named `max_sols`. This maximum value is 45, which is found with `subs`.

```
> with(simplex):
  Z:=6*x+8*y:
  constraints:={5*x+2*y<=20,x+2*y<=10}:
  max_sols:=maximize(Z,constraints,NONNEGATIVE);
```

$$max\_sols := \{y = \frac{15}{4}, x = \frac{5}{2}\}$$

```
> subs(max_sols,Z);
```

45

In this case, we can graph the feasibility set determined by the constraints with `inequal`, which is contained in the `plots` package.

```
> with(plots):
  inequal( {5*x+2*y<=20,x+2*y<=10,x>=0,y>=0},
  x=-1..10,y=-1..10,
```

```
optionsexcluded=(color=white,thickness=2));
```

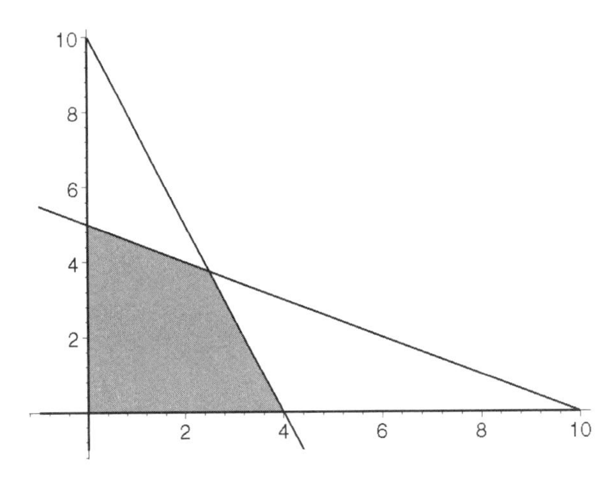

We consider the dual problem next. Since `y` is specified in the `dual` command, the two dual variables are given by `y1` and `y2`. The resulting problem is assigned the name `dual_problem`. Then, the solution of the dual problem is easily found with `minimize`. The point at which this minimum value occurs is called `min_sols`. By extracting the objective function of the dual problem with `dual_problem[1]`, we are able to substitute the values in `min_sols` into this function with `subs` to see that we obtain the same optimal value as that found for the original problem.

```
> dual_problem:=dual(Z,constraints,y);
```

$$dual\_problem := 20\,y1 + 10\,y2, \{6 \le 5\,y1 + y2, 8 \le 2\,y1 + 2\,y2\}$$

```
> min_sols:=minimize(dual_problem,NONNEGATIVE);
```

$$min\_sols := \{y2 = \frac{7}{2}, y1 = \frac{1}{2}\}$$

```
> dual_problem[1];
```

$$20\,y1 + 10\,y2$$

```
> subs(min_sols,dual_problem[1]);
```

$$45$$

■

Of course, linear programming models can involve numerous variables. Consider the following: Given the standard-form linear programming problem: Minimize $Z = c_1x_1 + c_2x_2 + \ldots + c_nx_n$, subject to the restrictions.

$$\begin{cases} a_{11}x_1 + a_{12}x_2 + \ldots + a_{1n}x_n \geq b_1 \\ a_{21}x_1 + a_{22}x_2 + \ldots + a_{2n}x_n \geq b_2 \\ \vdots \\ a_{m1}x_1 + a_{m2}x_2 + \ldots + a_{mn}x_n \geq b_m \end{cases}, \text{ and } x_1 \geq 0,\ x_2 \geq 0,\ \ldots,\ x_n \geq 0.$$

Let $\mathbf{x} = \begin{pmatrix} x_1 \\ x_2 \\ \vdots \\ x_n \end{pmatrix}$, $\mathbf{b} = \begin{pmatrix} b_1 \\ b_2 \\ \vdots \\ b_m \end{pmatrix}$, $c = (c_1c_2\ldots c_n)$, and $\mathbf{A}$ denote the $m \times n$ matrix.

$$\begin{pmatrix} a_{11} & a_{12} & \cdots & a_{1n} \\ a_{21} & a_{22} & \cdots & a_{2n} \\ \vdots & \vdots & \ddots & \vdots \\ a_{m1} & a_{m2} & \cdots & a_{mn} \end{pmatrix}.$$

Then the standard form of the linear programming problem is equivalent to finding the vector $\mathbf{x}$ that maximizes $Z = \mathbf{c} \bullet \mathbf{x}$ subject to the restrictions $\mathbf{A}_\mathbf{x} \geq \mathbf{b}$ and $x_1 \geq 0, x_2 \geq 0, \ldots, x_n \geq 0$. The dual problem of: Maximize the number $Z = \mathbf{c} \bullet \mathbf{x}$ subject to the restrictions $\mathbf{A}_\mathbf{x} \geq \mathbf{b}$ (componentwise) and $x_1 \geq 0, x_2 \geq 0, \ldots, x_n \geq 0$: Minimize the number $Y = \mathbf{y} \bullet \mathbf{b}$ where $\mathbf{y} = \begin{pmatrix} y_1 \\ y_2 \\ \vdots \\ y_m \end{pmatrix}$ subject to the restrictions $\mathbf{yA} \leq \mathbf{c}$ (componentwise) and $y_1 \geq 0, y_2 \geq 0, \ldots, y_m \geq 0$.

EXAMPLE 4: Maximize $Z = 5x_1 - 7x_2 + 7x_3 + 5x_4 + 6x_5$ subject to the constraints $2x_1 + 3x_2 + 3x_3 + 2x_4 + 2x_5 \geq 10$, $6x_1 + 5x_2 + 4x_3 + x_4 + 4x_5 \geq 30$, $-3x_1 - 2x_2 - 3x_3 - 4x_4 \geq -5$, $-x_1 - x_2 - x_4 \geq -10$, and $x_i \geq 0$ for $i = 1, 2, 3, 4$, and 5. State the dual problem. What is its solution?

SOLUTION: For this problem, $\mathbf{x} = \begin{pmatrix} x_1 \\ x_2 \\ x_3 \\ x_4 \\ x_5 \end{pmatrix}$, $\mathbf{b} = \begin{pmatrix} 10 \\ 30 \\ -5 \\ -10 \end{pmatrix}$, $\mathbf{c} = (5 \ \ -7 \ \ 7 \ \ 5 \ \ 6)$, and $\mathbf{A} = \begin{pmatrix} 2 & 3 & 3 & 2 & 2 \\ 6 & 5 & 4 & 1 & 4 \\ -3 & -2 & -3 & -4 & 0 \\ -1 & -1 & 0 & -1 & 0 \end{pmatrix}$.

We will use some of the commands located in the `linalg` package to manipulate matrices and vectors, so we begin by loading this package. The coefficient matrix is then defined in `A`. Notice that the fifth row, which is a zero vector, is entered with `[0$5]`. The vectors `b` and `c` are also defined as is the solution vector of five components `x`.

```
> with(linalg):
  Z:='Z':A:='A':c:='c':b:='b':x:='x':
  A:=array([[2,3,3,2,2],[6,5,4,1,4],
          [-3,-2,-3,-4,0],[-1,-1,0,-1,0],
          [0$5]]);
```

$$A := \begin{bmatrix} 2 & 3 & 3 & 2 & 2 \\ 6 & 5 & 4 & 1 & 4 \\ -3 & -2 & -3 & -4 & 0 \\ -1 & -1 & 0 & -1 & 0 \\ 0 & 0 & 0 & 0 & 0 \end{bmatrix}$$

```
> c:=array([[5,-7,7,5,6]]):
  b:=vector([10,30,-5,-10,0]):
  x:=vector(5):
```

The objective function Z is defined by finding the scalar (dot) product of the solution vector x and the vector c with multiply, which is found in the linalg package. This objective function is extracted from the output list that results with Z[1].

```
> Z:=multiply(c,x);
```

$$Z := [5x_1 - 7x_2 + 7x_3 + 5x_4 + 6x_5]$$

```
> Z[1];
```

$$5\,x_1 - 7\,x_2 + 7\,x_3 + 5\,x_4 + 6\,x_5$$

The left-hand sides of the contraints are then defined in prod by multiplying the coefficient matrix A by the solution vector x. Notice that the output list is made up of four components that can be extracted with prod[1],prod[2],prod[3],and prod[4]. In constraints, the set of constraints for this problem are constructed by substituting $i = 1, 2, 3, 4$ into the inequality prod[i]>=b[i]. Hence, a constraint is produced for each value of.

```
> prod:=multiply(A,x);
```

$$prod := [2x_1 + 3x_2 + 3x_3 + 2x_4 + 2x_5,$$
$$6x_1 + 5x_2 + 4x_3 + x_4 + 4x_5, -3x_1 - 2x_2 - 3x_3 - 4x_4$$
$$-x_1 - x_2 - x_4, 0]$$

```
> constraints:={seq(prod[i]>=b[i],i=1..4)};
```

$$constraints := \{30 \le 6x_1 + 5x_2 + 4x_3 + x_4 + 4x_5,$$
$$-5 \le -3x_1 - 2x_2 - 3x_3 - 4x_4, -10 \le x_1 - x_2 - x_4,$$
$$10 \le 2x_1 + 3x_2 + 3x_2 + 2x_4 + 2x_5\}$$

The point at which the minimum value of the objective function Z[1] subject to the constraints given in constraints is found with minimize with the NONNEGATIVE setting so that the nonnegativity constraints are considered by the simplex method. Of course, the simplex package must be loaded before this command is used. The solution is assigned the name vals. Finally, the components in vals are assigned to the solution vector x so that the minimum value of the objective function is determined by taking the scalar product of c and x which is equivalent to substituting the components of x into the objective function. The scalar product of 34/5 is found with multiply.

```
> with(simplex):
  vals:=minimize(Z[1],constraints,NONNEGATIVE);
```

$$vals := \{x_1 = 0, x_3 = 0, x_4 = 0, x_5 = \frac{35}{8}, x_2 = \frac{5}{2}\}$$

```
> assign(vals):
  multiply(c,x);
```

$$\left[\frac{35}{4}\right]$$

Next, we solve the corresponding dual problem. The vector Y, which is defined below, has as its components the dual variables y1, y2, y3, y4, and y5. The objective function of the dual problem is then defined by multiplying Y by the vector b. Similarly, the left-hand sides of the constraints are determined by multiplying the vector **Y** by the matrix of coefficients **A**.

```
> Y:=array([[y1,y2,y3,y4,y5]]):
  multiply(Y,b);
```

$$[10\,y1 + 30\,y2 - 5\,y3 - 10\,y4]$$

```
> multiply(Y,A);
```

$$\left[\begin{array}{c}[2y1 + 6y2 - 3y3 - y4, 3y1 + 5y2 - 2y3 - y4. \\ 3y1 + 4y2 - 3y3, 2y1 - y2 - 4y_3 - 4y, 2y1 + 4y2]\end{array}\right]$$

Hence, we may state the dual problem as:

Minimize $Y = 10y_1 + 30y_2 - 5y_3 - 10y_4$ subject to the constraints
$2y_1 + 6y_2 - 3y_3 - y_4 \le 5$,
$3y_1 + 5y_2 - 2y_3 - y_4 \le -7$,
$3y_1 + 4y_2 - 3y_3 \le 7$,
$2y_1 + y_2 - 4y_3 - y_4 \le 5$,
$2y_1 + 4y_2 \le 6$ and $y_i \ge 0$ for $i = 1, 2, 3$, and 4.

■

Application: A Transportation Problem

A certain company has two factories, F1 and F2, each producing two products, P1 and P2, that are to be shipped to three distribution centers, D1, D2, and D3. The following table illustrates the cost associated with shipping each product from the factory to the distribution center, the minimum number of each product each distribution center needs, and the maximum output of each factory. How much of each product should be shipped from each plant to each distribution center to minimize the total shipping costs?

	F1/P1	F1/P2	F2/P1	F2/P2	Minimum
D1/P1	\$0.75		\$0.80		500
D1/P2		\$0.50		\$0.40	400
D2/P1	\$1.00		\$0.90		300
D2/P2		\$0.75		\$1.20	500
D3/P1	\$0.90		\$0.85		700
D3/P2		\$0.80		\$0.95	300
Maximum Output	1000	400	800	900	

SOLUTION: Let x_1 denote the number of units of P1 shipped from F1 to D1; x_2 the number of units of P2 shipped from F1 to D1; x_3 the number of units of P1 shipped from F1 to D2; x_4 the number of units of P2 shipped from F1 to D2; x_5 the number of units of P1 shipped from F1 to D3; x_6 the number of units of P2 shipped from F1 to D3; x_7 the number of units of P1 shipped from F2 to D1; x_8 the number of units of P2 shipped from F2 to D1; x_9 the number of units of P1 shipped from F2 to D2; x_{10} the number of units of P2 shipped from F2 to D2; x_{11} the number of units of P1 shipped from F2 to D3; and x_{12} the number of units of P2 shipped from F2 to D3.

Then, it is necessary to minimize the number

$$Z = .75x_1 + .5x_2 + x_3 + .75x_4 + .9x_5 + .8x_6 + .8x_7 + .4x_8 + .9x_9 + 1.2x_{10} + .85x_{11} + .95x_{12}$$

subject to the constraints $x_1 + x_3 + x_5 \leq 1000$, $x_2 + x_4 + x_6 \leq 400$, $x_7 + x_9 + x_{11} \leq 800$, $x_8 + x_{10} + x_{12} \leq 900$, $x_1 + x_7 \geq 500$, $x_3 + x_9 \geq 300$, $x_5 + x_{11} \geq 700$, $x_2 + x_8 \geq 400$, $x_4 + x_{10} \geq 500$, $x_6 + x_{12} \geq 300$, and x_i nonnegative for $i=1, 2, \ldots, 12$.

In order to solve this linear programming problem, we must enter the objective function that computes the total cost, the twelve variables, and the set of inequalities. The coefficients of the objective function are given in the vector `c`. We will use several of the commands in the `linalg` package, and we begin by loading this package. The objective function is defined by computing the dot product of the vectors `x` and `c` with `dotprod`. (Recall that a similar computation was performed with `multiply` in a previous example.) The list of constraints are entered explicitly in `constraints`.

```
> with(linalg):
  c:=vector(
      [.75,.5,1,.75,.9,.8,.8,.4,.9,1.2,.85,.95]):
  x:=vector(12):
> Z:=dotprod(x,c);
```

$$Z := .75x_1 + .5x_2 + x_3 + .75x_4 + .9x_5 + .8x_6 + .8x_7 + .4x_8 + .9x_9 + 1.2x_{10} + .85x_{11} + .95x_{12}$$

```
> constraints:={x[1]+x[3]+x[5]<=1000,
      x[2]+x[4]+x[6]<=400,x[7]+x[9]+x[11]<=800,
      x[8]+x[10]+x[12]<=900,x[1]+x[7]>=500,
      x[3]+x[9]>=300,x[5]+x[11]>=700,
      x[2]+x[8]>=400,x[4]+x[10]>=500,
      x[6]+x[12]>=300}:
```

The `simplex` package is then loaded so that the `minimize` command can be used. This is done in `min_vals`, which determines the variable values at which the minimum occurs. These values are assigned to the components of the vector `x` with `assign`. Therefore, the total number of units produced of each product at each factory is easily found by entering `x[1]+x[3]+x[5]`, `x[2]+x[4]+x[6]`, ..., `x[6]+x[12]`. Also, the minimum value of the objective function 2115 is determined by entering `Z`.

```
> with(simplex):
  min_vals:=minimize(Z,constraints,NONNEGATIVE);
```

$$min\_vals := \{x_{10} = 100, x_3 = 0, x_2 = 0, x_6 = 0, x_8 = 400, x_{12} = 300, x_4 = 400, x_7 = 0, x_9 = 300, x_1 = 500, x_5 = 200, x_{11} = 500\}$$

```
> assign(min_vals):
  Z;
  x[1]+x[3]+x[5];
  x[2]+x[4]+x[6];
  x[7]+x[9]+x[11];
  x[3]+x[9];
  x[5]+x[11];
  x[2]+x[8];
  x[4]+x[10];
  x[6]+x[12];
```

```
2115.00
700
400
800
300
700
400
500
300
```

■

5.4 Vector Calculus

Definitions and Notation

The terminology and notation used in *Maple V By Example*, second edition, is standard. Nevertheless, we review basic definitions briefly.

A **scalar field** is a function with domain a set of ordered triples and range a subset of the real numbers:

$$\mathrm{f}:\mathrm{U} \to \mathrm{V} \text{ is a scalar field means } \mathrm{U} \subseteq \Re^3 \text{ and } \mathrm{V} \subseteq \Re .$$

A **scalar field** is a function with domain a set of ordered triples and range a subset of the real numbers:

$$f:U \to V \text{ is a scalar field means } U \subseteq \Re^3 \text{ and } V \subseteq \Re .$$

A **vector field f** is a vector-valued function:

$\mathbf{f}:V \to U$, $U \subseteq \Re^3$ and $V \subseteq \Re$, is a vector field means that **f** can be written in the form

$\mathbf{f}(x, y, z) = f_1(x, y, z)\mathbf{i} + f_2(x, y, z)\mathbf{j} + f_3(x, y, z)\mathbf{k} = \langle f_1(x, y, z), f_2(x, y, z), f_3(x, y, z)\rangle$

for each (x, y, z) in the domain of **f**.

The **gradient of the scalar field f** is defined to be the vector field,

$$\operatorname{grad} f = \nabla f = \frac{\partial f}{\partial x}\mathbf{i} + \frac{\partial f}{\partial y}\mathbf{j} + \frac{\partial f}{\partial z}\mathbf{k} = \langle \frac{\partial f}{\partial x}, \frac{\partial f}{\partial y}, \frac{\partial f}{\partial z}\rangle = \langle f_x, f_y, f_z\rangle,$$

where $\mathbf{i} = \langle 1, 0, 0\rangle$, $\mathbf{j} = \langle 0, 1, 0\rangle$, and $\mathbf{k} = \langle 0, 0, 1\rangle$.

A **conservative vector field f** is a vector field that is the gradient of a scalar field: **f** is a conservative vector field means that there is a scalar field g satisfying $\mathbf{f} = \nabla g$. In this case, g is usually called a **potential function** for **f**.

The **divergence of the vector field f** is defined to be the scalar

$$\operatorname{div}\mathbf{f} = \operatorname{div}\mathbf{f}(x, y, z) = \operatorname{div}\langle f_1(x, y, z), f_2(x, y, z), f_3(x, y, z)\rangle$$

$$= \frac{\partial f_1}{\partial x} + \frac{\partial f_2}{\partial x} + \frac{\partial f_3}{\partial x} = \nabla \bullet \mathbf{f}.$$

The **laplacian of the scalar field *f*** is defined to be div(grad *f*)):.

$$\operatorname{laplacian}(f) = \nabla^2 f = \Delta f = \frac{\partial^2 f}{\partial x^2} + \frac{\partial^2 f}{\partial y^2} + \frac{\partial^2 f}{\partial z^2} = f_{xx}, f_{yy}, f_{zz}.$$

The `linalg` package contains several commands that can be used to compute the gradient, laplacian, and divergence, like `grad`, `laplacian`, and `diverge`.

EXAMPLE 1: Let $f(x, y, z) = \cos(xyz)$. Compute ∇f, $\nabla^2 f$, and $\text{div}(\nabla f)$.

SOLUTION: First, the `linalg` package is loaded and then the function `f` of three variables is defined. In general, the gradient of the function $f(x_1, x_2, \ldots, x_n)$ is computed with

```
grad(f(x1,x2,...,xn),[x1,x2,...,xn])
```

so ∇f is computed with `grad(f(x,y,z),[x,y,z])`.

```
> with(linalg):
  f:=(x,y,z)->cos(x*y*z):
> grad_f:=grad(f(x,y,z),[x,y,z]);
```

$$grad\_f := [-\sin(x\,y\,z)\,y\,z, -\sin(x\,y\,z)\,x\,z, -\sin(x\,y\,z)\,x\,y]$$

The laplacian of $f(x_1, x_2, \ldots, x_n)$ is found with

```
laplacian(f(x1,x2,...,xn),[x1,x2,...,xn]).
```

```
> laplacian(f(x,y,z),[x,y,z]);
```

$$-\cos(x\,y\,z)\,y^2\,z^2 - \cos(x\,y\,z)\,x^2\,z^2 - \cos(x\,y\,z)\,x^2\,y^2$$

The divergence of the vector `[f1,f2,...,fn]`, where each component of the vector is a function of the variables in `[x1,x2,...,xn]`, is determined with

```
diverge([f1,f2,...,fn],[x1,x2,...,xn]).
```

This is computed for the gradient of f. Hence, the result is the same as that obtained for the laplacian of f.

```
> diverge(grad_f,[x,y,z]);
```

$$-\cos(x\,y\,z)\,y^2\,z^2 - \cos(x\,y\,z)\,x^2\,z^2 - \cos(x\,y\,z)\,x^2\,y^2$$

■

The **curl of the vector field f**(x,y,z) is defined to be the vector field:

$$\text{curl } \mathbf{f} = \text{curl } \mathbf{f}(x, y, z) = \text{curl}\langle f_1(x, y, z), f_2(x, y, z), f_3(x, y, z)\rangle$$
$$= \left(\frac{\partial f_3}{\partial y} - \frac{\partial f_2}{\partial z}\right)i + \left(\frac{\partial f_1}{\partial z} - \frac{\partial f_3}{\partial x}\right)j + \left(\frac{\partial f_2}{\partial x} - \frac{\partial f_1}{\partial y}\right)k$$
$$= \det\begin{pmatrix} \mathbf{i} & \mathbf{j} & \mathbf{k} \\ \frac{\partial}{\partial x} & \frac{\partial}{\partial y} & \frac{\partial}{\partial z} \\ f_1 & f_2 & f_3 \end{pmatrix},$$

and determined with

```
curl(f(x,y,z),[x,y,z]).
```

Note that curl is contained in the linalg package.

EXAMPLE 2: Let $\mathbf{f}(x, y, z) = xy\mathbf{i} + xyz^2\mathbf{j} - e^{2z}\mathbf{k} = \langle xy, xyz^2, -e^{2z}\rangle$. Compute curl **f**; div **f**, laplacian(div **f**), and grad(laplacian(div **f**)) = $\text{grad}(\nabla^2(\text{div}\mathbf{f}))$.

SOLUTION: First, we define the vector-valued function **f**.

```
> with(linalg):
  f:=(x,y,z)->[x*y,x*y*z^2,exp(-2*z)]:
```

Then, we use `curl` to compute the curl of **f**.

```
> curl(f(x,y,z),[x,y,z]);
```

$$[-2\,x\,y\,z, 0, y\,z^2 - x]$$

We use `diverge` to compute **f** and name the result `divf`.

```
> divf:=diverge(f(x,y,z),[x,y,z]);
```

$$div\ f := y + x\,z^2 - 2\,\mathbf{e}^{(-2\,z)}$$

Then, `laplacian` is used to compute laplacian(div **f**), and we name the result `ladivf`.

```
> ladivf:=laplacian(divf,[x,y,z]);
```

$$ladivf := 2\,x - 8\,\mathbf{e}^{(-2\,z)}$$

Finally, `grad` is used to compute grad(laplacian(div **f**))= grad(∇^2(div**f**)) .

```
> grad(ladivf,[x,y,z]);
```

$$[2, 0, 16\,\mathbf{e}^{(-2\,z)}]$$

■

Note that Maple supports a wide range of coordinate systems. For a complete list of all coordinate systems supported by Maple, enter `?coords`.

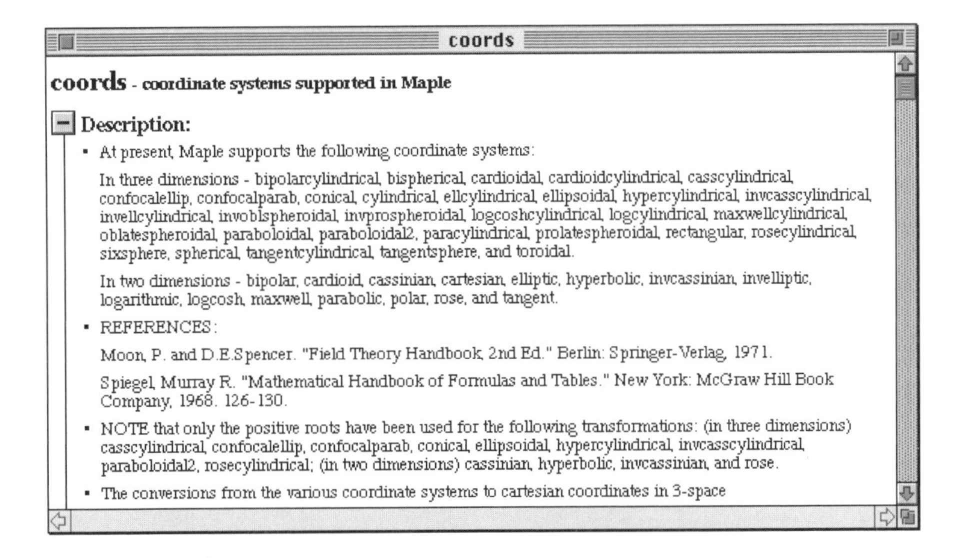

coords

coords - coordinate systems supported in Maple

Description:

- At present, Maple supports the following coordinate systems:

  In three dimensions - bipolarcylindrical, bispherical, cardioidal, cardioidcylindrical, casscylindrical, confocalellip, confocalparab, conical, cylindrical, ellcylindrical, ellipsoidal, hypercylindrical, invcasscylindrical, invellcylindrical, invoblspheroidal, invprospheroidal, logcoshcylindrical, logcylindrical, maxwellcylindrical, oblatespheroidal, paraboloidal, paraboloidal2, paracylindrical, prolatespheroidal, rectangular, rosecylindrical, sixsphere, spherical, tangentcylindrical, tangentsphere, and toroidal.

  In two dimensions - bipolar, cardioid, cassinian, cartesian, elliptic, hyperbolic, invcassinian, invelliptic, logarithmic, logcosh, maxwell, parabolic, polar, rose, and tangent.

- REFERENCES:

  Moon, P. and D.E.Spencer. "Field Theory Handbook, 2nd Ed." Berlin: Springer-Verlag, 1971.

  Spiegel, Murray R. "Mathematical Handbook of Formulas and Tables." New York: McGraw Hill Book Company, 1968. 126-130.

- NOTE that only the positive roots have been used for the following transformations: (in three dimensions) casscylindrical, confocalellip, confocalparab, conical, ellipsoidal, hypercylindrical, invcasscylindrical, paraboloidal2, rosecylindrical; (in two dimensions) cassinian, hyperbolic, invcassinian, and rose.

- The conversions from the various coordinate systems to cartesian coordinates in 3-space

Thus, many of these calculations (as well as plots generated with plot commands like `plot3d`) can be carried out in different coordinate systems by setting the `coords` option appropriately. Generally, cartesian is the default. For example, entering

```
> with(linalg):
  f:='f':
  grad(f[x,y,z],[x,y,z]);
```

$$\left[\frac{\partial}{\partial x} f_{x,\,y,\,z},\ \frac{\partial}{\partial y} f_{x,\,y,\,z},\ \frac{\partial}{\partial z} f_{x,\,y,\,z}\right]$$

computes the gradient of $f(x, y, z)$ in cartesian coordinates,

```
> grad(f[r,theta,phi],[r,theta,phi],coords=spherical);
```

$$\left[\frac{\partial}{\partial r} f_{r,\theta,\phi}, \frac{\frac{\partial}{\partial \theta} f_{r,\theta,\phi}}{r}, \frac{\frac{\partial}{\partial \phi} f_{r,\theta,\phi}}{r \sin(\theta)}\right]$$

computes the gradient of $f(r, \theta, \phi)$ in spherical coordinates, and

```
> grad(f[r,theta,z],[r,theta,z],coords=cylindrical);
```

$$\left[\frac{\partial}{\partial r} f_{r,\theta,z}, \frac{\frac{\partial}{\partial \theta} f_{r,\theta,z}}{r}, \frac{\partial}{\partial z} f_{r,\theta,z}\right]$$

computes the gradient of $f(r, \theta, z)$ in cylindrical coordinates. Similarly,

```
> curl([x[r,theta,phi],y[r,theta,phi],z[r,theta,phi]],
    [r,theta,phi],coords=spherical);
```

$$\left[\frac{r\cos(\theta)\, z_{r,\theta,\phi} + r\sin(\theta)\left(\frac{\partial}{\partial \theta} z_{r,\theta,\phi}\right) - r\left(\frac{\partial}{\partial \phi} y_{r,\theta,\phi}\right)}{r^2 \sin(\theta)},\right.$$

$$\frac{\left(\frac{\partial}{\partial \phi} x_{r,\theta,\phi}\right) - \sin(\theta)\, z_{r,\theta,\phi} - r\sin(\theta)\left(\frac{\partial}{\partial r} z_{r,\theta,\phi}\right)}{r\sin(\theta)},$$

$$\left.\frac{y_{r,\theta,\phi} + r\left(\frac{\partial}{\partial r} y_{r,\theta,\phi}\right) - \left(\frac{\partial}{\partial \theta} x_{r,\theta,\phi}\right)}{r}\right]$$

computes the curl of $\mathbf{R}(r, \theta, \phi) = x(r, \theta, \phi)\mathbf{i} + y(r, \theta, \phi)\mathbf{j} + z(r, \theta, \phi)\mathbf{k}$ in spherical coordinates.

If S is the graph of $f(x, y)$ and $g(x, y, z) = z - f(x, y)$, then the gradient $\nabla g(x, y, z)$ is a normal vector to the graph of $g(x, y, z) = 0$. At the point (x, y, z), a **unit normal vector, n,** can be obtained via:

$$\mathbf{n} = \frac{\nabla g(x, y, z)}{\|\nabla g(x, y, z)\|} = \frac{-f_x(x, y)\mathbf{i} - f_y(x, y)\mathbf{j} + \mathbf{k}}{\sqrt{(f_x(x, y))^2 + (f_y(x, y))^2 + 1}} = \frac{\langle -f_x(x, y), -f_y(x, y), 1\rangle}{\sqrt{(f_x(x, y))^2 + (f_y(x, y))^2 + 1}}$$

EXAMPLE 3: Let $w(x, y) = \cos(4x^2 + 9y^2)$. Let $\mathbf{n}(x, y)$ denote a unit vector normal to the graph of w at the point $(x, y, w(x, y))$. Find a formula for $\mathbf{n}$.

SOLUTION: In order to visualize the unit normal vector at points $(x, y, w(x, y))$ to the surface $w(x, y)$ this function is plotted using several of the options available with `plot3d`. The option `axes=BOXED` specifies that a box be displayed around the resulting `graphics3d` object while the option `grid=[40,45]` specifies that the number of sample points used consist of 40 along the x-axis and 45 along the y-axis for a total of $40 \times 45 = 1800$ sample points used in the generation of the graph The graph is named plotw and is displayed with `display3d`, which is contained in the `plots` package.

```
> w:=(x,y)->cos(4*x^2+9*y^2):
```

```
> with(plots):
  plotw:=plot3d(w(x,y),x=-1..1,y=-1..1,
       axes=BOXED,grid=[40,45]):
  display(plotw);
```

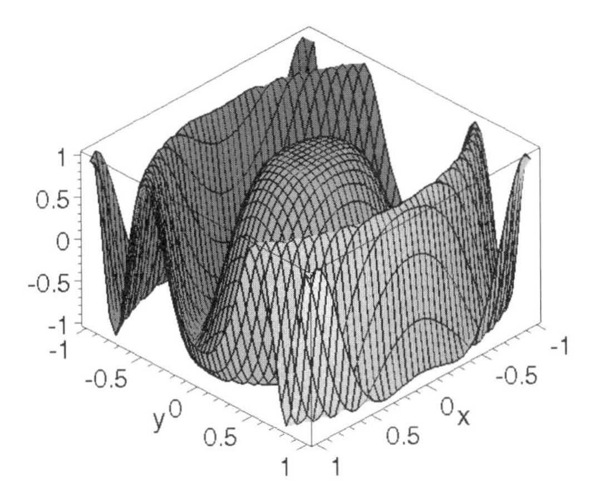

The equation $z = w(x, y)$ is written as $z - w(x, y) = 0$. The left-hand side of this equation is a function of x, y, and z, and it is defined as `wz`. Then, the `linalg` package is loaded. The gradient of `wz` is found with `grad` in `grad_w`.

```
> wz:=(x,y,z)->z-w(x,y):
  with(linalg):
  grad_w:=grad(wz(x,y,z),[x,y,z]);
```

$$grad\_w := [8\sin(4x^2+9y^2)\,x, 18\sin(4x^2+9y^2)\,y, 1]$$

Next $\|\nabla g(x, y, z)\|$ is computed and named `norm_grad_w`, using `norm` together with the frobenius option. Finally, by using this value, the ratio $\dfrac{\nabla g(x, y, z)}{\|\nabla g(x, y, z)\|}$ is calculated in `norm_w`. Notice that the command `scalarmul` is used. In its general form, `scalarmul(A,k)` calculates the product of the scalar `k` and the matrix (or vector) `A`. Also, notice that the result involves the expression `%1`, which is given below as $\sin(4x^2+9y^2)$.

```
> norm_grad_w:=norm(grad_w,frobenius);
```

$$\text{norm\_grad\_w} := \sqrt{1+64\left|\sin(4x^2+9y^2)\,x\right|^2+324\left|\sin(4x^2+9y^2)\,y\right|^2}$$

```
> norm_w:=scalarmul(grad_w,1/norm_grad_w);
```

$$norm\_w := \left[8\frac{\%1\,x}{\sqrt{\%2}}, 18\frac{\%1\,y}{\sqrt{\%2}}, \frac{1}{\sqrt{\%2}}\right]$$

$$\%1 = \sin(4x^2+9y^2)$$

$$\%2 = 1+64\left|\%1\,x\right|^2+324\left|\%1\,y\right|^2$$

We can graph various normal vectors with the command `fieldplot3d`, which is contained in the `plots` package. (Note that we do not have to reload the `plots` package, because we loaded it at the beginning of the example.) We graph `norm_w` in the cube given by $[-1,1] \times [-1,1] \times [-1,1]$ and name the resulting graph `plotn`.

```
> plotn:=fieldplot3d(norm_w,x=-1..1,y=-1..1,z=-1..1,
      color=black,axes=BOXED):
  display3d(plotn);
```

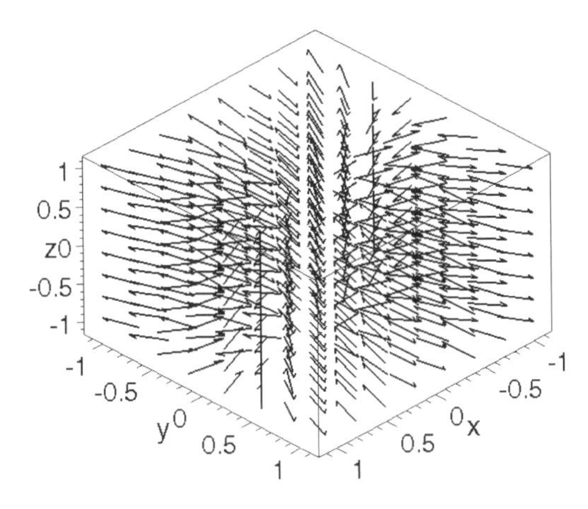

We then use `display3d` to show plotw and `plotn` together. Notice the way the normal vectors appear to vary continuously over the surface, which indicates that this is an oriented surface.

```
> display3d({plotw,plotn});
```

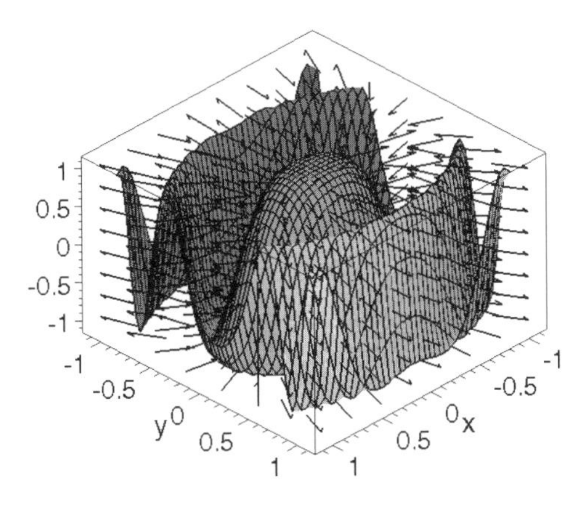

Similarly, we may graph the gradient vector field of the scalar function w with the command `gradplot3d`, which is contained in the `plots` package. In the following command, we include the option `color=BLACK` to instruct Maple to draw the vectors in black, and the option `axes=BOXED` to indicate that a box is to be placed around the resulting graphic.

```
> gradplot3d(w(x,y,z),x=-1..1,y=-1..1,z=-1..1,
      color=BLACK,axes=BOXED);
```

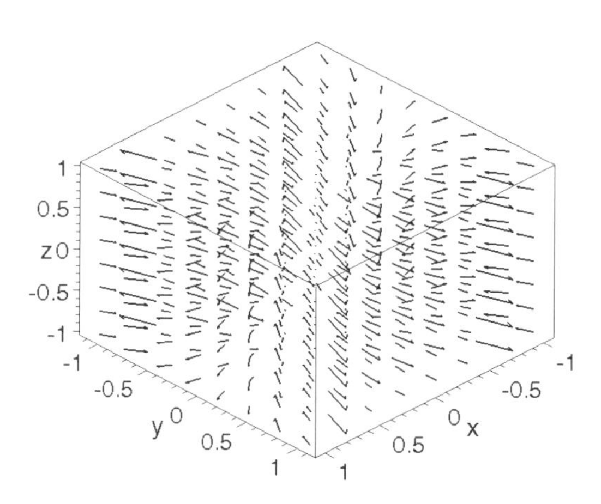

■

If, on the other hand, the surface S is defined parametrically by $\mathbf{r} = \mathbf{r}(u, v) = x(u, v)\mathbf{i} + y(u, v)\mathbf{j} + z(u, v)\mathbf{k}$ and is orientable, then $\frac{\partial \mathbf{r}}{\partial u} = \frac{\partial x}{\partial u}\mathbf{i} + \frac{\partial y}{\partial u}\mathbf{j} + \frac{\partial z}{\partial u}\mathbf{k}$, $\frac{\partial \mathbf{r}}{\partial v} = \frac{\partial x}{\partial v}\mathbf{i} + \frac{\partial y}{\partial v}\mathbf{j} + \frac{\partial z}{\partial v}\mathbf{k}$ and the vector $\frac{\partial \mathbf{r}}{\partial u} \times \frac{\partial \mathbf{r}}{\partial v}$ is normal to the surface. Graphically, you can tell if a surface is orientable if it has two sides.

EXAMPLE 4: A parametrization of **Umbilic Torus NC** is given by

$$\mathbf{r}_1(u, v) = x_1(u, v)\mathbf{i} + y_1(u, v)\mathbf{j} + z_1(u, v)\mathbf{k},$$

where $x_1(u, v) = \sin u\,(7 + \cos(u/3 - 2v) + 2\cos(u/3 + v))$,
$y_1(u, v) = \cos u\,(7 + \cos(u/3 - 2v) + 2\cos(u/3 + v))$, and
$z_1(u, v) = \sin(u/3 - 2v) + 2\sin(u/3 + v)$, $-\pi \le u \le \pi$, $-\pi \le v \le \pi$.
A parametrization of the **Mobius strip** is given by

$$\mathbf{r}_2(u, v) = x_2(u, v)\mathbf{i} + y_2(u, v)\mathbf{j} + z_2(u, v)\mathbf{k},$$

where $x_2(u, v) = (4 - v\sin u)\cos 2u$, $y_2(u, v) = (4 - v\sin u)\sin 2u$, and $z_2(u, v) = v\cos u$, $0 \le u \le \pi$, $-1 \le v \le 1$. Graphically show that these surfaces are nonorientable. Is the **torus** with parametrization

$$\mathbf{r}_3(u, v) = x_3(u, v)\mathbf{i} + y_3(u, v)\mathbf{j} + z_3(u, v)\mathbf{k},$$

where $x_3(u, v) = \cos u\,(3 + \cos v)$, $y_3(u, v) = \sin u\,(3 + \cos v)$, and $z_3(u, v) = \sin v$, $-\pi \le u \le \pi$, $-\pi \le v \le \pi$, orientable?

SOLUTION: After defining x_1, y_1, and z_1, we use plot3D to graph the umbilic torus. From the graph, we see that it has only one side (click and drag the graphic so you can see it from different viewpoints), so is not orientable.

```
> x[1]:=(u,v)->sin(u)*(7+cos(u/3-2*v)+2*cos(u/3+v)):
  y[1]:=(u,v)->cos(u)*(7+cos(u/3-2*v)+2*cos(u/3+v)):
  z[1]:=(u,v)->sin(u/3-2*v)+2*sin(u/3+v):
  plot3d([x[1](u,v),y[1](u,v),z[1](u,v)],
      u=-Pi..Pi,v=-Pi..Pi,grid=[45,45]);
```

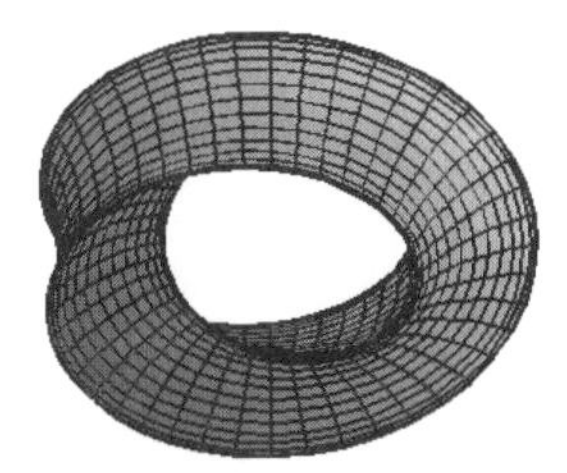

In the same way, we see that the Mobius strip is not orientable.

```
> x[2]:=(u,v)->(4-v*sin(u))*cos(2*u):
  y[2]:=(u,v)->(4-v*sin(u))*sin(2*u):
  z[2]:=(u,v)->v*cos(u):
  plot3d([x[2](u,v),y[2](u,v),z[2](u,v)],
      u=0..Pi,v=-1..1,grid=[30,30]);
```

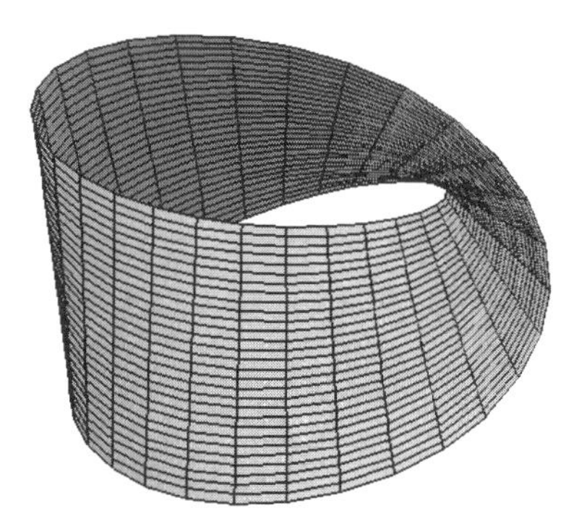

On the other hand, the torus is orientable. It has two distinct sides: the inside and the outside.

```
> x[3]:=(u,v)->cos(u)*(3+cos(v)):
  y[3]:=(u,v)->sin(u)*(3+cos(v)):
  z[3]:=(u,v)->sin(v):with(plots):
  plot3d([x[3](u,v),y[3](u,v),z[3](u,v)],
      u=-Pi..Pi,v=-Pi..Pi,grid=[40,40]);
```

■

Another interesting nonorientable surface is the Klein bottle. Unlike the Mobius strip and the umbilic torus, it cannot be **embedded** into 3-space, but it can be **immersed** into 3-space. The "usual" immersion of the Klein bottle has parametrization $\mathbf{r}(u, v) = x(u, v)\mathbf{i} + y(u, v)\mathbf{j} + z(u, v)\mathbf{k}$, where

$$x = \begin{cases} 6(1+\sin u)\cos u + r\cos u\cos v, & 0 \le u \le \pi \\ 6(1+\sin u)\cos u + r\cos v(v+\pi), & \pi \le u \le 2\pi \end{cases}, \quad y = \begin{cases} 16\sin u + r\sin u\cos v, & 0 \le u \le \pi \\ 16\sin u, & \pi \le u \le 2\pi \end{cases},$$

$z = r\sin v$, $r = 4\left(1 - \frac{1}{2}\cos u\right)$, $0 \le u \le 2\pi$, and $0 \le v \le 2\pi$.

```
> with(plots):
  r:=4*(1-cos(u)/2):
  p1:=plot3d([6*(1+sin(u))*cos(u)+r*cos(u)*cos(v),
```

```
    16*sin(u)+r*sin(u)*cos(v),r*sin(v)],
      u=0..Pi,v=0..2*Pi,grid=[40,40]):
p2:=plot3d([6*(1+sin(u))*cos(u)+r*cos(v+Pi),
    16*sin(u),r*sin(v)],u=Pi..2*Pi,v=0..2*Pi,
      grid=[40,40]):
display3d({p1,p2});
```

Another immersion of the Klein bottle, the "figure 8" immersion, has parametrization

$$\begin{cases} x = \left(a + \cos\left(\frac{1}{2}u\right)\sin v - \sin\left(\frac{1}{2}u\right)\sin 2v\right)\cos u \\ y = \left(a + \cos\left(\frac{1}{2}u\right)\sin v - \sin\left(\frac{1}{2}u\right)\sin 2v\right)\sin u, \ -\pi \le u \le \pi, \ -\pi \le v \le \pi. \\ z = \sin\left(\frac{1}{2}u\right)\sin v + \cos\left(\frac{1}{2}u\right)\sin 2v \end{cases}$$

```
> a:=3:
  x:=(a+cos(u/2)*sin(v)-sin(u/2)*sin(2*v))*cos(u):
  y:=(a+cos(u/2)*sin(v)-sin(u/2)*sin(2*v))*sin(u):
  z:=sin(u/2)*sin(v)+cos(u/2)*sin(2*v):
  plot3d([x,y,z],u=-Pi..Pi,v=-Pi..Pi,grid=[40,40]);
```

Application: Green's Theorem

Green's Theorem states: Let C be a piecewise smooth simple closed curve and let R be the region consisting of C and its interior. If f and g are functions that are continuous and have continuous first partial derivatives throughout an open region D containing R, then

$$\oint_C (m(x, y)dx + n(x, y)dy) = \iint_R \left(\frac{\partial n}{\partial x} - \frac{\partial m}{\partial y}\right) dA.$$

EXAMPLE 5: Use Green's Theorem to evaluate $\oint_C (x + e^{\sqrt{y}})dx + (2y + \cos x)dy$, where C is the boundary of the region enclosed by the parabolas $y = x^2$ and $x = y^2$.

SOLUTION: To calculate the limits of integration, we use Maple to graph the functions x^2 and $\sqrt{x}$. Note that the two functions intersect at the points (0,0) and (1,1).

```
> plot({x^2,sqrt(x)},x=0..1.1,0..1.2);
```

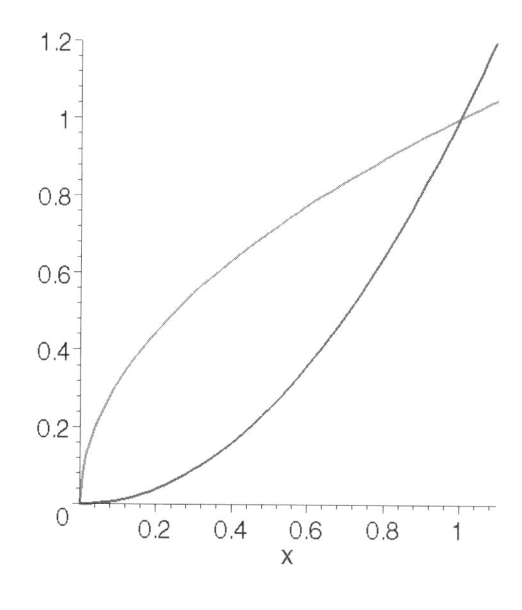

In this example, $m(x, y) = x + e^{\sqrt{y}}$ and $n(x, y) = 2y + \cos x$. Therefore, applying Green's Theorem,

$$\oint_C (x + e^{\sqrt{y}}) + (2y + \cos x)\text{dy} = \oint_C m(x, y)dx + n(x, y)dy$$

$$= \iint_R \left(\frac{\partial n}{\partial x} - \frac{\partial m}{\partial y}\right)dA = \int_0^1 \int_{x^2}^{\sqrt{x}} \left(\frac{\partial n}{\partial x} - \frac{\partial m}{\partial y}\right)dydx.$$

Next, we will use Maple to define $m(x, y)$, $n(x, y)$, and to compute $\partial n/\partial x$, $\partial m/\partial y$, and

$$\int_0^1 \int_{x^2}^{\sqrt{x}} \left(\frac{\partial n}{\partial x} - \frac{\partial m}{\partial y}\right)dydx.$$

First, the functions $m(x, y)$, and $n(x, y)$ are defined. Recall that when the partial derivatives are computed, the variable of differentiation must be specified. These partial derivatives are calculated in `nx` and `my`, respectively.

```
> m:=(x,y)->x+exp(sqrt(y)):
  n:=(x,y)->2*y+cos(x):

> nx:=diff(n(x,y),x);
  my:=diff(m(x,y),y);
```

$$nx := -\sin(x)$$

$$my := \frac{1}{2}\frac{e^{(\sqrt{y})}}{\sqrt{y}}$$

`int` is used to find an exact value of the integral $\int_0^1 \int_{x^2}^{\sqrt{x}} \left(\frac{\partial n}{\partial x} - \frac{\partial m}{\partial y}\right)dydx$.

```
> int(int(nx-my,y=x^2..sqrt(x)),x=0..1);
```

$$2\cos(1) - \frac{1}{2}\sqrt{2}\sqrt{\pi}\,\text{FresnelC}\left(\frac{\sqrt{2}}{\sqrt{\pi}}\right) - 27 + 2\sin(1) + 9\mathbf{e}$$

although the result contains the `FresnelC` function, which is defined by $\int_0^x \cos\left(\frac{\pi}{2}t^2\right)dt$. We approximate the integral with `evalf`.

```
> evalf(int(int(nx-my,y=x^2..sqrt(x)),x=0..1));
                    -.6764412004
```

■

Application: The Divergence Theorem

The **Divergence Theorem** states: Let Q be any domain with the property that each line through any interior point of the domain cuts the boundary in exactly two points, and such that the boundary S is a piecewise smooth closed, oriented surface with unit outer normal **n**. If **f** is a vector field that has continuous partial derivatives on Q, then

$$\iint_S \mathbf{f} \bullet \mathbf{n} dS = \iiint_Q \mathrm{div}\mathbf{f} dV = \iiint_Q \nabla \bullet \mathbf{f} dV.$$

$\iint_S \mathbf{f} \bullet \mathbf{n} dS$ is called the **outward flux** of the vector field **f** across the surface S. If S is a portion of the level curve $g(x, y) = c$ for some g, then a unit normal vector **n** may be taken to be either $\mathbf{n} = \frac{\nabla g}{\|\nabla g\|}$ or $\mathbf{n} = \frac{-\nabla g}{\|\nabla g\|}$.

Recall the following formulas for the evaluations of surface integrals: Let S be the graph of $z = (f(x, y)$ $(y = h(x, z)$ or $x = k(y, z))$ and let $R_{xy}(R_{xz}$ or $R_{yz})$ be the projection of S on the xy- (xz- or yz-) plane. Then,

$$\iint_S g(x, y, z) dS = \begin{cases} \iint_{R_{xy}} g(x, y, f(x, y)) \sqrt{(f_x(x, y))^2 + (f_y(x, y))^2 + 1} dA \\ \iint_{R_{xz}} g(x, h(x, z), z) \sqrt{(h_x(x, z))^2 + (h_z(x, z))^2 + 1} dA. \\ \iint_{R_{yz}} g(k(y, z), y, z) \sqrt{(k_y(y, z))^2 + (k_z(y, z))^2 + 1} dA \end{cases}$$

EXAMPLE 6: Use the Divergence Theorem to compute the outward flux of the field

$$\mathbf{vf}(x, y, z) = \langle xy + x^2yz, yz + xy^2z, xz + xyz^2 \rangle$$

through the surface of the cube cut from the first octant by the planes $x = 2$, $y = 2$, and $z = 2$.

SOLUTION: By the Divergence Theorem,

$$\underset{\text{Cube Surface}}{\iint \mathbf{vf} \bullet \mathbf{n}\, dA} = \underset{\text{Cube Interior}}{\iiint \nabla \bullet \mathbf{vf}\, dV}.$$

Notice that without the Divergence Theorem, calculating $\underset{\text{Cube Surface}}{\iint \mathbf{vf} \bullet \mathbf{n}\, dA}$ would require six separate integrals. However, with the Divergence Theorem, the flux can be calculated by integrating the divergence. Because we need the command `diverge`, we load the `linalg` package. The vector field is defined in `vf` as a list of three elements, the x, y, and z components so that the divergence can be determined in `div_v`. The divergence is then integrated over cube $[0,2] \times [0,2] \times [0,2]$ to yield a value of 72.

```
> with(linalg):
  vf:=(x,y,z)
       ->[x*y+x^2*y*z,y*z+x*y^2*z,x*z+x*y*z^2];
```

$$vf := (x,y,z) \rightarrow [xy + x^2yz, yz + xy^2z, xz + xyz^2]$$

```
> div_v:=diverge(vf(x,y,z),[x,y,z]);
```

$$div\_v = y + 6xyz + z + x$$

```
> int(int(int(div_v,z=0..2),y=0..2),x=0..2);
```

72

In the same manner as in Example 3, we can use the command `fieldplot3d`, contained in the plots package, to graph the vector field `vf`. After loading the `plots` package, we graph `vf` in the cube$[0,2] \times [0,2] \times [0,2]$.

```
> with(plots):
  fieldplot3d(vf(x,y,z),x=0..2,y=0..2,z=0..2,
```

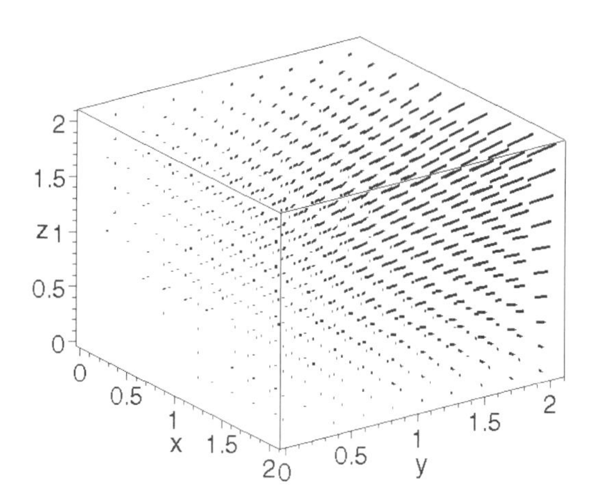

■

Application: Stoke's Theorem

Stoke's Theorem says: Let S be an oriented surface with finite surface area, unit normal $\mathbf{n}$, and boundary C. Let $\mathbf{F}$ be a continuous vector field defined on S such that the component functions of $\mathbf{F}$ have continuous partial derivatives at each nonboundary point of S. Then, $\oint_C \mathbf{F} \bullet d\mathbf{r} = \iint_S (\text{curl } \mathbf{F}) \bullet \mathbf{n} dS$. In other words, the surface integral of the normal component of the curl of $\mathbf{F}$ taken over S equals the line integral of the tangential component of the field taken over C: $\oint_C \mathbf{F} \bullet \mathbf{T} ds = \iint_S (\text{curl } \mathbf{F}) \bullet \mathbf{n} dS$. In particular, if $\mathbf{F} = \langle M, N, P \rangle = M\mathbf{i} + N\mathbf{j} + P\mathbf{k}$, then

$$\int_C M(x, y, z)dx + N(x, y, z)dy + P(x, y, z)dz = \iint_S (\text{curl } \mathbf{F}) \bullet \mathbf{n} dS.$$

EXAMPLE 7: Verify Stoke's Theorem for the vector field $\mathbf{vf}(x, y, z) = \langle y^2 - z, z^2 + x, x^2 - y \rangle$ and S the paraboloid $z = f(x, y) = 4 - (x^2 + y^2)$, z nonnegative.

SOLUTION: Because we must show $\oint_C \mathbf{vf} \bullet d\mathbf{r} = \iint_S (\text{curl } \mathbf{vf}) \bullet \mathbf{n} dS$, we must compute curl **vf**, **n**, $\iint_S (\text{curl } \mathbf{vf}) \bullet \mathbf{n} dS$, **r**, dr, **r**, dr, and $\oint_C \mathbf{vf} \bullet d\mathbf{r}$. We begin by loading the `linalg` package and defining the the vector field `vf` and the function `f`. The curl of `vf` is then computed in `curl_vf`. The function $h(x, y, z) = z - f(x, y)$ which will be used in the computation of the unit normal vector is also defined. Hence, the normal vector to the surface is given by ∇h which is found in `norm_to_surf`.

```
> with(linalg):
  vf:=(x,y,z)->[y^2-z,z^2+x,x^2-y]:
  f:=(x,y)->4-(x^2+y^2):
```

```
> curl_vf:=curl(vf(x,y,z),[x,y,z]);
```

$$curl\_vf := [-1-2z, -1-2, x1-2y]$$

```
> h:=(x,y,z)->z-f(x,y);
  norm_to_surf:=grad(h(x,y,z),[x,y,z]);
```

$$h = (x,y,z) \to z - f(x, y)$$

$$norm\_to\_surf := [2x, 2y, 1]$$

Because `norm_to_surf` is a normal vector to the surface, $\frac{\nabla(\text{norm\_to\_surf})}{|\nabla(\text{norm\_to\_surf})|}$ represents a unit normal vector. This vector is found in `un` with `norm` using the `frobenius` option because the denominator of $\frac{\nabla(\text{norm\_to\_surf})}{|\nabla(\text{norm\_to\_surf})|}$ is a scalar while the numerator is a vector.

```
> un:=scalarmul(norm_to_surf,
      1/norm(norm_to_surf,frobenius));
```

$$un := \left[2\frac{x}{\sqrt{1+4|x|^2+4|y|^2}}, 2\frac{y}{\sqrt{1+4|x|^2+4|y|^2}}, \frac{1}{\sqrt{1+4|x|^2+4|y|^2}}\right]$$

Notice that Maple does not automatically assume that x and y are real numbers, so we use assume to instruct Maple to assume that x and y are real numbers.

```
> assume(x,real):assume(y,real):
  un:=simplify(scalarmul(norm_to_surf,
      1/norm(norm_to_surf,frobenius)));
```

$$un := \left[2\frac{x\sim}{\sqrt{1+4x\sim^2+4y\sim^2}}, 2\frac{y\sim}{\sqrt{1+4x\sim^2+4y\sim^2}}, \frac{1}{\sqrt{1+4x\sim^2+4y\sim^2}}\right]$$

The dot product of `curl_vf` and `un` is calculated, simplified, and named `g`. Note that `simplify` writes `dotprod(curl_vf,un)` as a single fraction.

```
> g:=simplify(dotprod(curl_vf,un));
```

$$g := -\frac{2x\sim+4x\sim z\sim+4y\sim+4y\sim x\sim-1}{\sqrt{1+4x\sim^2+4y\sim^2}}$$

By the surface integral evaluation formula,

$$\iint_S (\text{curl } \mathbf{vf}) \bullet \mathbf{n}\, dS = \iint_S g(x,y,z)\,dS = \iint_R g(x,y,f(x,y))\sqrt{(f_x(x,y))^2+(f_y(x,y))^2+1}\,dA,$$

where R is the projection of $f(x, y)$ on the xy-plane. Hence, in this example, R is the region bounded by the graph of the circle $x^2 + y^2 = 4$. Thus,

$$\iint_R g(x,y,f(x,y))\sqrt{(f_x(x,y))^2+(f_y(x,y))^2+1}\,dA =$$

$$\int_{-2}^{2}\int_{-\sqrt{4-x^2}}^{\sqrt{4-x^2}} g(x,y,f(x,y))\sqrt{(f_x(x,y))^2+(f_y(x,y))^2+1}\,dA.$$

This surface integral is computed to yield a value of 4π.

```
> to_integrate:=simplify(subs(z=f(x,y),g)*
  sqrt((diff(f(x,y),x)^2+diff(f(x,y),y)^2+1)));
```

$$to\_integrate := -18x\sim+4x\sim^3+4x\sim y\sim^2-4y\sim-4y\sim x\sim+1$$

```
> int(int(to_integrate,
      y=-sqrt(4-x^2)..sqrt(4-x^2)),x=-2..2);
```

$$4\pi$$

Notice that the integral

$$\int_{-2}^{2}\int_{-\sqrt{4-x^2}}^{\sqrt{4-x^2}} g(x, y, f(x, y))\sqrt{(f_x(x, y))^2 + (f_y(x, y))^2 + 1}\,dA$$

can be easily evaluated using polar coordinates. To do so, replace each occurrence of x and y in $g(x, y, f(x, y))\sqrt{(f_x(x, y))^2 + (f_y(x, y))^2 + 1}$ by $r\cos t$ and $r\sin t$, respectively. In `polar_approach`, we substitute the polar coordinates `x=r*cos(theta)` and `y=r*sin(theta)` into the expression calculated above in `to_integrate`. Notice that the result involves powers of the trigonometric functions. Hence, in `polar_int`, we use `simplify` with the `trig` option setting to simplify the expression.

```
> polar_approach:=
  subs(x=r*cos(t),y=r*sin(t),to_integrate);
```

$$polar\_approach := -18r\cos(\theta) + 4r^3\cos(\theta)^3 + 4r^3\cos(\theta)\sin(\theta)^2 - 4r\sin(\theta) - 4r^2\sin(\theta)\cos(\theta) + 1$$

```
> polar_int:=simplify(polar_approach,trig);
```

$$polar\_int := -18r\cos(\theta) + 4r^3\cos(\theta) - 4r\sin(\theta) - 4r^2\sin(\theta)\cos(\theta) + 1$$

The expression in `polar_int` is then integrated over the circular region R: $0 \le r \le 2$, $0 \le t \le 2\pi$ to yield the value of 4π, which was obtained in the integral in Cartesian coordinates above.

```
> int(int(polar_int*r,theta=0..2*Pi),r=0..2);
```

$$4\pi$$

Now, to verify Stoke's Theorem, we must compute the associated line integral. We begin by noticing that the boundary of $z = f(x, y) = 4 - (x^2 + y^2)$, $z \ge 0$ is the circle $x^2 + y^2 = 4$, which has parameterization `x=2*cos(s)`, `y=2*sin(s)`, and `z=0` for $0 \le s \le 2\pi$. This parameterization is substituted into `vf` and named `pvf`. In order to evaluate the line integral along the circle, we must define the parameterization of the circle and calculate `dr`. This is done below in `r` and `dr`, respectively. The dot product of `pvd` and `dr` represents the integrand of the line integral.

```
> pvf:=vf(2*cos(s),2*sin(s),0);
```

$$pvf := [4\sin(s)^2, 2\cos(s), 4\cos(s)^2 - 2\sin(s)]$$

```
> r:=s->[2*cos(s),2*sin(s),0]:
  dr:=diff(r(s),s);
```

$$dr = [-2\sin(s), 2\cos(s), 0]$$

After clearing prior definitions of `to_integrate`, we calculate this dot product. The resulting expression in `to_integrate` is then integrated along the circle to yield a value of 4π, which verifies Stoke's Theorem.

```
> assume(s,real):
  to_integrate:='to_integrate':
  to_integrate:=dotprod(pvf,dr);
```

$$to\_integrate := -8\sin(s\sim)^3 + 4\cos(s\sim)^2$$

```
> int(to_integrate,s=0..2*Pi);
```

$$4\pi$$

■

5.5 Using Results in Later Maple Sessions

Beginning users of Maple quickly notice that in order for results from a previous Maple session to be used, they must first be recalculated. The purpose of this example is to illustrate how results can be saved for future use. We place this topic here since it is frequently necessary to construct a set or array of numbers during one Maple session and then use the table or array in a subsequent Maple session.

Application: Constructing a Table of Zeros of Bessel Functions

The zeros of the Bessel functions play an important role in the generalized Fourier series involving Bessel functions, use Maple to find the first eight zeros of the Bessel functions of the first kind, $J_n(x)$, of order $n = 0, 1, \ldots, 8$.

SOLUTION: The **Bessel function of the first kind of order** n, $J_n(x)$, is represented by `BesselJ(n,x)`. We graph the Bessel functions of the first kind of order n for $n = 0, 1, \ldots, 8$ on the interval [0,40] with `animate` and then display the graphs as an array with `display`.

```
> with(plots):
  A:=animate(BesselJ(i,x),x=0..40,i=0..8,
       frames=9,color=BLACK,numpoints=100):
  display(A);
```

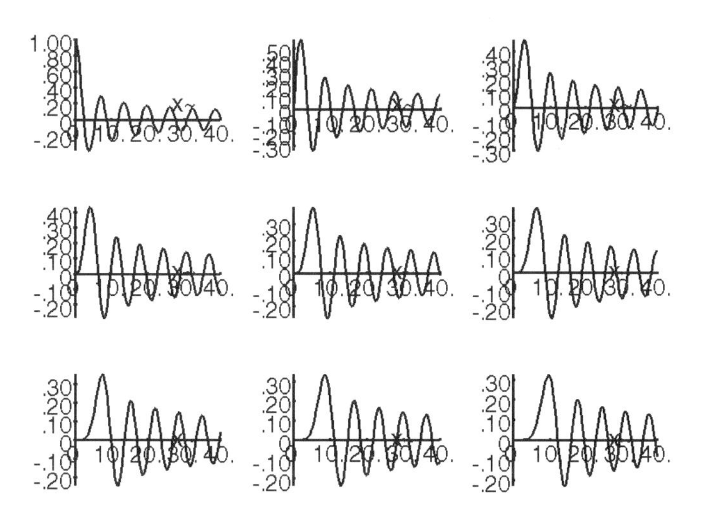

In order to approximate the zeros of the Bessel functions, we will use the command fsolve. Recall that fsolve(equation,x,a..b) attempts to locate an approximation of the solution to equation, which represents an equation in x, on the interval (a,b). We use these graphs to approximate (open) intervals containing the zeros to be approximated. For example, we see that the first zero of the Bessel function of order zero occurs in the interval (2,3), the second in (5,6), the third in (8,9), the fourth in (11,12), the fifth in (14,15), the sixth in (18,19), the seventh in (21,22), and the eighth in (24,25). Hence, these initial intervals are entered in the array guesses. Note that guesses[1] corresponds to the list of eight intervals containing the first eight zeros of the Bessel function of order 0. Hence, guesses[i] corresponds to the list of eight intervals containing the first eight zeros of the Bessel function of order i–1; guesses[i,j] corresponds to the interval containing the jth zero of the Bessel function of order i–1.

```
> guesses:=array(
         [[2..3,5..6,8..9,11..12,14..15,
         18..19,21..22,24..25],
         [3..4,7..8,10..11,13..14,16..17,
            19..20,22..23,25..26],
```

```
[5..6,8..9,11..12,14..15,17..18,
   21..22,24..25,27..28],
[6..7,9..10,13..14,16..17,19..20,
   22..23,25..26,28..29],
[7..8,11..12,14..15,17..18,20..21,
   24..25,27..28,30..31],
[8..9,12..13,15..16,18..19,22..23,
   25..26,28..29,31..32],
[9..10,13..14,17..18,20..21,23..24,
   26..27,30..31,33..34]]):
```

The function `alpha` uses `fsolve` and the intervals in `guesses` to approximate the zeros of the Bessel functions. In the following two `for` loops, `i` corresponds to the Bessel function of order i and `j` represents the jth zero. For example, for $i = 0$, the first eight zeros of the Bessel function of order 0 are computed by using `fsolve` to approximate the zero on each of the intervals in the first list in the array `guesses`. This is carried out for $i = 0$ to $i = 6$ to yield a table of zeros of the Bessel functions of order 0 to order 6. This table is then printed. Notice that each zero corresponds to an orderd pair (i,j) that represents the jth zero of the Bessel function of order i.

```
> i:='i':j:='j':
  alpha:=table():
  for i from 0 to 6 do
   for j to 8 do
      alpha[i,j]:=
         fsolve(BesselJ(i,x)=0,x,guesses[i+1,j])
         od od:


> print(alpha);
```

table([

(0, 6) = 18.07106397	(1, 1) = 3.831705970
(2, 6) = 21.11699705	(0, 1) = 2.404825558
(4, 7) = 27.19908777	(3, 2) = 9.761023130
(6, 5) = 23.58608444	(5, 3) = 15.70017408

(1, 6) = 19.61585851	(2, 2) = 8.417244140
(3, 7) = 25.74816670	(4, 3) = 14.37253667
(0, 8) = 24.35247153	(6, 1) = 9.936109524
(5, 5) = 22.21779990	(1, 2) = 7.015586670
(2, 7) = 24.27011231	(3, 3) = 13.01520072
(4, 8) = 30.37100767	(0, 2) = 5.520078110
(6, 6) = 26.82015198	(2, 3) = 11.61984117
(1, 7) = 22.76008438	(4, 4) = 17.61596605
(3, 8) = 28.90835078	(6, 2) = 13.58929017
(5, 6) = 25.43034115	(1, 3) = 10.17346814
(2, 8) = 27.42057355	(3, 4) = 16.22346616
(6, 7) = 30.03372239	(0, 7) = 21.21163663
(1, 8) = 25.90367209	(2, 4) = 14.79595178
(5, 7) = 28.62661831	(4, 5) = 20.82693296
(0, 5) = 14.93091771	(6, 3) = 17.00381967
(6, 8) = 33.23304176	(0, 4) = 11.79153444
(5, 8) = 31.81171672	(1, 4) = 13.32369194
(5, 1) = 8.771483816	(3, 5) = 19.40941523
(4, 1) = 7.588342435	(2, 5) = 17.95981950
(5, 4) = 18.98013388	(4, 6) = 24.01901952
(3, 1) = 6.380161896	(6, 4) = 20.32078921
(5, 2) = 12.33860420	(1, 5) = 16.47063005
(2, 1) = 5.135622302	(3, 6) = 22.58272959
(4, 2) = 11.06470949	(0, 3) = 8.653727913])

```
> ALPHA:=array(1..7,1..8);
> for i from 1 to 7 do
  for j from 1 to 8 do
      ALPHA[i,j]:=evalf(alpha[i-1,j],5) od od:
```

```
print(ALPHA);
```

$$\begin{bmatrix} 2.4048 & 5.5201 & 8.6537 & 11.792 & 14.931 & 18.071 & 21.212 & 24.352 \\ 3.8217 & 7.0156 & 10.173 & 13.3242 & 16.471 & 19.616 & 22.760 & 25.904 \\ 5.1356 & 8.4172 & 11.620 & 14.796 & 17.960 & 21.117 & 24.270 & 27.421 \\ 6.3802 & 9.7610 & 13.015 & 16.223 & 19.409 & 22.583 & 25.748 & 28.908 \\ 7.5883 & 11.065 & 14.373 & 17.616 & 20.827 & 24.019 & 27.199 & 30.371 \\ 8.7715 & 12.339 & 15.700 & 18.980 & 22{,}218 & 25.430 & 28.627 & 31.812 \\ 9.9361 & 13.589 & 17.004 & 20.321 & 23.586 & 26.820 & 30.034 & 33.233 \end{bmatrix}$$

We then save this table of numbers, for later use, in a text editor and name the resulting file **besselzeros**. In this particular case, a word processor was used to manipulate the list so that the result is an array of numbers. The array of numbers is displayed in the following screenshot. Note that the first row corresponds to the zeros of the Bessel function of order zero, the second row corresponds to the zeros of the Bessel function of order one, and so on. An alternative approach would be to save a Maple scratchpad as **besselzeros**.

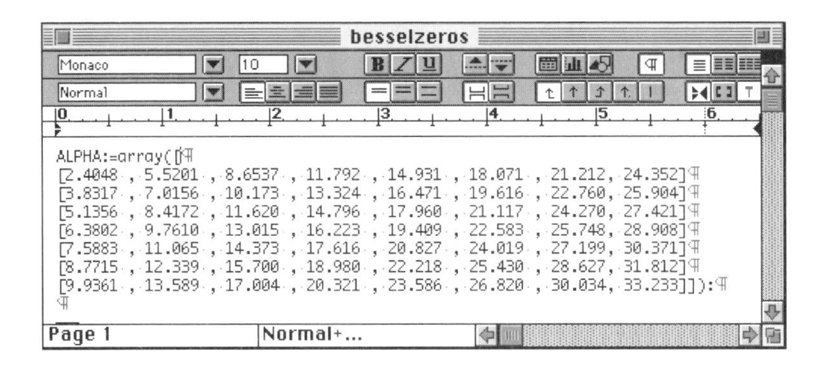


```
ALPHA:=array([
[2.4048 , 5.5201 , 8.6537 , 11.792 , 14.931 , 18.071 , 21.212, 24.352]
[3.8317 , 7.0156 , 10.173 , 13.324 , 16.471 , 19.616 , 22.760, 25.904]
[5.1356 , 8.4172 , 11.620 , 14.796 , 17.960 , 21.117 , 24.270, 27.421]
[6.3802 , 9.7610 , 13.015 , 16.223 , 19.409 , 22.583 , 25.748, 28.908]
[7.5883 , 11.065 , 14.373 , 17.616 , 20.827 , 24.019 , 27.199, 30.371]
[8.7715 , 12.339 , 15.700 , 18.980 , 22.218 , 25.430 , 28.627, 31.812]
[9.9361 , 13.589 , 17.004 , 20.321 , 23.586 , 26.820 , 30.034, 33.233]]):
```

CHAPTER 6

Related Topics from Linear Algebra

Although many calculations encountered in elementary linear algebra were discussed in Chapter 5, Maple V has an extensive library of linear algebra functions. In this chapter, we discuss some of the additional commands found in the linear algebra package `linalg` that perform computations similar to those encountered in elementary linear algebra classes. In each case, be sure to load the package by entering `with(linalg)` before you enter any of the commands contained in the linear algebra package. With the following command, we load the linear algebra package `linalg`; the commands contained in the package are displayed. Remember that if a colon (:) is included at the end of the command, the resulting output is suppressed. Throughout this chapter, we assume that you have loaded the `linalg` package before completing any example.

```
> with(linalg);
```

[BlockDiagonal, GramSchmidt, JordanBlock, LUdecomp, QRdecomp, Wronskian, addcol, addrow, adj, adjoint, angle, augment, backsub, band, basis, bezout, blockmatrix, charmat, charpoly, cholesky, col. coldim, colspace, colspan, companion, concat, cond, copyinto, crossprod, curl, definite, delcols, delrows, det, diag, diverge, dotprod, eigenvals, eigenvalues, eigenvectors, eigenvects, entermatrix, equal, exponential, extend, ffgausselim, fibonacci, forwardsub, frobenius, gausselim, gaussjord, geneqns, genmatrix, grad, hadamard, hermite, hesssian, hilbert, htranspose, ihermite, indexfunc, innerprod, intbasis, inverse, ismith, issimilar, iszero, jacobian, jordan, kernel, laplacian, leastsqrs, linsolve, matadd, matrix, minor, minpoly, mulcol, mulrow, multiply, norm, normalize, nullspace, orthog, permanent, pivot, potential, randmatrix,

randvector, rank, ratform, row, rowdim, rowspace, rowspan, rref, scalarmul, singularvals, smith, stackmatrix, submatrix, subvector, sumbasis, swapcol, swaprow, sylvester, toeplitz, trace, transpose, vandermonde, vecpotent, vectdim, vector, wronskian]

In addition, detailed information regarding the linalg package is obtained by entering `?linalg`.

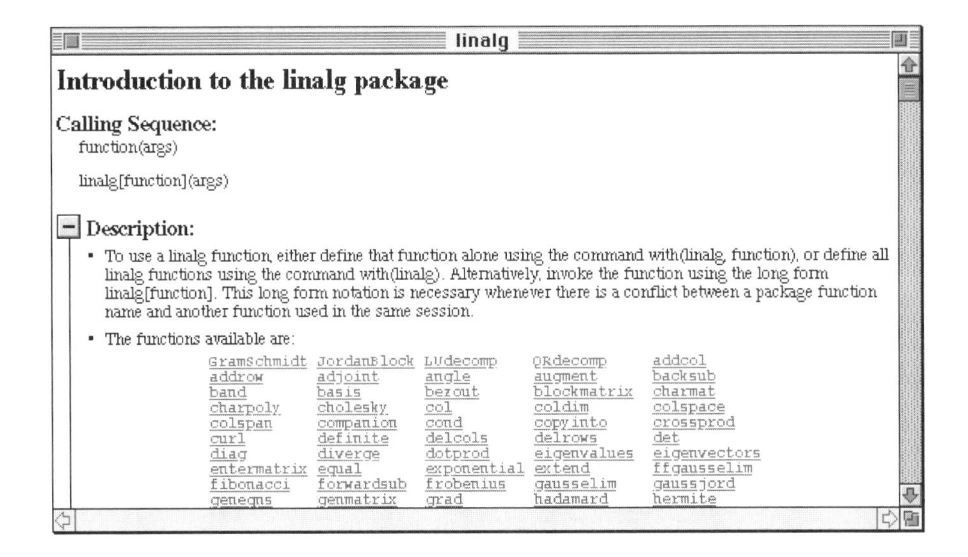

6.1 Elementary Row and Column Operations and Related Matrix Operations

Elementary Row and Column Operations

The linalg package contains several commands that perform elementary row (column) operations:

1. interchanging two rows (columns) of a matrix
2. multiplying a row (column) by a nonzero number
3. adding a constant multiple of one row (column) to another

These and related commands, contained in the linear algebra package `linalg`, are discussed in the following examples. Be sure that you have loaded the linalg package before you complete any of them.

`addcol`

`addcol(mat,coli,colj,m)` replaces column `colj` of the matrix `mat` with `m*coli+colj` and returns the resulting matrix.

EXAMPLE 1: Add 3 times column one of the matrix $\begin{pmatrix} -1 & 3 \\ -3 & 5 \\ 3 & -5 \\ 4 & 4 \end{pmatrix}$ to column two.

SOLUTION: The 4×2 matrix is defined in `cm`. In `stepone`, 3 times column one of `cm` is added to column two of `cm`.

```
> cm:=matrix(4,2,[-1,3,-3,5,3,-5,4,4]);
```

$$cm := \begin{bmatrix} -1 & 3 \\ -3 & 5 \\ 3 & -5 \\ 4 & 4 \end{bmatrix}$$

```
> stepone:=addcol(cm,1,2,3);
```

$$stepone := \begin{bmatrix} -1 & 0 \\ -3 & -4 \\ 3 & 4 \\ 4 & 16 \end{bmatrix}$$

■

`addrow`

`addrow(mat,rowi,rowj,m)` is similar to `addcol` in that it replaces `rowj` of matrix `mat` with `rowj+m*rowi`, and returns the matrix that results.

EXAMPLE 2: Add 3 times row one of the matrix $\begin{pmatrix} -1 & 3 \\ -3 & 5 \\ 3 & -5 \\ 4 & 4 \end{pmatrix}$ to row two.

SOLUTION: We define the matrix in cm. The desired matrix is obtained with addrow.

```
> cm:=matrix(4,2,[-1,3,-3,5,3,-5,4,4]);
```

$$cm := \begin{bmatrix} -1 & 3 \\ -3 & 5 \\ 3 & -5 \\ 4 & 4 \end{bmatrix}$$

```
> steptwo:=addrow(cm,1,2,3);
```

$$steptwo := \begin{bmatrix} -1 & 3 \\ -6 & 14 \\ 3 & -5 \\ 4 & 4 \end{bmatrix}$$

■

col

col(mat,n) extracts the *n*th column from the matrix mat; col(mat,n..m) extracts the *n*th though *m*th columns of the matrix mat.

row

row(mat,n) extracts the *n*th row from the matrix mat; row(mat,n..m) extracts the *n*th through *m*th rows from the matrix mat.

EXAMPLE 3: Extract the (a) second and third columns and (b) second and third rows from $A=\begin{pmatrix} -4 & 1 & -4 & 3 \\ -2 & -4 & -2 & -3 \\ 0 & -1 & -1 & 1 \\ -4 & 1 & -3 & 1 \end{pmatrix}$.

SOLUTION: The 4×4 matrix `A` is defined so that the second and third columns can be extracted with `col` in the list `twocolumns`.

```
> A:=matrix(4,4,[-4,1,-4,3,-2,-4,-2,-3,
       0,-1,-1,1,-4,1,-3,1]);
```

$$A := \begin{bmatrix} -4 & 1 & -4 & 3 \\ -2 & -4 & -2 & -3 \\ 0 & -1 & -1 & 1 \\ -4 & 1 & -3 & 1 \end{bmatrix}$$

```
> twocolumns:=col(A,2..3);
```

$$[1,-4,-1,1],[-4,-2,-1,-3]$$

Note that the same result is obtained with the command `twocolumns:=[col(A,2),col(A,3)];`. Similarly,

```
> row(A,2..3);
```

$$[-2,-4,-2,-3],[0,-1,-1,1]$$

extracts the second through third rows of **A**.

■

`gaussjord`

`gaussjord(mat)` performs Gauss-Jordan elimination on the matrix `mat` so that the matrix that results is of reduced row-echelon form. Note that `gaussjord(mat)` produces the same result as `rref(mat)`.

EXAMPLE 4: Find the augmented matrix that corresponds to the system $\begin{cases} -2x+y-2z=4 \\ 2x-4y-2z=-4 \\ x-4y-2z=3 \end{cases}$ and reduce the matrix to reduced row-echelon form.

SOLUTION: This system is equivalent to the matrix equation **Ax=b,** where A= $\begin{pmatrix} -2 & 1 & -2 \\ 2 & -4 & -2 \\ 1 & -4 & -2 \end{pmatrix}$ is the coefficient matrix and b= $\begin{pmatrix} 4 \\ -4 \\ 3 \end{pmatrix}$. After defining the system of equations in `sys`, we use `genmatrix` to generate **A** and **b**.

```
> sys:={-2*x+y-2*z=4,2*x-4*y-2*z=-4,x-4*y-2*z=3}:
  A:=genmatrix(sys,[x,y,z],'
```

The augmented matrix that corresponds to the system of equations is defined in `C` with `augment`, where `augment(A,b)` forms the augmented matrix [A:**b**].

```
> C:=augment(A,b);
```

$$C:=\begin{bmatrix} 2 & -4 & -2 & -4 \\ 1 & -4 & -2 & 3 \\ -2 & 1 & -2 & 4 \end{bmatrix}$$

Gauss-Jordan elimination is performed with `gaussjord` in `solmatrix`.

```
> solmatrix:=gaussjord(C);
```

$$solmatrix:=\begin{bmatrix} 1 & 0 & 0 & -7 \\ 0 & 1 & 0 & -4 \\ 0 & 0 & 1 & 3 \end{bmatrix}$$

The result means that the solution to the system is $(x,y,z)=(-7,-4,3)$, which we confirm with `solve`.

```
> solve(sys);
```

$$\{z = 3, y = -4, x = -7\}$$

■

rref

rref(mat) uses elementary row operations to place the matrix mat in reduced row-echelon form. This command may be entered as rref(mat,'r'), which assigns the value of the rank of mat the name r. Also, rref(mat,'r','d') assigns the value of the determinant of mat the name d. Finally, rref(mat,val) indicates that the elimination procedure ceases at column val. Note that gaussjord(matrix) produces the same result as rref(matrix).

EXAMPLE 5: Use rref to place the matrix A= $\begin{pmatrix} -1 & -1 & 2 & 0 & -1 \\ -2 & 2 & 0 & 0 & -2 \\ 2 & -1 & -1 & 0 & 1 \\ -1 & -1 & 1 & 2 & 2 \\ 1 & -2 & 2 & -2 & 0 \end{pmatrix}$ in reduced row-echelon form. Then use this command to find the rank of A and the determinant of A.

SOLUTION: We begin by defining the matrix **A**. Then, rref is used to place **A** in reduced row-echelon form. Notice that since the arguments 'r' and 'd' are included in this command, the rank and determinant of **A** are computed. These values are obtained by entering these two variable names. Hence, the rank of **A** is 4 and the determinant of **A** is 0.

```
> A:=array([[-1,-1,2,0,-1],[-2,2,0,0,-2],
        [2,-1,-1,0,1],[-1,-1,1,2,2],
        [1,-2,2,-2,0]]):
```

```
> rref(A,'r','d');
```

$$\begin{bmatrix} 1 & 0 & 0 & -2 & 0 \\ 0 & 1 & 0 & -2 & 0 \\ 0 & 0 & 1 & -2 & 0 \\ 0 & 0 & 0 & 0 & 1 \\ 0 & 0 & 0 & 0 & 0 \end{bmatrix}$$

```
> r;
  d;
```

$$4$$

$$0$$

■

`mulcol`

`mulcol(mat,cnum,expr)` multiplies column `cnum` of the matrix `mat` by the expression `expr`.

EXAMPLE 6: Multiply column two of $A = \begin{pmatrix} -1 & 2 & 0 & 0 \\ 1 & -2 & 1 & 0 \\ 2 & 2 & 1 & 0 \\ 1 & 1 & 2 & -1 \end{pmatrix}$ by $\frac{1}{2}$.

SOLUTION: After the 4×4 matrix `A` is entered, the second column of `A` is multiplied by `1/2` with `mulcol` to yield the matrix that results from this operation.

```
> A:=array([[-1,2,0,0],[1,-2,1,0],
      [2,2,1,0],[1,1,2,-1]]):
  mulcol(A,2,1/2);
```

$$\begin{bmatrix} -1 & 1 & 0 & 0 \\ 1 & -1 & 1 & 0 \\ 2 & 1 & 1 & 0 \\ 1 & \frac{1}{2} & 2 & -1 \end{bmatrix}$$

■

mulrow

`mulrow(mat,rnum,expr)` multiplies row `rnum` of the matrix `mat` by the expression `expr`.

EXAMPLE 7: Multiply row three of $A = \begin{pmatrix} -1 & 2 & 0 & 0 \\ 1 & -2 & 1 & 0 \\ 2 & 2 & 1 & 0 \\ 1 & 1 & 2 & -1 \end{pmatrix}$ by $\frac{1}{2}$.

SOLUTION: After the 4×4 matrix `A` is entered, the third row of `A` is multiplied by `1/2` with `mulrow` to yield the matrix that results from this operation.

```
> A:=array([[-1,2,0,0],[1,-2,1,0],
      [2,2,1,0],[1,1,2,-1]]):
  mulrow(A,3,1/2);
```

$$\begin{bmatrix} -1 & 2 & 0 & 0 \\ 1 & -2 & 1 & 0 \\ 1 & 1 & \frac{1}{2} & 0 \\ 1 & 1 & 2 & -1 \end{bmatrix}$$

■

pivot

`pivot(mat,i,j)` pivots the matrix `mat` about the nonzero element `(i,j)` of `mat`. Hence, every element of the `j`th column of `mat` except that in row `i` is converted to zero through elementary row operations. Likewise, `pivot(mat,i,j,m...n)` causes the pivoting to be carried out over rows `m` through `n` only.

EXAMPLE 8: Find a basis for the row space of $A=\begin{pmatrix} -1 & -1 & 2 & 0 & -1 \\ -2 & 2 & 0 & 0 & -2 \\ 2 & -1 & -1 & 0 & 1 \\ -1 & -1 & 1 & 2 & 2 \\ 1 & -2 & 2 & -2 & 0 \end{pmatrix}$.

SOLUTION: First, the 5×5 matrix `A` is defined. Then, in `stepone`, the 5×5 matrix obtained through multiplication of the first row of `A` by `-1` is given. Also, in `steptwo`, the element of `stepone` in position (1,1) is selected as the pivot element. Hence, every element in the first column of the matrix that results is zero, except for the element in position (1,1).

```
> A:=array([[-1,-1,2,0,-1],[-2,2,0,0,-2],
        [2,-1,-1,0,1],[-1,-1,1,2,2],
        [1,-2,2,-2,0]]):
  stepone:=mulrow(A,1,-1):
  steptwo:=pivot(stepone,1,1);
```

$$steptwo := \begin{bmatrix} 1 & 1 & -2 & 0 & 1 \\ 0 & 4 & -4 & 0 & 0 \\ 0 & -3 & 3 & 0 & -1 \\ 0 & 0 & -1 & 2 & 3 \\ 0 & -3 & 4 & -2 & -1 \end{bmatrix}$$

In `stepthree` next, the second row of `steptwo` is multiplied by `1/4` to yield a value of 1, and the element in position (2,2) of the resulting matrix is selected as the pivot element. Hence, all elements of the second column of the matrix that results are zero, except the element in position (2,2).

```
> stepthree:=pivot(mulrow(steptwo,2,1/4),2,2);
```

$$stepthree := \begin{bmatrix} 1 & 0 & -1 & 0 & 1 \\ 0 & 1 & -1 & 0 & 0 \\ 0 & 0 & 0 & 0 & -1 \\ 0 & 0 & -1 & 2 & 3 \\ 0 & 0 & 1 & -2 & -1 \end{bmatrix}$$

Continuing in the same manner in `stepfour`, the fourth row of `stepthree` is multiplied by `-1`, and the element in position (4,3) of the resulting matrix is selected as the pivot element. Hence, all elements of the fourth column of the matrix that results are zero, except the element in position (4,3).

```
> stepfour:=pivot(mulrow(stepthree,4,-1),4,3);
```

$$stepfour := \begin{bmatrix} 1 & 0 & 0 & -2 & -2 \\ 0 & 1 & 0 & -2 & -3 \\ 0 & 0 & 0 & 0 & -1 \\ 0 & 0 & 1 & -2 & -3 \\ 0 & 0 & 0 & 0 & 2 \end{bmatrix}$$

In `stepfive`, the fifth row of `stepfour` is multiplied by `1/2`, and the element in position (5,5) of the resulting matrix is selected as the pivot element. Hence, all elements of the fifth column of the matrix that results are zero, except the element in position (5,5). Hence, the nonzero row vectors that result from this process form a basis for the row space of `A`.

```
> stepfive:=pivot(mulrow(stepfour,5,1/2),5,5);
```

$$stepfive := \begin{bmatrix} 1 & 0 & 0 & -2 & 0 \\ 0 & 1 & 0 & -2 & 0 \\ 0 & 0 & 0 & 0 & 0 \\ 0 & 0 & 1 & -2 & 0 \\ 0 & 0 & 0 & 0 & 1 \end{bmatrix}$$

■

`scalarmul`

`scalarmul(mat,expr)` multiplies each element of the matrix `mat` by the expression `expr`. Hence, the result is a matrix of the same dimensions as that of `mat`.

EXAMPLE 9: If x denotes a scalar and A= $\begin{pmatrix} -4 & 1 & -4 & 3 \\ -2 & -4 & -2 & -3 \\ 0 & -1 & -1 & 1 \\ -4 & 1 & -3 & 1 \end{pmatrix}$, compute $-2x^2A$.

SOLUTION: First, the 4×4 matrix A is entered. Then, each element of A is multiplied by $-2x^2A$. to obtain the desired result with scalarmul. Note that the same result is obtained with the command evalm(-2*x^2*A);.

```
> A:=matrix(4,4,[-4,1,-4,3,-2,-4,-2,-3,0,-1,-1,
   1,-4,1,-3,1]):
  scalarmul(A,-2*x^2);
```

$$\begin{bmatrix} 8\,x^2 & -2\,x^2 & 8\,x^2 & -6\,x^2 \\ 4\,x^2 & 8\,x^2 & 4\,x^2 & 6\,x^2 \\ 0 & 2\,x^2 & 2\,x^2 & -2\,x^2 \\ 8\,x^2 & -2\,x^2 & 6\,x^2 & -2\,x^2 \end{bmatrix}$$

■

swapcol

swapcol(mat,coli,colj) interchanges columns coli and colj in the matrix mat and returns the matrix that results.

EXAMPLE 10: If A= $\begin{pmatrix} -3 & -1 & 1 \\ 1 & 3 & -2 \\ 0 & 1 & 3 \end{pmatrix}$, interchange columns one and three of **A**.

SOLUTION: First, the 3×3 matrix A is defined. Then, columns one and three of A are reversed with swapcol.

```
> A:=matrix(3,3,[-3,-1,1,1,3,-2,0,1,3]):
  swapcol(A,1,3);
```

$$\begin{bmatrix} 1 & -1 & -3 \\ -2 & 3 & 1 \\ 3 & 1 & 0 \end{bmatrix}$$

■

`swaprow`

`swaprow(mat,rowi,rowj)` interchanges rows `rowi` and `rowj` in the matrix `mat` and returns the matrix that results.

EXAMPLE 11: If A= $\begin{pmatrix} -3 & -1 & 1 \\ 1 & 3 & -2 \\ 0 & 1 & 3 \end{pmatrix}$, use interchange rows one and two of **A**.

SOLUTION: After defining the 3×3 matrix `A`, the first and second rows of `A` are reversed with `swaprow`.

```
> A:=matrix(3,3,[-3,-1,1,1,3,-2,0,1,3]):
  swaprow(A,1,2);
```

$$\begin{bmatrix} 1 & 3 & -2 \\ -3 & -1 & 1 \\ 0 & 1 & 3 \end{bmatrix}$$

■

Related Matrix Operations

In addition to providing the commands that perform elementary row and column operations and combinations of them, Maple V provides a variety of commands to help construct and manipulate matrices. As in the previous examples, the following commands are contained in the package `linalg`, so you must enter `with(linalg)` to load this package before you use the commands given in the following examples.

`augment`

`augment(matrixlist)` yields the matrix that results when the rows of the matrices in `matrixlist` are joined. Example 12 again illustrates the use of this command, which was included earlier in the `gaussjord` example.

EXAMPLE 12: Both $\left\{\begin{pmatrix}-5\\1\\-2\end{pmatrix},\begin{pmatrix}-3\\0\\-2\end{pmatrix},\begin{pmatrix}-1\\2\\-2\end{pmatrix}\right\}$ and $\left\{\begin{pmatrix}1\\-1\\-1\end{pmatrix},\begin{pmatrix}5\\-5\\-4\end{pmatrix},\begin{pmatrix}3\\-2\\5\end{pmatrix}\right\}$ are bases of $\Re^3$. Find the transition matrix from $\left\{\begin{pmatrix}-5\\1\\-2\end{pmatrix},\begin{pmatrix}-3\\0\\-2\end{pmatrix},\begin{pmatrix}-1\\2\\-2\end{pmatrix}\right\}$ to $\left\{\begin{pmatrix}1\\-1\\-1\end{pmatrix},\begin{pmatrix}5\\-5\\-4\end{pmatrix},\begin{pmatrix}3\\-2\\5\end{pmatrix}\right\}$.

SOLUTION: First, the matrices `A` and `B` are defined with columns vectors from the two sets of basis vectors. The rows of these two matrices are joined with `augment` in `C`. Then, the transition matrix is determined with `rowspace`, where `rowspace(mat)` calculates a basis for the row space of the matrix `mat`. The result means that the transition matrix is $\begin{pmatrix}-145 & 35 & -54\\-162 & 39 & -60\\22 & -5 & 8\end{pmatrix}$.

```
> A:=matrix(3,3,[-5,1,-2,-3,0,-2,-1,2,-2]):
  B:=matrix(3,3,[1,-1,-1,5,-5,-4,3,-2,5]):
> C:=augment(B,A);
```

$$C := \begin{bmatrix}1 & -1 & -1 & -5 & 1 & -2\\5 & -5 & -4 & -3 & 0 & -2\\3 & -2 & 5 & -1 & 2 & -2\end{bmatrix}$$

```
> rowspace(C);
```

$$\{[0,0,1,22,-5,8],[0,1,0,-162,39,-60],[1,0,0,-145,35,-54]\}$$

■

concat

concat(matrix) produces the same result as augment(matrix).

EXAMPLE 13: Use concat to form the augmented matrix for the system of equations $\begin{cases} 5x+y-2z = 1 \\ -3x-2z = -1. \\ -x+2y-2z = -1 \end{cases}$

SOLUTION: After defining the system of equations in sys, we determine the matrix of coefficients A and the vector b with genmatrix. Then, concat is used to form the augmented matrix.

```
> sys:={5*x+y-2*z=1,-3*x-2*z=-1,-x+2*y-2*z=-1}:
> A:=genmatrix(sys,[x,y,z],'b');
```

$$A := \begin{bmatrix} 5 & 1 & -2 \\ -3 & 0 & -2 \\ -1 & 2 & -2 \end{bmatrix}$$

```
> concat(A,b);
```

$$\begin{bmatrix} 5 & 1 & -2 & 1 \\ -3 & 0 & -2 & -1 \\ -1 & 2 & -2 & -1 \end{bmatrix}$$

■

`BlockDiagonal`

Maple can also be used in the formation of block diagonal matrices through the use of the command `BlockDiagonal`. `BlockDiagonal(matrixlist)` yields a matrix with diagonal elements listed as matrices or scalars in `matrixlist`. Note that this command gives the same result as `diag`.

EXAMPLE 14: Let $A=\begin{pmatrix} -1 & 5 \\ -3 & -1 \end{pmatrix}$ and $B=\begin{pmatrix} -1 & 1 & 3 \\ -3 & 4 & 2 \\ -2 & -4 & 1 \end{pmatrix}$ Construct the matrix $\begin{pmatrix} A & 0 \\ 0 & B \end{pmatrix}$.

SOLUTION: The two matrices are defined in `A` and `B`. Then, the matrix with these matrices along the diagonal is generated with `BlockDiagonal`.

```
> A:=matrix(2,2,[-1,5,-3,-1]):
  B:=matrix(3,3,[-1,1,3,-3,4,2,-2,-4,1]):
  BlockDiagonal(A,B);
```

$$\begin{bmatrix} -1 & 5 & 0 & 0 & 0 \\ -3 & -1 & 0 & 0 & 0 \\ 0 & 0 & -1 & 1 & 3 \\ 0 & 0 & -3 & 4 & 2 \\ 0 & 0 & -2 & -4 & 1 \end{bmatrix}$$

■

`extend`

`extend(mat,m,n,val)` yields a new matrix in which `m` rows and `n` columns are added to the matrix `mat`. The entries of these new rows and columns are assigned the value `val`. If this initialization value is not indicated, then the entry in position [`m`,`n`] is displayed as *?m,n*. Hence, the command `extend` is useful in adding rows and columns to a matrix, as is shown in Example 15.

EXAMPLE 15: If $A=\begin{pmatrix} 4 & 0 \\ -1 & 2 \end{pmatrix}$, construct the matrix $\begin{pmatrix} 4 & 0 & 0 & 0 \\ -1 & 2 & 0 & 0 \\ 0 & 0 & 0 & 0 \\ 0 & 0 & 0 & 0 \end{pmatrix}$. Also, use `extend` to add a row and column of arbitrary elements to A.

SOLUTION: First, the 2×2 matrix `A` is defined. The desired matrix is then constructed by adding two rows and two columns of zeros to `A`. Hence, a 4×4 matrix results. Since no initialization value is indicated in the second `extend` command, arbitrary elements appear in the 3×3 matrix that results.

```
> A:=matrix(2,2,[4,0,-1,2]):
  extend(A,2,2,0);
```

$$\begin{bmatrix} 4 & 0 & 0 & 0 \\ -1 & 2 & 0 & 0 \\ 0 & 0 & 0 & 0 \\ 0 & 0 & 0 & 0 \end{bmatrix}$$

```
> extend(A,1,1);
```

$$\begin{bmatrix} 4 & 0 & ?_{1,3} \\ -1 & 2 & ?_{2,3} \\ ?_{3,1} & ?_{3,2} & ?_{3,3} \end{bmatrix}$$

■

`copyinto`

`copyinto(mat1,mat2,m,n)` creates a new matrix by copying `mat1` into `mat2` beginning at position (m,n) in `mat2`. Hence, `mat1[1,1]` is assigned the position `mat2[m,n]`, `mat1[2,1]` (if one exists) is assigned the position `mat2[m+1,n]`, etc. Hence, `copyinto` can be used to form a new matrix from existing matrices. This command is demonstrated in Example 16.

EXAMPLE 16: If $A=\begin{pmatrix} 1 & 2 & -1 \\ 1 & 5 & 3 \\ -2 & 5 & 3 \end{pmatrix}$ and $C=\begin{pmatrix} 4 & -4 \\ -5 & 3 \end{pmatrix}$, construct

$$B=\begin{pmatrix} 1 & 2 & -1 & 0 & 0 \\ 1 & 5 & 3 & 0 & 0 \\ -2 & 5 & 3 & 0 & 0 \\ 0 & 0 & 0 & 0 & 0 \\ 0 & 0 & 0 & 0 & 0 \end{pmatrix} \text{ and } C=\begin{pmatrix} 1 & 2 & -1 & 0 & 0 \\ 1 & 5 & 3 & 0 & 0 \\ -2 & 5 & 3 & 0 & 0 \\ 0 & 0 & 0 & 4 & -4 \\ 0 & 0 & 0 & -5 & 3 \end{pmatrix}.$$

SOLUTION: First, the 3×3 matrix `A` is defined. In order to construct `B`, two rows and two columns of zeros are needed. Hence, `extend` is employed in `B` to accomplish this. Next, the 2×2 matrix `C` is defined. Finally, matrix `C` is copied into matrix `B`, with `C11` assigned to position (4,4) in matrix `B` to produce the desired result.

```
> A:=matrix(3,3,[1,2,-1,1,5,3,-2,5,3]):
  B:=extend(A,2,2,0);
  C:=matrix(2,2,[4,-4,-5,3]);
```

$$B := \begin{bmatrix} 1 & 2 & -1 & 0 & 0 \\ 1 & 5 & 3 & 0 & 0 \\ -2 & 5 & 3 & 0 & 0 \\ 0 & 0 & 0 & 0 & 0 \\ 0 & 0 & 0 & 0 & 0 \end{bmatrix}$$

$$C := \begin{bmatrix} 4 & -4 \\ -5 & 3 \end{bmatrix}$$

```
> copyinto(C,B,4,4);
```

$$\begin{bmatrix} 1 & 2 & -1 & 0 & 0 \\ 1 & 5 & 3 & 0 & 0 \\ -2 & 5 & 3 & 0 & 0 \\ 0 & 0 & 0 & 4 & -4 \\ 0 & 0 & 0 & -5 & 3 \end{bmatrix}$$

■

In addition to the command `extend`, which is used to add rows and columns to a matrix, Maple also has commands for deleting rows and columns from matrices.

`delcols`

`delcols(mat,m..n)` deletes columns `m` through `n` from the matrix `mat` and yields the matrix that results after the deletion.

EXAMPLE 17: Delete the second and third columns from the matrix

$$A=\begin{pmatrix} -3 & -2 & 1 & 2 \\ 1 & 2 & 3 & 2 \\ -4 & 5 & 4 & -2 \end{pmatrix}.$$

SOLUTION: First, the 3×4 matrix is entered. Then, the second and third columns are deleted from `A` with `delcols`. Hence, only the first and last columns remain from the original matrix.

```
> A:=matrix(3,4,[-3,-2,1,2,1,2,3,2,-4,5,4,-2]);
  delcols(A, 2..3);
```

$$A:=\begin{bmatrix} -3 & -2 & 1 & 2 \\ 1 & 2 & 3 & 2 \\ -4 & 5 & 4 & -2 \end{bmatrix}$$

$$\begin{bmatrix} -3 & 2 \\ 1 & 2 \\ -4 & -2 \end{bmatrix}$$

■

`delrows`

`delrows(mat,m..n)` deletes rows `m` through `n` from the matrix `mat` and yields the matrix that results after the deletion.

EXAMPLE 18: Use Maple to delete the second and third rows from the matrix $A = \begin{pmatrix} -3 & -2 & 1 & 2 \\ 1 & 2 & 3 & 2 \\ -4 & 5 & 4 & -2 \end{pmatrix}$.

SOLUTION: The 3×4 matrix `A` is entered, and then the second and third rows are deleted with `delrows`. This leaves the first row only.

```
> A:=matrix(3,4,[-3,-2,1,2,1,2,3,2,-4,5,4,-2]);
  delrows(A, 2..3);
```

$$A := \begin{bmatrix} -3 & -2 & 1 & 2 \\ 1 & 2 & 3 & 2 \\ -4 & 5 & 4 & -2 \end{bmatrix}$$

$$\begin{bmatrix} -3 & -2 & 1 & 2 \end{bmatrix}$$

■

`diag`

`diag(C1,C2,...,Cn)`, where the `Ci`, `i=1,..., n`, represent square matrices or scalars, yields a matrix with the `Ci` as diagonal elements. Many applications of linear algebra require the use of diagonal matrices. Maple easily forms these matrices through the use of this command as well as `BlockDiagonal`.

EXAMPLE 19: Construct the matrix $\begin{pmatrix} A & 0 \\ 0 & B \end{pmatrix}$ if $A = \begin{pmatrix} 5 & 3 & -2 \\ 4 & 4 & 5 \\ 3 & 2 & -3 \end{pmatrix}$ and $B = \begin{pmatrix} 3 & -1 \\ 3 & 5 \end{pmatrix}$.

SOLUTION: First, the 3×3 matrix `A` and the 2×2 matrix `B` are defined. Then, the desired matrix is constructed with `diag`.

```
> A:=matrix(3,3,[5,3,-2,4,4,5,3,2,-3]):
  B:=matrix(2,2,[3,-1,3,5]):
  diag(A,B);
```

$$\begin{bmatrix} 5 & 3 & -2 & 0 & 0 \\ 4 & 4 & 5 & 0 & 0 \\ 3 & 2 & -3 & 0 & 0 \\ 0 & 0 & 0 & 3 & -1 \\ 0 & 0 & 0 & 3 & 5 \end{bmatrix}$$

■

`hilbert`

`hilbert(n,x)` yields the $n \times n$ Hilbert matrix in which the (i,j)th element is given by the formula $\frac{1}{i+j-x}$ If `hilbert(n)` is entered, the Hilbert matrix with x=1 results. Since Hilbert matrices are useful in studying many topics in linear algebra, this command proves to be quite convenient in the construction of these well-known matrices.

EXAMPLE 20: Compute the 8×8 Hilbert matrix with x=1. Also, compute the 3×3 Hilbert matrix with the parameter x^2.

SOLUTION: The desired 8×8 Hilbert matrix with x=1 is found with `hilbert`.

```
> hilbert(8);
```

$$\begin{bmatrix} 1 & \frac{1}{2} & \frac{1}{3} & \frac{1}{4} & \frac{1}{5} & \frac{1}{6} & \frac{1}{7} & \frac{1}{8} \\ \frac{1}{2} & \frac{1}{3} & \frac{1}{4} & \frac{1}{5} & \frac{1}{6} & \frac{1}{7} & \frac{1}{8} & \frac{1}{9} \\ \frac{1}{3} & \frac{1}{4} & \frac{1}{5} & \frac{1}{6} & \frac{1}{7} & \frac{1}{8} & \frac{1}{9} & \frac{1}{10} \\ \frac{1}{4} & \frac{1}{5} & \frac{1}{6} & \frac{1}{7} & \frac{1}{8} & \frac{1}{9} & \frac{1}{10} & \frac{1}{11} \\ \frac{1}{5} & \frac{1}{6} & \frac{1}{7} & \frac{1}{8} & \frac{1}{9} & \frac{1}{10} & \frac{1}{11} & \frac{1}{12} \\ \frac{1}{6} & \frac{1}{7} & \frac{1}{8} & \frac{1}{9} & \frac{1}{10} & \frac{1}{11} & \frac{1}{12} & \frac{1}{13} \\ \frac{1}{7} & \frac{1}{8} & \frac{1}{9} & \frac{1}{10} & \frac{1}{11} & \frac{1}{12} & \frac{1}{13} & \frac{1}{14} \\ \frac{1}{8} & \frac{1}{9} & \frac{1}{10} & \frac{1}{11} & \frac{1}{12} & \frac{1}{13} & \frac{1}{14} & \frac{1}{15} \end{bmatrix}$$

Similarly, the 3×3 Hilbert matrix with the parameter x^2 is found with the following command:

```
> hilbert(3,x^2);
```

$$\begin{bmatrix} \frac{1}{2-x^2} & \frac{1}{3-x^2} & \frac{1}{4-x^2} \\ \frac{1}{3-x^2} & \frac{1}{4-x^2} & \frac{1}{5-x^2} \\ \frac{1}{4-x^2} & \frac{1}{5-x^2} & \frac{1}{6-x^2} \end{bmatrix}$$

■

`JordanBlock`

`JordanBlock(x,n)` yields the Jordan block matrix using the constant, rational number, or algebraic number `val` and the integer `n`. This matrix is made of elements *Jij* of the form

$$Jij = \begin{cases} x,\ i = j,\ i = 1, \ldots, n-1 \\ 1,\ j = i+1,\ i = 1, \ldots, n = 1. \\ 0, \text{otherwise} \end{cases}$$

Since Jordan block matrices arise in numerous applications of linear algebra, this command is quite useful. We show how `JordanBlock` is used to place a matrix in Jordan canonical form in Example 21.

EXAMPLE 21: Construct the Jordan canonical form of the matrix with Jordan blocks (3), $\begin{pmatrix} -5 & 1 \\ 0 & -5 \end{pmatrix}$, and $\begin{pmatrix} 4 & 1 & 0 \\ 0 & 4 & 1 \\ 0 & 0 & 4 \end{pmatrix}$.

SOLUTION: The desired Jordan blocks are determined with `JordanBlock` in the array `blocks`. In `stepone`, five rows and five columns of zeros are added to the first element of `blocks` through the use of `extend`. In `steptwo`, the second element of blocks is copied into `stepone`, with the element in position (1,1) of `blocks[2]` placed in position (2,2) of `stepone`. Then, in `stepthree`, the third element of `blocks` is copied into `steptwo`, with the element of `blocks[2]` in position (1,1) placed in position (4,4) of `steptwo`. This yields the desired matrix.

```
> blocks:=[JordanBlock(3,1),JordanBlock(-5,2),
           JordanBlock(4,3)];
```

$$blocks := \left[[\ 3], \begin{bmatrix} -5 & 1 \\ 0 & -5 \end{bmatrix}, \begin{bmatrix} 4 & 1 & 0 \\ 0 & 4 & 1 \\ 0 & 0 & 4 \end{bmatrix} \right]$$

```
> stepone:=extend(blocks[1],5,5,0);
```

$$stepone := \begin{bmatrix} 3 & 0 & 0 & 0 & 0 & 0 \\ 0 & 0 & 0 & 0 & 0 & 0 \\ 0 & 0 & 0 & 0 & 0 & 0 \\ 0 & 0 & 0 & 0 & 0 & 0 \\ 0 & 0 & 0 & 0 & 0 & 0 \\ 0 & 0 & 0 & 0 & 0 & 0 \end{bmatrix}$$

```
> steptwo:=copyinto(blocks[2],stepone,2,2):
  stepthree:=copyinto(blocks[3],steptwo,4,4);
```

$$stepthree := \begin{bmatrix} 3 & 0 & 0 & 0 & 0 & 0 \\ 0 & -5 & 1 & 0 & 0 & 0 \\ 0 & 0 & -5 & 0 & 0 & 0 \\ 0 & 0 & 0 & 4 & 1 & 0 \\ 0 & 0 & 0 & 0 & 4 & 1 \\ 0 & 0 & 0 & 0 & 0 & 4 \end{bmatrix}$$

■

`matrix`

`matrix(m,n,vlist)` generates the m × n rectangular array using the `m*n` values given in `vlist`. (The first `n` values in `vlist` are placed in the first row of the matrix, the next `n` values in the second row, and so on.) `matrix(veclist)` creates a matrix with the first element of `veclist` assigned to the first row, the second element of `veclist` to the second row, and so on. Also, `matrix(m,n)` generates an m × n matrix in which the elements are not specified. In this case, the result is given in the form `array(1 .. m, 1 .. n, [])`. A function `func` can be used as the third argument of `matrix` in order that element values be determined according to `func(m,n)`. We illustrate several of the forms of this command that can be used to define a matrix in Example 22.

EXAMPLE 22: Use Maple to construct the matrices $A=\begin{pmatrix} 2 & 0 \\ -1 & -1 \end{pmatrix}$,

$B=\begin{pmatrix} -2 & 1 & 0 & 2 \\ -3 & 1 & -2 & 1 \\ -3 & 3 & 0 & 3 \end{pmatrix}$, and C, where C represents an arbitrary 2×2 matrix.

SOLUTION: These three matrices are defined using `matrix`. Because no specific entries are indicated in `C`, the result is given in terms of `array`.

```
> A:=matrix(2,2,[2,0,-1,-1]);
```

$$A := \begin{bmatrix} 2 & 0 \\ -1 & -1 \end{bmatrix}$$

```
> B:=matrix(3,4,[-2,1,0,2,-3,1,-2,1,-3,3,0,3]);
```

$$B := \begin{bmatrix} -2 & 1 & 0 & 2 \\ -3 & 1 & -2 & 1 \\ -3 & 3 & 0 & 3 \end{bmatrix}$$

```
> C:=matrix(2,2);
```

$$C := \mathrm{array}(1..2, 1..2, [\ ])$$

■

`stackmatrix`

`stackmatrix(list)` vertically joins the elements of `list` (which may be matrices or vectors). Hence, the elements of `list` must have the same number of columns.

EXAMPLE 23: If A= $\begin{pmatrix} -3 & -1 \\ 0 & 2 \end{pmatrix}$ and B= $\begin{pmatrix} 2 & -2 \\ 2 & 2 \end{pmatrix}$, use Maple to construct $\begin{pmatrix} -3 & -1 \\ 0 & 2 \\ 2 & -2 \\ 2 & 2 \end{pmatrix}$.

SOLUTION: First, the two 2×2 matrices are entered in `A` and `B`. Then, the two are joined vertically with `stackmatrix` to obtain the desired result.

```
> A:=array([[-3,-1],[0,2]]):
  B:=array([[2,-2],[2,2]]):
  stackmatrix(A,B);
```

$$\begin{bmatrix} -3 & -1 \\ 0 & 2 \\ 2 & -2 \\ 2 & 2 \end{bmatrix}$$

■

`toeplitz`

`toeplitz(vlist)` gives the Toeplitz matrix associated with the list of values in `vlist`. Hence, if `vlist` has n elements, the result is a square matrix of dimension $n \times n$. The first element of `vlist` becomes the diagonal elements of the Toeplitz matrix, while the jth element of `vlist` makes up the (j–1)st sub and super diagonals of the result.

EXAMPLE 24: Construct the matrix $\begin{pmatrix} -3 & -1 & 0 & 2 \\ -1 & -3 & -1 & 0 \\ 0 & -1 & -3 & -1 \\ 2 & 0 & -1 & -3 \end{pmatrix}$.

SOLUTION: The desired matrix with main diagonal elements -3, upper and lower diagonal elements -1, second sub and super diagonal elements 0, and third sub and super diagonal elements 2 is constructed with the following command. Notice that these elements are read from the list in toeplitz from left to right.

```
> toeplitz([-3,-1,0,2]);
```

$$\begin{bmatrix} -3 & -1 & 0 & 2 \\ -1 & -3 & -1 & 0 \\ 0 & -1 & -3 & -1 \\ 2 & 0 & -1 & -3 \end{bmatrix}$$

■

vandermonde

vandermonde(vlist) calculates the Vandermonde matrix that corresponds to the values in vlist. Hence, the (i,j)th element of this square matrix is given by vlist[i]^(j-1). Note that if vlist contains n elements, then the result is an $n \times n$ matrix.

EXAMPLE 25: Find the Vandermonde matrix associated with the list {–4,3,1,2}.

SOLUTION: This matrix is easily found with the following command. Hence, a 4×4 matrix results because there are four elements in the indicated list.

```
> vandermonde([-4,3,1,2]);
```

$$\begin{bmatrix} 1 & -4 & 16 & -64 \\ 1 & 3 & 9 & 27 \\ 1 & 1 & 1 & 1 \\ 1 & 2 & 4 & 8 \end{bmatrix}$$

■

Application: Computing the Adjacency Matrix of a Graph

An application of these commands is the manipulation of the adjacency matrix of a graph. Recall that two vertices of a graph are said to be **adjacent** if there is at least one edge joining them. Consider the graph G with no loops and n vertices labeled 1, 2, ... , n. The **adjacency matrix of G** is the $n \times n$ matrix in which the entry in row i and column j is the number of edges joining the vertices i and j. For example, suppose that a graph has the adjacency matrix A= $\begin{pmatrix} 0&1&0&1 \\ 1&0&1&2 \\ 0&1&0&1 \\ 1&2&1&0 \end{pmatrix}$. This matrix is represented as `adj_matrix` below. Then, suppose that two more vertices are added to the graph, with vertex 5 adjacent to vertices 2 and 3, and vertex 6 adjacent to vertices 1 and 4. Instead of defining a new adjacency matrix for the revised graph, which can be quite cumbersome in many cases, these additions are made with `stackmatrix` and `augment`.

This is done below in `addtocols` and `addtorows`. The rows added in `addtocols` represent the edges from vertices 5 and 6 to vertices 1, 2, 3, and 4 while the columns added in `addtorows` give these same edges from the original set of vertices to vertices 5 and 6.

```
> with(linalg):
  adj_matrix:=matrix(4,4,[0,1,0,1,1,0,1,2,
      0,1,0,1,1,2,1,0]);
```

$$adj\_matrix := \begin{bmatrix} 0&1&0&1 \\ 1&0&1&2 \\ 0&1&0&1 \\ 1&2&1&0 \end{bmatrix}$$

```
> b:=matrix(2,4,[0,1,1,0,1,0,0,1]):
  c:=matrix(6,2,[0,1,1,0,1,0,0,1,0,1,1,0]):
> addtocols:=stackmatrix(adj_matrix,b);
```

$$addtocols := \begin{bmatrix} 0&1&0&1 \\ 1&0&1&2 \\ 0&1&0&1 \\ 1&2&1&0 \\ 0&1&1&0 \\ 1&0&0&1 \end{bmatrix}$$

```
> addtorows:=augment(addtocols,c);
```

$$addtorows := \begin{bmatrix} 0 & 1 & 0 & 1 & 0 & 1 \\ 1 & 0 & 1 & 2 & 1 & 0 \\ 0 & 1 & 0 & 1 & 1 & 0 \\ 1 & 2 & 1 & 0 & 0 & 1 \\ 0 & 1 & 1 & 0 & 0 & 1 \\ 1 & 0 & 0 & 1 & 1 & 0 \end{bmatrix}$$

An interesting fact concerning an adjacency matrix **M** is that the (i,j)th element of the kth power of **M** represents the number of walks of length k from vertex i to vertex j. A **walk of length k** in a graph is a succession of k edges. This is important in problems in which the number of ways to travel between two locations must be determined. Matrix multiplication is carried out with `multiply` or `evalm` and `&*`. Hence, the product **AB**, when defined, is obtained with `multiply(A,B)`. Like the previous command discussed, the command `multiply` is contained in the `linalg` package. The command `evalm(A&*B)` produces the same results as `multiply(A,B)`. We illustrate this command in the following example.

Using the matrix given in `addtorows`, the number of walks of length 2 between every vertex pair is determined from `twowalks`. For example, there are 4 walks of length 2 from vertex 4 to vertex 5, as seen with `twowalks[4,5]`. The number of walks of length 3 are found in `threewalks`.

```
> twowalks:=multiply(addtorows,addtorows);!
```

$$twowalks := \begin{bmatrix} 3 & 2 & 2 & 3 & 2 & 1 \\ 2 & 7 & 3 & 2 & 1 & 4 \\ 2 & 3 & 3 & 2 & 1 & 2 \\ 3 & 2 & 2 & 7 & 4 & 1 \\ 2 & 1 & 1 & 4 & 3 & 0 \\ 1 & 4 & 2 & 1 & 0 & 3 \end{bmatrix}$$

```
> twowalks[4,5];
```

$$4$$

```
> threewalks:=multiply(twowalks,addtorows);
```

$$threewalks := \begin{bmatrix} 6 & 13 & 7 & 10 & 5 & 8 \\ 13 & 10 & 10 & 23 & 14 & 5 \\ 7 & 10 & 6 & 13 & 8 & 5 \\ 10 & 23 & 13 & 10 & 5 & 14 \\ 5 & 14 & 8 & 5 & 2 & 9 \\ 8 & 5 & 5 & 14 & 9 & 2 \end{bmatrix}$$

The command `multiply` is used below to find the number of walks of length 10.

```
> multiply(adj_matrix$10);
```

$$\begin{bmatrix} 17408 & 28160 & 17408 & 28160 \\ 28160 & 46080 & 28160 & 45056 \\ 17408 & 28160 & 17408 & 28160 \\ 28160 & 45056 & 28160 & 46080 \end{bmatrix}$$

6.2 The Euclidean Vector Space $\Re_n$

The Row and Column Space of a Matrix

Let $\{\mathbf{v}_1,\mathbf{v}_2,\ldots,\mathbf{v}_n\}$ be a set of n vectors and $\{c_1,c_2,\ldots,c_n\}$ a set of n numbers. Then $c_1\mathbf{v}_1 + c_2\mathbf{v}_2+\ldots c_n\mathbf{v}_n$ is called a **linear combination** of $\mathbf{v}_1,\mathbf{v}_2,\ldots,\mathbf{v}_n$. If a set of vectors $\{\mathbf{v}_1,\mathbf{v}_2,\ldots,\mathbf{v}_n\}$ is said to be **linearly independent**, this means that if

$$c_1\mathbf{v}_1 + c_2\mathbf{v}_2+\ldots c_n\mathbf{v}_n= \mathbf{0}$$

then $c_1=c_2=\ldots=c_n=0$.. A set of vectors $\{\mathbf{v}_1,\mathbf{v}_2,\ldots,\mathbf{v}_n\}$ is **linearly dependent** if it is not linearly independent. Equivalently, a set of vectors $\{\mathbf{v}_1,\mathbf{v}_2,\ldots,\mathbf{v}_n\}$ is linearly dependent if there are numbers $c_1,c_2,\ldots,c_n$ not all zero such that $c_1\mathbf{v}_1 + c_2\mathbf{v}_2+\ldots c_n\mathbf{v}_n= \mathbf{0}$.

If $\{\mathbf{v}_1,\mathbf{v}_2,\ldots,\mathbf{v}_n\}$ is a set of vectors in $\Re^m$, then the **spanning set** of $\{\mathbf{v}_1,\mathbf{v}_2,\ldots,\mathbf{v}_n\}$ span$\{\mathbf{v}_1,\mathbf{v}_2,\ldots,\mathbf{v}_n\}$ is the set of all linear combinations of the vectors $\mathbf{v}_1,\mathbf{v}_2,\ldots,\mathbf{v}_n$:

span$\{\mathbf{v}_1,\mathbf{v}_2,\ldots,\mathbf{v}_n\}=\{c_1\mathbf{v}_1 + c_2\mathbf{v}_2+\ldots c_n\mathbf{v}_n : c_i$ is a real number for all values of $i\}$.

If $\Re^m =$span$\{\mathbf{v}_1,\mathbf{v}_2,\ldots,\mathbf{v}_n\}$ and the set of vectors $\{\mathbf{v}_1,\mathbf{v}_2,\ldots,\mathbf{v}_n\}$ is linearly independent then $\{\mathbf{v}_1,\mathbf{v}_2,\ldots,\mathbf{v}_n\}$ is called a **basis** for $\Re^n$. More generally, if V is said to be a **subspace** of $\Re^m$, this means that if $\mathbf{v}_1$ and $\mathbf{v}_2$ are vectors in V and a and b are real numbers, then $a\mathbf{v}_1+b\mathbf{v}_2$ is also an element of V. Note that span$\{\mathbf{v}_1,\mathbf{v}_2,\ldots,\mathbf{v}_n\}$ is a subspace of $\Re^m$.

Maple's linear algebra package contains many commands that can be used to investigate vector spaces and avoid the computational difficulties. We describe and illustrate these commands in the following examples. Recall that `with(linalg)` must be entered before an attempt to use these commands

Let **A** denote the $n \times m$ matrix $\mathbf{A} = \begin{pmatrix} a_{11} & a_{12} & \cdots & a_{1m} \\ a_{21} & a_{22} & \cdots & a_{2m} \\ \vdots & \vdots & \ddots & \vdots \\ a_{n1} & a_{n2} & \cdots & a_{nm} \end{pmatrix}$. The **row space of A**, row(**A**), is the spanning set of the rows of **A**; the **column space of A**, col(**A**), is the spanning set of the columns of **A**.

If **A** is any matrix, then the dimension of the column space of **A** is equal to the dimension of the row space of **A**. The dimension of the row space (column space) of a matrix **A** is called the **rank** of **A**. The **nullspace** of **A** is the set of solutions to the system of equations **Ax=0**. The nullspace of **A** is a subspace and its dimension is called the **nullity** of **A**. In the same manner as the rank of **A** is equal to the number of nonzero rows in the row-echelon form of **A**, the nullity of **A** is equal to the number of zero rows in the row-echelon form of **A**. Thus, if **A** is a square matrix, the sum of the rank of **A** and the nullity of **A** is equal to the number of rows (columns) of **A**.

The following examples illustrate how Maple can be used to determine the column space, rank, null space, and nullity of matrices.

`colspace`

`colspace(mat)` yields a basis for the column space of the matrix `mat`. The vectors returned have lead coefficient one.

EXAMPLE 1: Find a basis for the column space of A= $\begin{pmatrix} 1 & -2 & 2 & 1 & -2 \\ 1 & 1 & 2 & -2 & -2 \\ 1 & 0 & 0 & 2 & -1 \\ 0 & 0 & 0 & -2 & 0 \\ -2 & 1 & 0 & 1 & 2 \end{pmatrix}$.

SOLUTION: First, the matrix `A` is defined. Then, the four basis vectors for the column space of `A` are found with `colspace`.

```
> A:=matrix(5,5,[1,-2,2,1,-2,1,1,2,-2,-2,1,0,0,2,
          -1,0,0,0,-2,0,-2,1,0,1,2]);
```

$$A := \begin{bmatrix} 1 & -2 & 2 & 1 & -2 \\ 1 & 1 & 2 & -2 & -2 \\ 1 & 0 & 0 & 2 & -1 \\ 0 & 0 & 0 & -2 & 0 \\ -2 & 1 & 0 & 1 & 2 \end{bmatrix}$$

```
> colspace(A);
```

$$\{[0, 0, 0, 1, -3], [0, 0, 1, 0, -2], \left[1, 0, 0, 0, \frac{-1}{3}\right], \left[0, 1, 0, 0, \frac{1}{3}\right]\}$$

■

`rowspace`

`rowspace(mat)` yields the basis for the vector space spanned by the rows of the matrix `mat`.

EXAMPLE 2: Find a basis for the row space of $A=\begin{pmatrix} 0 & 1 & 1 & -1 \\ 3 & 2 & -1 & -1 \\ -2 & -1 & -3 & -1 \\ 0 & -1 & 2 & 2 \end{pmatrix}$.

SOLUTION: Below, the 4×4 matrix `A` is defined so that a basis for the row space of `A`, which consists of a set of three vectors, is given with `rowspace`.

```
> A:=array([[0,1,1,-1],[3,2,-1,-1],[-2,-1,-3,-1],
[0,-1,2,2]]):
  rowspace(A);
```

$$\{\left[0, 1, 0, \frac{-4}{3}\right], \left[1, 0, 0, \frac{2}{3}\right], \left[0, 0, 1, \frac{1}{3}\right]\}$$

■

`rank`

`rank(mat)` computes the rank of the matrix `mat`. This value is found through Gaussian elimination and is equal to the number of nonzero rows in the matrix that results.

EXAMPLE 3: Calculate the rank of $A=\begin{pmatrix} 3 & 3 & 1 & 0 \\ 3 & 3 & 3 & 3 \\ 1 & 1 & 1 & 1 \end{pmatrix}$.

SOLUTION: After the 3×4 matrix `A` is defined, the rank is found to be 2.

```
> A:=array([[3,3,1,0],[3,3,3,3],[1,1,1,1]]):
  rank(A);
```

$$2$$

■

`nullspace`

`nullspace(mat)` and `nullspace(mat,'nd')` yield the same result as `kernel(mat)` and `kernel(mat,'nd')`. In both cases, the value of the nullity (dimension of the null space) of the matrix `mat` is named `nd`.

EXAMPLE 4: Find a basis for the nullspace of $A=\begin{pmatrix} 0 & 2 & -2 & 1 & 1 \\ -1 & 3 & -3 & 3 & -1 \\ 0 & -1 & 1 & 1 & 2 \end{pmatrix}$.

SOLUTION: After the 3×5 matrix `A` is defined the basis for the nullspace of `A` is found with `nullspace`. Note that this result is a set of two vectors.

```
> A:=array([[0,2,-2,1,1],[-1,3,-3,3,-1],
      [0,-1,1,1,2]]):
```

```
nullspace(A);
```

$$\left\{\left[-5, \frac{1}{3}, 0, \frac{-5}{3}, 1\right], [0, 1, 1, 0, 0]\right\}$$

■

colspan

colspan(mat) determines the spanning set of the columns of the matrix mat. Unlike with colspace, the lead entry of each of the vectors returned will not usually be 1.

EXAMPLE 5: Find a basis for the spanning set of the columns of the matrix

$$A = \begin{pmatrix} 1 & -2 & 2 & 1 & -2 \\ 1 & 1 & 2 & -2 & -2 \\ 1 & 0 & 0 & 2 & -1 \\ 0 & 0 & 0 & -2 & 0 \\ -2 & 1 & 0 & 1 & 2 \end{pmatrix}.$$

SOLUTION: The 5×5 matrix is entered so that the spanning set of the columns of A can be found with colspan. The result means that a basis for the column space of **A** is $\left\{\begin{pmatrix} 0 \\ 3 \\ 2 \\ 0 \\ -3 \end{pmatrix}\begin{pmatrix} 1 \\ 1 \\ 1 \\ 0 \\ -2 \end{pmatrix}\begin{pmatrix} 0 \\ 0 \\ -6 \\ 0 \\ 12 \end{pmatrix}\begin{pmatrix} 0 \\ 0 \\ 0 \\ 12 \\ -36 \end{pmatrix}\right\}$.

```
> A:=matrix(5,5,[1,-2,2,1,-2,1,1,2,-2,-2,1,0,0,2,
            -1,0,0,0,-2,0,-2,1,0,1,2]);
```

$$A := \begin{bmatrix} 1 & -2 & 2 & 1 & -2 \\ 1 & 1 & 2 & -2 & -2 \\ 1 & 0 & 0 & 2 & -1 \\ 0 & 0 & 0 & -2 & 0 \\ -2 & 1 & 0 & 1 & 2 \end{bmatrix}$$

```
> colspan(A);
```

$$\{[0, 3, 2, 0, -3], [1, 1, 1, 0, -2], [0, 0, 0, 12, -36], [0, 0, -6, 0, 12]\}$$

■

`rowspan`

`rowspan(mat)` determines the spanning set of the row space of the matrix `mat`, where `mat` is a made up of multivariate polynomials over the rationals. This command can be entered as `rowspan(mat,'d')` so that the rank of `mat` is found and assigned the name `d`. Unlike with `rowspace`, the lead entry of each of the vectors returned may not be 1.

EXAMPLE 6: Find a spanning set for the row space of $A = \begin{pmatrix} 0 & 1 & 1 & -1 \\ 3 & 2 & -1 & -1 \\ -2 & -1 & -3 & -1 \\ 0 & -1 & 2 & 2 \end{pmatrix}$.

SOLUTION: After the 4×4 matrix `A` is defined, the spanning set for the row space of `A`, which is made up of three vectors, is determined with `rowspan`. The result means that a basis for the row space of **A** is {(0 0 9 3),(3 2 -1 -1),(0 3 3 -3)}.

```
> A:=array([[0,1,1,-1],[3,2,-1,-1],[-2,-1,-3,-1],
        [0,-1,2,2]]):
  rowspan(A);
```

$$\{[3, 2, -1, -1], [0, 3, 3, -3], [0, 0, 9, 3]\}$$

■

Application: Transition Matrices

Let $\{\mathbf{v}_1,\mathbf{v}_2,\ldots,\mathbf{v}_n\}$ and $\{\mathbf{u}_1,\mathbf{u}_2,\ldots,\mathbf{u}_n\}$ both be (ordered) bases for $\Re^n$. If $\mathbf{v}=c_1\mathbf{v}_1 + c_2\mathbf{v}_2+\ldots+c_n\mathbf{v}_n$, there are unique numbers $a_{11},a_{12},\ldots,a_{1n},a_{21},a_{22},\ldots a_{2n},\ldots,a_{n1},a_{n2},\ldots a_{nn}$ so that $\mathbf{v}_1=a_{11}\mathbf{u}_1 + a_{21}\mathbf{u}_2+\ldots+a_{n1}\mathbf{u}_n$, $\mathbf{v}_2=a_{12}\mathbf{u}_1 + a_{22}\mathbf{u}_2+\ldots+a_{n2}\mathbf{u}_n$, ... , $\mathbf{v}_n=a_{1n}\mathbf{u}_1 + a_{2n}\mathbf{u}_2+\ldots+a_{nn}\mathbf{u}_n$. Then, if $\mathbf{v}=c_1\mathbf{v}_1 + c_2\mathbf{v}_2+\ldots+c_n\mathbf{v}_n$ is expressed in the form $\mathbf{v}=\begin{pmatrix} c_1 \\ c_2 \\ \vdots \\ c_n \end{pmatrix}$, computing

$$\begin{pmatrix} a_{11} & a_{12} & \ldots & a_{1n} \\ a_{21} & a_{22} & \ldots & a_{2n} \\ \vdots & \vdots & \ddots & \vdots \\ a_{n1} & a_{n2} & \ldots & a_{nn} \end{pmatrix}\begin{pmatrix} c_1 \\ c_2 \\ \vdots \\ c_n \end{pmatrix} = \begin{pmatrix} d_1 \\ d_2 \\ \vdots \\ d_n \end{pmatrix}$$

shows that $\mathbf{v}=d_1\mathbf{u}_1 + d_2\mathbf{u}_2+\ldots+d_n\mathbf{u}_n$. The matrix $\begin{pmatrix} a_{11} & a_{12} & \ldots & a_{1n} \\ a_{21} & a_{22} & \ldots & a_{2n} \\ \vdots & \vdots & \ddots & \vdots \\ a_{n1} & a_{n2} & \ldots & a_{nn} \end{pmatrix}$ is called the **transition matrix** from $\{\mathbf{v}_1,\mathbf{v}_2,\ldots,\mathbf{v}_n\}$ to $\{\mathbf{u}_1,\mathbf{u}_2,\ldots,\mathbf{u}_n\}$.

The transition matrix from $\{\mathbf{v}_1,\mathbf{v}_2,\ldots,\mathbf{v}_n\}$ to $\{\mathbf{u}_1,\mathbf{u}_2,\ldots,\mathbf{u}_n\}$, $\begin{pmatrix} a_{11} & a_{12} & \ldots & a_{1n} \\ a_{21} & a_{22} & \ldots & a_{2n} \\ \vdots & \vdots & \ddots & \vdots \\ a_{n1} & a_{n2} & \ldots & a_{nn} \end{pmatrix}$, can be obtained by row reducing the matrix $(\mathbf{u}_1 \ \mathbf{u}_2 \ \ldots \ \mathbf{u}_n | \mathbf{v}_1 \ \mathbf{v}_2 \ \ldots \ \mathbf{v}_n)$ to the form

$$\left(\begin{array}{cccc|cccc} 1 & 0 & \ldots & 0 & a_{11} & a_{12} & \ldots & a_{1n} \\ 0 & 1 & \ldots & 0 & a_{21} & a_{22} & \ldots & a_{2n} \\ \vdots & \vdots & \ddots & \vdots & \vdots & \vdots & \ddots & \vdots \\ 0 & 0 & 0 & 1 & a_{n1} & a_{n2} & \ldots & a_{nn} \end{array}\right).$$

Many of the commands that have been introduced can be used to find the transition matrix.

EXAMPLE 7: Both $B_1 = \left\{ \begin{pmatrix} 2 \\ 1 \\ 2 \\ 1 \end{pmatrix}, \begin{pmatrix} 0 \\ 0 \\ 2 \\ 2 \end{pmatrix}, \begin{pmatrix} 1 \\ 2 \\ 0 \\ 0 \end{pmatrix}, \begin{pmatrix} 0 \\ 1 \\ 2 \\ 2 \end{pmatrix} \right\}$ and

$B_2 = \left\{ \begin{pmatrix} 1 \\ 2 \\ 1 \\ 2 \end{pmatrix}, \begin{pmatrix} 0 \\ 2 \\ 1 \\ 1 \end{pmatrix}, \begin{pmatrix} 0 \\ 2 \\ 2 \\ 2 \end{pmatrix}, \begin{pmatrix} 2 \\ 1 \\ 1 \\ 1 \end{pmatrix} \right\}$ are bases of $\Re^4$. If $v = \begin{pmatrix} 1 \\ -1 \\ 5 \\ -1 \end{pmatrix}$ with respect to the basis B_1, calculate **v** with respect to the basis B_2.

SOLUTION: We begin by computing the transition matrix from B_1 to B_2. Row reducing the matrix $\left(\underbrace{\begin{matrix} 1 & 0 & 0 & 2 \\ 2 & 2 & 2 & 1 \\ 1 & 1 & 2 & 1 \\ 2 & 1 & 2 & 1 \end{matrix}}_{\text{New Basis}} \quad \underbrace{\begin{matrix} 2 & 0 & 1 & 0 \\ 1 & 0 & 2 & 1 \\ 2 & 2 & 0 & 2 \\ 1 & 2 & 0 & 2 \end{matrix}}_{\text{Old Basis}} \right)$ yields $\left(\begin{array}{cccc|cccc} 1 & 0 & 0 & 0 & -1 & 0 & 0 & 0 \\ 0 & 1 & 0 & 0 & 0 & -2 & 2 & -1 \\ 0 & 0 & 1 & 0 & 3/4 & 2 & -5/4 & 3/2 \\ 0 & 0 & 0 & 1 & 3/2 & 0 & 1/2 & 0 \end{array} \right)$

Therefore, the transition matrix from B_1 to B_2 is $\begin{pmatrix} -1 & 0 & 0 & 0 \\ 0 & -2 & 2 & -1 \\ 3/4 & 2 & -5/4 & 3/2 \\ 3/2 & 0 & 1/2 & 0 \end{pmatrix}$ These steps are carried out with Maple. First, the matrices `B1` and `B2` are defined. The augmented matrix `B1_B2` is determined with `augment`. This matrix is row-reduced with `rref` in `red_B1_B2`. Finally, the transition matrix is obtained with `delcols`.

```
> with(linalg):
  B1:=array([[1,0,0,2],[2,2,2,1],
```

```
    [1,1,2,1],[2,1,2,1]]):
  B2:=array([[2,0,1,0],[1,0,2,1],
    [2,2,0,2],[1,2,0,2]]):
> B1_B2:=augment(B1,B2);
```

$$B1\_B2 := \begin{bmatrix} 1 & 0 & 0 & 2 & 2 & 0 & 1 & 0 \\ 2 & 2 & 2 & 1 & 1 & 0 & 2 & 1 \\ 1 & 1 & 2 & 1 & 2 & 2 & 0 & 2 \\ 2 & 1 & 2 & 1 & 1 & 2 & 0 & 2 \end{bmatrix}$$

```
> red_B1_B2:=rref(B1_B2);
```

$$red\_B1\_B2 := \begin{bmatrix} 1 & 0 & 0 & 0 & -1 & 0 & 0 & 0 \\ 0 & 1 & 0 & 0 & 0 & -2 & 2 & -1 \\ 0 & 0 & 1 & 0 & \frac{3}{4} & 2 & \frac{-5}{4} & \frac{3}{2} \\ 0 & 0 & 0 & 1 & \frac{3}{2} & 0 & \frac{1}{2} & 0 \end{bmatrix}$$

```
> tran_matrix:=delcols(red_B1_B2,1..4);
```

$$tran\_matrix := \begin{bmatrix} -1 & 0 & 0 & 0 \\ 0 & -2 & 2 & -1 \\ \frac{3}{4} & 2 & \frac{-5}{4} & \frac{3}{2} \\ \frac{3}{2} & 0 & \frac{1}{2} & 0 \end{bmatrix}$$

Since $\mathbf{v}=\begin{pmatrix} 1 \\ -1 \\ 5 \\ -1 \end{pmatrix}$ with respect to the basis B_1, $\mathbf{v}=\begin{pmatrix} -1 & 0 & 0 & 0 \\ 0 & -2 & 2 & -1 \\ 3/4 & 2 & -5/4 & 3/2 \\ 3/2 & 0 & 1/2 & 0 \end{pmatrix}\times$ $\times\begin{pmatrix} 1 \\ -1 \\ 5 \\ -1 \end{pmatrix}=\begin{pmatrix} -1 \\ 13 \\ -9 \\ 4 \end{pmatrix}$ with respect to the basis B_2. This is also illustrated with Maple.

```
> v:=vector([1,-1,5,-1]);
```

$$v := [1, -1, 5, -1]$$

```
> multiply(tran_matrix,v);
```

$$[-1, 13, -9, 4]$$

■

The Gram-Schmidt Process

The set of vectors $\{\mathbf{v}_1,\mathbf{v}_2,\ldots,\mathbf{v}_n\}$ is an **orthogonal set** of vectors means that $\mathbf{v}_i$ and $\mathbf{v}_j$ are orthogonal if i and j are not equal; $\{\mathbf{v}_1,\mathbf{v}_2,\ldots,\mathbf{v}_n\}$ is an **orthonormal set** of vectors means that the norm of $\mathbf{v}_i$ is one for each i and $\mathbf{v}_i$ and $\mathbf{v}_j$ are orthogonal if i and j are not equal. Given a set of linearly independent vectors $\{\mathbf{v}_1,\mathbf{v}_2,\ldots,\mathbf{v}_n\}$, the set of all linear combinations of the elements of $\{\mathbf{v}_1,\mathbf{v}_2,\ldots,\mathbf{v}_n\}$, $V=\text{span}\{\mathbf{v}_1,\mathbf{v}_2,\ldots,\mathbf{v}_n\}$, is a vector space. Note that if $\{\mathbf{u}_1,\mathbf{u}_2,\ldots,\mathbf{u}_n\}$ is an orthonormal set and $\mathbf{v}\in\text{span}\{\mathbf{u}_1,\mathbf{u}_2,\ldots,\mathbf{u}_n\}$, then $\mathbf{v}=(\mathbf{v}\bullet\mathbf{u}_1)\mathbf{u}_1+(\mathbf{v}\bullet\mathbf{u}_2)\mathbf{u}_2+\ldots+(\mathbf{v}\bullet\mathbf{u}_n)\mathbf{u}_n$. Thus, we may easily express $\mathbf{v}$ as a linear combination of the vectors in $\{\mathbf{u}_1,\mathbf{u}_2,\ldots,\mathbf{u}_n\}$. Consequently, if we are given any vector space, V, it is frequently convenient to be able to find an orthonormal basis of V. We may use the Gram-Schmidt process to find an orthonormal basis of the vector space $V=\text{span}\{\mathbf{v}_1,\mathbf{v}_2,\ldots,\mathbf{v}_n\}$.

We summarize the algorithm of the Gram-Schmidt Process so that given a set of linearly independent vectors $\{\mathbf{v}_1,\mathbf{v}_2,\ldots,\mathbf{v}_n\}$, where $V=span\{\mathbf{v}_1,\mathbf{v}_2,\ldots,\mathbf{v}_n\}$, we can construct a set of orthonormal vectors $\{\mathbf{u}_1,\mathbf{u}_2,\ldots,\mathbf{u}_n\}$. so that $V=span\{\mathbf{u}_1,\mathbf{u}_2,\ldots,\mathbf{u}_n\}$.

1. Let $\mathbf{u}_1=\dfrac{\mathbf{v}_1}{\|\mathbf{v}_1\|}$;;
2. Compute $\text{proj}_{\{\mathbf{u}_1\}}\mathbf{v}_2=(\mathbf{u}_1\bullet\mathbf{v}_2)\mathbf{u}_1$,, $v_2-\text{proj}_{\{\mathbf{u}_1\}}\mathbf{v}_2$, and let $\mathbf{u}_2=\dfrac{\mathbf{v}_2-\text{proj}_{\mathbf{u}_1}\mathbf{v}_2}{\|\mathbf{v}_2-\text{proj}_{\mathbf{u}_1}\mathbf{v}_2\|}$. Then, span$\{\mathbf{u}_1,\mathbf{u}_2\}$ = span$\{\mathbf{v}_1,\mathbf{v}_2\}$ and span$\{\mathbf{u}_1, \mathbf{u}_2, \mathbf{v}_3,\ldots\mathbf{v}_n\}$= span$\{\mathbf{v}_1,\mathbf{v}_2,\mathbf{v}_3,\ldots,\mathbf{v}_n\}$;
3. Generally, for $3\le i\le n$,, compute

$$\text{proj}_{\{\mathbf{u}_1,\mathbf{u}_2,\ldots,\mathbf{u}_{i-1}\}}\mathbf{v}_i=(\mathbf{u}_1\bullet\mathbf{v}_i)\mathbf{u}_1+(\mathbf{u}_2\bullet\mathbf{v}_i)\mathbf{u}_2+\ldots+(\mathbf{u}_{i-1}\bullet\mathbf{v}_i)\mathbf{u}_{i-1},$$

$\mathbf{v}_i - \text{proj}_{\{\mathbf{u}_1, \mathbf{u}_2, \ldots, \mathbf{u}_{i-1}\}} \mathbf{v}_i$, and let $\mathbf{u}_i = \dfrac{\mathbf{v}_i - \text{proj}_{\{\mathbf{u}_1, \mathbf{u}_2, \ldots, \mathbf{u}_{i-1}\}} \mathbf{v}_i}{\left\| \mathbf{v}_i - \text{proj}_{\{\mathbf{u}_1, \mathbf{u}_2, \ldots, \mathbf{u}_{i-1}\}} \mathbf{v}_i \right\|}$. Then,

$\text{span}\{\mathbf{u}_1, \mathbf{u}_2, \ldots, \mathbf{u}_i\} = \text{span}\{\mathbf{v}_1, \mathbf{v}_2, \ldots, \mathbf{v}_i\}$ and

$\text{span}\{\mathbf{u}_1, \mathbf{u}_2, \ldots, \mathbf{u}_i, \mathbf{v}_{i+1}, \ldots, \mathbf{v}_n\} = \text{span}\{\mathbf{v}_1, \mathbf{v}_2, \mathbf{v}_3, \ldots, \mathbf{v}_n\}$; and

4. Because sspan$\{\mathbf{u}_1, \mathbf{u}_2, \ldots, \mathbf{u}_n\} = \text{span}\{\mathbf{v}_1, \mathbf{v}_2, \ldots, \mathbf{v}_n\}$ and $\{\mathbf{u}_1, \mathbf{u}_2, \ldots, \mathbf{u}_n\}$ is an orthonormal set, $\{\mathbf{u}_1, \mathbf{u}_2, \ldots, \mathbf{u}_n\}$ is an orthonormal basis of $\{\mathbf{v}_1, \mathbf{v}_2, \ldots, \mathbf{v}_n\}$.

The Gram-Schmidt procedure is well suited to computer arithmetic, and Maple contains the command `GramSchmidt` (which is described and demonstrated in the following example) to perform the Gram-Schmidt Process.

`GramSchmidt`

`GramSchmidt(veclist)` uses the Gram-Schmidt process to produce a list of orthogonal vectors from those given in `veclist`. Note that the vectors that result are not normalized, and if the vectors in `veclist` are linearly dependent, then so are those given in the result of `GramSchmidt`.

EXAMPLE 8: Use the Gram-Schmidt process to transform the basis $\left\{ \begin{pmatrix} -2 \\ -1 \\ -2 \end{pmatrix}, \begin{pmatrix} 0 \\ -1 \\ 2 \end{pmatrix}, \begin{pmatrix} 1 \\ 3 \\ -2 \end{pmatrix} \right\}$ of $\Re^3$ into an orthonormal basis.

SOLUTION: The three vectors are defined as `v1`, `v2`, and `v3`. Then, the orthogonal basis is obtained with `GramSchmidt`.

```
> v1:=vector([-2,-1,-2]):
  v2:=vector([0,-1,2]):
  v3:=vector([1,3,-2]):
```

```
GramSchmidt([v1,v2,v3]);
```

$$\left[[-2,-1,-2],\left[\frac{-2}{3},\frac{-4}{3},\frac{4}{3}\right],\left[\frac{-4}{9},\frac{4}{9},\frac{2}{9}\right]\right]$$

■

Linear Transformations

When a function $(T:\Re^n \to \Re^m)$ is said to be a **linear transformation**, this means that T satisfies the properties $T(\mathbf{u} + \mathbf{v})=T(\mathbf{u})+T(\mathbf{v})$ and $T(c\mathbf{u})=cT(\mathbf{u})$ for all vectors **u** and **v** in $\Re^n$ and all real numbers c. Let $(T:\Re^n \to \Re^m)$ be a linear transformation, and suppose $T(\mathbf{e}_1) = \mathrm{v}_1, T(\mathbf{e}_2) = \mathrm{v}_2, \ldots, T(\mathbf{e}_n) = \mathbf{v}_n$, where $\{\mathbf{e}_1,\mathbf{e}_2,\ldots,\mathbf{e}_n\}$ represents the standard basis of $\Re^n$ and $\mathbf{v}_1,\mathbf{v}_2,\ldots,\mathbf{v}_n$ are (column) vectors in $\Re^m$. The **associated matrix** of T is the $\mathrm{m} \times \mathrm{n}$ matrix

$$\mathrm{A}=(\mathbf{v}_1\ \mathbf{v}_2\ \ldots\ \mathbf{v}_n)\text{: if } \mathbf{x}=\begin{pmatrix} x_1 \\ x_2 \\ \vdots \\ x_n \end{pmatrix}, T(\mathbf{x})= T\left(\begin{pmatrix} x_1 \\ x_2 \\ \vdots \\ x_2 \end{pmatrix}\right) = \mathbf{A}\mathbf{x}= (\mathbf{v}_1\ \ \mathbf{v}_2\ \ \ldots\ \ \mathbf{v}_n)\begin{pmatrix} x_1 \\ x_2 \\ \vdots \\ x_n \end{pmatrix}.$$

Moreover, if **A** is any $\mathrm{m} \times \mathrm{n}$ matrix, then **A** is the associated matrix of the linear transformation defined by $T(\mathbf{x}) = \mathbf{A}\mathbf{x}$. In fact, a linear transformation T is completely determined by its action on any basis. We show how a transformation matrix is determined with Maple in Example 9.

EXAMPLE 9: Let $T:\Re^4 \to \Re^3$ be the linear transformation defined by

$T(\mathbf{e}_1)=\begin{pmatrix}1\\-2\\1\end{pmatrix}$, $T(\mathbf{e}_2)=\begin{pmatrix}-4\\5\\-4\end{pmatrix}$, $T(\mathbf{e}_3)=\begin{pmatrix}1\\-1\\-3\end{pmatrix}$, and $T(\mathbf{e}_4)=\begin{pmatrix}-3\\-2\\3\end{pmatrix}$, where **e**1, **e**2, **e**3, and **e**4 represent the standard basis vectors $\begin{pmatrix}1\\0\\0\\0\end{pmatrix}$, $\begin{pmatrix}0\\1\\0\\0\end{pmatrix}$, $\begin{pmatrix}0\\0\\1\\0\end{pmatrix}$, and $\begin{pmatrix}0\\0\\0\\1\end{pmatrix}$, respectively. Find the associated matrix of T, and calculate $T\left(\begin{pmatrix}3\\4\\0\\-2\end{pmatrix}\right)$.

SOLUTION: The associated matrix of T is $\begin{pmatrix}1&-4&1&-3\\-2&5&-1&-2\\-1&-4&-3&3\end{pmatrix}$ so that $T(\mathbf{x})=\begin{pmatrix}1&-4&1&-3\\-2&5&-1&-2\\-1&-4&-3&3\end{pmatrix}\mathbf{x}$, where $\mathbf{x}=\begin{pmatrix}x_1\\x_2\\x_3\\x_4\end{pmatrix}$. We define the 3×4 transformation matrix in `a_T` so that we can determine the value of $T(\mathbf{x})$ with `multiply` in `T`. Notice that this product is defined for all $4 \times n$ vectors **x**. The product corresponding to $T\left(\begin{pmatrix}3\\4\\0\\-2\end{pmatrix}\right)$ is then found with `T([3,4,0,-2])`.

The product is then defined for computing the product of T and a 4×1 vector `[x1, x2, x3, x4]`. Notice the difference in these two definitions. In the first case, the product is found with `T([3,4,0,-2])`, while in the second it is determined with `T(3,4,0,-2)`. Both functions yield the same value.

```
> with(linalg):
  a_T:=array([[1,-3,1,-3],[-2,5,-1,-2],
        [-1,-4,-3,3]]):
> T:=x->multiply(a_T,x);
```

$$T := x \to \text{multiply}(\,a\_T, x\,)$$

```
> T([3,4,0,-2]);
```

$$[\,-3, 18, -25\,]$$

```
> T:='T':
  T:=(x1,x2,x3,x4)->multiply(a_T,[x1,x2,x3,x4]);
```

$$T := (\,x1, x2, x3, x4\,) \to \text{multiply}(\,a\_T, [\,x1, x2, x3, x4\,]\,)$$

```
> T(3,4,0,-2);
```

$$[\,-3, 18, -25\,]$$

■

The **kernel** of the linear transformation *T*, ker(*T*), is the set of all vectors **x** in $\Re^n$ such that *T*(**x**)= **0**: ker(T)= $\{\mathbf{x} \in \Re^n : T(\mathbf{x}) = \mathbf{0}\}$. The kernel of *T* is a subspace of $\Re^n$. Because *T*(x)=**Ax** for all **x** in $\Re^n$, (T)= $\{\mathbf{x} \in \Re^n : T(\mathbf{x}) = \mathbf{0}\} = \{\mathbf{x} \in \Re^n : \mathbf{Ax} = \mathbf{0}\}$ so the kernel of *T* is the same as the nullspace of **A**. Maple's linear algebra package contains the following command, which can be used to find the kernel of a linear transformation.

`kernel`

`kernel(mat)` yields the basis for the null space for the linear transformation associated with the matrix `mat`. This command can be entered as `kernel(mat,´nd´)`, so that the dimension of the nullspace (nullity) is also determined and assigned the name `nd`. `kernel` yields the same results as `nullspace`.

EXAMPLE 10: Let $T:\Re^5 \to \Re^3$ be the linear transformation defined by

$T(\mathbf{x}) = \begin{pmatrix} 0 & -3 & -1 & -3 & -1 \\ -3 & 3 & -3 & -3 & -1 \\ 2 & 2 & -1 & 1 & 2 \end{pmatrix}\mathbf{x}$. (a) Calculate a basis for the kernel of the linear transformation. (b) Determine which of the vectors $\begin{pmatrix} 4 \\ 2 \\ 0 \\ 0 \\ -6 \end{pmatrix}$ and $\begin{pmatrix} 1 \\ 2 \\ -1 \\ -2 \\ 3 \end{pmatrix}$ is in the kernel of T.

SOLUTION: (a) After the matrix `A` is defined, a basis for the kernel of the associated linear transformation is found with `kernel` to be a set of two vectors.

```
> A:=array([[0,-3,-1,-3,-1],[-3,3,-3,-3,-1],
            [2,2,-1,1,2]]):
  kernel(A);
```

$$\left\{[2, 1, 0, 0, -3], \left[\frac{10}{13}, 0, \frac{-15}{13}, 1, \frac{-24}{13}\right]\right\}$$

(b) First, we define the linear transformation T and then compute $T\left(\left(\begin{array}{c}4\\2\\0\\0\\-6\end{array}\right)\right)$ and $T\left(\left(\begin{array}{c}1\\2\\-1\\-2\\3\end{array}\right)\right)$ to see that $\left(\left(\begin{array}{c}4\\2\\0\\0\\-6\end{array}\right)\right)$ is in the kernel, but $\left(\left(\begin{array}{c}1\\2\\-1\\-2\\3\end{array}\right)\right)$ is not.

```
> t:=x->evalm(A&*x):
  t([4,2,0,0,-6]);
```

$$[0, 0, 0]$$

```
> t([1,2,-1,-2,3]);
```

$$[-2, 9, 11]$$

■

Application: Rotations

Let $\mathbf{x} = \left(\begin{array}{c}x_1\\x_2\end{array}\right)$ be a vector in $\Re^2$ and θ an angle. Then, there are numbers r and ϕ given by

$r = \sqrt{x_1^2 + x_2^2}$ and $\phi = \tan^{-1}\frac{x_2}{x_1}$, so that $x_1 = r\cos\phi$ and $x_2 = r\sin\phi$. When we rotate

$$\mathbf{x} = \left(\begin{array}{c}x_1\\x_2\end{array}\right) = \left(\begin{array}{c}r\cos\phi\\r\sin\phi\end{array}\right)$$

through the angle θ, we obtain the vector $\mathbf{x}' = \left(\begin{array}{c}r\cos(\theta+\phi)\\r\sin(\theta+\phi)\end{array}\right)$.

Using the trigonometric identities

$$\sin(\theta \pm \phi) = \sin\theta\cos\phi \pm \sin\phi\cos\theta \text{ and } \cos(\theta \pm \phi) = \cos\theta\cos\phi \mp \sin\theta\sin\phi$$

we rewrite

$$\mathbf{x}' = \begin{pmatrix} r\cos(\theta+\phi) \\ r\sin(\theta+\phi) \end{pmatrix} = \begin{pmatrix} r\cos\theta\cos\phi - r\sin\theta\sin\phi \\ r\sin\theta\cos\phi + r\sin\phi\cos\theta \end{pmatrix}$$
$$= \begin{pmatrix} \cos\theta & -\sin\theta \\ \sin\theta & \cos\theta \end{pmatrix}\begin{pmatrix} r\cos\phi \\ r\sin\phi \end{pmatrix} = \begin{pmatrix} \cos\theta & -\sin\theta \\ \sin\theta & \cos\theta \end{pmatrix}\begin{pmatrix} x_1 \\ x_2 \end{pmatrix}.$$

Thus, the vector $\mathbf{x}'$ is obtained from $\mathbf{x}$ by computing $\begin{pmatrix} \cos\theta & -\sin\theta \\ \sin\theta & \cos\theta \end{pmatrix}\mathbf{x}$. Generally, if θ represents an angle, the linear transformation $T:\Re^2 \to \Re^2$ defined by $T(\mathbf{x}) = \begin{pmatrix} \cos\theta & -\sin\theta \\ \sin\theta & \cos\theta \end{pmatrix}\mathbf{x}$ is called the **rotation of $\Re^2$ through the angle θ**.

We can use the `rotate` command that is contained in the `plottools` package to rotate two- and three-dimensional graphics objects.

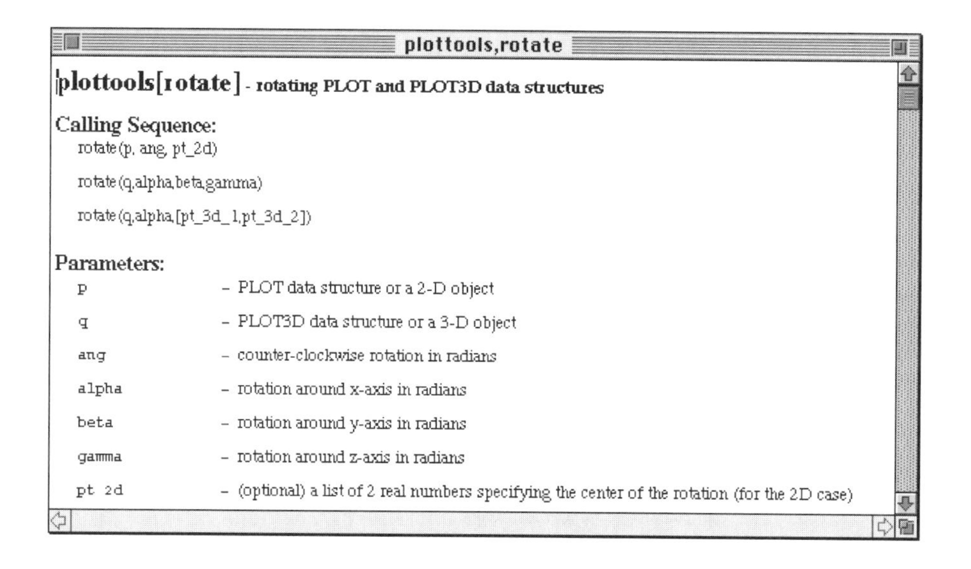

plottools,rotate

plottools[rotate] - **rotating PLOT and PLOT3D data structures**

Calling Sequence:

rotate(p, ang, pt_2d)

rotate(q,alpha,beta,gamma)

rotate(q,alpha,[pt_3d_1,pt_3d_2])

Parameters:

p	- PLOT data structure or a 2-D object
q	- PLOT3D data structure or a 3-D object
ang	- counter-clockwise rotation in radians
alpha	- rotation around x-axis in radians
beta	- rotation around y-axis in radians
gamma	- rotation around z-axis in radians
pt_2d	- (optional) a list of 2 real numbers specifying the center of the rotation (for the 2D case)

As an illustration of `rotate`, we first load the `plots` and `plottools` packages

```
> with(plots):
  with(plottools);
```

[*arc, arrow, circle, cone, cuboid, curve, cutin, cutout, cylinder, disk, dodecahedron, ellipse, ellipticArc, hemisphere, hexahedron, homothety, hyperbola, icosahedron line, octahedron, pieslice, point, polygon, project, rectangle, reflect, rotate, scale, semitorus, sphere, stellate, tetrahedron, torus, transform, translate, vrml*]

and then use `rectangle` to create a gray square with corners at (–1/2,–1/2) and (1/2,1/2). The resulting graphics object named `sq` is displayed with `display`.

```
> sq:=rectangle([-1/2,-1/2],[1/2,1/2],color=GRAY):
  display(sq,axes=NONE);
```

Next, we use `rotate` to rotate the square counterclockwise $\pi/3$ radians. Notice that Maple returns the coordinates of the vertices.

```
> rotate(sq, Pi/3);
```

POLYGONS([[.1830127020, -.6830127020], [.6830127020, .1830127020], [-.1830127020, .6830127020], [-.6830127020, -.1830127020]], COLOUR(*RGB*, .75294118, .75294118, .75294118))

We use `display` to see the rotated square.

```
> display(rotate(sq,Pi/3),axes=NONE);
```

We can rotate the square through various angles and either animate the result or display the result as an array with `display`, together with the option `insequence=true`. We begin by defining `thetavals` to be nine equally spaced numbers between 0 and $\pi/2$. Then we use `seq` and `display` to rotate the square about each of the angles in `thetavals` and name the resulting list of graphics objects `toanimate`.

```
> thetavals:=seq(j*Pi/16,j=0..8):
  toanimate:=[seq(display(rotate(sq,theta),axes=NONE),
      theta=thetavals)]:
```

When we use `display` to display `toanimate`, all nine graphs are shown together.

```
> display(toanimate);
```

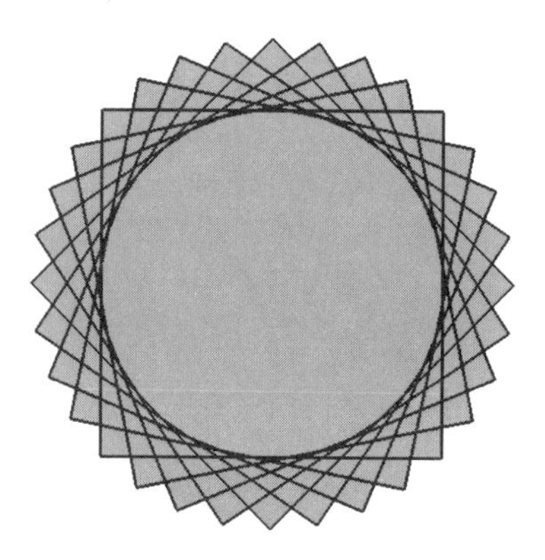

To display toanimate as an array or as an animation, we use display together with the option insequence=true.

```
> p:=display(toanimate,insequence=true):
```

Entering

```
> display(p)
```

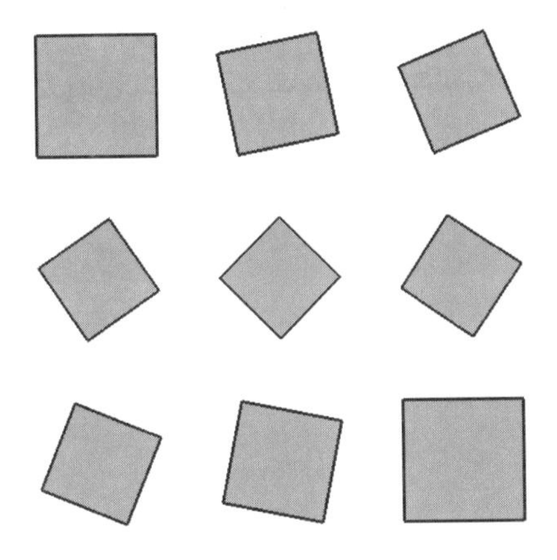

displays p as an array of graphics objects, while entering

```
> display(p,insequence=true)
```

animates the graphics in p. One frame from the resulting animation is shown in the following screenshot.

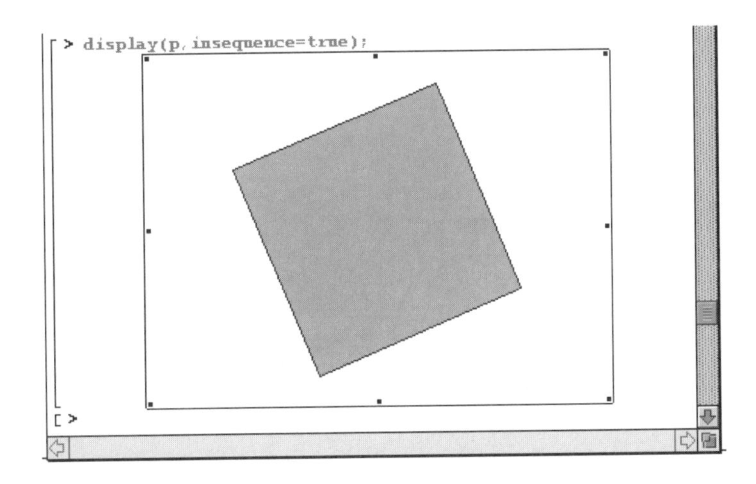

6.3 Determinants

In Chapter 5, we briefly discussed how to use the command `det` to compute the determinant of a square matrix. In this section, we review the definition of the determinant and related Maple functions that can be used to help find the determinant of a square matrix.

Let $\mathbf{A} = (a)$ be a 1×1 matrix. Then the determinant of **A**, det(**A**), is det(**A**) = a. If $\mathbf{A} = \begin{pmatrix} a_{11} & a_{12} \\ a_{21} & a_{22} \end{pmatrix}$ is a 2 x 2 matrix, then the determinant of **A**, det(**A**), is det(**A**) = $a_{11}a_{22} - a_{12}a_{21}$. We use induction to define the determinant of an $n \times n$ matrix. Let **A** be an $n \times n$ matrix. The ***ijth minor*** of **A**, denoted M_{ij}, is the $(n-1)\times(n-1)$ matrix obtained by deleting the ith row and jth column of **A**. The ***ijth cofactor*** of **A**, denoted A_{ij}, is determined by the formula $A_{ij} = (-1)^{i+j}\det(\mathbf{M}_{ij})$.

With Maple, `minor(mat,i,j)` yields the ijth minor of the matrix `mat` that is obtained by removing the ith row and the jth column of `mat`. As with most of the command discussed earlier in the chapter, `minor` is contain in the `linalg` package.

EXAMPLE 1: If $\mathbf{A} = \begin{pmatrix} 2 & 0 & -1 & -1 \\ 3 & -1 & -3 & -1 \\ -3 & 0 & 0 & -1 \\ 2 & -2 & -3 & 1 \end{pmatrix}$, calculate the minors $\mathbf{M}_{13}$, $\mathbf{M}_{23}$, and $\mathbf{M}_{33}$.

SOLUTION: After the 4×4 matrix `A` is defined, the indicated minors of `A` are found in the form of a list. The first entry of this list is determined by eliminating the first row and third column of `A`, the second entry by eliminating the second row and third column of `A`, and the third entry by eliminating the third row and third column of `A`.

```
> with(linalg):
  A:=array([[2,0,-1,-1],[3,-1,-3,-1],[-3,0,0,-1],
     [2,-2,-3,1]]);
```

```
[minor(A,1,3),minor(A,2,3),minor(A,3,3)];
```

$$A := \begin{bmatrix} 2 & 0 & -1 & -1 \\ 3 & -1 & -3 & -1 \\ -3 & 0 & 0 & -1 \\ 2 & -2 & -3 & 1 \end{bmatrix}$$

$$\left[\begin{bmatrix} 3 & -1 & -1 \\ -3 & 0 & -1 \\ 2 & -2 & 1 \end{bmatrix}, \begin{bmatrix} 2 & 0 & -1 \\ -3 & 0 & -1 \\ 2 & -2 & 1 \end{bmatrix}, \begin{bmatrix} 2 & 0 & -1 \\ 3 & -1 & -1 \\ 2 & -2 & 1 \end{bmatrix}\right]$$

■

If **A** is an $n \times n$ matrix where $n \geq 2$, the **determinant** of **A** is

$$\det(\mathbf{A}) = |\mathbf{A}| = a_{11}A_{11} + a_{12}A_{12} + \cdots + a_{1n}A_{1n} = \sum_{j=1}^{n} a_{1j}A_{1j}.$$

Fortunately, the cofactor expansion can be carried out along any row or column of the matrix. These alternate forms of expansion are given below and are especially useful for finding the determinant of matrices having rows or columns with numerous zeros: Let **A** be an $n \times n$ matrix where $n \geq 2$. Then the determinant of **A** may be calculated by expanding along any row or column of **A**. Hence, expansion along the kth row yields

$$\det(\mathbf{A}) = |\mathbf{A}| = a_{k1}A_{k1} + a_{k2}A_{k2} + \cdots + a_{kn}A_{kn} = \sum_{j=1}^{n} a_{kj}A_{kj},$$

and expansion along the kth column yields

$$\det(\mathbf{A}) = |\mathbf{A}| = a_{1k}A_{1k} + a_{2k}A_{2k} + \cdots + a_{nk}A_{nk} = \sum_{j=1}^{n} a_{jk}A_{jk}.$$

Of course with Maple, the cofactor expansion to find the value of the determinant of the matrix `mat` can be avoided with the command `det(mat)`.

EXAMPLE 2: If $\mathbf{A} = \begin{pmatrix} 2 & 0 & -1 & -1 \\ 3 & -1 & -3 & -1 \\ -3 & 0 & 0 & -1 \\ 2 & -2 & -3 & 1 \end{pmatrix}$, calculate the determinant of **A**, det(**A**).

SOLUTION: In the first case, we compute the determinant by expanding along the first row. `Det` represents the inert form of the `det` function. Of course, the same value is obtained if we expand along any row.

```
> with(linalg):
  A:=array([[2,0,-1,-1],[3,-1,-3,-1],
      [-3,0,0,-1],[2,-2,-3,1]]);
```

$$A := \begin{bmatrix} 2 & 0 & -1 & -1 \\ 3 & -1 & -3 & -1 \\ -3 & 0 & 0 & -1 \\ 2 & -2 & -3 & 1 \end{bmatrix}$$

```
> stepone:=sum(
   '(-1)^(1+i)*A[1,i]*Det(minor(A,1,i))','i'=1..4);
```

$$stepone := 2\,\mathrm{Det}\left(\begin{bmatrix} -1 & -3 & -1 \\ 0 & 0 & -1 \\ -2 & -3 & 1 \end{bmatrix}\right) - \mathrm{Det}\left(\begin{bmatrix} 3 & -1 & -1 \\ -3 & 0 & -1 \\ 2 & -2 & 1 \end{bmatrix}\right) + \mathrm{Det}\left(\begin{bmatrix} 3 & -1 & -3 \\ -3 & 0 & 0 \\ 2 & -2 & -3 \end{bmatrix}\right)$$

We evaluate the previous expression by replacing each occurrence of `Det` in `stepone` by `det` and then using `eval` to evaluate the result.

```
> steptwo:=subs(Det=det,stepone);
```

$$steptwo := 2\,\det\left(\begin{bmatrix} -1 & -3 & -1 \\ 0 & 0 & -1 \\ -2 & -3 & 1 \end{bmatrix}\right) - \det\left(\begin{bmatrix} 3 & -1 & -1 \\ -3 & 0 & -1 \\ 2 & -2 & 1 \end{bmatrix}\right) + \det\left(\begin{bmatrix} 3 & -1 & -3 \\ -3 & 0 & 0 \\ 2 & -2 & -3 \end{bmatrix}\right)$$

```
> eval(steptwo);
```

$$-2$$

In the second case, we use the command `det` to compute the determinant of A.

```
> det(A);
```

$$2$$

■

6.4 Eigenvalues and Eigenvectors

Let **A** be an $n \times n$ matrix with real components. Then the number λ is called an **eigenvalue** of **A** if there is a nonzero vector **v** that satisfies $\mathbf{A}\mathbf{v} = \lambda\mathbf{v}$. This nonzero vector is called the **eigenvector** of **A** that corresponds to λ. The **characteristic matrix** of **A** is the matrix $\mathbf{A} - \lambda\mathbf{I}$, where **I** represents the identity matrix. The eigenvalues are roots of the **characteristic equation** $|\mathbf{A} - \lambda\mathbf{I}| = 0$, which may have at most n real distinct roots; it may also have repeated roots and roots that are complex conjugates. The left-hand side of the characteristic equation, $\mathbf{A} - \lambda\mathbf{I}$, is called the **characteristic polynomial**. (Therefore, the eigenvalues of **A** are the roots of the characteristic polynomial.) After the eigenvalues are obtained, the corresponding eigenvectors are found by substituting the eigenvalues into the homogeneous system of equations $(\mathbf{A} - \lambda\mathbf{I})\mathbf{v} = \mathbf{0}$ and solving for **v**.

Maple contains several commands that can be used to determine the characteristic matrix, characteristic polynomial, eigenvalues, and eigenvectors of an $n \times n$ matrix. We begin by discussing the commands contained in the `linalg` package that we will use for determining the characteristic matrix and characteristic polynomial below.

`charmat`

> `charmat(A,lambda)` computes the characteristic matrix, $\mathbf{A} - \lambda\mathbf{I}$, where **I** represents the identity matrix with the same dimensions as that of the square matrix `A`.

`charpoly`

> `charpoly(A,lambda)` calculates the characteristic polynomial, the determinant of the characteristic matrix, $\mathbf{A} - \lambda\mathbf{I}$, where **I** represents the identity matrix with the same dimensions as that of the square matrix `A`.

EXAMPLE 1: If $A = \begin{pmatrix} 4 & -3 & 4 \\ 1 & -3 & 1 \\ -1 & -1 & 5 \end{pmatrix}$, find the characteristic polynomial of **A** (with respect to the variable x).

SOLUTION: The characteristic matrix is determined in terms of the variable `lambda` with `charmat`, and the characteristic polynomial is found in `polyA` in terms of the variable `x`. This polynomial is the same as the determinant of the characteristic matrix with `lambda` replaced by `x`.

```
> with(linalg):
  A:=matrix(3,3,[4,-3,4,1,-3,1,-1,-1,-5]);
```

$$A := \begin{bmatrix} 4 & -3 & 4 \\ 1 & -3 & 1 \\ -1 & -1 & -5 \end{bmatrix}$$

```
> charmat(A,lambda);
```

$$\begin{bmatrix} \lambda - 4 & 3 & -4 \\ -1 & \lambda + 3 & -1 \\ 1 & 1 & \lambda + 5 \end{bmatrix}$$

```
> polyA:=charpoly(A,x);
```

$$polyA := x^3 + 4\,x^2 - 9\,x - 36$$

■

Maple also contains commands in the linear algebra `linalg`,package that compute the eigenvalues and eigenvectors of a matrix.

`eigenvals`

`eigenvals(A)` lists the eigenvalues `lambda` of the matrix `A`. If the characteristic polynomial associated with `A` is of degree greater than four, then `eigenvals(A,'implicit')` should be used, in which case the eigenvalues are given in terms of the `RootOf` command. The generalized eigenvalue problem is solved with `eigenvals(A,B)`, which solves `det(lambda*A-B)=0`.

EXAMPLE 2: (a) If $\mathbf{A} = \begin{pmatrix} 5 & 2/3 & 1 & -4/3 & -4 & -4/3 \\ 0 & -1/6 & -2 & -1/6 & 7 & 23/6 \\ -1/2 & -1/4 & 5/2 & -1/4 & 1 & 3/4 \\ 4 & 1/2 & 0 & 1/2 & -3 & -1/2 \\ 0 & -1 & 0 & 0 & 4 & 1 \\ -1 & 1/6 & -1 & 1/6 & 2 & 19/6 \end{pmatrix}$, find the eigenvalues of **A**. (b) If $\mathbf{B} = \begin{pmatrix} 1 & 5 & 1 \\ -3 & 6 & -6 \\ -4 & 1 & 7 \end{pmatrix}$, find both exact and approximate values for the eigenvalues of **B**.

SOLUTION: We define **A** and then use `eigenvals` to compute the eigenvalues of **A**. We see that **A** has two distinct eigenvalues, $\lambda_{1,2,3} = 2$ and $\lambda_{4,5,6} = 3$, each with multiplicity three.

```
> A:=array([[5,2/3,1,-4/3,-4,-4/3],
      [0,-1/6,-2,-1/6,7,23/6],
      [-1/2,-1/4,5/2,-1/4,1,3/4],
      [4,1/2,0,1/2,-3,-1/2],
      [0,-1,0,0,4,1],[-1,1/6,-1,1/6,2,19/6]]):
  with(linalg):
  eigenvals(A);
```

$$2, 3, 2, 3, 2, 3$$

In the same manner, we define **B** and then use `eigenvals` to obtain the exact value of the eigenvalues of **B** in `eigs_B`.

```
> B:=array([[1,5,1],[-3,6,-6],[-4,1,7]]):
  eigs_B:=eigenvals(B);
```

$$eigs\_B := \frac{1}{3}(1673+3\sqrt{320457})^{(1/3)} - \frac{44}{3}\frac{1}{(1673+3\sqrt{320457})^{(1/3)}} + \frac{14}{3},$$

$$-\frac{1}{6}(1673+3\sqrt{320457})^{\left(\frac{1}{3}\right)}+\frac{22}{3}+\frac{1}{(1673+3\sqrt{320457})^{\left(\frac{1}{3}\right)}}+\frac{14}{3}$$

$$+\frac{1}{2}I\sqrt{3}\left(\frac{1}{3}(1673+3\sqrt{320457})^{(1/3)}+\frac{44}{3}\frac{1}{(1673+3\sqrt{320457})^{(1/3)}}\right),$$

$$-\frac{1}{6}(1673+3\sqrt{320457})^{(1/3)}+\frac{22}{3}\frac{1}{(1673+3\sqrt{320457})^{(1/3)}}+\frac{14}{3}$$

$$-\frac{1}{2}I\sqrt{3}\left(\frac{1}{3}(1673+3\sqrt{320457})^{(1/3)}+\frac{44}{3}\frac{1}{(1673+3\sqrt{320457})^{(1/3)}}\right)$$

The first number in the set of numbers `eigs_B` is obtained below with `eigs_B[1]`. We then use `evalf` to compute approximations of the eigenvalues of B by first using the command `evalf({eigs_B})` and then by using `map` to apply `evalf` to the elements of `{eigs_B}`.

```
> eigs_B[1];
```

$$\frac{1}{3}(1673+3\sqrt{320457})^{(1/3)}-\frac{44}{3}\frac{1}{(1673+3\sqrt{320457})^{(1/3)}}+\frac{14}{3}$$

```
> evalf({eigs_B});
```

$$\{8.686684370, 2.656657815-5.175623000\,I, 2.656657815+5.175623000\,I\}$$

```
> map(evalf,{eigs_B});
```

$$\{8.686684370, 2.656657815-5.175623000\,I, 2.656657815+5.175623000\,I\}$$

■

The built-in command `Eigenvals` is the inert form of the command `eigenvals`. Note that `eigenvals` is contained in the package `linalg`, so it must be loaded by entering `with(linalg)` or the command must be entered in the form `linalg[eigenvals]`. On the other hand, `Eigenvals` is a built-in function and need not be loaded before it is used.

Approximations of eigenvalues and the corresponding eigenvectors of a given matrix can be approximated with `Eigenvals`. When the command is entered in the form `Eigenvals(mat,'vecs')`, then the eigenvectors of the matrix `mat` are given in the array `vecs`.

EXAMPLE 3: If $\mathbf{A}=\begin{pmatrix} 3 & -5 & -4 \\ -5 & 6 & 3 \\ -3 & 2 & -2 \end{pmatrix}$, approximate the eigenvalues and corresponding eigenvectors of **A**.

SOLUTION: As in the previous examples, we begin by defining **A**, and then we use `Eigenvals` and `evalf` to approximate the eigenvalues of **A** in `stepone` and `vals_A`.

```
> A:=array([[3,-5,-4],[-5,6,3],[-3,2,-2]]);
  stepone:=Eigenvals(A,'vecs_A');
```

$$A := \begin{bmatrix} 3 & -5 & -4 \\ -5 & 6 & 3 \\ -3 & 2 & -2 \end{bmatrix}$$

$$stepone := \text{Eigenvals}\left(\begin{bmatrix} 3 & -5 & -4 \\ -5 & 6 & 3 \\ -3 & 2 & -2 \end{bmatrix}, vecs\_A\right)$$

```
> vals_A:=evalf(stepone);
```

$$vals\_A := [\,10.98792900, -.217222039, -3.770706957\,]$$

The matrix with columns corresponding to approximations of the eigenvectors of A is named `vecs_A`. The *i*th eigenvalue in the list `vals_A` has eigenvector corresponding to the *i*th column of `vecs_A`. We display `vecs_A` with `print`.

```
> print(vecs_A);
```

$$\begin{bmatrix} .609587416 & .5581271084 & .5851919459 \\ -.7667655370 & .5852417795 & -.0075842680 \\ -.2588783261 & -.2826475183 & 1.000021129 \end{bmatrix}$$

We illustrate that the third column of vecs_A is the eigenvector corresponding to the third eigenvalue given in vals_A by verifying that $\mathbf{Av} = \lambda\mathbf{v}$ for the corresponding eigenvalue and eigenvector. We first load the package linalg, and we then use the command col to extract the third column of vecs_A.

```
> with(linalg):
  col(vecs_A,3);
```

$$[.5851919459, -.0075842680, 1.000021129]$$

In the same manner, the third eigenvalue is extracted from vals_A with vals_A[3]:

```
> vals_A[3];
```

$$-3.770706959$$

We then use evalm and &* to compare the results of multiplying **A** by the third column of vecs_A and multiplying the third column of vecs_A by the third number in the list vals_A. The results are the same, so we conclude that the third column of vecs_A is the eigenvector corresponding to the third eigenvalue given in vals_A.

```
> evalm(A&*col(vecs_A,3));
```

$$[-2.206587338, .028598049, -3.770786632]$$

```
> evalm(col(vecs_A,3)*vals_A[3]);
```

$$[-2.206587342, .02859805211, -3.770786628]$$

With the following command, we use an alternative method to verify that the second column of vecs_A is the eigenvector corresponding to the second eigenvalue given in vals_A. The result indicates that the value of $(\mathbf{A} - \lambda\mathbf{I})\mathbf{v} = \mathbf{0}$ for each the eigenvalue λ in the second column of vals_A and its corresponding eigenvector **v** in vecs_A. Notice that the values in the vectors are approximately equal to zero.

```
> evalm(A&*col(vecs_A,2)-vals_A[2]*col(vecs_A,2));
```

$$[.85\ 10^{-8}, -.72\ 10^{-8}, .36\ 10^{-9}]$$

Finally, we use a `for` loop to show that the ith eigenvalue in `vals_A` has eigenvector corresponding to the ith column of `vecs_A`. Due to the precision of computer arithmetic, we interpret all results to be 0.

```
> for i to 3 do
    evalm(A&*col(vecs_A,i)-
          vals_A[i]*col(vecs_A,i))
          od;
```

$$[-.9\ 10^{-8}, 0, -.3\ 10^{-8}]$$
$$[.85\ 10^{-8}, -.72\ 10^{-8}, .36\ 10^{-9}]$$
$$[.4\ 10^{-8}, -.311\ 10^{-8}, -.4\ 10^{-8}]$$

■

In addition to `Eigenvals`, `eigenvects`, which is contained in the `linalg` package, can be used to compute eigenvalues and eigenvectors.

`eigenvects`

`eigenvects(A)` computes the exact value of the eigenvectors of the square matrix `A`. The entries of the output list of this command have the form `[eigenval[i],mult[i], {vec[1,i],..., vec[m,i]}]` where `eigenval[i]` represents the eigenvalue, `mult[i]` the multiplicity of the eigenvalue, and `{vec[1,i],..., vec[m,i]}` the set of linearly independent eigenvectors associated with `eigenval[i]`. Eigenvectors are determined by finding and factoring the characteristic polynomial associated with `mat`. If nonlinear factors are encountered in this factoring process, then results are given in terms of the `RootOf` command.

EXAMPLE 4: Find the eigenvalues and corresponding eigenvectors of
$\mathbf{A} = \begin{pmatrix} 0 & 4 \\ 2 & -2 \end{pmatrix}$.

SOLUTION: First, the 2×2 matrix `A` is entered. Then, the eigenvectors [2,1], which corresponds to the eigenvalue 2 of multiplicity 1, and [-1,1], which corresponds to the eigenvalue -4 of multiplicity 1, are found with `eigenvects`.

```
> with(linalg):
  A:=matrix(2,2,[0,4,2,-2]):
  eigs_A:=eigenvects(A);
```

$$eigs\_A := [-4, 1, \{[1, -1]\}], [2, 1, \{[2, 1]\}]$$

The eigenvalue –4 is extracted from `eigs_A`, with

```
> eigs_A[1,1];
```

$$-4$$

while the corresponding eigenvector is extracted from `eigs_A`, with

```
> eigs_A[1,3,1];
```

$$[1, -1]$$

The eigenvalue 2 and corresponding eigenvector are extracted from `eigs_A` in the same way.

```
> eigs_A[2,1];
```

$$2$$

```
> eigs_A[2,3,1];
```

$$[2, 1]$$

Finally, we verify that the numbers and vectors given in `eigsb` are eigenvalues and corresponding eigenvectors. We first verify that $\mathbf{A}\begin{pmatrix}-1\\1\end{pmatrix} = 4\begin{pmatrix}-1\\1\end{pmatrix}$ and

```
> evalm(A&*eigs_A[1,3,1]-eigs_A[1,1]*eigs_A[1,3,1]);
```

$$[0, 0]$$

then we verify that

$$\mathbf{A}\begin{pmatrix} 2 \\ 1 \end{pmatrix} = 2\begin{pmatrix} 2 \\ 1 \end{pmatrix}$$

$$(\mathbf{A} - 2\mathbf{I})\begin{pmatrix} 2 \\ 1 \end{pmatrix} = \begin{pmatrix} 0 \\ 0 \end{pmatrix}.$$

In this case, notice how we use the symbol `&*()` to represent the identity matrix.

```
> evalm((A-eigs_A[2,1]*&*())&*eigs_A[2,3,1]);
                        [0,0]
```

■

Let λ be an eigenvalue of the $n \times n$ matrix $\mathbf{A}$ and let $E_\lambda = \{\mathbf{v} \neq 0 : \mathbf{A}\mathbf{v} = \lambda\mathbf{v}\}$. Then E_λ is called the **eigenspace** of **A** associated with the eigenvalue λ. Hence, if $\mathbf{v}_1, \mathbf{v}_2, \ldots, \mathbf{v}_n$ is a set of n linearly independent eigenvectors corresponding to the eigenvalue λ, then the associated eigenspace is given by $E_\lambda = \text{span}\{\mathbf{v}_1, \mathbf{v}_2, \ldots, \mathbf{v}_n\}$.

EXAMPLE 5: Find the eigenvalues and corresponding eigenvectors of $\mathbf{A} = \begin{pmatrix} 1 & 0 & -4 & 3 \\ -3/2 & -1/2 & 9/2 & 7/2 \\ -2 & 0 & 3 & -3 \\ -2 & 0 & 4 & -4 \end{pmatrix}$ and $\mathbf{B} = \begin{pmatrix} -3 & 3 & -2 & 2 \\ 5/2 & -5/2 & 2 & -2 \\ 5/2 & -9/2 & 4 & -2 \\ -2 & -1 & 1 & 1 \end{pmatrix}$. What is the dimension of the eigenspace of **A**? of **B**?

SOLUTION: As in the preceding examples, we begin by defining the matrices **A** and **B**. We then use `eigenvals` to compute the eigenvalues of **A** and **B**. Notice that both **A** and **B** have the same eigenvalues.

```
> A:='A':B:='B':
  with(linalg):
  A:=array([[1,0,-4,3],[-3/2,-1/2,9/2,7/2],
       [-2,0,3,-3],[-2,0,4,-4]]):
  B:=array([[-3,2,-2,2],[5/2,-5/2,2,-2],
       [5/2,-9/2,4,-2],[-2,-1,1,1]]):
```

```
> eigenvals(A);
```

$$\frac{-1}{}, 2, -1, -1$$

```
> eigenvals(B);
```

$$\frac{-1}{2}, 2, -1, -1$$

We use `eigenvects` to compute the eigenvalues and corresponding eigenvectors of **A** and **B**. Note that both **A** and **B** have the same eigenvalues, each with the same multiplicity. Also note that **A** has two linearly independent eigenvectors corresponding to −1, while **B** has only one. Hence, the dimension of the eigenspace of **A** is four; the dimension of the eigenspace of **B** is three.

```
> eigenvects(A);
```

$$\left[2, 1, \left\{\left[-1, \frac{19}{5}, 1, 1\right]\right\}\right], \left[\frac{-1}{2}, 1, \{[0, 1, 0, 0]\}\right],$$

$$\left[-1, 2, \left\{[2, -3, 1, 0], \left[\frac{-3}{2}, \frac{-23}{2}, 0, 1\right]\right\}\right]$$

```
> eigenvects(B);
```

$$[-1, 2, \{[1, -1, -1, 1]\}], [2, 1, \{[0, 0, 1, 1]\}], \left[\frac{-1}{2}, 1, \{[0, 1, 1, 0]\}\right]$$

■

Jordan Canonical Form

Let $\mathbf{N}_k = (n_{ij}) = \begin{cases} 1, j = i+1 \\ 0, \text{otherwise} \end{cases}$ represent a $k \times k$ matrix with indicated elements. The $k \times k$ **Jordan block matrix** is given by $\mathbf{B}(\lambda) = \lambda\mathbf{I} + \mathbf{N}_k$ where λ is a constant. These matrices are defined by the following matrices:

$$\mathbf{N}_k = \begin{pmatrix} 0 & 1 & 0 & \dots & 0 \\ 0 & 0 & 1 & \dots & 0 \\ \vdots & \vdots & \vdots & & \vdots \\ 0 & 0 & 0 & \dots & 1 \\ 0 & 0 & 0 & \dots & 0 \end{pmatrix} \text{ and } \mathbf{B}(\lambda) = \lambda\mathbf{I} + \mathbf{N}_k = \begin{pmatrix} \lambda & 1 & 0 & \dots & 0 & 0 \\ 0 & \lambda & 0 & \dots & 0 & 0 \\ \vdots & \vdots & \vdots & & \vdots & \vdots \\ 0 & 0 & 0 & \dots & \lambda & 1 \\ 0 & 0 & 0 & \dots & 0 & \lambda \end{pmatrix}.$$

Hence, $\mathbf{B}(\lambda)$ can be defined as $\mathbf{B}(\lambda) = (b_{ij}) = \begin{cases} \lambda, i = j \\ 1, j = i+1 \\ 0, \text{otherwise} \end{cases}$. A **Jordan matrix** has the

form $\mathbf{J} = \begin{pmatrix} \mathbf{B}_1(\lambda) & 0 & \dots & 0 \\ 0 & \mathbf{B}_2(\lambda) & \dots & 0 \\ \vdots & \vdots & & \vdots \\ 0 & 0 & \dots & \mathbf{B}_n(\lambda) \end{pmatrix}$ where the entries $\mathbf{B}_j(\lambda)$, $j = 1,2,\dots,n$ represent Jordan block matrices. Suppose that $\mathbf{A}$ is an $n \times n$ matrix. Then there is an invertible $n \times n$ matrix $\mathbf{C}$ such that $\mathbf{C}^{-1}\mathbf{AC} = \mathbf{J}$ where $\mathbf{J}$ is a Jordan matrix with the eigenvalues of $\mathbf{A}$ as diagonal elements. The matrix $\mathbf{J}$ is called the **Jordan canonical form** of $\mathbf{A}$. From the Jordan canonical form of a matrix, we can determine the eigenvalues and the minimal and characteristic polynomials of a matrix.

Maple contains the jordan to find the Jordan canonical form of a matrix.

`jordan`

`jordan(A)` calculates the Jordan form of the matrix `A`. Note that the diagonal elements of the Jordan form are the eigenvalues of `A`. Hence, if the eigenvalues cannot be determined exactly, then a floating point entry must be included in `A` so that a result can be obtained. This command may be entered as `jordan(A,'T')`, so that the transition matrix which satisfies the equation `inverse(T)*jordan(A)*T=A` is assigned the name `T`.

EXAMPLE 6: Find the Jordan canonical form, $\mathbf{J_A}$ and $\mathbf{J_B}$, of $\mathbf{A}=\begin{pmatrix} 2 & 9 & -9 \\ 0 & 8 & -6 \\ 0 & 9 & -7 \end{pmatrix}$ and

$$\mathbf{B} = \begin{pmatrix} 21 & 27 & -13 & -11 & -32/3 & -20/3 \\ -17 & -19 & 9 & 7 & 8 & 4 \\ -49/2 & -27 & 25/2 & 11 & 77/6 & 35/6 \\ 7/2 & 5 & -5/2 & -4 & -5/2 & -3/2 \\ 55/2 & 34 & -33/2 & -14 & -91/6 & -49/6 \\ -8 & -4 & 3 & 2 & 11/3 & -1/3 \end{pmatrix}.$$ Find matrices $\mathbf{C_1}$ and $\mathbf{C_2}$ so that $\mathbf{A} = (\mathbf{C}_1)^{-1}\mathbf{J_A}\mathbf{C}_1$ and $\mathbf{B} = (\mathbf{C}_2)^{-1}\mathbf{J_B}\mathbf{C}_2$.

SOLUTION: We begin by defining **A** and using `jordan` to compute the Jordan canonical form of **A**. The resulting output is named `j_A`. The result shows us that the characteristic polynomial of **A** is $(x+1)(x-2)^2$, while the minimal polynomial of **A** is $(x+1)(x-2)$; the eigenvalues of **A** are 2 and −1. The matrix `C1` that satisfies the relationship `inverse(C1)*jordan(A)*C1=A` is also determined. The matrix `C1` is then displayed, and the above relationship is verified with `evalm`.

```
> A:='A':B:='B':C1:='C1':C2:='C2':
  with(linalg):
  A:=array([[2,9,-9],[0,8,-6],[0,9,-7]]);
```

$$A := \begin{bmatrix} 2 & 9 & -9 \\ 0 & 8 & -6 \\ 0 & 9 & -7 \end{bmatrix}$$

```
> j_A:=jordan(A,'C1');
```

$$j\_A := \begin{bmatrix} -1 & 0 & 0 \\ 0 & 2 & 0 \\ 0 & 0 & 2 \end{bmatrix}$$

```
> print(C1);
```

$$\begin{bmatrix} -3 & 3 & 0 \\ -2 & 4 & 1 \\ -3 & 4 & 1 \end{bmatrix}$$

```
> evalm(inverse(C1) &* j_A &* C1);
```

$$\begin{bmatrix} 2 & 0 & 0 \\ 3 & -1 & 0 \\ -12 & 12 & 2 \end{bmatrix}$$

In a similar manner, the matrix **B** is defined and the Jordan canonical form of **B** is found in `j_B`. In this case, we see that the eigenvalues of **B** are 3, –1, and –2, and the minimal and characteristic polynomials of **B** are both $(x-3)(x+1)^2(x+2)^3$. In this case, the matrix `C2` that satisfies `inverse(C2)*jordan(B)*C2=B` is determined with `jordan`. This matrix is then displayed, and the relationship `inverse(C2)*jordan(B)*C2=B` is verified with the `linalg` command `multiply`.

```
> B:=array([[21,27,-13,-11,-32/3,-20/3],
      [-17,-19,9,7,8,4],
      [-49/2,-27,25/2,11,77/6,35/6],
      [7/2,5,-5/2,-4,-5/2,-3/2],
      [55/2,34,-33/2,-14,-91/6,-49/6],
      [-8,-4,3,2,11/3,-1/3]]):
> j_B:=jordan(B,'C2');
```

$$j\_B := \begin{bmatrix} 3 & 0 & 0 & 0 & 0 & 0 \\ 0 & -1 & 1 & 0 & 0 & 0 \\ 0 & 0 & -1 & 0 & 0 & 0 \\ 0 & 0 & 0 & -2 & 1 & 0 \\ 0 & 0 & 0 & 0 & -2 & 1 \\ 0 & 0 & 0 & 0 & 0 & -2 \end{bmatrix}$$

```
> print(C2);
```

$$\begin{bmatrix} \frac{9}{2} & 0 & \frac{-5}{2} & 0 & 3 & -1 \\ \frac{-9}{2} & 0 & \frac{5}{2} & 0 & 3 & 2 \\ \frac{-9}{2} & \frac{-5}{2} & \frac{-7}{2} & 6 & 4 & 8 \\ 0 & \frac{-5}{2} & 4 & -6 & 2 & -4 \\ \frac{9}{2} & \frac{5}{2} & \frac{-3}{2} & -3 & 4 & -3 \\ \frac{-9}{2} & 5 & \frac{9}{2} & 3 & 5 & 0 \end{bmatrix}$$

```
> multiply(inverse(C2),j_B,C2);
```

$$\begin{bmatrix} \frac{79}{6} & \frac{-265}{54} & \frac{-79}{6} & \frac{29}{9} & \frac{442}{27} & 3 \\ \frac{-93}{25} & \frac{-37}{15} & \frac{27}{25} & 2 & \frac{88}{75} & \frac{66}{25} \\ \frac{153}{10} & \frac{-11}{2} & \frac{-157}{10} & 3 & \frac{82}{5} & \frac{17}{5} \\ \frac{44}{9} & \frac{95}{162} & \frac{-22}{27} & \frac{-130}{27} & \frac{-227}{81} & \frac{-110}{27} \\ \frac{2}{3} & \frac{35}{54} & \frac{-1}{9} & \frac{-1}{9} & \frac{-41}{27} & \frac{-5}{9} \\ \frac{19}{2} & \frac{-115}{18} & \frac{-77}{6} & \frac{20}{3} & \frac{172}{9} & \frac{19}{3} \end{bmatrix}$$

■

Companion Matrices

Let $p(x) = a_0 + a_1 x + a_2 x^2 + \ldots + a_{n-1} x^{n-1} + x^n$ be a monic polynomial. The **companion matrix** of $p(x)$ is the matrix $\mathbf{A} = \begin{pmatrix} 0 & 0 & \ldots & 0 & 0 & -a_0 \\ 1 & 0 & \ldots & 0 & 0 & -a_1 \\ 0 & 1 & \ldots & 0 & 0 & -a_2 \\ \vdots & \vdots & \ddots & \vdots & \vdots & \vdots \\ 0 & 0 & 0 & 1 & 0 & -a_{n-1} \\ 0 & 0 & 0 & 0 & 1 & -a_{n-2} \end{pmatrix}$. Maple contains the command `companion`, which can be used to determine the companion matrix.

`companion`

`companion(poly,var)` yields the companion matrix associated with the nth order univariate monic polynomial `poly` in the variable `var`.

EXAMPLE 7: Find the companion matrix for the polynomial

$$p(x) = -30 - 35x - 6x^2 - 2x^3 + x^5.$$

SOLUTION: First, the polynomial is defined as `p`. The 5×5 companion matrix is then found with `companion`, according to the definition above.

```
> with(linalg):
  p:=-30-35*x-6*x^2-2*x^3+x^5;
  companion(p,x);
```

$$p := -30 - 35\,x - 6\,x^2 - 2\,x^3 + x^5$$

$$\begin{bmatrix} 0 & 0 & 0 & 0 & 30 \\ 1 & 0 & 0 & 0 & 35 \\ 0 & 1 & 0 & 0 & 6 \\ 0 & 0 & 1 & 0 & 2 \\ 0 & 0 & 0 & 1 & 0 \end{bmatrix}$$

■

EXAMPLE 8: Find the Jordan canonical form for the companion matrix of

$$q(x) = x^6 - 5x^5 - 41x^4 + 257x^3 + 212x^2 - 3280x + 4800.$$

SOLUTION: The polynomial of degree six is entered as `q`. The companion matrix of `q` is then found in `cq`. Finally, this matrix is placed in Jordan canonical form with `jordan`.

```
> with(linalg):
  q:=x^6-5*x^5-41*x^4+257*x^3+212*x^2-3280*x+4800;
```

$$q := x^6 - 5\,x^5 - 41\,x^4 + 257\,x^3 + 212\,x^2 - 3280\,x + 4800$$

```
> cq:=companion(q,x);
```

$$cq := \begin{bmatrix} 0 & 0 & 0 & 0 & 0 & -4800 \\ 1 & 0 & 0 & 0 & 0 & 3280 \\ 0 & 1 & 0 & 0 & 0 & -212 \\ 0 & 0 & 1 & 0 & 0 & -257 \\ 0 & 0 & 0 & 1 & 0 & 41 \\ 0 & 0 & 0 & 0 & 1 & 5 \end{bmatrix}$$

```
> jordan(cq);
```

$$\begin{bmatrix} 3 & 0 & 0 & 0 & 0 & 0 \\ 0 & -5 & 1 & 0 & 0 & 0 \\ 0 & 0 & -5 & 0 & 0 & 0 \\ 0 & 0 & 0 & 4 & 1 & 0 \\ 0 & 0 & 0 & 0 & 4 & 1 \\ 0 & 0 & 0 & 0 & 0 & 4 \end{bmatrix}$$

From the resulting output, we see that the eigenvalues of `cq` are 3, –5, and 4, and the characteristic and minimal polynomials of `cq` are both $q(x) = (x-3)(x+5)^2$ $(x-4)^3$.

```
> expand((x-3)*(x+5)^2*(x-4)^3);
```

$$x^6 - 5\,x^5 - 41\,x^4 + 257\,x^3 + 212\,x^2 - 3280\,x + 4800$$

■

For a given matrix **A**, the unique monic polynomial p of least degree satisfying $p(\mathbf{A}) = \mathbf{0}$ is called the **minimal polynomial of A**. Let q denote the characteristic polynomial of **A**. Because $q(\mathbf{A}) = \mathbf{0}$, it follows that p divides q. Several of the commands that have been introduced in this section can be used to determine the minimal polynomial of a matrix.

EXAMPLE 9: Let $\mathbf{A} = \begin{pmatrix} 3 & 8 & -6 & -1 \\ -3 & 2 & 0 & 3 \\ 3 & -3 & -1 & -3 \\ 4 & 8 & 6 & -2 \end{pmatrix}$ and $\mathbf{B} = \begin{pmatrix} 1 & -1 & -6 & -12 \\ -2 & 0 & -12 & -30 \\ 2 & -1 & 2 & 8 \\ 0 & 0 & 0 & -1 \end{pmatrix}$. Find the characteristic and minimal polynomials of **A** and **B**.

SOLUTION: Proceeding as in the previous examples, we clear all prior definitions of **A** and **B**, load the linear algebra package `linalg`, and define **A** and **B**.

```
> A:='A':B:='B':C1:='C1':C2:='C2':
  with(linalg):
  A:=array([[3,8,6,-1],[-3,2,0,3],
       [3,-3,-1,-3],[4,8,6,-2]]):
  B:=array([[1,-1,-6,-12],[-2,0,-12,-30],
       [2,-1,2,8],[0,0,0,-1]]):
```

Later we will use the identity matrix. We use `alias` to define `Id` to represent the identity matrix.

```
> alias(Id=&*()):
```

We use charpoly and factor to compute and factor the characteristic polynomial of **A**.

```
> p_A:=charpoly(A,x);
  factor(p_A);
```

$$p\_A := x^4 - 2\,x^3 - 3\,x^2 + 4\,x + 4$$
$$(x+1)^2\,(x-2)^2$$

To find the minimal polynomial of **A**, we compute the Jordan canonical form of **A**.

```
> jordan(A);
```

$$\begin{bmatrix} -1 & 0 & 0 & 0 \\ 0 & 2 & 1 & 0 \\ 0 & 0 & 2 & 0 \\ 0 & 0 & 0 & -1 \end{bmatrix}$$

The result means that the minimal polynomial of **A** is $(x+1)(x-2)^2$. We then define this to be q_A and verify that q_A divides p_A with divide.

```
> q_A:=(x+1)*(x-2)^2;
```

$$q\_A := (x+1)\,(x-2)^2$$

```
> divide(p_A,q_A);
```

$$true$$

We then use evalm to verify that $(x+1)(x-2)^2$ is the minimal polynomial of **A**. Notice that we use the alias Id that was defined earlier to represent the identity matrix.

```
> evalm((A+Id)&*(A-2*Id)^2);
```

$$\begin{bmatrix} 0 & 0 & 0 & 0 \\ 0 & 0 & 0 & 0 \\ 0 & 0 & 0 & 0 \\ 0 & 0 & 0 & 0 \end{bmatrix}$$

In the same manner, we use `charpoly` to compute the characteristic polynomial of **B** and name the resulting output `p_B`. Note that both **A** and **B** have the same characteristic polynomial.

```
> p_B:=charpoly(B,x);
```

$$p\_B := x^4 - 2\,x^3 - 3\,x^2 + 4\,x + 4$$

However, when we use `jordan` to find the Jordan canonical form of **B**, we see that the minimal polynomial of **B** is $(x+1)^2(x-2)^2$. This is also verified with `evalm`.

```
> jordan(B);
```

$$\begin{bmatrix} -1 & 1 & 0 & 0 \\ 0 & -1 & 0 & 0 \\ 0 & 0 & 2 & 1 \\ 0 & 0 & 0 & 2 \end{bmatrix}$$

```
> evalm((B+1)^2 &* (B-2)^2);
```

$$\begin{bmatrix} 0 & 0 & 0 & 0 \\ 0 & 0 & 0 & 0 \\ 0 & 0 & 0 & 0 \\ 0 & 0 & 0 & 0 \end{bmatrix}$$

■

CHAPTER 7

Applications Related to Ordinary and Partial Differential Equations

Maple can perform the calculations necessary to compute solutions of various differential equations. In some cases, it can be used to find the exact solution of certain differential equations using the built-in command dsolve. We obtain basic information about the dsolve command with ?dsolve.

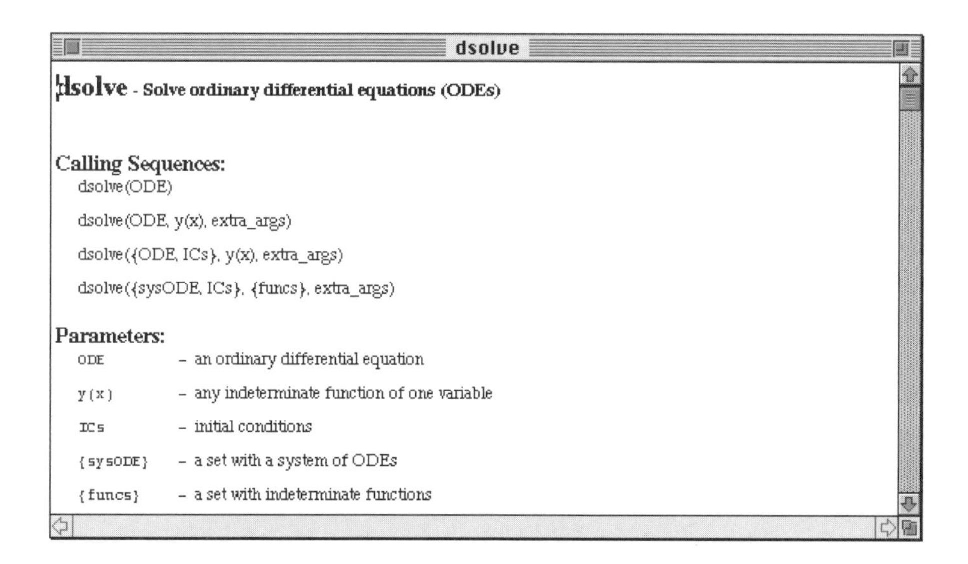
dsolve

dsolve - Solve ordinary differential equations (ODEs)

Calling Sequences:
dsolve(ODE)
dsolve(ODE, y(x), extra_args)
dsolve({ODE, ICs}, y(x), extra_args)
dsolve({sysODE, ICs}, {funcs}, extra_args)

Parameters:
ODE - an ordinary differential equation
y(x) - any indeterminate function of one variable
ICs - initial conditions
{sysODE} - a set with a system of ODEs
{funcs} - a set with indeterminate functions

In addition to the `dsolve` command, the `DEtools` package contains a large number of functions for working with differential equations.

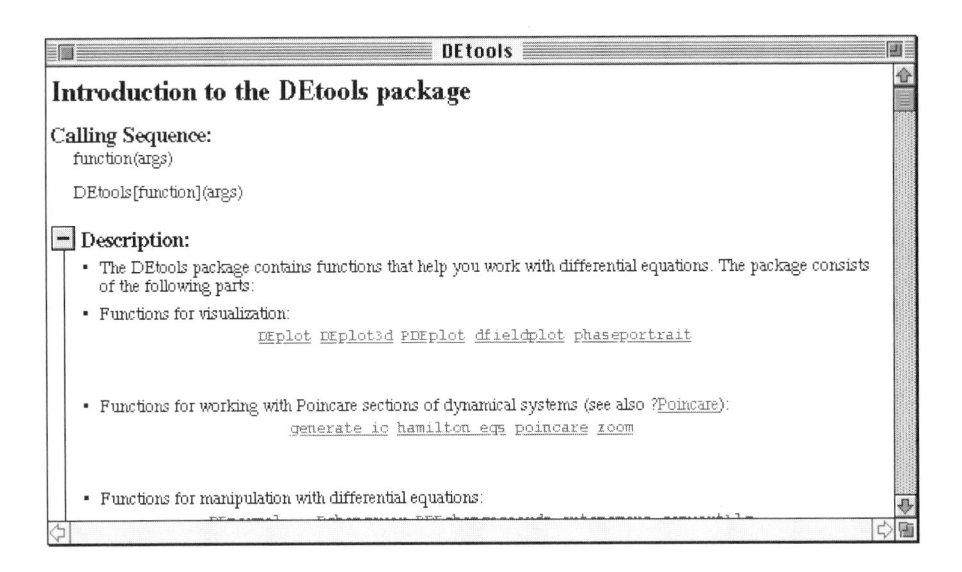

The purpose of Chapter 7 is to illustrate various computations Maple can perform when solving differential equations. Applications discussed in this chapter include the Falling Bodies Problem, Spring Problems, Classification of Equilibrium Points, and the Wave Equation.

7.1 First-Order Ordinary Differential Equations

Separable Differential Equations

A differential equation that can be written in the form $g(y)y' = f(x)$ is called a **separable differential equation**. Rewriting $g(y)y' = f(x)$ in the form $g(y)\frac{dy}{dx} = f(x)$ yields $g(y)dy = f(x)dx$ so that $\int g(y)dy = \int f(x)dx + C$, where C represents an arbitrary constant. An equation of this type is solved with Maple in Example 1.

EXAMPLE 1: Solve the initial-value problem $y\cos x\,dx - (1 + y^2)dy = 0$, $y(0) = 1$.

SOLUTION: Note that this equation can be rewritten as $\frac{dy}{dx} = \frac{y\cos x}{1+y^2}$. We first use dsolve to solve the solution. We are careful to include the argument, x, each time we type the dependent variable $y = y(x)$.

```
> sol1:=dsolve(diff(y(x),x)=y(x)*cos(x)/(1+y(x)^2),
  y(x));
```

$$sol1 := y(x) = -I\sqrt{-\text{LambertW}(\mathbf{e}^{(2\sin(x)+2\_C1)})}$$

(Note that entering dsolve(D(y)(x)=y(x)*cos(x)/(1+y(x)^2),y(x)) produces the same result.) In this case, we see that dsolve is able to solve the nonlinear equation, although the result contains the LambertW function. Given z, the Lambert W function returns the value of w that satisfies $z = we^w$. If we include the implicit option in the dsolve command, a more familiar form of the solution is found.

```
> sol2:=dsolve(D(y)(x)=y(x)*cos(x)/(1+y(x)^2),y(x),
     implicit);
```

$$sol2 := \sin(x) - \frac{1}{2}y(x)^2 - \ln(y(x)) + \_C1 = 0$$

We can also use Maple to implement the steps necessary to solve the equation by hand. We see that it is separable using odeadvisor

```
> odeadvisor(D(y)(x)=y(x)*cos(x)/(1+y(x)^2));
```

$$[\_separable]$$

and rewrite the equation as $\cos x dx = \frac{1+y^2}{y}dy$. To solve the equation, we must integrate both the left- and right-hand sides, which we do with int, naming the resulting output LHS and RHS, respectively.

```
> LHS:=int(cos(x),x);
  RHS:=int((1+y^2)/y,y);
```

$$LHS := \sin(x)$$

$$RHS := \frac{1}{2}y^2 + \ln(y)$$

Alternatively, the separablesol command that is contained in the DEtools package, which is illustrated next, can be used to determine if an equation is separable and, if so, can solve it.

```
> with(DEtools):
  sol2:=separablesol(
      D(y)(x)=y(x)*cos(x)/(1+y(x)^2),y(x));
```

$$sol2 := \{\frac{1}{2}\mathrm{y}(x)^2 + \ln(\mathrm{y}(x)) - \sin(x) = \_C_1\}$$

Therefore, a general solution to the equation is $\sin x = \ln|y| + \frac{1}{2}y^2 + C$. We now use contourplot to graph $\sin x = \ln|y| + \frac{1}{2}y^2 + C$ for various values of C by observing that the level curves of $\sin x - \ln|y| + \frac{1}{2}y^2$ correspond to the graph of $\sin x = \ln|y| + \frac{1}{2}y^2 + C$ for various values of C. First, we replace each occurrence of $y(x)$ in $\frac{1}{2}y(x)^2 + \ln(y(x)) - \sin(x)$ with y

```
  toplot:=subs(y(x)=y,lhs(sol2[1]));
```

$$toplot := \frac{1}{2}y^2 + \ln(y) - \sin(x)$$

and then we use contourplot to generate the graph.

```
> with(plots):
  contourplot(toplot,x=0..10,y=0..10,
```

```
color=BLACK,contours=10,grid=[60,60]);
```

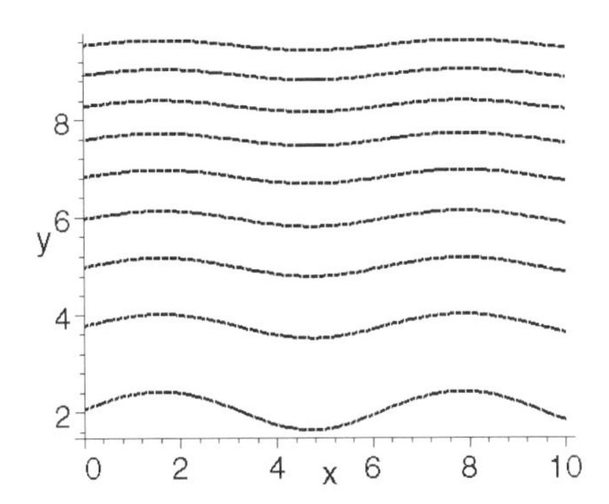

By substituting $y(0) = 1$ into this equation, we find that $C = 1/2$, so the implicit solution is given by $\sin x + \frac{1}{2} = \ln|y| + \frac{1}{2}y^2$.

```
> eval(subs({x=0,y=1},toplot));
```

$$\frac{1}{2}$$

■

Homogeneous Differential Equations

A differential equation that can be written in the form $M(x,y)dx + N(x,y)dy = 0$, where $M(tx,ty) = t^n M(x,y)$ and $N(tx,ty) = t^n N(x,y)$, is called a **homogeneous differential equation** (of degree n). This type of equation can be solved with the help of Maple, as illustrated in Example 2.

EXAMPLE 2: Solve the equation $(x^2 - y^2)dx + xydy = 0$.

SOLUTION: Proceeding as in the previous example, we first define `eq` to be the equation $(x^2 - y^2)dx + xydy = 0$, and then we attempt to use `dsolve` to solve the equation.

```
> eq:='eq':
  eq:=x*y(x)*D(y)(x)=y(x)^2-x^2:
  sol:=dsolve(eq,y(x))
```

$$sol := y(x) = \sqrt{-2\ln(x) + \_C1}\, x,\ y(x) = -\sqrt{-2\ln(x) + \_C1}\, x$$

Alternatively, we use Maple to implement the steps needed to solve this homogeneous equation. Let $M(x,y) = x^2 - y^2$ and $N(x,y) = xy$. Then, $M(tx,ty) = t^2 M(x,y)$ and $M(tx,ty) = t^2 M(x,y)$, which means that $N(tx,ty) = t^2 N(x,y)$ is a homogeneous equation of degree 2.

```
> m:='m':n:='n':x:='x':y:='y':
  m:=(x,y)->x^2-y^2:
  n:=(x,y)->x*y:
  factor(m(t*x,t*y));
  factor(n(t*x,t*y));
```

$$t^2 (x - y)(x + y)$$

$$t^2 x\, y$$

This is confirmed with `odeadvisor`.

```
> odeadvisor(eq);
```

$$[[\_homogeneous, class\ A], \_rational, \_Bernoulli]$$

Assume $x = vy$. Then, $dx = vdy + ydv$ and directly substituting into the equation and simplifying yields $y^2 v^3 dy + y^3 (v^2 - 1) dv = 0$ as obtained in `leqthree`.

```
> leqone:=m(x,y)*D(x)+n(x,y)*D(y);
```

$$leqone := (x^2 - y^2)\, D(x) + x\, y\, D(y)$$

```
> x:=v*y:
  leqtwo:=expand(leqone);
```

$$leqtwo := v^2 y^3 D(v) + v^3 y^2 D(y) - y^3 D(v)$$

```
> leqthree:=collect(leqtwo,[D(v),D(y)]);
```

$$leqthree := (v^2\,y^3 - y^3)\,\mathrm{D}(v) + v^3\,y^2\,\mathrm{D}(y)$$

Dividing this equation by y^3v^3 yields the separable differential equation $\frac{dy}{y} + \frac{(v^2-1)dv}{v^3} = 0$.

```
> leqfour:=collect(normal(leqthree/(y^3*v^3)),
      [D(v),D(y)]);
```

$$leqfour := \frac{(v^2\,y - y)\,\mathrm{D}(v)}{y\,v^3} + \frac{\mathrm{D}(y)}{y}$$

We solve this equation by rewriting it in the form

$$\frac{dy}{y} = \frac{(1-v^2)dv}{v^3} = \left(\frac{1}{v^3} - \frac{1}{v}\right)dv$$

and integrating each side with int.

```
> LHS:=int(1/y,y);
```

$$LHS := \ln(y)$$

```
> RHS:=int((1/v^3-1/v),v);
```

$$RHS := -\frac{1}{2}\frac{1}{v^2} - \ln(v)$$

This yields $\ln y = -\frac{1}{2v^2} - \ln v + C_1$, which can be simplified as ,

$$\ln(vy) = -\frac{1}{2v^2} + C_1$$

$$vy = Ce^{-1/(2v^2)}.$$

Because $x = vy$, $v = x/y$ and resubstituting into the above equation yields

$$\frac{x}{y} \cdot y = Ce^{-1/(2(x/y)^2)}$$

$$x = Ce^{-y^2/(2x^2)}$$

as a general solution of the equation $(x^2 - y^2)dx + xydy = 0$. Of course, the same results are obtained by substituting $v = x/y$ into RHS.

```
> x:='x':
  subs(v=x/y,RHS);
```

$$-\frac{1}{2}\frac{y^2}{x^2} - \ln\left(\frac{x}{y}\right)$$

To graph the solution for various values of C, we find an implicit solution using dsolve together with the implicit option.

```
> implicitsol:=dsolve(eq,y(x),implicit);
```

$$implicitsol := \mathrm{y}(x)^2 + 2x^2 \ln(x) - x^2 \_C1 = 0$$

Then, we use seq and subs to replace _C1 by 0.5i for various values of i, and we name the resulting set of equations toplot.

```
> toplot:={seq(subs(_C1=0.5*i,implicitsol),i=-5..5)};
```

$$toplot := \{ \mathrm{y}(x)^2 + 2x^2\ln(x) + 1.5x^2 = 0, \mathrm{y}(x)^2 + 2x^2\ln(x) + 2.0x^2 = 0, \mathrm{y}(x)^2 + 2x^2\ln(x) = 0,$$
$$\mathrm{y}(x)^2 + 2x^2\ln(x) + 2.5x^2 = 0, \mathrm{y}(x)^2 + 2x^2\ln(x) - .5x^2 = 0, \mathrm{y}(x)^2 + 2x^2\ln(x) + .5x^2 = 0,$$
$$\mathrm{y}(x)^2 + 2x^2\ln(x) + 1.0x^2 = 0, \mathrm{y}(x)^2 + 2x^2\ln(x) - 1.5x^2 = 0, \mathrm{y}(x)^2 + 2x^2\ln(x) - 1.0x^2 = 0,$$
$$\mathrm{y}(x)^2 + 2x^2\ln(x) - 2.5x^2 = 0, \mathrm{y}(x)^2 + 2x^2\ln(x) - 2.0x^2 = 0 \}$$

The set of equations toplot is then graphed with implicitplot.

```
> with(plots):
  implicitplot(toplot,x=0..6,y=-3..3,
      grid=[60,60],color=BLACK);
```

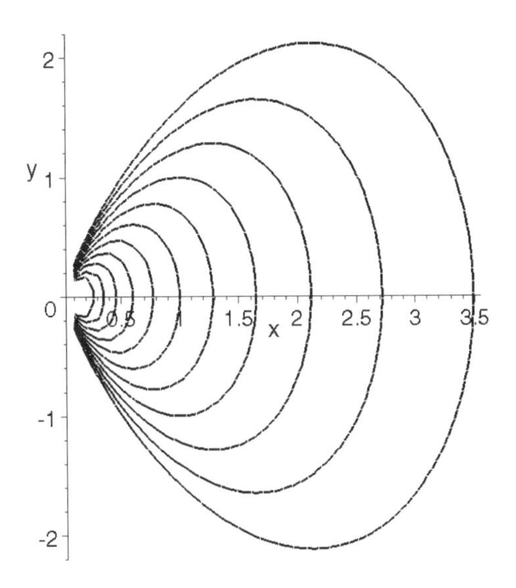

The command

```
DEplot(differential equation,y(x),x=xmin..xmax,
        [initial conditions],y=ymin..ymax)
```

graphs the solutions to the differential equation `differential equation` for the initial conditions specified in `initial conditions` on the rectangle [`xmin`,`xmax`] × [`ymin`,`ymax`], along with the direction field for the equation, if the equation has a direction field. Note that initial conditions should be a set or list of the form `[[x0,y0], [x1,y1], ... ]`, `{[x0,y0], [x1,y1], ... }`, `[[y(x0)=y0],[y(x1)=y1],...]` or `{[y(x0)=y0],[y(x1)=y1],...}`. Note that `DEplot` is contained in the `DEtools` package.

We use `DEplot` to graph the direction field for the equation on the rectangle [0,6] x [–3,3].

```
> with(DEtools):
  DEplot(eq,y(x),x=0.1..6,y=-3..3);
```

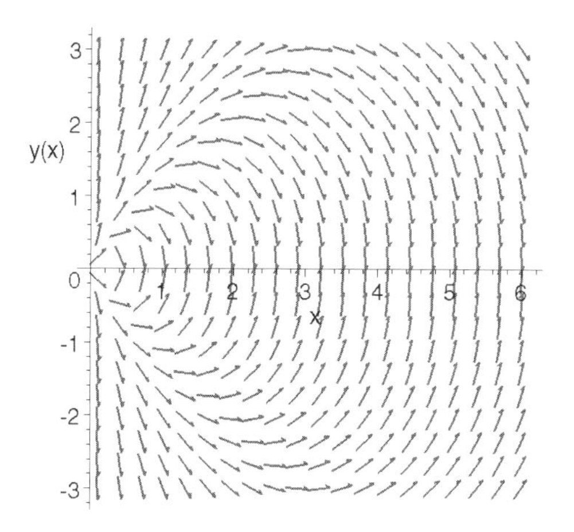

■

Exact Equations

A differential equation that can be written in the form $M(x,y)dx + N(x,y)dy = 0$, where $\partial N/\partial x = \partial M/\partial y$, is called an **exact differential equation**. If $M(x,y)dx + N(x,y)dy = 0$ is exact, there is a function F such that the total differential of F, dF, satisfies the equation $dF = M(x,y)dx + N(x,y)dy$. The solution of the exact differential equation is $F(x,y) = C$ where C is a constant. The method by which $F(x,y)$ is determined with Maple is illustrated in Example 3.

EXAMPLE 3: Find a general solution of the equation $(-1 + e^{xy}y + y\cos(xy))dx + (1 + e^{xy}x + x\cos(xy))dy = 0$.

SOLUTION: We begin by defining $M(x,y) = -1 + e^{xy}y + y\cos(xy)$, $N(x,y) = 1 + e^{xy}x + x\cos(xy)$, and then defining `capm` and `capn`. We then use `dsolve` to solve the equation $M(x,y)dx + N(x,y)dy = 0$.

```
> capm:=(x,y)->-1+y*exp(x*y)+y*cos(x*y):
  capn:=(x,y)->1+x*exp(x*y)+x*cos(x*y):
> dsolve(capm(x,y(x))+capn(x,y(x))*diff(y(x),x)=0
      ,y(x));
```

$$y(x) = \frac{\text{RootOf}(\sin(\_Z)\,x + \_C1\,x - x^2 + \mathbf{e}^{\_Z}\,x + \_Z)}{x}$$

An implicit solution is found is found when we include the `implicit` option in the `dsolve` command.

```
> implicitsol:=dsolve(capm(x,y(x))+
      capn(x,y(x))*diff(y(x),x)=0,y(x),implicit);
```

$$implicitsol := \_C1 - x + \mathbf{e}^{(x\,y(x))} + \sin(x\,y(x)) + y(x) = 0$$

Alternatively, we can use Maple to implement the steps we encounter when we solve exact equations. First, we verify that $\partial N/\partial x = \partial M/\partial y$ and consequently that the equation is exact by entering

```
> testeq(diff(capm(x,y),y),diff(capn(x,y),x));
```

$$true$$

We can also use `odeadvisor` to determine that the equation is exact.

```
> odeadvisor(capm(x,y(x))+capn(x,y(x))*diff(y(x),x)=0);
```

$$[\_exact]$$

We then define `StepOne` to be $\int M(x, y)dx + g(y)$ and `StepTwo` to be $\frac{\partial}{\partial y} + (\int M(x, y)dx + g(y))$.

```
> StepOne:=int(capm(x,y),x)+g(y);
```

$$StepOne := -x + \mathbf{e}^{(xy)} + \sin(xy) + g(y)$$

```
> StepTwo:=diff(StepOne,y);
```

$$StepTwo := x\mathbf{e}^{(xy)} + \cos(xy)x + \left(\frac{\partial}{\partial y}g(y)\right)$$

Because the equation is exact, we must have that $N(x, y) = \frac{d}{dy}(\int M(x, y)dx + g(y))$. We use `solve` to solve the equation $N(x, y) = \frac{d}{dy}(\int M(x, y)dx + g(y))$ for $g'(y)$.

```
> solve(capn(x,y)=StepTwo,diff(g(y),y));
```

$$1$$

We could also have used the command `isolate` to determine $g'(y)$. The command `isolate` must be loaded with the command `readlib(isolate)` before use. We use `isolate` in the same manner as we use `solve`.

```
> readlib(isolate):
  isolate(capn(x,y)=StepTwo,diff(g(y),y));
```

$$\frac{\partial}{\partial y} \mathrm{g}(y) = 1$$

In this case, because $g'(y) = 1$, $g(y) = y + C$, where C denotes any constant. Note that if the previous results had been complicated, we could have computed $g(y)$ using the `int` command to integrate $g'(y)$. We conclude that a general solution is $-x + e^{xy} + sin(xy) + y = C$.

In Section 5.4, we learned that **f** is a conservative vector field if that there is a scalar field F satisfying $\mathbf{f} = \Delta F$. Thus, the equation $M(x,y)dx + N(x,y)dy = 0$ is exact if $\langle M(x,y), N(x\ y)\rangle$ is a conservative vector field. We can determine if **f** is a conservative vector field and, if so, its potential function F, with `potential`, which is contained in the `linalg` package. Thus, entering

```
> with(linalg):
  potential([capm(x,y),capn(x,y)],[x,y],'F');
```

$$true$$

```
> F;
```

$$-x + \mathbf{e}^{(x\,y)} + \sin(x\,y) + y - 1$$

also shows that a general solution of the equation is $-x + e^{xy} + sin(xy) + y = C$.

We can graph various solutions with the command `contourplot` by observing that level curves of the function $f(x,y) = -x + e^{xy} + \sin(x\ y) + y$ correspond to graphs of $-x + e^{xy} + \sin(x\ y) + y = C$ for various values of C. We load the `plots` package, and then use `contourplot` to graph several level curves of F on the rectangle $[\ \pi,\pi] \times [-\pi,\pi]$.

```
> with(plots):
  contourvals:=[seq(evalf(subs({x=0,y=0.5*i},F)),
      i=-4..4)];
  contourplot(F,x=-Pi..Pi,y=-Pi..Pi,
```

```
grid=[70,70],contours=contourvals,axes=BOXED);
```

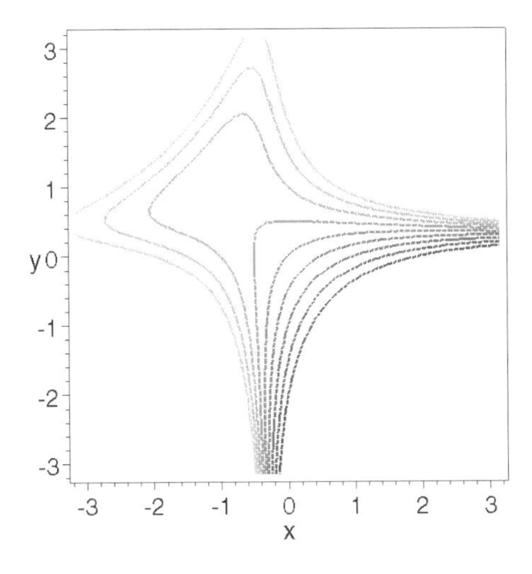

As in the previous example, we can graph the direction field for the equation using `DEplot`, contained in the `DEtools` package, to graph the direction field for the equation. We first load the `DEtools` package and then use `DEplot` to graph the direction field for the equation on the rectangle $[-\pi,\pi] \times [-\pi,\pi]$. DEplotDEplot

```
> with(DEtools):
  DEplot(capm(x,y(x))+capn(x,y(x))*diff(y(x),x)=0,
      y(x),x=-Pi..Pi,y=-Pi..Pi);
```

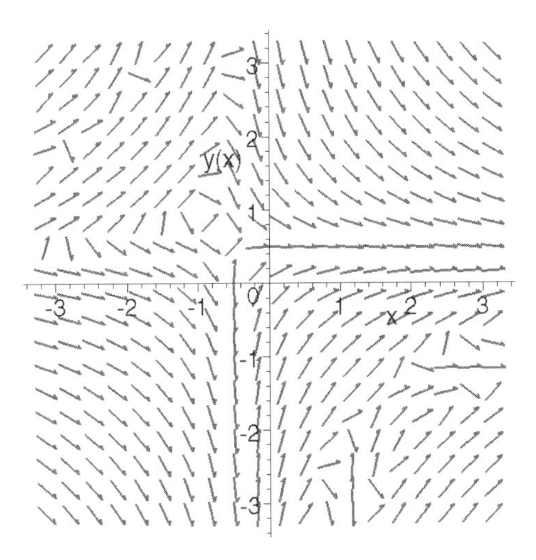

■

Linear Equations

A differential equation that can be written in the form $\frac{dy}{dx}+p(x)y=q(x)$ is called a **first-order linear differential equation**. If $\frac{dy}{dx}+p(x)y=q(x)$, then multiplying by $e^{\int p(x)dx}$ results in .

$$e^{\int p(x)dx}\frac{dy}{dx}+e^{\int p(x)dx}p(x)y=e^{\int p(x)dx}q(x).$$

The term $e^{\int p(x)dx}$ is called the **integrating factor**. Applying the product rule and Fundamental Theorem of Calculus yields

$$\frac{d}{dx}\left(e^{\int p(x)dx}y\right)=e^{\int p(x)dx}\frac{dy}{dx}+e^{\int p(x)dx}p(x)y,$$

so the equation $e^{\int p(x)dx}\frac{dy}{dx}+e^{\int p(x)dx}p(x)y=e^{\int p(x)dx}q(x)$ is equivalent to the equation

$$\frac{d}{dx}\left(e^{\int p(x)dx}y\right)+e^{\int p(x)dx}q(x).$$

Integrating, we obtain $e^{\int p(x)dx}y=\int e^{\int p(x)dx}q(x)dx$. Dividing by $e^{\int p(x)dx}$ results in the solution

$$y = e^{-\int p(x)dx}\int e^{\int p(x)dx}q(x)dx.$$

Maple's `dsolve` command can solve all first-order linear differential equations, although the result might contain unevaluated integrals.

```
> x:='x':y:='y':
  dsolve(diff(y(x),x)+p(x)*y(x)=q(x),y(x));
```

$$y(x)=\mathbf{e}^{\left(-\int p(x)\,dx\right)}\int q(x)\,\mathbf{e}^{\left(\int p(x)\,dx\right)}dx+\mathbf{e}^{\left(-\int p(x)\,dx\right)}\_C1$$

In addition, the command `linearsol`, which is contained in the `DEtools` package, can be used to find solutions of first-order linear equations.

EXAMPLE 4: (a) Find a general solution of $x dy/dx + 3y = x \sin x$. (b) Graph the solution for the values of, C = –3,-2, -1, 0, 1, 2, and 3. (c) Graph the solutions that satisfy the conditions $y(1) = 2$ and $y(1) = -1/2$.

SOLUTION: After clearing prior definitions of `Equation` and y, if any, we define `Equation` to be the first-order linear differential equation $x dy/dx + 3y = x \sin x$. We then use `dsolve` to find a general solution of `Equation` and name the resulting output `gen_sol`. Note that the symbol `_C1` represents the arbitrary constant in the solution.

```
> x:='x':y:='y':
  Equation:=x*diff(y(x),x)+3*y(x)=x*sin(x):
> gen_sol:=dsolve(Equation,y(x));
```

$$gen\_sol := \mathrm{y}(x) = \frac{-x^3 \cos(x) + 3x^2 \sin(x) - 6\sin(x) + 6x\cos(x) + \_C1}{x^3}$$

Because `gen_sol` does not automatically assign $y(x)$ to be the explicit solution of the equation, we use `assign` to define $y(x)$ to be the explicit solution of the equation and then compute $y(x)$.

```
> assign(gen_sol):
  y(x);
```

$$\frac{-x^3 \cos(x) + 3x^2 \sin(x) - 6\sin(x) + 6x\cos(x) + \_C1}{x^3}$$

To graph the solutions for the values $C = -3$, -2, -1, 0, 1, 2, and 3, we define `tograph` to be the set of functions obtained by replacing `_C1` in $y(x)$ by i for $i = -3, -2, -1, 0,$ 1, 2, and 3. Note that the set of functions is enclosed in braces ({...}). (Using square brackets ([...]) in the subsequent `plot` command would produce an error.) After defining `tograph`, we graph the set of functions in `tograph` on the interval [0, 3/π]. In this case, we specify that the range displayed correspond to the interval [–4, 4], and we use 150 plot points to help insure that the resulting graphs are smooth.

```
> tograph:={seq(subs(_C1=i,y(x)),i=-3..3)}:
```

```
plot(tograph,x=.2..3/2*Pi,-4..4,numpoints=150);
```

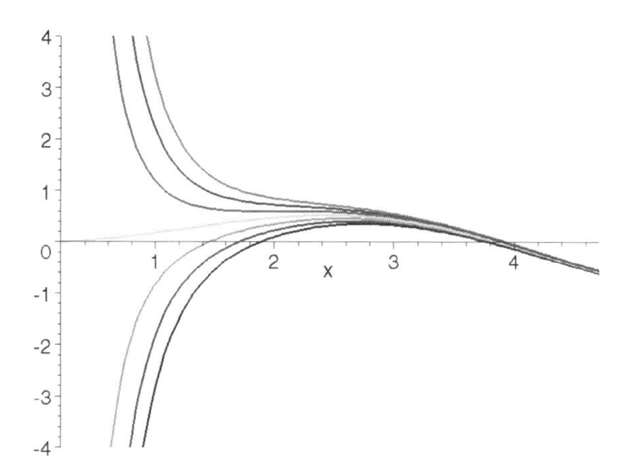

The solutions that satisfy the conditions $y(1) = 2$ and $y(1) = -1/2$ are determined with `dsolve`.

```
> y:='y':
  sol1:=dsolve({Equation,y(1)=2},y(x));
  sol2:=dsolve({Equation,y(2)=-1/2},y(x));
```

$$sol1 := \mathrm{y}(x) = \frac{(-x^3\cos(x) + 3x^2\sin(x) - 6\sin(x) + 6x\cos(x) - 5\cos(1) + 3\sin(1) + 2)}{x^3}$$

$$sol2 := \mathrm{y}(x) = \frac{(-x^3\cos(x) + 3x^2\sin(x) - 6\sin(x) + 6x\cos(x) - 4\cos(2) - 6\sin(2) - 4)}{x^3}$$

To graph the solutions that satisfy the initial conditions $y(1) = 2$ and $y(2) = -1/2$, we use the `DEplot` command together with the option `arrows=NONE` so the the graph of the direction field for the equation is not displayed.

```
> with(DEtools):
  DEplot(Equation,y(x),
      x=0..3/2*Pi,{[1,2],[2,-1/2]},y=-4..4,
      arrows=NONE,linecolor=[BLACK,GRAY]);
```

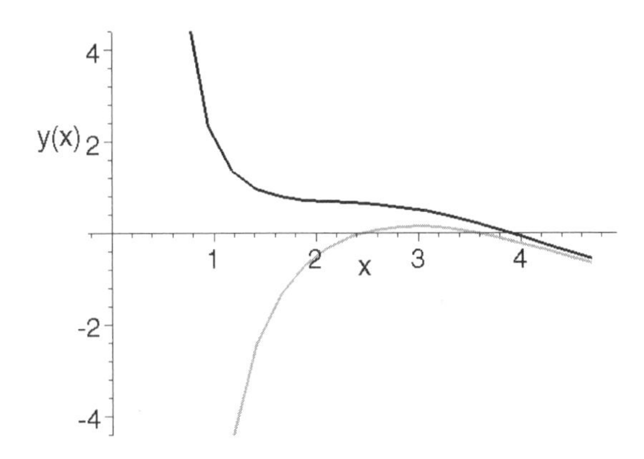

■

EXAMPLE 5: Compare the solutions of $dy/dx + y = f(x)$ subject to $y(0) = 0$, where $f(x) = x$, $\sin x$, $\cos x$, e^x, $e^{-x} \sin x$, $x \cos x$, and xe^{-x}.

SOLUTION: To solve this problem, we define a function `ex` that given f, (1) declares `sol` to be local to the procedure `ex`; (2) solves the differential equation $dy/dy + y = f(x)$ subject to the initial condition $y(0) = 0$ and names the result `sol` (the expression `sol` contains two parts, the second part of which is the explicit solution to the equation and is extracted with `op`); and (3) returns the explicit solution to the differential equation.

```
> ex:=proc(f)
      local sol;
      sol:=dsolve({diff(y(x),x)+y(x)=f,y(0)=0},
         y(x));
      op(2,sol)
      end:
```

We then define `funcs` to be the set of functions x, $\sin x$, $\cos x$, e^x, $e^{-x} \sin x$, $x \cos x$, and xe^{-x} and use `map` to compute `ex` for each function in the list `funcs`. The resulting set of functions is named `set_of_sols`.

```
> funcs:={x,sin(x),cos(x),exp(x),exp(-x),
      exp(-x)*sin(x),x*cos(x),x*exp(-x)};
```

$$funcs := \{x, \cos(x), \mathbf{e}^{(-x)}\sin(x), \mathbf{e}^{x}, x\,\mathbf{e}^{(-x)}, \mathbf{e}^{(-x)}, \sin(x), x\cos(x)\}$$

```
> set_of_sols:=map(ex,funcs);
```

$$set\_of\_sols := \{\frac{1}{2}\mathbf{e}^{(-x)}x^2, x\,\mathbf{e}^{(-x)}, -\mathbf{e}^{(-x)}\cos(x)+\mathbf{e}^{(-x)}, \frac{1}{2}x\cos(x)+\frac{1}{2}x\sin(x)-\frac{1}{2}\sin(x),$$

$$\frac{1}{2}\cos(x)+\frac{1}{2}\sin(x)-\frac{1}{2}\mathbf{e}^{(-x)}, -\frac{1}{2}\cos(x)+\frac{1}{2}\sin(x)+\frac{1}{2}\mathbf{e}^{(-x)}, x-1+\mathbf{e}^{(-x)}, \frac{1}{2}\mathbf{e}^{x}-\frac{1}{2}\mathbf{e}^{(-x)}\}$$

In order to graph the set of functions `set_of_sols`, we note that `set_of_sols` contains eight parts, obtained with `nops`, which correspond to the functions to be graphed. We then use `seq` to graph each member of `set_of_sols` on the interval [−π/4, 3π] and name the resulting list of graphs `toshow`.

```
> nops(set_of_sols);
```

$$8$$

```
> toshow:=[seq(plot(set_of_sols[i],x=-Pi/4..3*Pi,
      color=BLACK),i=1..8)]:
```

After loading the `plots` package, we use `display` together with the option `insequence=true` to generate an array of graphs that is displayed with `display`.

```
> with(plots):
> arrayofgraphs:=display(toshow,insequence=true):
> display(arrayofgraphs);
```

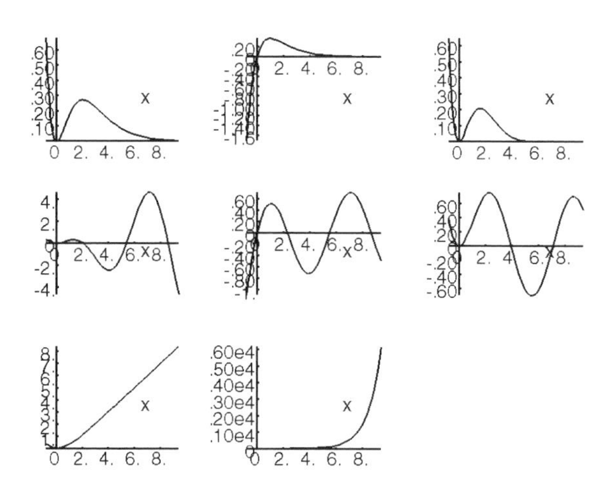

■

Numerical Solutions of First-Order Ordinary Differential Equations

Numerical approximations of solutions to differential equations can be obtained with `dsolve` together with the `numeric` option. This command is particularly useful for work with nonlinear equations for which `dsolve` alone is unable to find an explicit solution. This command is entered in the form

```
dsolve({deq,ics},fun,numeric),
```

where `deq` is solved for `fun`. Note that the number of initial conditions in `ics` must equal the order of the differential equation indicated in `deq`.

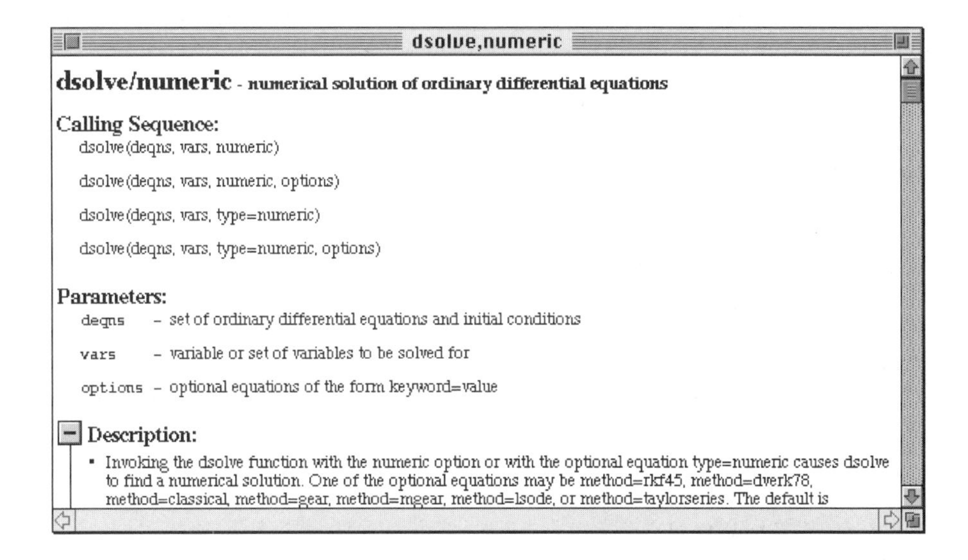

EXAMPLE 6: Graph the approximate solution of $dy/dx = \sin(2x - y)$, subject the initial condition $y(0) = 0.5$ on the interval [0, 15].

SOLUTION: First, we define `Eq` to be the equation $dy/dx = \sin(2x - y)$, and then we use `dsolve` to approximate the solution of `Eq` subject to the initial condition $y(0) = 0.5$, naming the resulting output `Sol`. The resulting output is a procedure that represents an approximate function obtained through interpolation.

```
> x:='x':y:='y':
  Eq:=D(y)(x)=sin(2*x-y(x)):
> Sol:=dsolve({Eq,y(0)=.5},y(x),numeric);
```

$$Sol := \mathbf{proc}(rkf45\_x) \ \ldots \ \mathbf{end}$$

We can use this result to approximate the solution to the initial-value problem for values of x that satisfy $0 \le x \le 15$. For example, entering

```
> Sol(1);
```

$$[x = 1, y(x) = .8758947797345141]$$

returns an ordered pair corresponding to x and $y(x)$ if $x = 1$. Entering

```
> Sol(1)[2];
```

$$y(x) = .8758947797345141$$

returns the second part of `Sol`. Entering

```
> rhs(Sol(1)[2]);
```

```
.8758947797345141
```

returns the value of the solution, $y(x)$, if $x = 1$.

We then graph the solution using `plot`. In the `plot` command, note how we enclose `rhs(Sol(x)[2]` in single quotes (`'  '`) to delay the evaluation of `rhs(Sol(x)[2]`.

```
> plot('rhs(Sol(x)[2])',x=0..15);
```

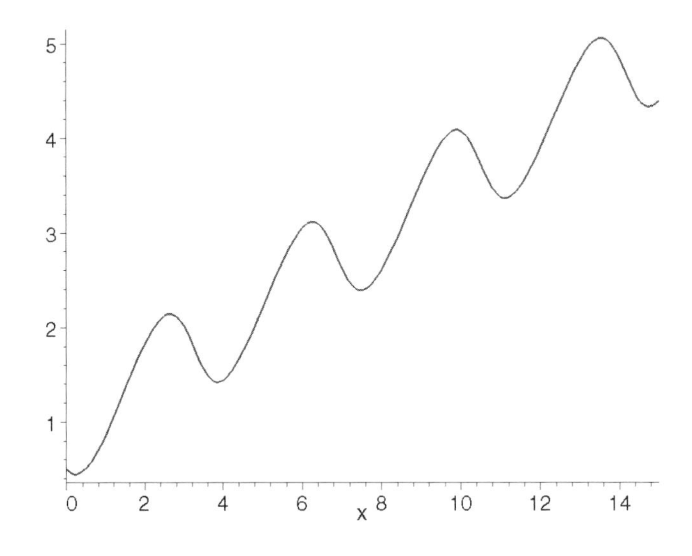

Alternatively, we can graph the solution by using the command odeplot contained in the plots package. Generally, we will graph numerical solutions of differential equations with odeplot. Entering

```
> with(plots):
  odeplot(Sol,[x,y(x)],0..15);
```

produces the same graph as that obtained above.

We can also use DEplot to graph solutions. However, in this case, a numerical solution that can be evaluated, like the one we obtain with dsolve, is not generated. For example, entering

```
> with(DEtools):
  DEplot(diff(y(x),x)=sin(2*x-y),y(x),
      x=0..15,{[0,1],[0,-1]},
      color=GRAY,linecolor=BLACK,stepsize=0.01);
```

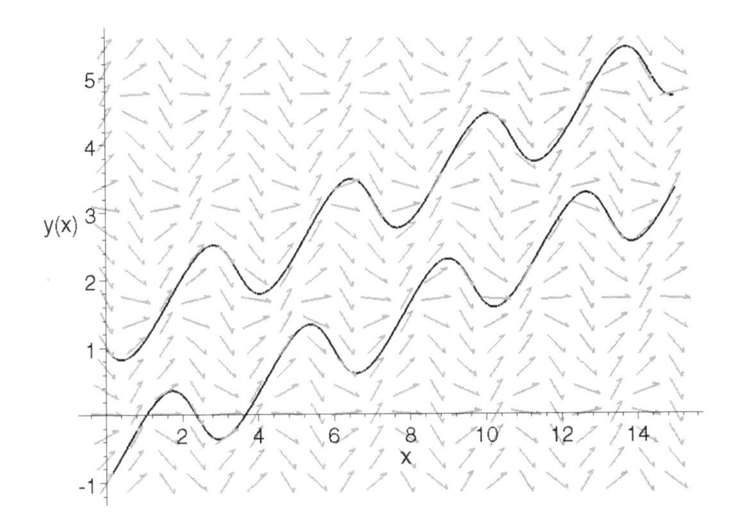

graphs the solutions of the equation satisfying $y(0) = 1$ and $y(0) = -1$ along with the direction field for the equation on the interval [0,15]. The direction field is not displayed if the options arrows=NONE is included in the DEplot command.

■

We can also use DEplot to graph solutions to a differential equation under changing initial conditions.

EXAMPLE 7: Graph the solution of $y' = \sin(xy)$ subject to the initial condition $y(0) = j$ on the interval [0, 7] for j = 0.5,1.0,1.5,2.0, and 2.5.

SOLUTION: We begin by using `seq` to define the set of ordered pairs $(0, i/2)$ for $i = 1, 2, \ldots, 5$. These correspond to the initial conditions $y(0) = i/2$ for $i = 1, 2, \ldots, 5$.

```
> inits:={seq([0,i/2],i=1..5)};
```

$$inits := \{\left[0, \frac{1}{2}\right], [0, 1], [0, 2], \left[0, \frac{3}{2}\right], \left[0, \frac{5}{2}\right]\}$$

Next, we use `DEplot` to graph the solutions to $y' = \sin(xy)$ for the initial conditions specified in `inits`. The option `arrows=NONE` is included so that the direction fields for the equation are not included in the graph.

```
> with(DEtools):
  DEplot(diff(y(x),x)=sin(x*y),y(x),x=0..7,inits,
      arrows=NONE,stepsize=0.05,linecolor=BLACK);
```

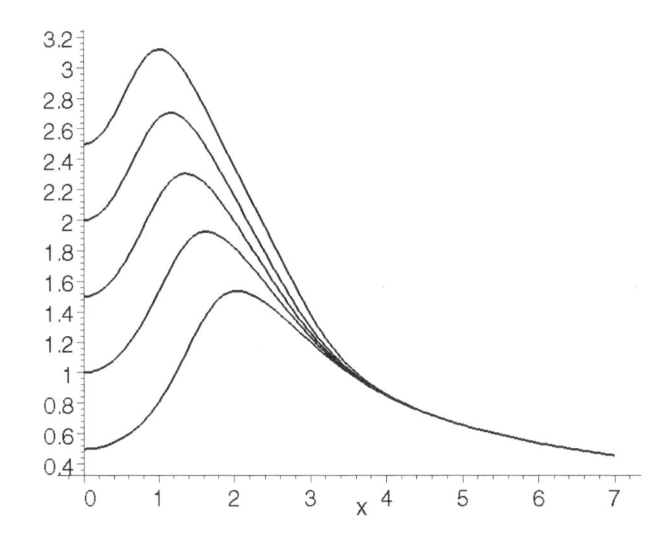

■

Application: Population Growth and the Logistic Equation

The **logistic equation** (or **Verhulst equation**), first introduced by the Belgian mathematician Pierre Verhulst to study population growth, is the equation

$$y'(t) = (r - ay(t))y(t),$$

where r and a are constants, subject to the condition $y(0) = y_0$. This equation can be written as $dy/dt = (r - ay)y = ry - ay^2$ where the term $-ay^2$ represents an inhibitive factor or "death rate." Hence, the population under these assumptions is not allowed to grow out of control as it is in some other models like the Malthus model. Also, the population does not grow or decay constantly.

The logistic equation is separable, so it can be solved by separation of variables. We proceed by using `dsolve` to solve the initial-value problem and simplify to `simplify` the result.

```
> y:='y':
  logeqn:=diff(y(t),t)=(r-a*y(t))*y(t):
  sol:=dsolve({logeqn,y(0)=y0},y(t));
```

$$sol := y(t) = \frac{r}{a - \frac{e^{(-r\,t)}\,(-r + y0\,a)}{y0}}$$

```
> sol:=simplify(rhs(sol));
```

$$sol := -\frac{r\,y0}{-y0\,a - e^{(-r\,t)}\,r + e^{(-r\,t)}\,y0\,a}$$

We define `y` as a function of `t`, `y0`, `r`, and `a` so that we can refer to this solution in other problems without solving the differential equation again.

```
> y:=(t0,y00,r0,a0)->subs({t=t0,y0=y00,r=r0,a=a0},sol):
  y(t,y0,r,a);
```

$$-\frac{r\,y0}{-y0\,a - e^{(-r\,t)}\,r + e^{(-r\,t)}\,y0\,a}$$

The solution can also be written as

$$y = \frac{ry_0}{ay_0 + (r - ay_0)e^{-rt}}.$$

Notice that if $r > 0$, $\lim_{t \to \infty} y(t) = (r/a)$.

EXAMPLE 8: Use the logistic equation to approximate the population of the United States using $r = 0.03$, $a = 0.0001$, and $y_0 = 5.3$. Compare this result with the actual census values given in the following table. Use the model obtained to predict the population of the United States in the year 2000.

Year (t)	Population (in millions)	Year (t)	Population (in millions)
1800 (0)	5.30	1900 (100)	76.21
1810 (10)	7.24	1910 (110)	92.23
1820 (20)	9.64	1920 (120)	106.02
1830 (30)	12.68	1930 (130)	123.20
1840 (40)	17.06	1940 (140)	132.16
1850 (50)	23.19	1950 (150)	151.33
1860 (60)	31.44	1960 (160)	179.32
1870 (70)	38.56	1970 (170)	203.30
1880 (80)	50.19	1980 (180)	226.54
1890 (90)	62.98	1990 (190)	248.71

SOLUTION: We substitute the indicated values of r, a, and y_0 into $y = \frac{ry_0}{ay_0 + (r - ay_0)e^{-rt}}$ to obtain the approximation of the population of the United States at time t, where t represents the number of years since 1800,

$$y(t) = \frac{0.03 \bullet 5.3}{0.0001 \bullet 5.3 + (0.03 - 0.0001 \bullet 5.3)e^{-0.03t}} = \frac{0.159}{0.00053 + 0.02947e^{-0.03t}}.$$

We compare the approximation of the population of the United States given by the approximation $y(t)$ with the actual population obtained from census figures. First, we enter the data represented in the table as `data` and and then graph the points in `data` with `plot` together with the option `style=POINT`, naming the resulting graphics object `p1`. We then graph `y(t,5.3,0.03,0.0001)` for the years corresponding to 1800 to 2000, and we name the resulting graph `p2`. Finally, `display` is used to display `p1` and `p2` together.

```
data:=[[0,5.30],[10,7.24],[20,9.64],[30,12.68],
      [40,17.06],[50,23.19],[60,31.44],[70,38.56],
      [80,50.19],[90,62.98],[100,76.21],[110,92.23],
      [120,106.02],[130,123.20],[140,132.16],
      [150,151.33],[160,179.32],[170,203.30],
      [180,226.54],[190,248.71]]:
  p1:=plot(data,style=POINT,color=BLACK):
> p2:=plot(y(t,5.3,0.03,0.0001),t=0..200,color=BLACK):
  with(plots):
  display({p1,p2});
```

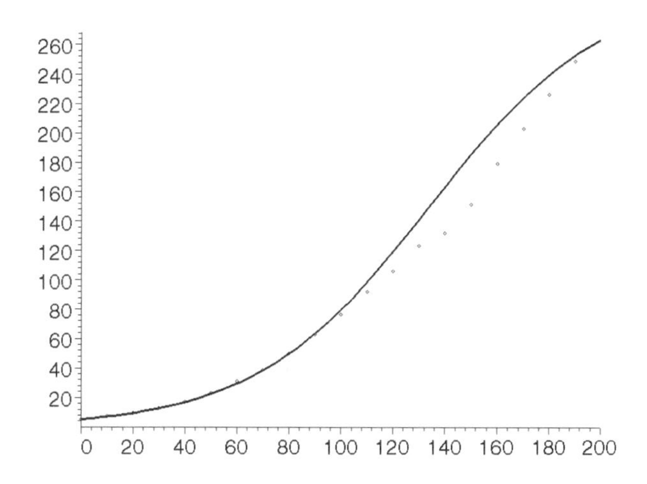

We can also compare the data by making a table of the year, actual population, and population predicted by the model with `seq` and `array`.

```
> array([seq([data[i,1]+1800,data[i,2],
      evalf(y(data[i,1],5.3,0.03,0.0001))],
```

```
i=1..20)]);
```

$$\begin{bmatrix} 1800 & 5.30 & 5.299999999 \\ 1810 & 7.24 & 7.110304038 \\ 1820 & 9.64 & 9.518975104 \\ 1830 & 12.68 & 12.70819879 \\ 1840 & 17.06 & 16.90375616 \\ 1850 & 23.19 & 22.37657266 \\ 1860 & 31.44 & 29.43704029 \\ 1870 & 38.56 & 38.41701652 \\ 1880 & 50.19 & 49.63386311 \\ 1890 & 62.98 & 63.33282516 \\ 1900 & 76.21 & 79.61046329 \\ 1910 & 92.23 & 98.33346770 \\ 1920 & 106.02 & 119.0805988 \\ 1930 & 123.20 & 141.1414901 \\ 1940 & 132.16 & 163.5937704 \\ 1950 & 151.33 & 185.4482322 \\ 1960 & 179.32 & 205.8170778 \\ 1970 & 203.30 & 224.0474428 \\ 1980 & 226.54 & 239.7815217 \\ 1990 & 248.71 & 252.9407939 \end{bmatrix}$$

To predict the population of the United States in the year 2000 with this model, we evaluate

```
evalf(y(200,5.3,0.03,0.0001));
```

263.6602427

Thus, we predict that the population will be approximately 263.66 million in the year 2000. Note that the projects of the population of the United States in the year 2000 made by the Bureau of the Census range from 259.57 million to 278.23 million.

■

Application: Newton's Law of Cooling

Newton's Law of Cooling states that the rate at which the temperature $T(t)$ changes in a cooling body is proportional to the difference between the temperature of the body and the constant temperature T_s of the surrounding medium. This situation is represented as the first-order initial-value problem

$$\begin{cases} \dfrac{dT}{dt} = k(T - T_s) \\ T(0) = T_0 \end{cases},$$

where T_0 is the initial temperature of the body and k is the constant of proportionality.

EXAMPLE 9: A pie is removed from a 350°F oven. In 15 minutes, the pie has a temperature of 150°F. Determine the time required to cool the pie to a temperature of 80°F so that it may be eaten.

SOLUTION: We use `dsolve` to solve the initial-value problem .

$$\begin{cases} \dfrac{dT}{dt} = k(T - T_s) \\ T(0) = T_0 \end{cases}.$$

```
> de1:=dsolve({diff(tp(t),t)=-k*(tp(t)-ts),
      tp(0)=t0},tp(t));
```

$$de1 := \mathrm{tp}(t) = ts + \mathbf{e}^{(-k\,t)}\,(-ts + t0)$$

We use the result to define `temp`.

```
> temp:=(ts,t0,k,t)->ts+exp(-k*t)*(-ts+t0);
```

$$temp := (ts, t0, k, t) \to ts + \mathbf{e}^{(-k\,t)}\,(t0 - ts)$$

The solution based on the data indicated in the example is then found.

```
> temp(75,350,k,15);
```

$$75 + 275\,\mathbf{e}^{(-15\,k)}$$

The constant k is then determined with `solve` and named `k1` for convenience.

```
> k1:=solve(temp(75,350,k,15)=150);
```

$$k1 := -\frac{1}{15}\ln\left(\frac{3}{11}\right)$$

We then use `k1` to determine the time at which the pie reaches the desired temperature. From the graph, we see that value of the function seems to equal 80 near $t = 45$.

```
> plot({temp(75,350,k1,t),80},t=0..120,
      color=[BLACK,GRAY]);
```

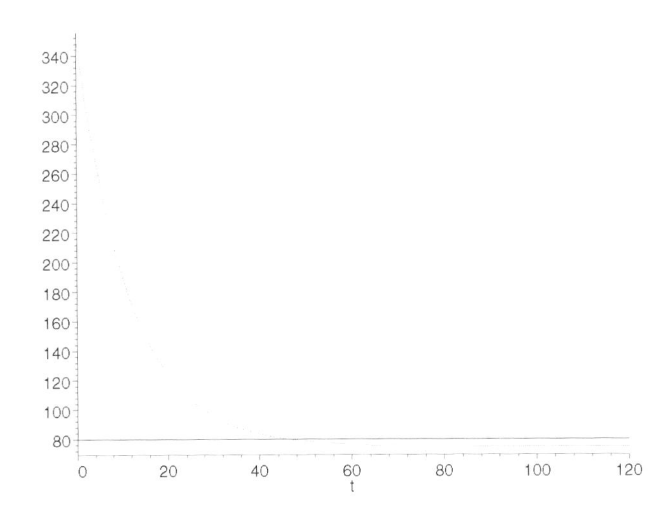

The exact value is found with `solve`:

```
> solve(temp(75,350,k1,t)=80,t);
```

$$-15\frac{\ln(55)}{\ln\left(\frac{3}{11}\right)}$$

and the approximation $t \approx 46.264$ is obtained with `evalf`.

```
> evalf(%);
```

$$46.26397676$$

■

Application: Free-Falling Bodies

The motion of objects can be determined through the solution of first-order initial-value problems. We begin by explaining some of the theory that is needed to set up the differential equation that models the situation.

Newton's Second Law of Motion states that the rate at which the momentum of a body changes with respect to time is equal to the resultant force acting on the body. Because the body's momentum is defined as the product of its mass and velocity, this statement is modeled as

$$\frac{d}{dt}(mv) = F,$$

where m and v represent the body's mass and velocity, respectively, and F is the sum of the forces (the resultant force) acting on the body. Because m is constant, differentiation leads to the well-known equation

$$m\frac{d}{dt} = F.$$

If the body is subjected only to the force due to gravity, then its velocity is determined by solving the differential equation

$$m\frac{dv}{dt} = mg \text{ or } \frac{dv}{dt} = g,$$

where $g \approx 32\ ft/s^2$ (English system) and $g \approx 9.8\ m/s^2$ (international system). This differential equation is applicable only when the resistive force due to the medium (such as air resistance) is ignored. If this offsetting resistance is considered, we must discuss all of the forces acting on the object. Mathematically, we write the equation as

$$m\frac{dv}{dt} = \sum \text{(forces acting on the object)},$$

where the direction of motion is taken to be the positive direction.

We use a force diagram to set up the differential equation that models the situation. Because air resistance acts against the object as it falls and g acts in the same direction of the motion, we state the differential equation in the form

$$m\frac{dv}{dt} = mg + (-F_R) \text{ or } m\frac{dv}{dt} = mg - F_R,$$

where F_R represents this resistive force. Note that down is assumed to be the positive direction. The resistive force is typically proportional to the body's velocity, v, or the square of its velocity, v^2. Hence, the differential equation is linear or nonlinear based on the resistance of the medium taken into account.

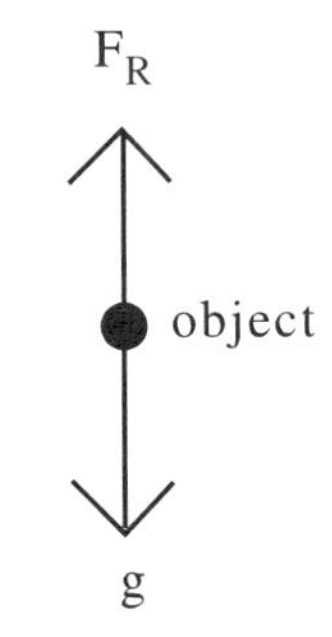

positive direction

Force Diagram

EXAMPLE 10: An object of mass $m = 1$ is dropped from a height of 50 feet above the surface of a small pond. While the object is in the air, the force due to air resistance is v. However, when the object is in the pond, it is subjected to a buoyancy force equivalent to $6v$. Determine how much time is required for the object to reach a depth of 25 feet in the pond.

SOLUTION: This problem must be broken into two parts: an initial-value problem for the object above the pond, and an initial-value problem for the object below the surface of the pond. The initial-value problem above the pond's surface is found to be

$$\begin{cases} \frac{dv}{dt} = (32 - v) \\ V(0) = 0 \end{cases}.$$

However, to define the initial-value problem to find the velocity of the object beneath the pond's surface, we must know the velocity of the object when it reaches the surface. Hence, the velocity of the object above the surface must be determined by solving the initial-value problem above. The equation $dv/dt = 32 - v$ is separable and solved with `dsolve` in `d1`.

```
> d1:=dsolve({diff(v(t),t)=32-v(t),v(0)=0},v(t));
```

$$d1 := v(t) = 32 - 32\,\mathbf{e}^{(-t)}$$

Note that `d1` consists of two parts: the left- and right-hand sides of the equation. The second part of `d1`, corresponding to the right-hand side of the equation, is extracted with `op` or, equivalently, `rhs(d1)`.

```
> op(2,d1);
```

$$32 - 32\,\mathbf{e}^{(-t)}$$

To find the velocity when the object hits the pond's surface, we must know the time at which the distance traveled by the object (or the displacement of the object) is 50. Thus, we must find the displacement function, by integrating the velocity function obtaining $s(t) = 32e^{-t} + 32t - 32$.

```
> p1:=dsolve({diff(y(t),t)=op(2,d1),y(0)=0},y(t));
```

$$p1 := y(t) = 32\,t + 32\,\mathbf{e}^{(-t)} - 32$$

The displacement function is graphed next. Note that the explicit solution for y is obtained with `op(2,p1)`. However, in this case, we use `assign` to define y to be the explicit solution and then graph y on the interval [0,5]. We need to know the value of t for which the object has traveled 50 feet. This time appears to be approximately 2.5 seconds.

```
> assign(p1);
  plot({y(t),50},t=0..5);
```

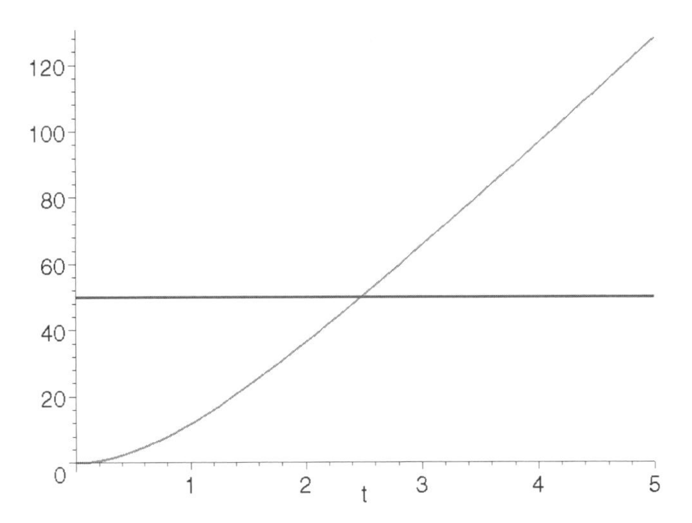

A more accurate value of the time at which the object hits the surface is found with `fsolve`. In this case, we obtain $t \approx 2.47864$. The velocity at this time is then determined by substitution into the velocity function resulting in $v(2.47864) \approx 29.3166$. Note that this value is the initial velocity of the object when it hits the surface of the pond. This result is called `v1` for convenience.

```
> t1:=fsolve(op(2,p1)=50,t);
```

$$t1 := 2.478643063$$

```
> v1:=evalf(subs(t=t1,op(2,d1)));
```

$$v1 := 29.31657802$$

Thus, the initial-value problem that determines the velocity of the object beneath the surface of the pond is given by

$$\begin{cases} dv/dt = 32 - 6v \\ V(0) = 29.3166 \end{cases}.$$

The solution of this initial-value problem is $v(t) = \frac{16}{3} + 23.9833e^{-t}$ and integrating to obtain the displacement function (the initial displacement is 0) we obtain $s(t) = 3.99722 - 3.99722e^{-6t} + \frac{16}{3}t$. These steps are carried out in d2 and p2. Note that because y was defined above, we first clear prior definitions of y before entering the dsolve command.

```
> d2:=dsolve({diff(v(t),t)=32-6*v(t),
      v(0)=v1},v(t));
```

$$d2 := \mathrm{v}(t) = \frac{16}{3} + 23.98324469\,\mathbf{e}^{(-6\,t)}$$

```
> y:='y':
  p2:=dsolve({diff(y(t),t)=op(2,d2),y(0)=0},y(t));
```

$$p2 := \mathrm{y}(t) = -3.997207448\,\mathbf{e}^{(-6.\,t)} + 5.333333334\,t + 3.997207448$$

In the same manner as above, this displacement function is then plotted to determine when the object is 25 feet beneath the surface of the pond. This time appears to be near 4 seconds.

```
> assign(p2):
  plot({y(t),25},t=0..5);
```

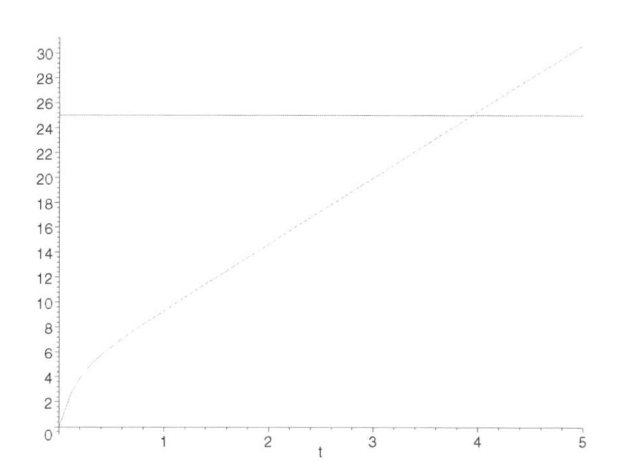

`fsolve` is used below to obtain a more accurate approximation of the value of t at which the object is 25 feet beneath the pond's surface. Finally, the time required for the object to reach the pond's surface is added to the time needed for it to travel 25 feet beneath the surface to see that approximately 6.41667 seconds are required for the object to travel from a height of 50 feet above the pond to a depth of 25 feet below the surface.

```
> t2:=fsolve(op(2,p2)=25,t);
```

$$t2 := 3.938023603$$

```
> t1+t2;
```

```
6.416666666
```

■

7.2 Higher-Order Ordinary Differential Equations

An ordinary differential equation of the form

$$a_n(x)y^{(n)}(x) + a_{n-1}(x)y^{(n-1)}(x) + \ldots\ a_1(x)y'(x) + a_0(x)y(x) = f(x),$$

where $a_n(x)$ is not identically the zero function, is called an ***n*th-order ordinary linear differential equation**. If $f(x)$ is identically the zero function, the equation is said to be **homogeneous**; if $f(x)$ is not the zero function, the equation is said to be **nonhomogeneous**; and if the functions $a_i(x)$, $i = 0, 1, 2, \ldots, n$ are constants, the equation is said to have **constant coefficients**.

Let $f_1(x), f_2(x), f_3(x), \ldots, f_{n-1}(x)$, and $f_n(x)$ be a set of n functions at least $n-1$ times differentiable. If S is said to be **linearly dependent** on an interval I, this means that there are constants $c_1, c_2, \ldots, c_n$, not all zero, so that $\sum_{k=1}^{n} c_k f_k(x) = 0$ for every value of x in the interval I. If S is said to be **linearly independent**, this means that S is not linearly dependent. The **Wronskian** of S, denoted by $W(S) = W(f_1(x), f_2(x), f_3(x), \ldots, f_{n-1}(x), f_n(x))$, is the determinant

$$W(S) = \begin{vmatrix} f_1(x) & f_2(x) & \ldots & f_n(x) \\ f_1'(x) & f_2'(x) & \ldots & f_n'(x) \\ \vdots & \vdots & \vdots & \vdots \\ f_1^{(n-1)}(x) & f_2^{(n-1)}(x) & \ldots & f_n^{(n-1)}(x) \end{vmatrix}.$$

The following theorem can help us determine if a set of functions is either linearly dependent or linearly independent: Let $S = \{f_1(x), f_2(x), \ldots, f_n(x)\}$ be a set of n solutions of

$$a_n(x)y^{(n)}(x) + a_{n-1}(x)y^{(n-1)}(x) + \ldots a_1(x)y'(x) + a_0(x)y(x) = f(x)$$

on an interval I. S is linearly independent if and only if $W(S) \neq 0$ for at least one value of x in the interval I.

EXAMPLE 1: Show that $S = \{e^x, xe^x, x^2ex\}$ is linearly independent.

SOLUTION: Note that all three functions in S are solutions of $y''' - 3y'' + 3y' - y = 0$. The Wronskian of S is

$$W(S) = \begin{vmatrix} e^x & xe^x & x^2e^x \\ \frac{d}{dx}(e^x) & \frac{d}{dx}(xe^x) & \frac{d}{dx}(x^2e^x) \\ \frac{d^2}{dx^2}(e^x) & \frac{d^{22}}{dx}(xe^x) & \frac{d^2}{dx^2}(x^2e^x) \end{vmatrix} = \begin{vmatrix} e^x & xe^x & x^2e^x \\ e^x & (x+1)e^x & (x^2+2x)e^x \\ e^x & (x+2)e^x & (x^2+4x+2)e^x \end{vmatrix}$$

We compute this with Maple by defining `sof` to be the list of functions consisting of e^x, xe^x, and x^2e^x. Note that we will be using the commands `det`, which computes the determinant of a square matrix. Therefore, we first load `linalg`; otherwise, this commands would have to be entered in the form `linalg[det]`.

```
> with(linalg):
  sof:=[exp(x),x*exp(x),x^2*exp(x)];
```

$$sof := [\mathbf{e}^x, x\,\mathbf{e}^x, x^2\,\mathbf{e}^x]$$

We must compute the determinant of the matrix with first row corresponding to sof, second row corresponding to the derivative of each member of sof, and third row corresponding to the second derivative of each member of sof. We define this matrix, M, using array and diff. Note that diff(sog,x$2) computes the second derivative of each member of sof with respect to x.

```
> diff(sof,x);
```

$$[\mathbf{e}^x, \mathbf{e}^x + x\,\mathbf{e}^x, 2\,x\,\mathbf{e}^x + x^2\,\mathbf{e}^x]$$

```
> M:=array([sof,diff(sof,x),diff(sof,x$2)]);
```

$$M := \begin{bmatrix} \mathbf{e}^x & x\,\mathbf{e}^x & x^2\,\mathbf{e}^x \\ \mathbf{e}^x & \mathbf{e}^x + x\,\mathbf{e}^x & 2\,x\,\mathbf{e}^x + x^2\,\mathbf{e}^x \\ \mathbf{e}^x & 2\,\mathbf{e}^x + x\,\mathbf{e}^x & 2\,\mathbf{e}^x + 4\,x\,\mathbf{e}^x + x^2\,\mathbf{e}^x \end{bmatrix}$$

Finally, we compute the determinant of M with det to conclude that the Wronskian of S is $2(e^x)^3 = 2e^{3x}$ and that the set of functions is linearly independent.

```
> det(M);
```

$$2\,(\mathbf{e}^x)^3$$

■

The linear algebra package linalg contains a Wronskian function that returns the Wronskian matrix for a set of functions. Thus, entering

```
> with(linalg):
  sof:=[exp(x),x*exp(x),x^2*exp(x)];
  W:=Wronskian(sof,x);
```

$$sof := [\mathbf{e}^x, x\,\mathbf{e}^x, x^2\,\mathbf{e}^x]$$

$$W := \begin{bmatrix} \mathbf{e}^x & x\,\mathbf{e}^x & x^2\,\mathbf{e}^x \\ \mathbf{e}^x & \mathbf{e}^x + x\,\mathbf{e}^x & 2\,x\,\mathbf{e}^x + x^2\,\mathbf{e}^x \\ \mathbf{e}^x & 2\,\mathbf{e}^x + x\,\mathbf{e}^x & 2\,\mathbf{e}^x + 4\,x\,\mathbf{e}^x + x^2\,\mathbf{e}^x \end{bmatrix}$$

defines sof to be the list of functions $S = \{e^x, xe^x, x^2e^x\}$ and W to be the Wronskian matrix for sof. Entering

```
> step1:=det(W);
```

$$step1 := 2\,(e^{x})^{3}$$

```
> simplify(step1);
```

$$2\,e^{(3\,x)}$$

computes and simplifies the determinant of `W`.

EXAMPLE 2: Determine if the set of functions $S = \{\cos x, \cos 2x, \cos 3x, \cos 4x\}$ is linearly independent.

SOLUTION: Note that all three functions in S are solutions of $y^{(8)} + 30y^{(6)} + 273y^{(4)} + 820y'' + 576y = 0$. The Wronskian of S is computed with `wronskian`. and then `det`.

```
> with(linalg):
  sof:=[cos(x),cos(2*x),cos(3*x),cos(4*x)]:
  W:=Wronskian(sof,x);
```

$$W := \begin{bmatrix} \cos(x) & \cos(2\,x) & \cos(3\,x) & \cos(4\,x) \\ -\sin(x) & -2\sin(2\,x) & -3\sin(3\,x) & -4\sin(4\,x) \\ -\cos(x) & -4\cos(2\,x) & -9\cos(3\,x) & -16\cos(4\,x) \\ \sin(x) & 8\sin(2\,x) & 27\sin(3\,x) & 64\sin(4\,x) \end{bmatrix}$$

```
> step1:=det(W);
```

$$\begin{aligned} step1 :=\ & 768\cos(x)\sin(2\,x)\cos(3\,x)\sin(4\,x) - 450\cos(x)\sin(2\,x)\cos(4\,x)\sin(3\,x) \\ & - 252\cos(x)\cos(2\,x)\sin(3\,x)\sin(4\,x) - 300\sin(x)\cos(2\,x)\cos(3\,x)\sin(4\,x) \\ & + 288\sin(x)\cos(2\,x)\cos(4\,x)\sin(3\,x) - 42\sin(x)\sin(2\,x)\cos(3\,x)\cos(4\,x) \end{aligned}$$

We simplify the result using `simplify` together with the `trig` option.

```
> step2:=simplify(step1,trig);
```

$$step2 := 288 + 4320\cos(x)^{6} - 5280\cos(x)^{4} + 1440\cos(x)^{2} - 768\cos(x)^{10}$$

To see that this is not the zero function, we generate a graph with `plot`.

```
> plot(step2,x=0..2*Pi);
```

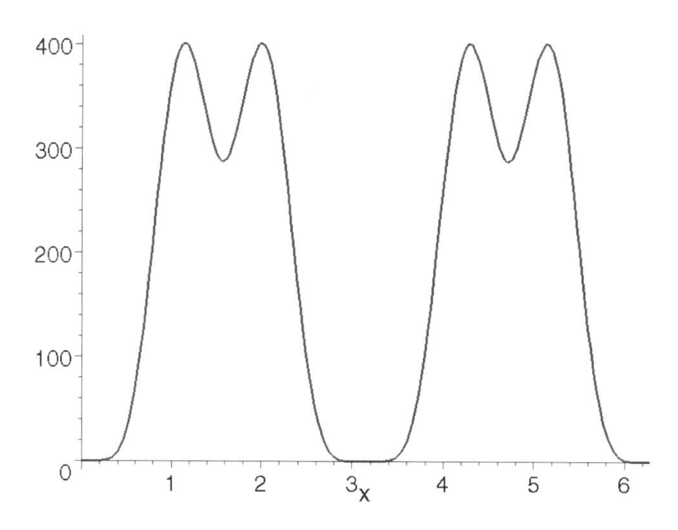

Alternatively, we can see the result is not identically the zero function by evaluating it for some value of x. For example, $W(S) = 288 \neq 0$ if $x = \pi/2$.

```
> eval(subs(x=Pi/2,step2));
                              288
```

■

A set $S = \{f_1(x), f_2(x), f_3(x), \ldots f_{n-1}(x), f_n(x)\}$ of n linearly independent nontrivial solutions of the nth-order linear homogeneous equation

$$a_n(x)y^{(n)}(x) + a_{n-1}(x)y^{(n-1)}(x) + \ldots a_1(x)y'(x) + a_0(x)y(x) = 0$$

is called a **fundamental set of solutions** of the equation. Observe that if $S = \{f_i(x)\}_{i=1}^{n}$ is a fundamental set of solutions of

$$a_n(x)y^{(n)}(x) + a_{n-1}(x)y^{(n-1)}(x) + \ldots a_1(x)y'(x) + a_0(x)y(x) = 0$$

and $\{c_i\}_{i=1}^{n}$ is a set of n numbers, then $f(x) = \sum_{i=1}^{n} c_i f_i(x)$ is also a solution of $\sum_{i=0}^{n} a_i(x)y^{(i)}(x) = 0$ (the **Principle of Superposition**). The following two theorems tell us that under reasonable conditions, the nth-order homogeneous equation

$$a_n(x)y^{(n)}(x) + a_{n-1}(x)y^{(n-1)}(x) + \ldots a_1(x)y'(x) + a_0(x)y(x) = 0$$

has a fundamental set of n solutions:

1. If $a_i(x)$ is continuous on an open interval I for $i = 0, 1, \ldots, n$, then the nth-order linear homogeneous equation

$$a_n(x)y^{(n)}(x) + a_{n-1}(x)y^{(n-1)}(x) + \ldots a_1(x)y'(x) + a_0(x)y(x) = 0$$

has a fundamental set of solutions on I.

2. Any set of $n + 1$ solutions of the nth-order linear homogeneous equation

$$a_n(x)y^{(n)}(x) + a_{n-1}(x)y^{(n-1)}(x) + \ldots a_1(x)y'(x) + a_0(x)y(x) = 0$$

is linearly dependent.

If $S = \{f_i(x)\}_{i=1}^{n}$ is a fundamental set of solutions of the nth-order linear homogeneous equation $a_n(x)y^{(n)}(x) + a_{n-1}(x)y^{(n-1)}(x) + \ldots a_1(x)y'(x) + a_0(x)y(x) = 0$, a **general solution** of the equation is $f(x) = c_1f_1(x) + c_2f_2(x) + \ldots + c_nf_n(x)$, where $\{c_i\}_{i=1}^{n}$ is a set of n arbitrary constants.

Given a linear homogeneous differential equation with constant coefficients, the command `constcoeffsols` that is contained in the `DEtools` package returns a fundamental set of solutions to the equation.

EXAMPLE 3: Find a fundamental set of solutions for $\frac{d^2y}{dx^2} + 4\frac{dy}{dx} + 20y = 0$. What is a general solution?

SOLUTION: After loading the `DEtools` package, we use `constcoeffsols` to find a fundamental set of solutions to the equation. The result indicates that a fundamental set of solutions is

$$S = \{e^{-2x}\cos 4\text{x}, e^{-2x}\sin 4x\};$$

a general solution is $y = e^{-2x}(c_1\cos 4x + c_2\sin 4x)$.

```
> y:='y':
  with(DEtools):
  constcoeffsols(diff(y(x),x$2)+4*diff(y(x),x)+
  20*y(x)=0,
```

```
y(x));
```

$$[e^{(-2x)}\cos(4x), e^{(-2x)}\sin(4x)]$$

■

The equation

$$a_n m^n + a_{n-1} m^{n-1} + \ldots + a_1 m + a_0 = \sum_{k=0}^{n} a_k m^k = 0$$

is called the **characteristic equation** of the nth-order homogeneous linear differential equation with constant coefficients $a_n y^{(n)}(x) + a_{n-1} y^{(n-1)}(x) + \ldots + a_1 y'(x) + a_0 y(x) = 0$. General solutions of an nth-order homogeneous linear differential equation with constant coefficients are determined by the solutions of its characteristic equation.

The Homogeneous Second-Order Equation with Constant Coefficients

Let $ay'' + by' + cy = 0$ be a homogeneous second-order equation with constant coefficients, and let m_1 and m_2 be the solutions of the characteristic equation $am^2 + bm + c = 0$.

1. If $m_1 \neq m_2$ and both m_1 and m_2 are real, a general solution of $ay'' + by' + cy = 0$ is
$$y(t) = c_1 e^{m_1 t} + c_2 e^{m_2 t}.$$
2. If $m_1 = m_2$ and both m_1 and m_2 are real, a general solution of $ay'' + by' + cy = 0$ is
$$y(t) = c_1 e^{m_1 t} + c_2 t e^{m_1 t}; \text{ and}$$
3. If $m_1 = \alpha + i\beta$, $\beta \neq 0$, and $m_1 = \overline{m_2}$, a general solution of $ay'' + by' + cy = 0$ is
$$y(t) = c_1 e^{\alpha t} \cos \beta t + c_2 e^{at} \sin \beta t$$

In (3), $\overline{m_2}$ is the **complex conjugate** of m_2: $\overline{m_2} = \overline{\alpha - i\beta} = \alpha + i\beta$.

EXAMPLE 4: Solve (a) $3y'' + 2y' - 5y = 0$; (b) $\begin{cases} 2y'' + 5y' + 5y = 0 \\ y(0) = 0, y'(0) = 1/2 \end{cases}$; and (c) $\begin{cases} y'' + 4y' + 4y = 0 \\ y(0) = 0, y'(0) = -1/2 \end{cases}$.

SOLUTION: (a) First, we use `dsolve` to find a general solution of the equation. Note that `diff(y(x),x$2)` represents $y''=d^2y/dx^2$.

```
sola:=dsolve(3*diff(y(x),x$2)+2*diff(y(x),x)-
   5*y(x)=0,y(x));
```

$$sola := y(x) = \_C1\, e^{x} + \_C2\, e^{(-5/3\,x)}$$

We graph the solution for various values of the arbitrary constants with `plot`.

```
> toplot:={seq(seq(subs({_C1=i,_C2=j},rhs(sola)),
     i=-1..1),j=-1..1)}:
  plot(toplot,x=-2..2,y=-10..10);
```

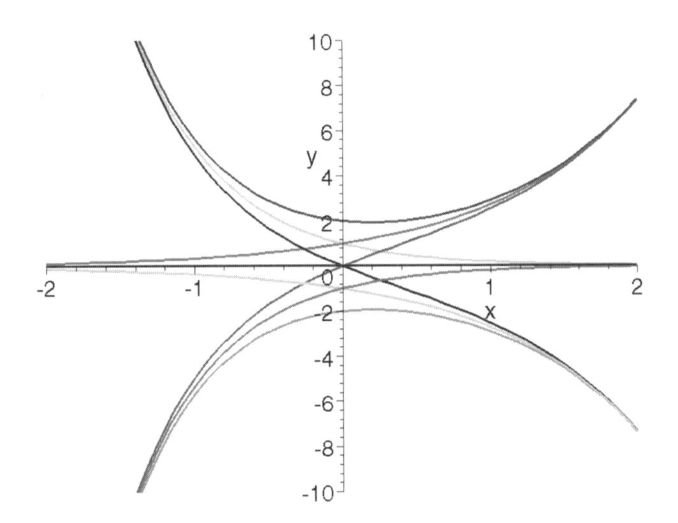

Alternatively, the characteristic equation for this equation is $3m^2 + 2m - 5 = 0$. We determine the solutions of the characteristic equation with `solve` or `factor`.

```
> solve(3*m^2+2*m-5=0);
```

$$\frac{-5}{3}, 1$$

```
> factor(3*m^2+2*m-5);
```

$$(3\,m + 5)\,(m - 1)$$

Because the roots of the characteristic equation are $m = -5/3$ and $m = 1$, a general solution of the equation is $y = c_1 e^{-5x/3} + c_2 e^{x}$.

(b) As in (a), we use `dsolve` to solve the initial-value problem directly.

```
solb:=dsolve({2*diff(y(x),x$2)+5*diff(y(x),x)+5*y(x)=0,
        y(0)=0,D(y)(0)=1/2},y(x));
```

$$solb := y(x) = \frac{2}{15}\sqrt{15}\,\mathbf{e}^{(-5/4\,x)}\sin\left(\frac{1}{4}\sqrt{15}\,x\right)$$

The solution is graph on the interval $[-\pi/2, \pi]$ with `plot`.

```
> plot(rhs(solb),x=-Pi/2..Pi,y=-0.3..0.2);
```

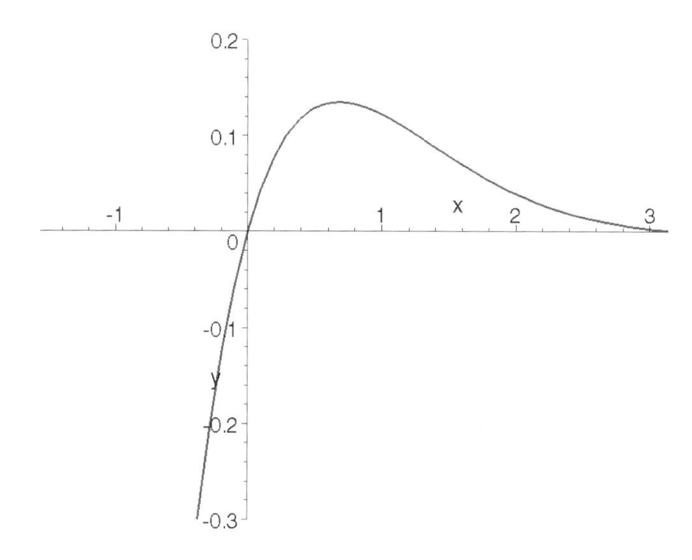

Alternatively, we can use Maple to assist us with the steps necessary to solve the initial-value problem. First, we use `solve` to solve the characteristic equation $2m^2 + 5m + 5 = 0$.

```
solve(2*m^2+5*m+5=0);
```

$$-\frac{5}{4}+\frac{1}{4}I\sqrt{15},\ -\frac{5}{4}-\frac{1}{4}I\sqrt{15}$$

Then, a general solution of the equation is $y = e^{-5x/4}\left(c_1 \cos\left(\frac{\sqrt{15}}{4}\right) + c_2 \sin\left(\frac{\sqrt{15}}{4}x\right)\right)$, defined next.

```
> y:='y':
  y:=x->exp(-5*x/4)*
       (c[1]*cos(sqrt(15)*x/4)+c[2]*sin(sqrt(15)*x/4)):
```

Application of the initial conditions

```
> sys:={y(0)=0,D(y)(0)=1/2};
```

$$sys := \{ c_1 = 0, -\frac{5}{4}c_1 + \frac{1}{4}c_2\sqrt{15} = \frac{1}{2}\}$$

yields the system of equations

$$\begin{cases} c_1 = 0 \\ -\frac{5}{4}c_1 + \frac{\sqrt{15}}{4}c_2 = \frac{1}{2} \end{cases},$$

which we solve with `solve`.

```
> cvals:=solve(sys);
```

$$cvals := \{ c_1 = 0, c_2 = \frac{2}{15}\sqrt{15}\}$$

Substitution into the general solution yields the solution to the initial-value problem.

```
> sol:=x->subs(cvals,y(x)):
  sol(x);
```

$$\frac{2}{15}\sqrt{15}\,\mathbf{e}^{(-5/4x)}\sin\left(\frac{1}{4}\sqrt{15}\,x\right)$$

(c) We use `dsolve` as in (a) and (b).

```
> y:='y':
```

```
solc:=dsolve({diff(y(x),x$2)+4*diff(y(x),x)+4*y(x)=0,
    y(0)=1,D(y)(0)=-1/2},y(x));
```

$$solc := \mathrm{y}(x) = \mathbf{e}^{(-2x)} + \frac{3}{2}\mathbf{e}^{(-2x)}\,x$$

The result is then graphed on the interval [–1,1].

```
> assign(solc):
  plot(y(x),x=-1..1);
```

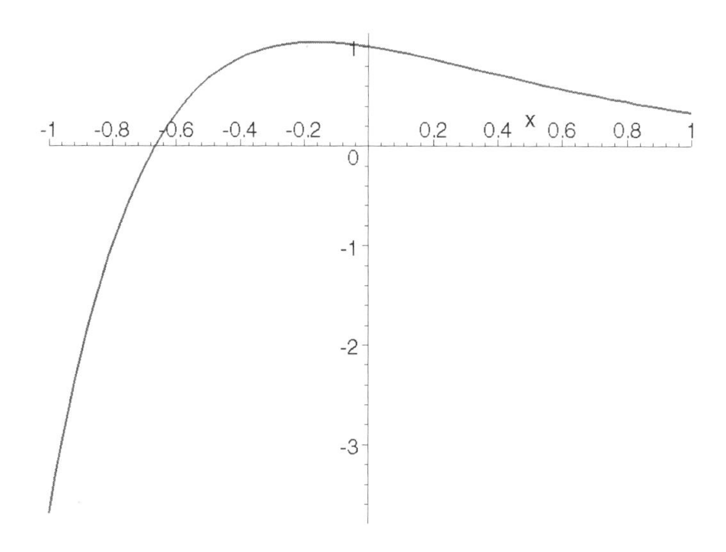

■

In the same manner as in the case for a second-order homogeneous equation with real constant coefficients, a general solution is also determined by the solutions of the characteristic equation. Instead of stating an exact rule for the numerous situations encountered, we illustrate how a general solution is found in the following examples.

The command `dsolve` can be used to solve nth-order linear homogeneous differential equations with constant coefficients as long as n is smaller than 5. In cases in which the roots of the characteristic equation are symbolically complicated, approximations of the roots of the characteristic equation can be computed with the command `fsolve`.

EXAMPLE 5: Find a general solution of $9y^{(4)} - 6y''' + 46y'' - 6y' + 37y = 0$.

SOLUTION: The characteristic equation of $9y^{(4)} - 6y''' + 46y'' - 6y' + 37y = 0$ is $9m^4 - 6m^3 + 46m^2 - 6m + 37 = 0$, solved with `solve`.

```
> solve(9*m^4-6*m^3+46*m^2-6*m+37=0)
```

$$I, -I, \frac{1}{3} + 2\,I, \frac{1}{3} - 2\,I$$

Because the solutions of the characteristic equation are $x = \pm i$ and $x = \frac{1}{3} \pm 2i$, a general solution of the equation is given by $y = c_1 \cos x + c_2 \sin x + e^{x/3}(c_3 \cos 2x + c_4 \sin 2x)$. Because the order of the equation is 4, DSolve can be used to find a general solution. First, we define eq to be the equation $9y^{(4)} - 6y''' + 46y'' - 6y' + 37y = 0$, and then we use `dsolve` to find a general solution of `eq`, naming the resulting output `sol`.

```
> eq:=9*diff(y(x),x$4)-6*diff(y(x),x$3)+
      46*diff(y(x),x$2)-6*diff(y(x),x)+37*y(x)=0:
  sol:=dsolve(eq,y(x));
```

$$sol := \mathrm{y}(x) = \_C1 \cos(x) + \_C2 \sin(x) + \_C3\, \mathbf{e}^{(1/3\,x)} \cos(2\,x) + \_C4\, \mathbf{e}^{(1/3\,x)} \sin(2\,x)$$

To graph the solution for various values of the constants, we define `tograph` to be the sequence of functions obtained by replacing each occurrence of `_C1`, `_C2`, `_C3`, and `_C4` in `rhs(sol)` by `i`, `j`, `k`, and `m`, respectively, for `i`=0 and 1, `j`=–1 and 0, `k`=0 and 1, and `m`=–1 and 0.

```
> tograph:=[seq(seq(seq(seq(subs({_C1=i,_C2=j,_C3=k,_
   C4=m},
      rhs(sol)),i=0..1),j=-1..0),k=0..2),m=-1..0)]:
```

```
> plot(tograph,x=0..2*Pi);
```

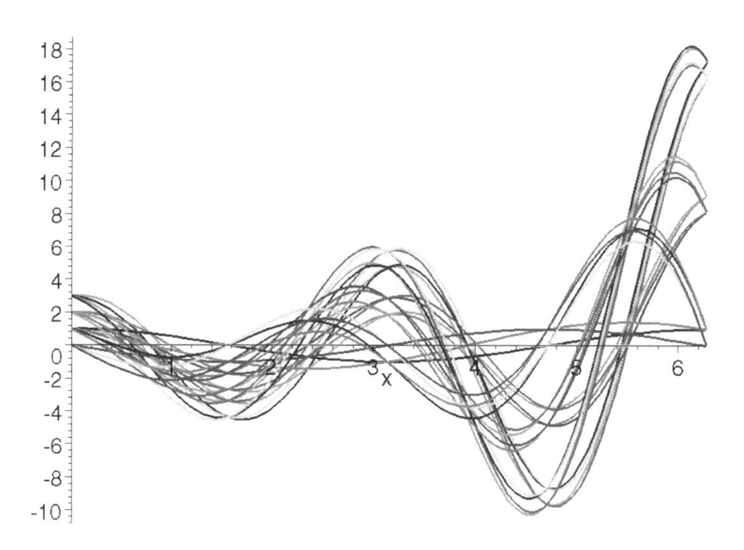

■

Nonhomogeneous Equations with Constant Coefficients: Undetermined Coefficients

Consider the nonhomogeneous linear nth order differential equation with constant coefficients

$$a_n y^{(n)}(x) + a_{n-1} y^{(n-1)}(x) + \dots + a_1 y'(x) + a_0 y(x) = f(x)$$

We know that a general solution of this differential equation is given by

$$(y(x) = y_h(x) + y_p(x)),$$

where $y_h(x)$ is a solution of the corresponding homogeneous equation

$$a_n y^{(n)}(x) + a_{n-1} y^{(n-1)}(x) + \dots + a_1 y'(x) + a_0 y(x) = 0,$$

and $y_p(x)$ is a particular solution involving no arbitrary constants of the nonhomogeneous equation

$$a_n y^{(n)}(x) + a_{n-1} y^{(n-1)}(x) + \dots + a_1 y'(x) + a_0 y(x) = f(x).$$

If $f_i(x)$ is a linear combination of the functions i = 1, x, x^2 ..., , e^{kx}, xe^{kx}, x^2xe^{kx}, ..., $e^{\alpha x}\cos\beta x$, $xe^{\alpha x}\cos\beta x$, $x^2e^{\alpha x}\cos\beta x$, ... $e^{\alpha x}\sin\beta x$, $xe^{\alpha x}\sin\beta x$, $x^2e^{\alpha x}\sin\beta x$, ... the Method of Undetermined Coefficients provides a method that we can use to determine a particular solution of the equation:

1. Solve the corresponding homogeneous equation for $y_h(x)$.
2. Determine the form of a particular solution $y_p(x)$.
3. Determine the unknown coefficients in $y_p(x)$ by substituting $y_p(x)$ into the nonhomogeneous equation and equating the coefficients of like terms.
4. Form a general solution with $y(x) = y_h(x) + y_p(x)$.

The Form of $y_p(x)$ (Step 2) is determined as follows :

Suppose that $f(x) = b_1f_1(x) + b_2f_2(x) + \ldots + b_jf_j(x)$, where b_1, b_2 ..., b_j are constants and each $f_i(x)$, i = 1, 2, ..., j is a function of the form x^m, x^me^{kx}, $x^me^{\alpha x}\cos\beta x$, or $x^me^{\alpha x}\cos\beta x$.

(a) If $f_i(x) = x^m$, the associated set of functions is

$$S = \left\{x^m, x^{m-1}, \ldots, x^2, x, 1\right\}.$$

(b) If $f_i(x) = x^me^{kx}$, the associated set of functions is

$$S = \left\{x^m, e^{kx}x^{m-1}e^{kx}, \ldots, x^2e^{kx}, xe^{kx}, e^{kx}\right\}.$$

(c) If $f_i(x) = x^me^{\alpha x}\cos\beta x$ or $f_i(x) = x^me^{\alpha x}\sin\beta x$, the associated set of functions is

$$S = (x^me^{ax}\cos\beta x, x^{m-1}e^{ax}\cos\beta x, \ldots, x^2e^{ax}\cos\beta x, xe^{ax}\cos\beta x, xe^{ax}\cos\beta x,$$
$$x^me^{ax}\sin\beta x, x^{m-1}e^{ax}\cos\beta x, \ldots, x^2e^{ax}\sin\beta x, xe^{ax}\sin\beta x, xe^{ax}\sin\beta x\}$$

For each function $f_i(x)$ in $f(x)$, determine the associated set of functions S. If any of the functions in S appears in the general solution to the corresponding homogeneous equation, $y_h(x)$, multiply each function in S by x^r to obtain a new set S', where r is the smallest positive integer so that each function in S' is not a function in $y_h(x)$. A particular solution is obtained by taking the linear combination of all functions in the associated sets, where repeated functions should appear only once in the particular solution.

EXAMPLE 6: If $\omega \geq 0$, solve the initial-value problem $\begin{cases} y'' + y = \sin \omega x \\ y(0) = 0, y'(0) = 1 \end{cases}$.

SOLUTION: A general solution of the corresponding homogeneous equation is

```
> yh:=dsolve(diff(y(x),x$2)+y(x)=0,y(x));
```

$$yh := y(x) = \_C1 \cos(x) + \_C2 \sin(x)$$

$y_h = c_1 \cos x + c_2 \sin x$. If $\omega \neq 1$, the associated set of functions is $S = \{\cos \omega x, \sin \omega x\}$, and because none of these are terms in the solution to the corresponding homogeneous equation, a particular solution to the nonhomogeneous equation has the form $y_p(x) = A \cos \omega x + B \sin(\omega x)$.

```
> yp:=x->A*cos(omega*x)+B*sin(omega*x);
```

$$yp := x \to A \cos(\omega x) + B \sin(\omega x)$$

We then substitute y_p into the nonhomogeneous equation. This equation is true for all values of x.

```
> eq:=diff(yp(x),x$2)+yp(x)=sin(x);
```

$$eq := -A \cos(\omega x)\, \omega^2 - B \sin(\omega x)\, \omega^2 + A \cos(\omega x) + B \sin(\omega x) = \sin(x)$$

In particular, evaluating eq for $x = 0$ and $x = \pi/2$ results in two linear equations

```
> eq1:=eval(subs(x=0,eq));
```

$$eq1 := -A\, \omega^2 + A = 0$$

```
> eq2:=eval(subs(x=Pi/2,eq));
```

$$eq2 := -A \cos\left(\frac{1}{2}\omega \pi\right) \omega^2 - B \sin\left(\frac{1}{2}\omega \pi\right) \omega^2 + A \cos\left(\frac{1}{2}\omega \pi\right) + B \sin\left(\frac{1}{2}\omega \pi\right) = 1$$

that we solve for A and B.

```
> abvals:=solve({eq1,eq2},{A,B});
```

$$abvals := \left\{ A = 0, B = -\frac{1}{\sin\left(\frac{1}{2}\,\omega\,\pi\right)(\omega^2 - 1)} \right\}$$

We substitute these values into y_p and form the general solution, $y = y_h + y_p$.

```
> y:=rhs(yh)+subs(abvals,yp(x));
```

$$y := \_C1\,\cos(x) + \_C2\,\sin(x) - \frac{\sin(\omega\,x)}{\sin\left(\frac{1}{2}\,\omega\,\pi\right)(\omega^2 - 1)}$$

Next, we apply the initial conditions

```
> sys:=eval({subs(x=0,y)=0,subs(x=0,diff(y,x))=1});
```

$$sys: = \left\{ \_C1 = 0, C2 - \frac{\omega}{\sin\left(\frac{1}{2}\omega\pi\right)(\omega^2 - 1)} = 1 \right\}$$

and solve for the arbitrary constants.

```
> cvals:=solve(sys,{_C1,_C2});
```

$$cvals := \left\{ \_C2 = \frac{\omega + \sin\left(\frac{1}{2}\,\omega\,\pi\right)\omega^2 - \sin\left(\frac{1}{2}\,\omega\,\pi\right)}{\sin\left(\frac{1}{2}\,\omega\,\pi\right)(\omega^2 - 1)}, \_C1 = 0 \right\}$$

Substituting these values into the general solution yields the solution to the initial-value problem if $\omega \neq 1$.

```
> y1:=subs(cvals,y);
```

$$y1 := \frac{\left(\omega + \sin\left(\frac{1}{2}\,\omega\,\pi\right)\omega^2 - \sin\left(\frac{1}{2}\,\omega\,\pi\right)\right)\sin(x)}{\sin\left(\frac{1}{2}\,\omega\,\pi\right)(\omega^2 - 1)} - \frac{\sin(\omega\,x)}{\sin\left(\frac{1}{2}\,\omega\,\pi\right)(\omega^2 - 1)}$$

In fact, when we use `dsolve` to solve this initial-value problem, Maple assumes that $\omega \neq 1$.

```
> y:='y':
  sola:=dsolve({diff(y(x),x$2)+y(x)=sin(omega*x),
      y(0)=0,D(y)(0)=1},y(x));
```

$$sola := \mathrm{y}(x) = \left(-\frac{1}{2}\frac{\sin((-1+\omega)\,x)}{-1+\omega} + \frac{1}{2}\frac{\sin((1+\omega)\,x)}{1+\omega}\right)\cos(x)$$
$$+\left(-\frac{1}{2}\frac{\cos((1+\omega)\,x)}{1+\omega} - \frac{1}{2}\frac{\cos((-1+\omega)\,x)}{-1+\omega}\right)\sin(x) + \frac{(\omega+\omega^2-1)\sin(x)}{\omega^2-1}$$

Note that if $\omega \neq 1$, the solution is bounded and periodic, as we see in the following graphs.

```
> omegavals:=[3/4,25/28,27/28,33/28]:
  somegraphs:=[seq(plot(subs(omega=i,y1),x=0..200,
      color=BLACK),i=omegavals)]:
> with(plots):
  anarray:=display(somegraphs,insequence=true):
  display(anarray);!
```

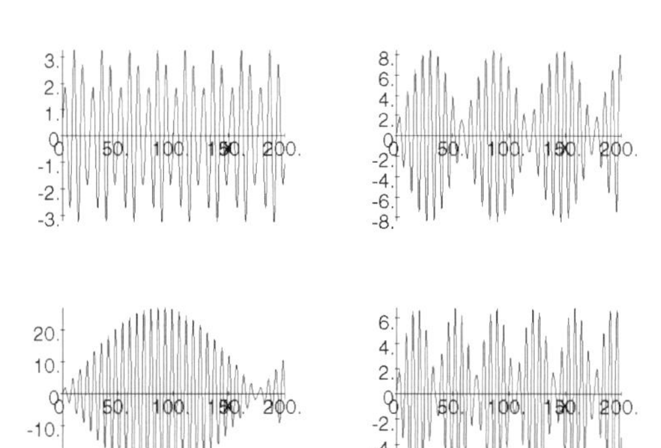

On the other hand, if $\omega = 1$, $S = \{x\cos x, x\sin x\}$, and because $\sin x$ appears as a term in y_h, we multiply each term in S by x to obtain $S' = \{x\cos x, x\sin x\}$, so a particular solution to the nonhomogeneous equation has the form $y_p(x) = Ax\cos x + Bx\sin x$. Proceeding as before, we substitute into the nonhomogeneous equation.

```
> yp:=x->A*x*cos(x)+B*x*sin(x);
```

$$yp := x \to A\,x\cos(x) + B\,x\sin(x)$$

Again, this equation is true for all values of x

```
> eq:=diff(yp(x),x$2)+yp(x)=sin(x);
```

$$eq := -2\,A\sin(x) + 2\,B\cos(x) = \sin(x)$$

so we evaluate it for two values of x and solve for A and B.

```
> eq1:=eval(subs(x=0,eq));
  eq2:=eval(subs(x=Pi/2,eq));
```

$$eq1 := 2\,B = 0$$
$$eq2 := -2\,A = 1$$

```
> abvals:=solve({eq1,eq2},{A,B});
```

$$abvals := \{A = \frac{-1}{2}, B = 0\}$$

The general solution is formed by computing $y = y_h + y_p$.

```
> y:=rhs(yh)+subs(abvals,yp(x));
```

$$y := \_C1\cos(x) + \_C2\sin(x) - \frac{1}{2}x\cos(x)$$

We then determine the values of the arbitrary constants so that the initial conditions are satisfied.

```
> sys:=eval({subs(x=0,y)=0,subs(x=0,diff(y,x))=1});
```

$$sys := \{-\frac{1}{2} + \_C2 = 1, \_C1 = 0\}$$

```
> cvals:=solve(sys,{_C1,_C2});
```

$$cvals := \{ \_C2 = \frac{3}{2}, \_C1 = 0 \}$$

We then form the solution to the initial-value problem.

```
> y2:=subs(cvals,y);
```

$$y2 := \frac{3}{2}\sin(x) - \frac{1}{2}x\cos(x)$$

Alternatively, we can also use `dsolve` to solve the initial-value problem $\begin{cases} y'' + y = \sin \omega x \\ y(0) = 0, y'(0) = 1 \end{cases}$.

```
> y:='y':
  solb:=dsolve({diff(y(x),x$2)+y(x)=sin(x),
       y(0)=0,D(y)(0)=1},y(x));
```

$$solb := y(x) = \left(\frac{1}{2}\cos(x)\sin(x) - \frac{1}{2}x\right)\cos(x) - \frac{1}{2}\cos(x)^2\sin(x) + \frac{3}{2}\sin(x)$$

Notice how the solution dramatically changes if $\omega = 1$: if $\omega = 1$, the solution is unbounded.

```
> plot(rhs(solb),x=0..200);
```

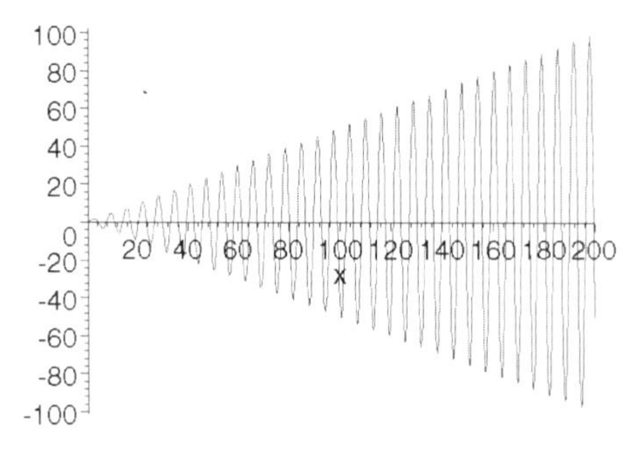

■

Nonhomogeneous Equations with Constant Coefficients: Variation of Parameters

Let $p(x)$, $q(x)$, and $f(x)$ be continuous on an interval I. The associated homogeneous equation of $y''(x)+p(x)y'(x)+q(x)y(x)=f(x)$ is $y''(x)+p(x)y'(x)+q(x)y(x)=0$. Let $S=\{y_1(x), y_2(x)\}$ be a fundamental set of solutions for the associated homogeneous equation of $y''(x)+p(x)y'(x)+q(x)y(x)=f(x)$. Let

$$u_1(x) = \int\frac{-y_2(x)f(x)}{y_1(x)y_2'(x)-y_1'(x)y_2(x)}dx = \int\frac{-y_2(x)f(x)}{W(S)}dx$$

and .

$$u_2(x) = \int\frac{y_1(x)f(x)}{y_1(x)y_2'(x)-y_1'(x)y_2(x)}dx = \int\frac{y_1(x)f(x)}{W(S)}dx.$$

Then, $y_p(x)=y_1(x)u_1(x)+y_2(x)u_2(x)$ is a particular solution of the nonhomogeneous equation $y''(x)+p(x)y'(x)+q(x)y(x)=f(x)$, and a general solution is $y(x)=c_1y_1(x)+c_2y_2(x)+y_p(x)$.

The `DEtools` package contains the `varparam` command that can be used to find a particular solution and, subsequently, the general solution of a nonhomogeneous equation. Given the nonhomogeneous equation

$$y^{(n)}+a_{n-1}(x)y^{(n-1)}+\ldots+a_2(x)y''+a_1(x)y'+a_0(x)y=f(x),$$

in standard form, and a fundamental set of solutions, S, to the corresponding homogeneous equation, ,

$$y^{(n)}+a_{n-1}(x)y^{(n-1)}+\ldots+a_2(x)y''+a_1(x)y'+a_0(x)y=0,$$

the command `varparam([S],f(x),x)` finds a particular solution and returns the general solution to the nonhomogeneous equation.

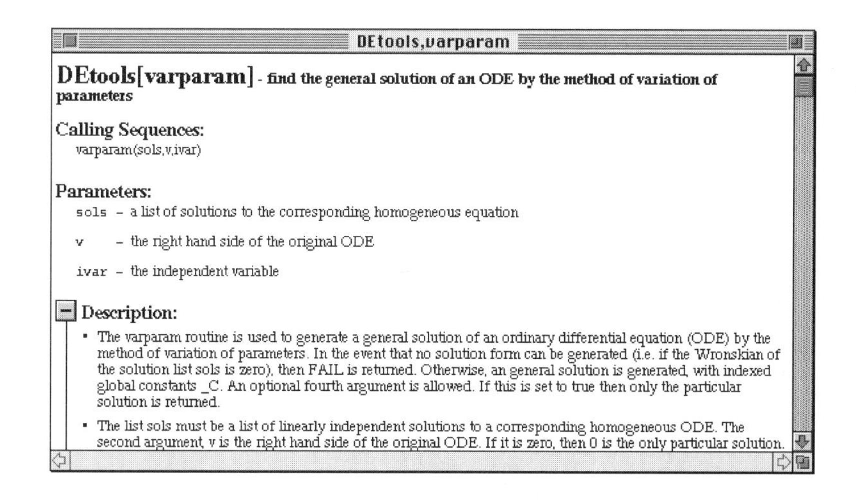

DEtools,varparam

DEtools[varparam] - find the general solution of an ODE by the method of variation of parameters

Calling Sequences:
varparam(sols,v,ivar)

Parameters:
sols - a list of solutions to the corresponding homogeneous equation
v - the right hand side of the original ODE
ivar - the independent variable

Description:
- The varparam routine is used to generate a general solution of an ordinary differential equation (ODE) by the method of variation of parameters. In the event that no solution form can be generated (i.e. if the Wronskian of the solution list sols is zero), then FAIL is returned. Otherwise, an general solution is generated, with indexed global constants _C. An optional fourth argument is allowed. If this is set to true then only the particular solution is returned.
- The list sols must be a list of linearly independent solutions to a corresponding homogeneous ODE. The second argument, v is the right hand side of the original ODE. If it is zero, then 0 is the only particular solution.

EXAMPLE 7: Solve (a) $y'' + 9y = \sec 3x$ and (b) the initial-value problem

$$\begin{cases} y'' + 9y = \sec 3x \\ y(0) = 0,\ y'(0) = 1 \end{cases}.$$

SOLUTION: (a) We illustrate three different approaches to the problem. First, we use Maple to assist us in implementing the method of variation of parameters when we search for a particular solution to the nonhomogeneous equation. In this case, we first find a general solution of the corresponding homogeneous equation, y_h, and see that a fundamental set of solutions is $S = \{y_1(x), y_2(x)\} = \{\cos 3x, \sin 3x\}$.

```
> y:='y':
  yh:=dsolve(diff(y(x),x$2)+9*y(x)=0,y(x));
```

$$yh := y(x) = \_C1 \cos(3x) + \_C2 \sin(3x)$$

We then define y_1, y_2, and $f(x) = \sec 3x$, and compute $W(S)$.

```
> y[1]:=x->cos(3*x):
  y[2]:=x->sin(3*x):
  with(linalg):
  ws:=simplify(det(
       Wronskian([y[1](x),y[2](x)],x)),trig);
  f:=x->sec(3*x):
```

$$ws := 3$$

We then compute $u'_1 = \dfrac{-y_2(x)f(x)}{W(X)}$, u'_1, $u'_2 = \dfrac{-y_1(x)f(x)}{W(S)}$, and u_2.

```
> u1prime:=-y[2](x)*f(x)/ws;
  u1:=int(u1prime,x);
  u2prime:=y[1](x)*f(x)/ws;
  u2:=int(u2prime,x);
```

$$u1prime := -\frac{1}{3}\sin(3x)\sec(3x)$$

$$u1 := \frac{1}{9}\ln(\cos(3x))$$

$$u2prime := \frac{1}{3}\cos(3x)\sec(3x)$$

$$u2 := \frac{1}{3}x$$

A particular solution of the nonhomogeneous equation is $y_p = u_1 y_1 + u_2 y_2$, and a general solution is $y = y_h + y_p$.

```
> yp:=u1*y[1](x)+u2*y[2](x);
  y:=rhs(yh)+yp;
```

$$yp := \frac{1}{9}\ln(\cos(3x))\cos(3x) + \frac{1}{3}x\sin(3x)$$

$$y := \frac{1}{9}\ln(\cos(3x))\cos(3x) + \frac{1}{3}x\sin(3x) + \_C1\cos(3x) + \_C2\sin(3x)$$

Alternatively, we can use `dsolve` to find a general solution of the equation

```
> y:='y':
  sola:=dsolve(diff(y(x),x$2)+9*y(x)=sec(3*x),y(x));
```

$$sola := y(x) = \frac{1}{9}\ln(\cos(3x))\cos(3x) + \frac{1}{3}x\sin(3x) + \_C1\cos(3x) + \_C2\sin(3x)$$

and the solution to the initial-value problem.

```
> solb:=dsolve({diff(y(x),x$2)+9*y(x)=sec(3*x),
     y(0)=0,D(y)(0)=1},y(x));
```

$$solb := y(x) = \frac{1}{9}\ln(\cos(3x))\cos(3x) + \frac{1}{3}x\sin(3x) + \frac{1}{3}\sin(3x)$$

We graph the solution obtained in `solb` on the interval $(-\pi/6, \pi/6)$ with `plot`.

```
> plot(rhs(solb),x=-Pi/6..Pi/6);
```

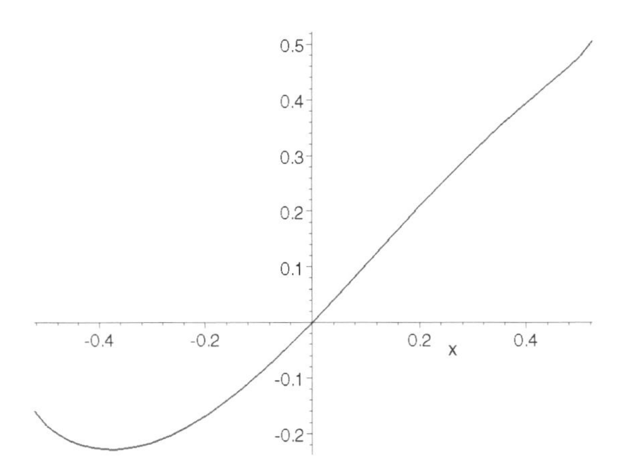

Last, we illustrate the use of `varparam`. First, we use `constcoeffsols` to find a fundamental set of solutions to the corresponding homogeneous equation.

```
> with(DEtools):
  y:='y':
  S:=constcoeffsols(diff(y(x),x$2)+9*y(x)=0,y(x));
```

$$S := [\cos(3x), \sin(3x)]$$

Then, entering

```
> varparam(S,sec(3*x),x);
```

$$\_C_1 \cos(3x) + \_C_2 \sin(3x) + \frac{1}{9} \ln(\cos(3x)) \cos(3x) + \frac{1}{3} x \sin(3x)$$

uses the method of variation of parameters to find a particular solution to the nonhomogeneous equation and returns the general solution to the equation.

■

Higher-order linear nonhomogeneous equations can be solved through variation of parameters as well.

EXAMPLE 8: Solve $y''' + 6y'' + 12y' + 8y = e^{-2x}\tan^{-1}x$.

SOLUTION: First, we use `dsolve` to find a general solution of this third-order linear nonhomogeneous equation.

```
odeadvisor(diff(y(x),x$3)+6*diff(y(x),x$2)+
>       12*diff(y(x),x)+8*y(x)=exp(-2*x)*arctan(x));
```

$$[[\_3rd\_order, \_linear, \_nonhomogeneous]]$$

```
> dsolve(diff(y(x),x$3)+6*diff(y(x),x$2)+
      12*diff(y(x),x)+8*y(x)=exp(-2*x)*arctan(x),y(x));
```

$$y(x) = \frac{1}{6}e^{(-2x)}x^3\arctan(x) + \frac{5}{12}e^{(-2x)}x^2 + \frac{1}{12}e^{(-2x)}\ln(x^2+1) - \frac{1}{4}e^{(-2x)}x^2\ln(x^2+1) - \frac{1}{2}e^{(-2x)}x\arctan(x) + \_C1\,e^{(-2x)} + \_C2\,e^{(-2x)}x^2 + \_C3\,e^{(-2x)}x$$

Alternatively, we illustrate the use of `varparam`. As in Example 7, we use `constcoeffsols` to find a fundamental set of solutions to the corresponding homogeneous equation, and then we use `varparam` to implement the method of variation of parameters to find a particular solution to the nonhomogeneous equation and obtain the general solution to the equation.

```
> with(DEtools):
  S:=constcoeffsols(diff(y(x),x$3)+6*diff(y(x),x$2)+
      12*diff(y(x),x)+8*y(x)=0,y(x));
```

$$S := [e^{(-2x)}, e^{(-2x)}x, e^{(-2x)}x^2]$$

```
> varparam(S,exp(-2*x)*arctan(x),x);
```

$$\_C_1\,e^{(-2x)} + \_C_2\,e^{(-2x)}x + \_C_3\,e^{(-2x)}x^2 + \left(\frac{1}{6}x^3\arctan(x) - \frac{1}{12}x^2 + \frac{1}{12}\ln(x^2+1)\right)e^{(-2x)} + \left(-\frac{1}{2}x^2\arctan(x) + \frac{1}{2}x - \frac{1}{2}\arctan(x)\right)e^{(-2x)}x + \left(\frac{1}{2}x\arctan(x) - \frac{1}{4}\ln(x^2+1)\right)e^{(-2x)}x^2$$

■

Application: Harmonic Motion

Suppose that a mass is attached to an elastic spring that is suspended from a rigid support such as a ceiling. According to Hooke's Law, the spring exerts a restoring force in the upward direction that is proportional to the displacement of the spring; that is, $F = ks$, where $k > 0$ is the constant of proportionality or spring constant, and s is the displacement of the spring.

Using Hooke's Law and assuming that $x(t)$ represents the displacement of the mass from the equilibrium position at time t, we obtain the initial-value problem .

$$\begin{cases} mx'' + kx = 0 \\ x(0) = \alpha, x'(0) = \beta \end{cases}$$

Note that the initial conditions give the initial displacement and velocity, respectively. This differential equation disregards all retarding forces acting on the motion of the mass and a more realistic model that takes these forces into account is needed. Studies in mechanics reveal that resistive forces due to damping are proportional to a power of the velocity of the motion. Hence, $F_R = cdx/dt$ or $F_R = c(dx/dt)^3$, where $c > 0$, are typically used to represent the damping force. Then, we have the following initial-value problem assuming that $F_R = cdx/dt$: .

$$\begin{cases} m\dfrac{d^2x}{dt^2} + c\dfrac{dx}{dt} + kx = 0 \\ x0 = \alpha, \dfrac{dx}{dt}0 = \beta \end{cases}$$

Problems of this type are characterized by the value of $c^2 - 4mk$ as follows:

1. $c^2 - 4mk > 0$. This situation is said to be **overdamped** because the damping coefficient c is large incomparison with the spring constant k.
2. $c^2 - 4mk = 0$. This situation is described as **critically damped** because the resulting motion is oscillatory with a slight decrease in the damping coefficient c.
3. $c^2 - 4mk < 0$. This situation is called **underdamped** because the damping coefficient c is small in comparison with the spring constant k.

EXAMPLE 9: Classify the following differential equations as overdamped, underdamped, or critically damped. Also, solve the corresponding initial-value problem using the given initial conditions, and investigate the behavior of the solutions.

(a) $\frac{d^2x}{dt^2} + 8\frac{dx}{dt} + 16x = 0$ subject to $x(0) = 0$ and $\frac{dx}{dt}(0) = 1$;

(b) $\frac{d^2x}{dt^2} + 5\frac{dx}{dt} + 4x = 0$ subject to $x(0) = 1$ and $\frac{dx}{dt}(0) = 1$; and

(c) $\frac{d^2x}{dt^2} + \frac{dx}{dt} + 16x = 0$ subject to $x(0) = 0$ and $\frac{dx}{dt}(0) = 1$.

SOLUTION: For (a), we identify $m = 1$, $c = 8$, and $k = 16$ so that $c^2 - 4mk = 0$, which means that the differential equation $x'' + 8x' + 16x = 0$ is critically damped. After defining DEOne, we solve the equation subject to the initial conditions and name the resulting output sola.

```
> m:=1:c:=8:k:=16:
  c^2-4*m*k;
```

$$0$$

```
> x:='x':
  DEOne:=diff(x(t),t$2)+8*diff(x(t),t)+16*x(t)=0:
> sola:=dsolve({DEOne,x(0)=0,D(x)(0)=1},x(t));
```

$$sola := x(t) = e^{(-4t)}\,t$$

We use assign to define $x(t)$ to be the explicit solution obtained in sola. We then graph $x(t)$ on the interval [0,4].

```
> assign(sola):
  plot(x(t),t=0..4);
```

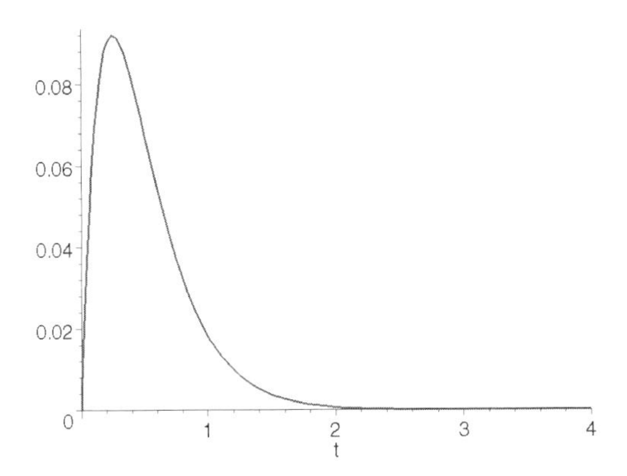

For (b), we proceed in the same manner. We identify $m = 1$, $c = 5$, and $k = 4$, so that $c^2 - 4mk = 9$ and the equation $x'' + 5x' + 4x = 0$ is overdamped. We then define `DETwo` to be the equation and the solution of the equation obtained with `dsolve`, `solb`. In the same manner, we use `assign` and `plot` to define $x(t)$ and then graph $x(t)$ on the interval [0,6].

```
> m:=1:c:=5:k:=4:
  c^2-4*m*k;
```

$$9$$

```
> x:='x':
  DETwo:=diff(x(t),t$2)+5*diff(x(t),t)+4*x(t)=0:
  solb:=dsolve({DETwo,x(0)=1,D(x)(0)=-1},x(t));
  assign(solb):
  plot(x(t),t=0..6);
```

$$solb := \mathrm{x}(t) = \mathbf{e}^{(-t)}$$

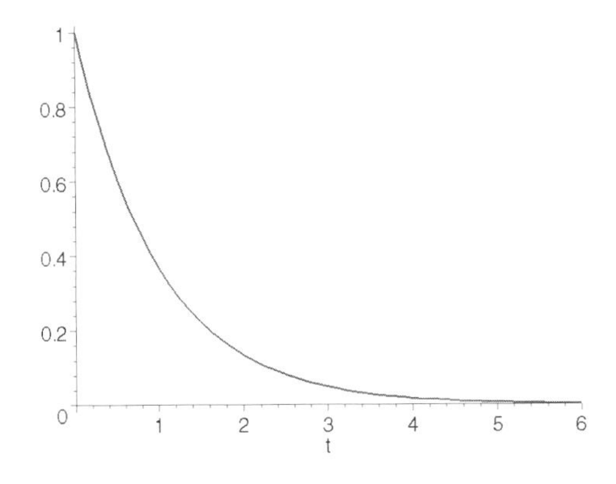

For (c), we proceed in the same manner as in (a) and (b) to show that the equation is underdamped because the value of $c^2 - 4mk$ is –63.

```
> m:=1:c:=1:k:=16:
  c^2-4*m*k;
  x:='x':
  DEThree:=diff(x(t),t$2)+diff(x(t),t)+16*x(t)=0:
  solc:=dsolve({DEThree,x(0)=0,D(x)(0)=1},x(t));
  assign(solc):
  plot(x(t),t=0..6);
```

$$-63$$

$$solc := \mathrm{x}(t) = \frac{2}{21}\sqrt{7}\,\mathbf{e}^{(-1/2\,t)}\sin\left(\frac{3}{2}\sqrt{7}\,t\right)$$

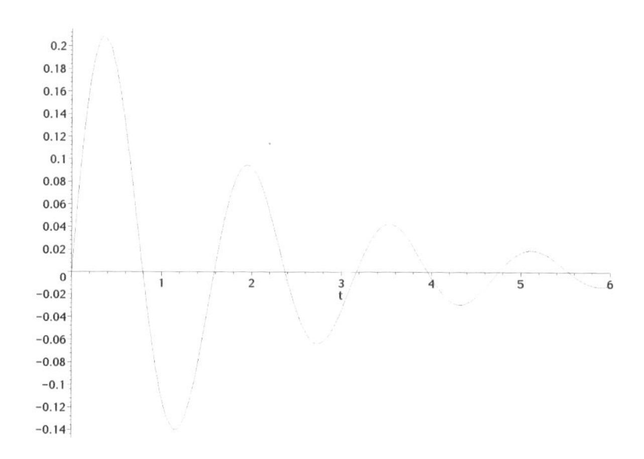

■

Numerical Solutions of Higher-Order Ordinary Differential Equations

Numerical approximations of solutions to higher-order ordinary differential equations can be obtained with `dsolve` together with the `numeric` option in the same way that numerical solutions of first-order equations are obtained with `dsolve` together with the `numeric` option. This command is particularly useful for work with nonlinear equations for which dsolve is unable to find an explicit solution. This command is entered in the form

```
dsolve({deq,ics},var,numeric),
```

where deq represents the differential equation, ics represents the initial conditions, and var represents the variable name of the solution of the differential equation. Note that the number of initial conditions in ics must equal the order of the differential equation indicated in deq.

EXAMPLE 10: Sketch the graph of $x(t)$ on the interval [0,30] if $x(t)$ satisfies the initial-value problem

$$\begin{cases} \dfrac{(d^2 x)}{(dt^2)} + 4e^{-t/4}x = 0 \\ x(0) = 1, \dfrac{dx}{dt}(0) = 0 \end{cases}.$$

What is the value of $x(15)$?

SOLUTION: After defining eq to be the equation $x'' + 4e^{-t/4}x = 0$, we see that dsolve is able both to find a general solution of the equation and to solve the initial-value problem, although the result is given in terms of the Bessel functions BesselJ and BesselY. Note that BesselJ, $J_n(x)$, and BesselY, $Y_n(x)$, are the Bessel functions of the first and second kind, respectively, and are linearly independent solutions to the differential equation $x^2y'' + xy' + (x^2 - n^2)y = 0$.

```
> eq:=diff(x(t),t$2)+4*exp(-t/4)*x(t)=0:
  dsolve(eq,x(t));
```

$$x(t) = \_C1 \text{ BesselJ}(0, 16\sqrt{e^{(-1/4\,t)}}) + \_C2 \text{ BesselY}(0, 16\sqrt{e^{(-1/4\,t)}})$$

```
> dsolve({eq,x(0)=1,D(x)(0)=0},x(t));
```

$$x(t) = \frac{\text{BesselY}(1, 16)\,\text{BesselJ}(0, 16\sqrt{e^{(-1/4\,t)}})}{-\text{BesselJ}(1, 16)\,\text{BesselY}(0, 16) + \text{BesselY}(1, 16)\,\text{BesselJ}(0, 16)} - \frac{\text{BesselJ}(1, 16)\,\text{BesselY}(0, 16\sqrt{e^{(-1/4\,t)}})}{-\text{BesselJ}(1, 16)\,\text{BesselY}(0, 16) + \text{BesselY}(1, 16)\,\text{BesselJ}(0, 16)}$$

We choose to proceed numerically. First, we use `dsolve` together with the `numeric` option to generate a numerical solution to the equation that we name `numsol`.

```
> numsol:=dsolve({eq,x(0)=1,D(x)(0)=0},x(t),numeric);
```

$$numsol := \mathbf{proc}(rkf45\_x) \ \ldots \ \mathbf{end}$$

Entering

```
> numsol(15);
```

$$t = 15, \mathrm{x}(t) = 1.258118099833521, \frac{\partial}{\partial t}\mathrm{x}(t) = -.6099409737401172$$

returns an ordered triple that corresponds to $t = 15$, the value of $x(t)$ if $t = 15$, and the value of dx/dt if $t = 15$. $x(15)$ is the second element of this ordered triple

```
> numsol(15)[2];
```

$$\mathrm{x}(t) = 1.258118099833521$$

and obtained with `rhs`.

```
> rhs(numsol(15)[2]);
```

1.258118099833521

We use the `odeplot` command that is contained in the `plots` package to graph the numerical solution t versus $x(t)$ for $0 \leq t \leq 30$.

```
> with(plots):
  odeplot(numsol,[t,x(t)],0..30);
```

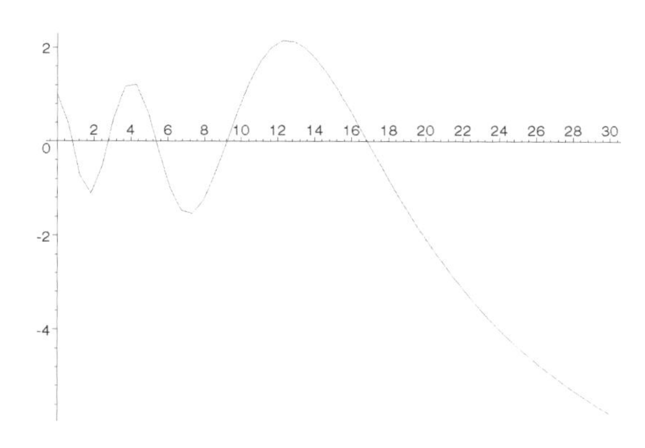

■

Application: The Simple Pendulum

Another situation that leads to a second-order ordinary differential equation is that of the simple pendulum. In this case, a mass m is attached to the end of a rod of length L, which is suspended from a rigid support. Because the motion is best described in terms of the angular displacement θ, we let $\theta = 0$ correspond to the rod hanging vertically. The objective is to find the motion of the mass as a function of θ, an initial displacement from the vertical (equilibrium) position, and an initial velocity. Assuming that the pendulum is allowed to rotate without friction, the only force acting on the pendulum is that due to gravity. Newton's Second Law and the relationship $s = L\theta$ are used to establish the initial-value problem that models this situation: .

$$\begin{cases} L\dfrac{d^2\Theta}{dt^2} + g\sin\Theta = 0 \\ \Theta(0) = \Theta_0, \dfrac{d\Theta}{dt}(0) = v_0 \end{cases}.$$

This differential equation is nonlinear. However, this nonlinear equation can be approximated with a linear differential equation by making use of the power series expansion for $\sin\theta$, which is given by: $\sin\Theta = \sum_{n=0}^{\infty} (-1)^n \dfrac{1}{(2n+1)!}\Theta^{2n+1} = \Theta - \dfrac{1}{3!}\Theta^3 + \dfrac{1}{5!}\Theta^5 + \ldots$.

Hence, for small displacements, we have the approximation $\sin\theta \approx \theta$. Therefore, the initial-value problem becomes

$$\begin{cases} L\dfrac{d^2\Theta}{dt^2} + g\Theta = 0 \\ \Theta(0) = \Theta_0, \dfrac{d\Theta}{dt}(0) = v_0 \end{cases}.$$

This is a linear second-order initial-value problem and is solved easily.

```
> dsolve({L*diff(theta(t),t$2)+g*theta(t)=0,
     theta(0)=theta[0],D(theta)(0)=v[0]},theta(t));
```

$$\theta(t) = \frac{v_0 \sin\left(\sqrt{\frac{g}{L}}\, t\right)}{\sqrt{\frac{g}{L}}} + \theta_0 \cos\left(\sqrt{\frac{g}{L}}\, t\right)$$

If the pendulum undergoes a damping force that is proportional to the instantaneous velocity, the force due to damping is given as $F_D = -bd\theta/dt$. Incorporating this force into the sum of the forces acting on the pendulum, we have

$$\begin{cases} L\dfrac{d^2\theta}{dt^2} + b\dfrac{d^2\theta}{dt} + g\sin\theta = 0 \\ \theta(0) = \theta_0, \theta'(0) = v_0 \end{cases}.$$

We now investigate the properties of this nonlinear differential equation.

EXAMPLE 11: Use dsolve to investigate the solutions to the damped pendulum problem $\dfrac{d^2\theta}{dt^2} + 0.50\dfrac{d\theta}{dt} + 16\sin\theta = 0$ subject to the initial conditions $\theta(0) = \theta_0$ and $\theta'(0) = \mathbf{v}_0$ using the following initial conditions: $\theta(0) = \mathrm{i}$ and $\theta'(0) = 0$ for $\mathrm{i} = 0.5, 1.0, 1.5, 2.0, 3.5, 3.0,$ and 3.5.

SOLUTION: Notice that, in this case, the damping coefficient is relatively small compared to the other coefficients. The differential equation is defined below as Eq. The solution to the problem is named Sol[1] and is obtained through the use of dsolve. Notice that the result of the dsolve command with the numeric option setting is in the form of a procedure. The values (t,3248136,[2]) within this procedure represent the independent variable of the equation, an arbitrary machine number, and the order of the equation.

```
> Eq:=diff(x(t),t$2)+.25*diff(x(t),t)+sin(x(t))=0:
  Sol:=table():
  Sol[1]:=dsolve({Eq,x(0)=1,D(x)(0)=0},x(t),
      numeric);
```

$$Sol_1 := \mathbf{proc}(rkf45\_x)\ \ldots\ \mathbf{end}$$

Note that Sol[1](t) yields an ordered triple: the first element indicates the value of t, the second corresponds to the value of $x(t)$, and the third the value of dx/dt. For example, we show the result if $t = 1$.

```
> Sol[1](1);
```

$$\left[t=1, \mathrm{x}(t)=.6310083041452602, \frac{\partial}{\partial t}\mathrm{x}(t)=-.6686014021833726\right]$$

To graph the solution, we take advantage of the `odeplot` command contained in the `plots` package. We first load the `plots` package and then use `odeplot` to graph the numerical solution obtained in `Sol[1]` for $0 \leq t \leq 15$.

```
> with(plots):
  Graph:=table():
  Graph[1]:=odeplot(Sol[1],[t,x(t)],0..15):
  display({Graph[1]});
```

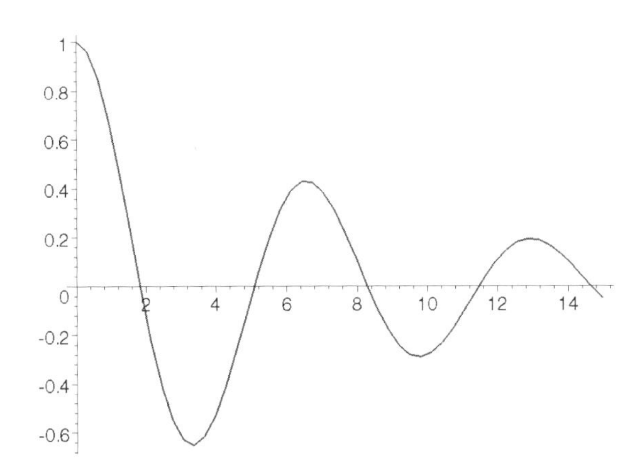

To graph the solution corresponding to various initial conditions, we first define `inits1` to be a list of ordered pairs. For each ordered pair in `inits1`, we will graph the solution satisfying $x(0)$ equal to the first coordinate of the ordered pair and $x'(0)$ equal to the second coordinate of the ordered pair. The number of ordered pairs in `inits1` is given by `nops(inits1)`. When entered, this command indicates that there are seven ordered pairs. This value is used in later calculations.

```
> inits1:=[[.5,0],[1,0],[1.5,0],[2,0],[2.5,0],
        [3,0],[3.5,0]];
```

$$inits1 := [[.5,0],[1,0],[1.5,0],[2,0],[2.5,0],[3,0],[3.5,0]]$$

```
> nops(inits1);
```

7

Then, for each ordered pair in inits1, we define Sol[i] to be the numerical solution of the differential equation subject to the initial conditions $x(0)$ inits[i][1] and $x'(0)$ inits[i][2]. Note that inits[i][1] corresponds to the first coordinate of the ith ordered pair of inits1; inits[i][2] corresponds to the second coordinate of the ith ordered pair of inits1. Graph[i] is the graph of Sol[i] on the interval [0, 15]. Note that Graph[i] is not displayed because a colon (:) is included at the end of the command. This is necessary because display, which is contained in Maple's plots package, must be used to show named graphs. Finally, we display the seven graphs simultaneously by using the display command. Note that the plots package, which was loaded above, is loaded by entering with(plots).

```
> i:='i':j:='j':
  for i to 7 do
      Sol[i]:=dsolve(
      {Eq,x(0)=inits1[i][1],D(x)(0)=inits1[i][2]},
         x(t),numeric):
      Graph[i]:=odeplot(Sol[i],[t,x(t)],0..15): od:
> display({seq(Graph[j],j=1..7)});
```

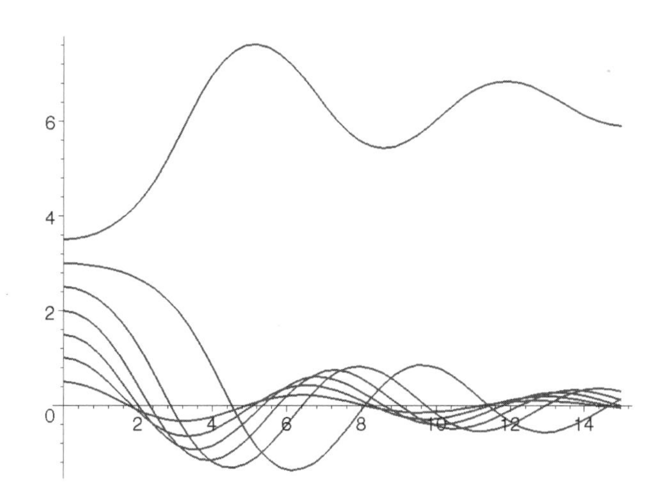

■

In the previous examples, we have illustrated how to evaluate the function returned by using the dsolve command together with the numeric option. Frequently, however, we may not need to use the function returned by the dsolve command. In those cases, as when we need to graph only the solution to an equation, Maple supplies other commands contained in the DEtools package that we can use to directly graph the solution to an equation.

Application: Hard Springs

In the case of a **hard spring,** the spring force strengthens with compression or extension. For springs of this type, we model the physical system with the nonlinear initial-value problem

$$\begin{cases} m\dfrac{d^2x}{dt^2} + c\dfrac{d^2x}{dt} + kx + jx^3 = f(t) \\ x(0) = \alpha, \dfrac{dx}{dt}(0) = \beta \end{cases},$$

where j is a positive constant.

EXAMPLE 12: Approximate the solution to

$$\begin{cases} \dfrac{d^2x}{dt^2} + 0.3x + 0.04x^3 = (0) \\ x(0) = \alpha, \dfrac{dx}{dt}(0) = \beta \end{cases},$$

for various values of α and β in the initial conditions.

SOLUTION: First, we define the undamped nonlinear differential equation in eq, and then we load the DEtools package so that we can use the DEplot command.

We graph the solution using the initial conditions defined in inits1 with DEplot. Notice that solutions with larger amplitudes have smaller periods, as expected with a hard spring.

```
> inits1:=[seq([x(0)=i,D(x)(0)=0],i=1..5)];
```

$inits1 := [[x(0) = 1, D(x)(0) = 0], [x(0) = 2, D(x)(0) = 0], [x(0) = 3, D(x)(0) = 0],$
$[x(0) = 4, D(x)(0) = 0], [x(0) = 5, D(x)(0) = 0]]$

```
> DEplot(eq,x(t),t=0..15,inits1,
      linecolor=BLACK,stepsize=0.1);
```

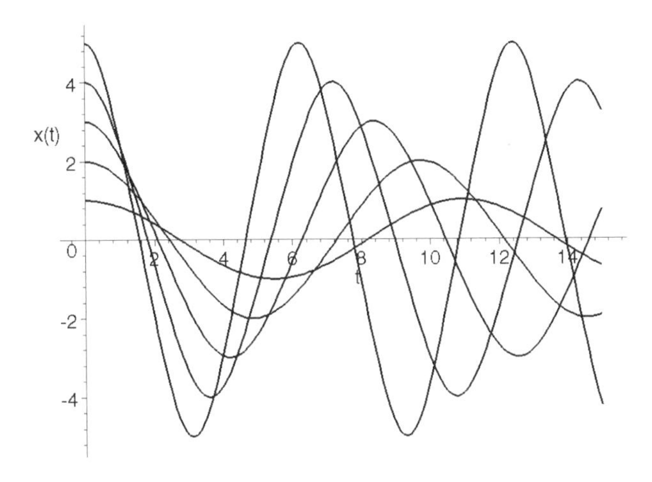

In a similar manner, we use the initial conditions in `inits2` as the initial velocity β. Again, we see that when the amplitude is large, the spring strengthens so that the period of the motion is decreased.

```
> inits2:=[seq([x(0)=0,D(x)(0)=i],i=1..5)]:
> DEplot(eq,x(t),t=0..15,inits2,
      linecolor=BLACK,stepsize=0.1);
```

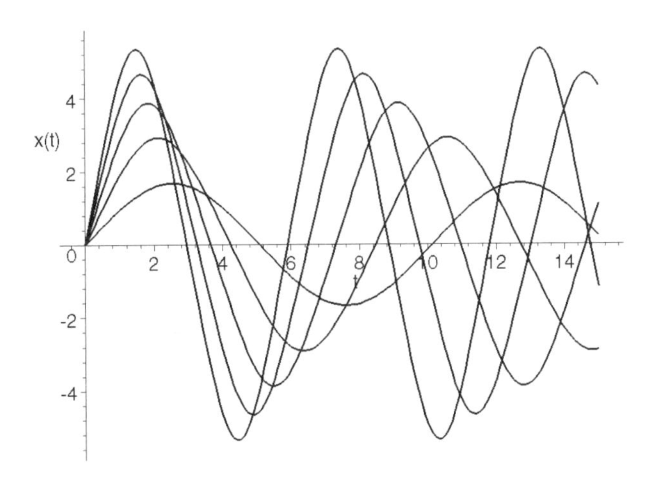

■

7.3 Series Solutions of Ordinary Differential Equations

Many users will find numerical solutions obtained using `dsolve` together with the `numeric` option adequate for most purposes. In some situations, however, a series solution may be useful, and in those cases, it can be obtained using `dsolve` together with the `'type=series'` option.

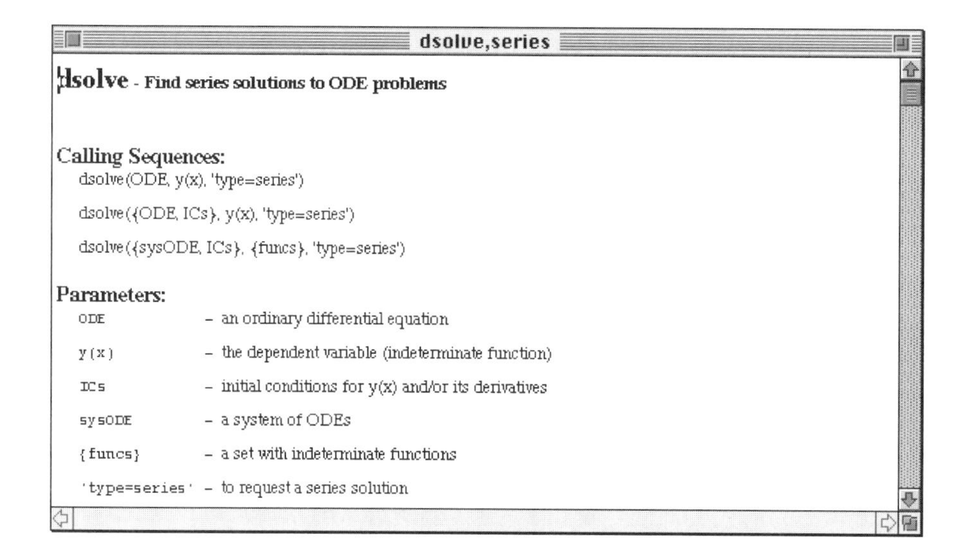
dsolve,series

dsolve - Find series solutions to ODE problems

Calling Sequences:

dsolve(ODE, y(x), 'type=series')

dsolve({ODE, ICs}, y(x), 'type=series')

dsolve({sysODE, ICs}, {funcs}, 'type=series')

Parameters:

ODE	- an ordinary differential equation
y(x)	- the dependent variable (indeterminate function)
ICs	- initial conditions for y(x) and/or its derivatives
sysODE	- a system of ODEs
{funcs}	- a set with indeterminate functions
'type=series'	- to request a series solution

Series Solutions about Ordinary Points

Consider the equation $a_2(x)y'' + a_1(x)y' + a_0(x)y = 0$ and let $p(x) = a_1/a_2(x)$ and $q(x) = a_0(x)/a_2(x)$. Then, $a_2(x)y'' + a_1(x)y' + a_0(x)y = 0$ is equivalent to $y'' + p(x)y' + q(x)y = 0$, which is called the **standard form** of the equation. If a number x_0 is said to be an **ordinary point** of the differential equation, this means that both $p(x)$ and $q(x)$ are analytic at x_0. If x_0 is not an ordinary point, x_0 is called a **singular point**.

If x_0 is an ordinary point of the differential equation $y'' + p(x)y' + q(x)y = 0$,, we can write

$p(x) = \sum_{n=0}^{\infty} b_n(x-x_0)^n$, where $b_n = \dfrac{p^{(n)}(x_0)}{n!}$, and $q(x) = \sum_{n=0}^{\infty} c_n(x-x_0)^n$, where

$c_n = \frac{q^{(n)}(x_0)}{n!}$. Substitution into the equation $y'' + p(x)y' + q(x)y = 0$ results in

$$y'' + y' + \sum_{n=0}^{\infty} b_n(x - x_0)^n + y\sum_{n=0}^{\infty} c_n(x - x_0)^n = 0.$$

If we assume that y is analytic at x_0, we can write $y(x) = \sum_{n=0}^{\infty} a_n(x - x_0)^n$. Because a power series can be differentiated term by term, we can compute the first and second derivatives of y and substitute back into the equation to calculate the coefficients a_n. Thus, we obtain a power series solution of the equation.

EXAMPLE 1: Let k and m be fixed integers. **Legendre's equation** is the equation

$$(1 - x^2)\frac{d^2y}{dx^2} - 2x\frac{dy}{dx} + \left[k(k+1) - \frac{m^2}{1 - x^2}\right]y = 0.$$

If $m = 0$, Legendre's equation becomes $(1 - x^2)\frac{d^2y}{dx^2} - 2x\frac{dy}{dx} + k(k+1)y = 0$. Find a general solution to Legendre's equation if $m = 0$.

SOLUTION: In standard form, the equation is $y'' - \frac{2x}{1 - x^2} + \frac{k(k+1)}{1 - x^2} = 0$. There is a solution to the equation of the form $y = \sum_{n=0}^{\infty} a_n x^n$ because $x = 0$ is an ordinary point. This solution will converge at least on the interval (–1,1) because the closest singular points to $x = 0$ are $x = \pm 1$. Substitution of this function and its derivatives $y' = \sum_{n=0}^{\infty} (n+1)a_{n+1}x^n$ and $y'' = \sum_{n=0}^{\infty} (n+2)a_{n+2}x^n$ into the differential equation and simplifying the results yields

$$[2a_2 + k(k+1)a_0]x^0 + [-2a_1 + k(k+1)a_1 + 6a_3]x.$$

$$+\sum_{n=4}^{\infty} n(n-1)a_n x^{n-2} - \sum_{n=2}^{\infty} n(n-1)a_n x^n - \sum_{n=2}^{\infty} 2na_n x^n + \sum_{n=2}^{\infty} k(k+1)a_n x^n = 0.$$

After substituting $n+2$ for each occurrence of n in the first series and simplifying, we have

$$[2a_2 + k(k+1)a_0]x^0 + [-2a_1 + k(k+1)a_1 + 6a_3]x$$

$$+\sum_{n=2}^{\infty} \{(n+2)(n+1)a_{n+2} + [-n(n-1) - 2n + k(k+1)]a_n\}x^n = 0.$$

Equating the coefficients to zero, we find a_2, a_3, and a_{n+2} with `solve`.

```
solve(2*a[2]+k*(k+1)*a[0]=0,a[2]);
```

$$-\frac{1}{2}k(k+1)a_0$$

```
> solve(-2*a[1]+k*(k+1)*a[1]+6*a[3]=0,a[3]);
```

$$\frac{1}{3}a_1 - \frac{1}{6}k^2 a_1 - \frac{1}{6}k\, a_1$$

```
> genform:=solve((n+1)*(n+1)*a[n+2]+
      (-n*(n-1)-2*n+k*(k+1))*a[n]=0,a[n+2]);
```

$$genform := -\frac{(-n^2 - n + k^2 + k)\, a_n}{(n+1)^2}$$

We obtain a formula for a_n by replacing each occurrence of n in a_{n+2} by $n-2$.

```
> subs(n=n-2,genform);
```

$$-\frac{(-(n-2)^2 - n + 2 + k^2 + k)\, a_{n-2}}{(n-1)^2}$$

Notice how we use the `remember` option in the procedure that defines the recursively defined function a.

```
> a:= proc(n) option remember;
        -(-(n-2)^2-n+2+k^2+k)*a(n-2)/((n-1)^2) end:
```

Using this formula, we find several coefficients with `seq`.

```
> a(1):=a[1]:
  a(0):=a[0]:
  array([seq([i,a(i)],i=2..10
```

$$\begin{bmatrix} 2 & -(k^2+k)\,a_0 \\ 3 & -\frac{1}{4}(-2+k^2+k)\,a_1 \\ 4 & \frac{1}{9}(-6+k^2+k)\,(k^2+k)\,a_0 \\ 5 & \frac{1}{64}(-12+k^2+k)\,(-2+k^2+k)\,a_1 \\ 6 & -\frac{1}{225}(-20+k^2+k)\,(-6+k^2+k)\,(k^2+k)\,a_0 \\ 7 & -\frac{1}{2304}(-30+k^2+k)\,(-12+k^2+k)\,(-2+k^2+k)\,a_1 \\ 8 & \frac{1}{11025}(-42+k^2+k)\,(-20+k^2+k)\,(-6+k^2+k)\,(k^2+k)\,a_0 \\ 9 & \frac{1}{147456}(-56+k^2+k)\,(-30+k^2+k)\,(-12+k^2+k)\,(-2+k^2+k)\,a_1 \\ 10 & -\frac{1}{893025}(-72+k^2+k)\,(-42+k^2+k)\,(-20+k^2+k)\,(-6+k^2+k)\,(k^2+k)\,a_0 \end{bmatrix}$$

Hence, we have the two linearly independent solutions

$$y_1(x) = a_0\Big(1 - \frac{k(k+1)}{2!}x^2 + \frac{(2-k)(3+k)k(k+1)}{4!}x^4$$

$$-\frac{(4-k)(5+k)(2-k)(3+k)k(k+1)}{6!}x^6 + \dots\Big)$$

and

$$y_2(\mathrm{x}) = a_1\left(x - \frac{(k-1)(k+2)}{6}x^3 + \frac{(3-k)(4+k)(k-1)(k+2)}{5!}x^5\right.$$
$$\left.\frac{(5-k)(6+k)(3-k)(4+k)(k-1)(k+2)}{7!}x^7 + \ldots\right),$$

so a general solution is

$$y = a_0\left(1 - \frac{k(k+1)}{2!}x^2 + \frac{(2-k)(3+k)k(k+1)}{4!}x^4 - \frac{(4-k)(5+k)(2-k)(3+k)k(k+1)}{6!}x^6 + \ldots\right)$$
$$+a_1\left(x - \frac{(k-1)(k+2)}{6}x^3 + \frac{(3-k)(4+k)(k-1)(k+2)}{5!}x^5\right.$$
$$\left.-\frac{(5-k)(6+k)(3-k)(4+k)(k-1)(k+2)}{7!}x^7 + \ldots\right).$$

We see that `dsolve` is able to find a general solution of the equation

```
> sola:=dsolve((1-x^2)*diff(y(x),x$2)-
     2*x*diff(y(x),x)+k*(k+1)*y(x)=0,y(x));
```

$$sola := \mathrm{y}(x) = \_C1\ \mathrm{LegendreP}(k, x) + \_C2\ \mathrm{LegendreQ}(k, x)$$

as well as a series solution when we include the option `'type=series'`.

```
> solb:=dsolve((1-x^2)*diff(y(x),x$2)-
     2*x*diff(y(x),x)+k*(k+1)*y(x)=0,y(x),
        'type=series');
```

$$solb := \mathrm{y}(x) = \mathrm{y}(0) + \mathrm{D}(y)(0)\,x + \left(-\frac{1}{2}k^2\,\mathrm{y}(0) - \frac{1}{2}k\,\mathrm{y}(0)\right)x^2 + \left(-\frac{1}{6}k\,\mathrm{D}(y)(0) - \frac{1}{6}k^2\,\mathrm{D}(y)(0) + \frac{1}{3}\mathrm{D}(y)(0)\right)x^3 +$$
$$\left(-\frac{5}{24}k^2\,\mathrm{y}(0) + \frac{1}{24}k^4\,\mathrm{y}(0) + \frac{1}{12}k^3\,\mathrm{y}(0) - \frac{1}{4}k\,\mathrm{y}(0)\right)x^4 +$$
$$\left(\frac{1}{60}k^3\,\mathrm{D}(y)(0) + \frac{1}{120}k^4\,\mathrm{D}(y)(0) - \frac{13}{120}k^2\,\mathrm{D}(y)(0) - \frac{7}{60}k\,\mathrm{D}(y)(0) + \frac{1}{5}\mathrm{D}(y)(0)\right)x^5 + \mathrm{O}(x^6)$$

In the first result, `LegendreP` and `LegendreQ` represent the Legendre functions of the first and second kind, respectively, and are linearly independent solutions to the equation.

■

An interesting observation from the general solution to Legendre's equation is that the series solutions terminate for integer values of k. If k is an even integer, the first series terminates, while if k is an odd integer the second series terminates. Therefore, polynomial solutions are found for integer values of k.

We compute these polynomial solutions for several values of k using a Maple spreadsheet inserted into the Maple notebook. Spreadsheets can be inserted into a Maple notebook by going to **Insert** followed by **Spreadsheet**.

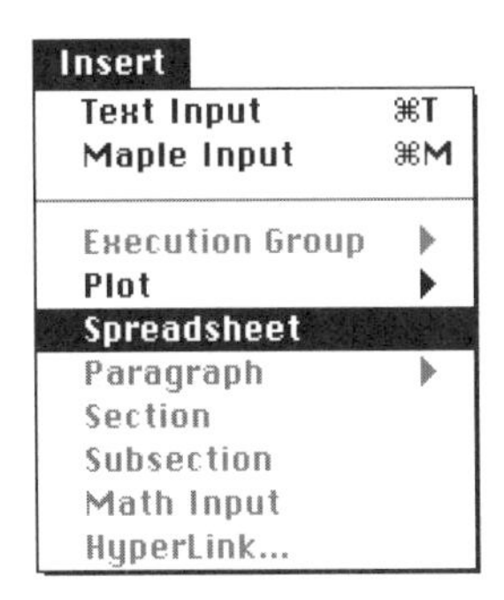

Once a spreadsheet has been inserted, input into cells can be entered manually. In addition, references can be made to other cells. For example, in the following spreadsheet, we manually entered n, 1, 2, 3, and 4 in the first five cells in column A.

07.3a.mbe.maple

Example 1

	A	B	C
1	n	$(1-x^2)\left(\frac{\partial^2}{\partial x^2}y(x)\right)-2x\left(\frac{\partial}{\partial x}y(x)\right)+n(n+1)y(x)=0$	$y(x)=\_C1\ \mathrm{LegendreP}(n,x)+\_C2\ \mathrm{LegendreQ}(n,x)$
2	1	$(1-x^2)\left(\frac{\partial^2}{\partial x^2}y(x)\right)-2x\left(\frac{\partial}{\partial x}y(x)\right)+2\,y(x)=0$	$y(x)=\_C1\,x+\_C2\left(\frac{1}{2}x\ln\left(-\frac{x+1}{-1+x}\right)-1\right)$
3	2	$(1-x^2)\left(\frac{\partial^2}{\partial x^2}y(x)\right)-2x\left(\frac{\partial}{\partial x}y(x)\right)+6\,y(x)=0$	$y(x)=\_C1\,(3x^2-1)+\_C2\left(\frac{3}{4}\ln\left(-\frac{x+1}{-1+x}\right)x^2-\frac{1}{4}\ln\left(-\frac{x+1}{-1+x}\right)-\frac{3}{2}x\right)$
4	3	$(1-x^2)\left(\frac{\partial^2}{\partial x^2}y(x)\right)-2x\left(\frac{\partial}{\partial x}y(x)\right)+12\,y(x)=0$	$y(x)=\_C1\,(5x^3-3x)+\_C2\left(\frac{5}{4}\ln\left(-\frac{x+1}{-1+x}\right)x^3-\frac{3}{4}x\ln\left(-\frac{x+1}{-1+x}\right)-\frac{5}{2}x^2+\frac{2}{3}\right)$
5	4	$(1-x^2)\left(\frac{\partial^2}{\partial x^2}y(x)\right)-2x\left(\frac{\partial}{\partial x}y(x)\right)+20\,y(x)=0$	$y(x)=\_C1\,(35x^4-30x^2+3)+\_C2\left(\frac{35}{16}\ln\left(-\frac{x+1}{-1+x}\right)x^4-\frac{15}{8}\ln\left(-\frac{x+1}{-1+x}\right)x^2+\frac{3}{16}\ln\left(-\frac{x+1}{-1+x}\right)-\frac{35}{8}x^3+\frac{55}{24}x\right)$

The contents of the cell in the row one and column B are

```
(1-x^2)*diff(y(x),`$`(x,2))-
2*x*diff(y(x),x)+~A1*(~A1+1)*y(x) = 0
```

Note that ~A1 is a relative reference to the first entry in the row. Similarly, the contents of the cell in the first row and column C are

```
dsolve(~B1,y(x))
```

and ~B1 is a relative reference to the second entry in the row.

When we select the cells in rows one through five and columns B and C and then go to **Spreadsheet** and select **Fill** followed by **Down**, the formulas are pasted into each cell, and the differential equation is solved.

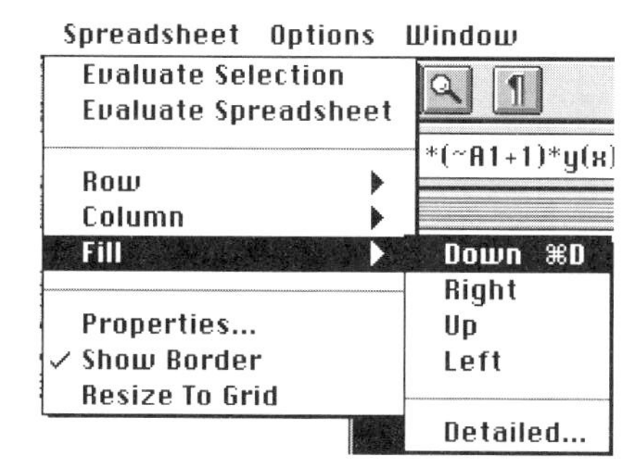

Thus, column A represents the value of n, Column B the equation $(1-x^2)\frac{d^2y}{dx^2} - 2x\frac{dy}{dx} + n(n+1)y = 0$, and column C its solution.

Because these polynomials are useful and are encountered in numerous applications, we have a special notation for them: $P_n(x)$ is called the **Legendre polynomial of degree *n*** and represents an nth-degree polynomial solution to Legendre's equation. The Maple command P(n,x), which is contained in the orthopoly package, returns $P_n(x)$. After loading the orthopoly package, we use seq together with P to list the first few Legendre polynomials.

```
> with(orthopoly):
  toplot:=seq(P(n,x),n=0..5);
```

$$toplot := 1, x, \frac{3}{2}x^2 - \frac{1}{2}, \frac{5}{2}x^3 - \frac{3}{2}x, \frac{35}{8}x^4 - \frac{15}{4}x^2 + \frac{3}{8}, \frac{63}{8}x^5 - \frac{35}{4}x^3 + \frac{15}{8}x$$

We graph these polynomials for $-2 \le x \le 2$.

```
> plot({toplot},x=-2..2,y=-2..2);
```

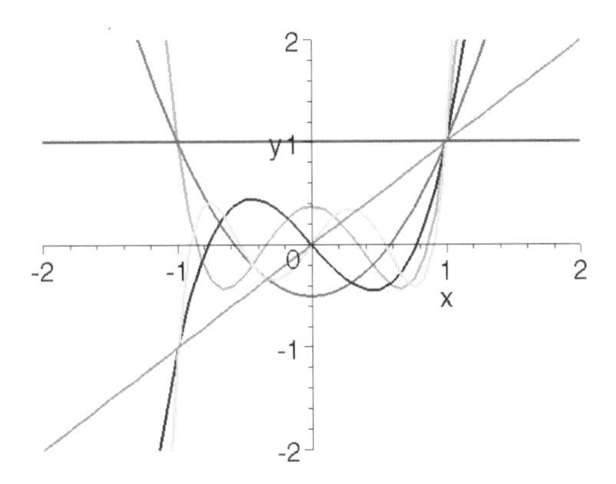

Another interesting observation is about the Legendre polynomials is that they satisfy the relationship $\int_{-1}^{1} P_m(x)P_n(x)dx = 0, m \neq n$, called an **orthogonality condition**, which we verify with `int` for $m,n = 0,1,\ldots,6$

```
array([seq([seq(int(P(n,x)*P(m,x),
    x=-1..1),n=0..6)],m=0..6)]);
```

$$\begin{bmatrix} 2 & 0 & 0 & 0 & 0 & 0 & 0 \\ 0 & \frac{2}{3} & 0 & 0 & 0 & 0 & 0 \\ 0 & 0 & \frac{2}{5} & 0 & 0 & 0 & 0 \\ 0 & 0 & 0 & \frac{2}{7} & 0 & 0 & 0 \\ 0 & 0 & 0 & 0 & \frac{2}{9} & 0 & 0 \\ 0 & 0 & 0 & 0 & 0 & \frac{2}{11} & 0 \\ 0 & 0 & 0 & 0 & 0 & 0 & \frac{2}{13} \end{bmatrix}$$

Note that the entries down the diagonal of this result correspond to the value of $\int_{-1}^{1} [P_n(x)]^2 dx$ for $n = 0, 1, \ldots, 6$ and indicate that $\int_{-1}^{1} [P_n(x)]^2 dx = \frac{2}{2n+1}$.

Regular and Irregular Singular Points and the Method of Frobenius

Let x_0 be singular point of $y'' + p(x)y' + q(x)y = 0$. x_0 is a **regular singular point** of the equation if both $(x - x_0)p(x)$ and $(x - x_0)^2 q(x)$ are analytic at $x = x_0$. If x_0 is not a regular singular point, then x_0 is called an **irregular singular point** of the equation. The **Method of Frobenius** states: Let x_0 be a regular singular point of $y'' + p(x)y' + q(x)y = 0$. Then this differential equation has at least one solution of the form

$$y = \sum_{n=0}^{\infty} a_n (x - x_0)^{n+r},$$

where r is a constant that must be determined. This solution is convergent at least on some interval $|x - x_0| < R, R > 0$.

Suppose that $x = 0$ is a regular singular point of the differential equation $y'' + p(x)y' + q(x)y = 0$. Then the functions $xp(x)$ and $x^2 q(x)$ are analytic, which means that both of these functions have a power series in x with a positive radius of convergence. Hence,

$$xp(x) = p_0 + p_1 x + p_2 x^2 \ldots \text{ and } x^2 q(\mathrm{x}) = q_0 + q_1 x + q_2 x^2 + \ldots.$$

Therefore,

$$p(x) = \frac{p_0}{x} + p_1 + p_2 x + p_3 x^2 + p_4 x^3 + \ldots q(x) = \frac{q_0}{x^2} + \frac{q_1}{x} + q_2 + q_3 x + q_4 x^2 + q_5 x^3 + \ldots$$

Substitution of these series into the differential equation $y'' + p(x)y' + q(x)y = 0$ and multiplying through by the first term in the power series for $p(x)$ and $q(x)$, we see that the lowest term in the series involves x^{n+r-2}:

$$\left(\sum_{n=0}^{\infty} a_n (n+r)(n+r-1)x^{n+r-2}\right) + \left(\sum_{n=0}^{\infty} a_n p_0 (n+r) x^{n+r-2}\right)$$
$$+ (p_1 + p_2 x + p_3 x^2 + p_4 x^3 + \ldots)\left(\sum_{n=0}^{\infty} a_n (n+r) x^{n+r-1}\right) + \left(\sum_{n=0}^{\infty} a_n q_0 x^{n+r-2}\right)$$
$$+ \left(\frac{q_1}{x} + q_2 + q_3 x + q_4 x^2 + q_5 x^3 + \ldots\right)\left(\sum_{n=0}^{\infty} a_n x^{n+r}\right) = 0.$$

Then, with $n = 0$, we find that the coefficient of x^{r-2} is

$$-ra_0 + r^2 a_0 + ra_0 p_0 + a_0 q_0 = a_0(r^2 + (p_0 - 1)r + q_0) = a_0(r(r-1) + p_0 r + q_0).$$

Thus, for any equation of the form $y'' + p(x)y' + q(x)y = 0$ with regular singular point $x = 0$, we have the **indicial equation** .

$$r(r-1) + p_0 r + q_0 = 0.$$

The command indicialeq, which is contained in the DEtools package, can be used to find the indicial equation.

The values of r that satisfy this equation, which are called the **exponents** or **indicial roots**, are

$$r1 = \frac{1}{2}\left(1 - p_0 + \sqrt{1 - 2p_0 + p_0^2 - 4q_0}\right) \text{ and } r2 = \frac{1}{2}\left(1 - p_0 + \sqrt{1 - 2p_0 + p_0^2 - 4q_0}\right).$$

Note that $r_1 \geq r_2$ and $r_1 - r_2 = \sqrt{1 - 2p_0 + p_0^2 - 4q_0}$.

Several situations can arise in the search for the roots of the indicial equation:

1. If $(r_1 \neq r_2)$ and $(r_1 - r_2 = \sqrt{1 - 2p_0 + p_0^2 - 4q_0})$ is not an integer, then there are two linearly independent solutions of the equation of the form $y_1(x) = x^{r_1} \sum_{n=0}^{\infty} a_n x^n$ and $y_2(x) = x^{r_2} \sum_{n=0}^{\infty} b_n x^n$.
2. If $r_1 \neq r_2$ and $r_1 - r_2 = \sqrt{1 - 2p_0 + p_0^2 - 4q_0}$ is an integer, then there are two linearly independent solutions of the equation of the form $y_1(x) = x^{r_1} \sum_{n=0}^{\infty} a_n x^n$ and $y_2(x) = c y_1(x) \ln x + x^{r_2} \sum_{n=0}^{\infty} b_n x^n$.
3. If $r_1 - r_2 = \sqrt{1 - 2p_0 + p_0^2 - 4q_0}$, then there are two linearly independent solutions of the problem of the form $y_1(x) = x^{r_1} \sum_{n=0}^{\infty} a_n x^n$ and $y_2(x) = c y_1(x) \ln x + x^{r_1} \sum_{n=0}^{\infty} b_n x^n$.

In any case, if $y_1(x)$ is a solution of the equation, a second linearly independent solution is given by

$$y_2(x) = y_1(x) \int \frac{e^{-\int p(x)dx}}{[y(x)]^2} dx,$$

which can be obtained through reduction of order.

EXAMPLE 2: Bessel's equation (of order μ), is the equation

$$(x^2 y'' + xy' + (x^2 - \mu^2)y = 0),$$

where $\mu \geq 0$ is a constant. Solve Bessel's equation.

SOLUTION: To use a power series method to solve Bessel's equation, we first write the equation in standard form as

$$y'' + \frac{1}{x}y' + \frac{x^2 - \mu^2}{x^2}y = 0,$$

so $x = 0$ is a regular singular point. Using the Method of Frobenius, we assume that there is a solution of the form $y = \sum_{n=0}^{\infty} a_n x^{n+r}$. We determine the value(s) of r with the indicial equation. Because $xp(x) = x \cdot 1/x = 1$ and $x^2 q(x) = x^2 \cdot (x^2 - \mu^2)/x^2 = x^2 - \mu^2$, $p_0 = 1$ and $q_0 = -\mu^2$. Hence, the indicial equation is

$$r(r-1) + p_0 r + q_0 = r(r-1) + r - \mu^2 = r^2 - \mu^2 = 0$$

with roots $r1 = \mu$ and $r2 = -\mu$, which we confirm using `indicialeq`.

```
> eq:=x^2*diff(y(x),x$2)+x*diff(y(x),x)+
      (x^2-mu^2)*y(x)=0:
  with(DEtools):
  indicialeq(eq,x,0,y(x));
```

$$x^2 - \mu^2 = 0$$

Therefore, we assume that $y = \sum_{n=0}^{\infty} a_n x^{n+\mu}$ with derivatives $y' = \sum_{n=0}^{\infty} (n+\mu) a_n x^{n+\mu-1}$ and $y'' = \sum_{n=0}^{\infty} (n+\mu)(n+\mu-1) a_n x^{n+\mu-2}$. Substitution into Bessel's equation and simplifying the result yields

$$\left[\mu(\mu-1) + \mu - \mu^2\right] a_0 x^{\mu} + \left[(1+\mu)\mu + (1+\mu) - \mu^2\right] a_1 x^{\mu+1}$$

$$+ \sum_{n=2}^{\infty} \left\{ \left[(n+\mu)(n+\mu-1) + (n+\mu) - \mu^2\right] a_n + a_{n-2} \right\} x^{n+\mu} = 0.$$

Notice that the coefficient of a_0x^μ is zero. After simplifying the other coefficients and equating them to zero, we have $(1+2\mu)a_1 = 0$ and $\left[(n+\mu)(n+\mu-1)+(n+\mu)-\mu^2\right]a_n + a_{n-2} = 0$, which we solve for a_n.

```
> a:='a':
  solve(((n+mu)*(n+mu-1)+
      (n+mu)-mu^2)*a[n]+a[n-2]=0,a[n]);
```

$$-\frac{a_{n-2}}{n\,(n+2\,\mu)}$$

From the first equation, $a_1 = 0$. Therefore, from $a_n = -\frac{a_{n-2}}{n(n-2\mu)}$, $n \geq 2$ for all odd n. We use the formula for a_n to calculate several of the coefficients that correspond to even indices.

```
> a:= proc(n) option remember;
      -a(n-2)/(n*(n+2*mu)) end:
  a(0):=a[0]:
> nvals:=seq(2*i,i=1..5):
  array([seq([n,a(n)],n=nvals)]);
```

$$\begin{bmatrix} 2 & -\frac{1}{2}\frac{a_0}{2+2\,\mu} \\ 4 & \frac{1}{8}\frac{a_0}{(2+2\,\mu)\,(4+2\,\mu)} \\ 6 & -\frac{1}{48}\frac{a_0}{(2+2\,\mu)\,(4+2\,\mu)\,(6+2\,\mu)} \\ 8 & \frac{1}{384}\frac{a_0}{(2+2\,\mu)\,(4+2\,\mu)\,(6+2\,\mu)\,(8+2\,\mu)} \\ 10 & -\frac{1}{3840}\frac{a_0}{(2+2\,\mu)\,(4+2\,\mu)\,(6+2\,\mu)\,(8+2\,\mu)\,(10+2\,\mu)} \end{bmatrix}$$

A general formula for these coefficients is given by $a_{2n} = \dfrac{(-1)^n a_0}{2^{2n}(1+\mu)(2+\mu)(3+\mu)\dots(n+\mu)}, n \geq 2$. Our solution can then be written as

$$y = \sum_{n=0}^{\infty} a_{2n} x^{2n+\mu} = \sum_{n=0}^{\infty} \frac{(-1)^n x^{2n+\mu}}{2^{2n}(1+\mu)(2+\mu)(3+\mu)\dots(n+\mu)}$$

$$= \sum_{n=0}^{\infty} \frac{(-1)^n 2^{\mu}}{(1+\mu)(2+\mu)(3+\mu)\dots(n+\mu)} \left(\frac{x}{2}\right)^{2n+\mu}.$$

If μ is an integer, then by using the gamma function $\Gamma(x)$, we can write this solution as

$$y = \sum_{n=0}^{\infty} \frac{(-1)^n}{n!\Gamma(1+\mu+n)} \left(\frac{x}{2}\right)^{2n+\mu}.$$

This function, denoted $J_{\mu}(x)$, is called the **Bessel function of the first kind of order** μ. The command `BesselJ(mu,x)` returns $J_{\mu}(x)$.

We use `BesselJ` to graph $J_{\mu}(x)$ for $\mu = 0, 1, 2, 3,$ and 4. Notice that these functions have numerous zeros. We will need to know these values in subsequent sections.

```
> plot({seq(BesselJ(mu,x),mu=0..4)},x=0..10);
```

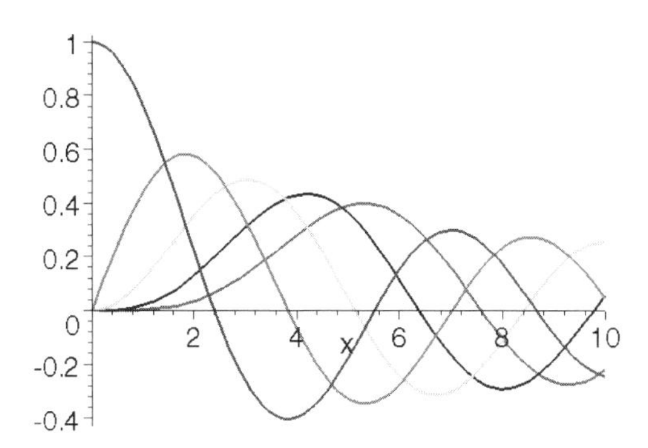

For the other root $r2 = -\mu$, a similar derivation yields a second solution

$$y = \sum_{n=0}^{\infty} \frac{(-1)^n}{n!\Gamma(1-\mu+n)}\left(\frac{x}{2}\right)^{2n-\mu},$$

which is the **Bessel function of the first kind of order** $-\mu$ and is denoted $J_{\mu}(x)$. Now, we must determine if the functions $J_{\mu}(x)$ and $J_{-\mu}(x)$ are linearly independent. Notice that if $\mu = 0$, then these two functions are the same. If $\mu > 0$, then $r_1 - r_2 = \mu - (-\mu) = 2\mu$. If 2μ is not an integer, then by the Method of Frobenius, the two solutions $J_{\mu}(x)$ and $J_{-\mu}(x)$ are linearly independent. Also, we can show that if 2μ is an odd integer, then $J_{\mu}(x)$ and $J_{-\mu}(x)$ are linearly independent.

If μ is not an integer, we define the **Bessel function of the second kind of order** μ by the linear combination of the functions $J_{\mu}(x)$ and $J_{-\mu}(x)$. This function, denoted by $Y_{\mu}(x)$, is given by

$$Y_{\mu}(x) = \frac{\cos\mu\pi J_{\mu}(x) - J_{-\mu}(x)}{\sin\mu\pi}.$$

The command `BesselY(m,x)` returns $Y_{\mu}(x)$. We can show that $J_{\mu}(x)$ and $Y_{\mu}(x)$ are linearly independent, so a general solution of Bessel's equation of order μ can be represented by $y = c_1 J_{\mu}(x) + c_2 Y_{\mu}(x)$. In fact, when we use Maple to solve the differential equation, it returns the solution $y = c_1 J_{\mu}(x) + c_2 Y_{\mu}(x)$.

```
> dsolve(eq,y(x));
```

$$y(x) = \_C1 \text{ BesselJ}(\mu, x) + \_C2 \text{ BesselY}(\mu, x)$$

We use `BesselY` to graph the functions $Y_{\mu}(x)$ for $\mu = 0, 1, 2, 3, \text{and } 4$. Notice that $\lim_{x\to 0^+} Y_{\mu}(x) = -\infty$. This property will be important in several applications in later chapters.

```
> plot({seq(BesselY(mu,x),mu=0..4)},x=0..10,y=-9..1);
```

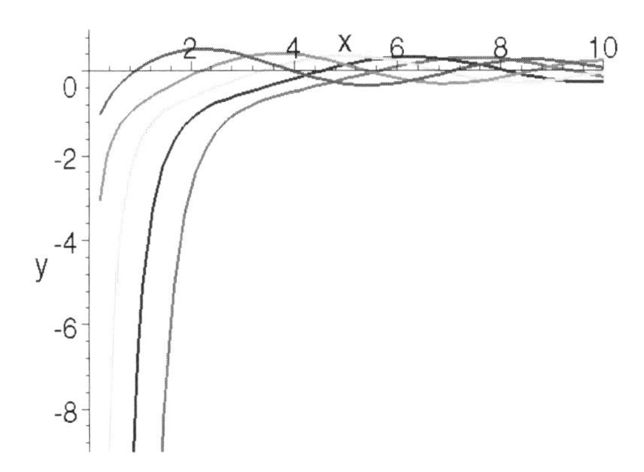

Alternatively, a series solution can be found using `dsolve` together with the option `'type=series'`.

```
> dsolve(eq,y(x),'type=series');
```

$$y(x) = \_C1\, x^{(-\mu)} \left(1 + \frac{1}{4\,\mu - 4} x^2 + \frac{1}{(8\,\mu - 16)\,(4\,\mu - 4)} x^4 + O(x^6) \right)$$
$$+ \_C2\, x^{\mu} \left(1 + \frac{1}{-4\,\mu - 4} x^2 + \frac{1}{(-8\,\mu - 16)\,(-4\,\mu - 4)} x^4 + O(x^6) \right)$$

■

7.4 Using the Laplace Transform to Solve Ordinary Differential Equations

Definition of the Laplace Transform

Let $f(t)$ be a function defined on the interval $[0, \infty)$. The **Laplace transform** of $f(t)$ is the function (of s)

$$\mathcal{L}\{f(t)\} = \int_0^{\infty} e^{-st} f(t)(dt),$$

provided the improper integral exists.

If $f(t)$ is said to be the **inverse Laplace transform** of $F(s)$, this means that $\mathcal{L}\{f(t)\} = F(s)$, and we write $\mathcal{L}^{-1}\{f(s)\} = f(t)$.

Commands that can be used to compute Laplace transforms and inverse Laplace transforms are located in the `intransforms` package. The command `laplace(f(t),t,s)` computes the Laplace transform of $f(t)$ as a function of s; the command `invlaplace(F(s),s,t)` computes the inverse Laplace transform of $F(s)$ as a function of t.

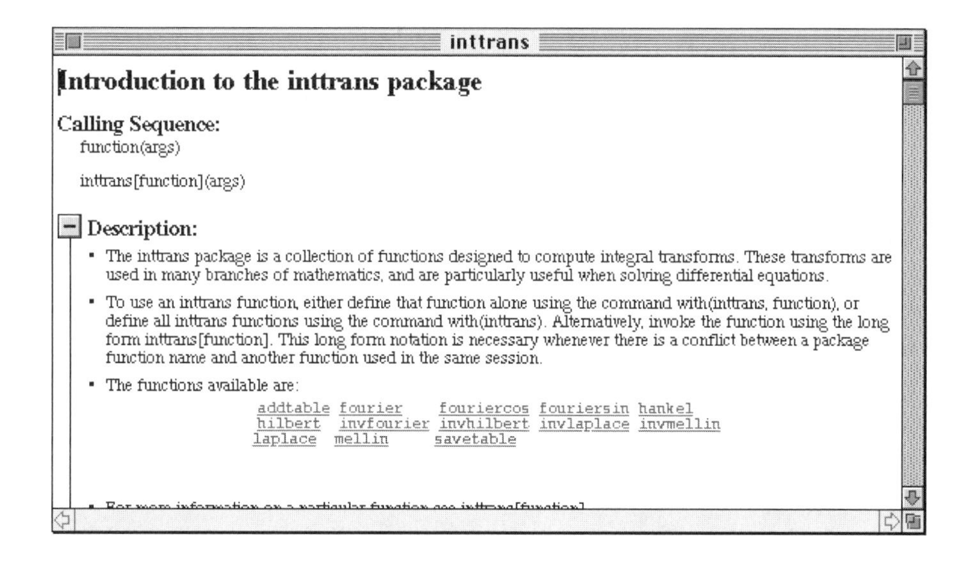

EXAMPLE 1: Find the Laplace transform of (a) $f(t) = t^3$; (b) $f(t) = \sin at$; and (c) $f(t) = \cos at$.

SOLUTION: For (a), we use the definition of the Laplace transform and compute $\int_0^L t^3 e^{-st} dt$ with `int`, naming the result `step1`.

```
> step1:=int(t^3*exp(-s*t),t=0..L);
```

$$step1 := -\frac{e^{(-s\,L)}\,s^3\,L^3 + 3\,e^{(-s\,L)}\,s^2\,L^2 + 6\,e^{(-s\,L)}\,s\,L + 6\,e^{(-s\,L)} - 6}{s^4}$$

Then, the Laplace transform of $f(t) = t^3$ is $\lim_{L\to\infty} \int_0^L t^3 e^{-st} dt = \frac{6}{s^4}$. However, we see that Maple is unable to compute this limit with `limit` because Maple does not assume that s is positive.

```
> limit(step1,L=infinity);
```

$$\lim_{L \to \infty} -\frac{e^{(-s\,L)}\,s^3\,L^3 + 3\,e^{(-s\,L)}\,s^2\,L^2 + 6\,e^{(-s\,L)}\,s\,L + 6\,e^{(-s\,L)} - 6}{s^4}$$

However, when we instruct Maple to assume that $s > 0$ with `assume`, Maple is able to evaluate the limit with `limit`.

```
> assume(s>0);
  limit(step1,L=infinity);
```

$$6\,\frac{1}{s\sim^4}$$

For (b) and (c) we use the command `laplace`.

```
> s:='s':
> laplace(sin(a*t),t,s);
```

$$\frac{a}{s^2 + a^2}$$

```
> laplace(cos(a*t),t,s);
```

$$\frac{s}{s^2 + a^2}$$

We see that `laplace` is successful and that the Laplace transform of $f(t) = \sin at$ is $\frac{a}{a^2 + s^2}$, while the Laplace transform of $f(t) = \cos at$ is $\frac{s}{a^2 + s^2}$.

■

Although `laplace` can be used to compute the Laplace transform of many "standard" functions, in other cases it is best to proceed directly and use Maple to perform the necessary calculations to compute the Laplace transform of a function.

EXAMPLE 2: Find the Laplace transform of the function *f* defined by $f(t) = \begin{cases} 1-t, 0 \le t < 1 \\ f(t-1), t \ge 1 \end{cases}$. Hence, *f* represents the periodic extension of the function $1 - t$ on $[0, 1]$ on [0,1].

SOLUTION: In this case, we illustrate how to use Maple to define and graph a recursively-defined function. We begin by using `piecewise` to define the function $g(t) = \begin{cases} 1-t, 0 \le t < 1 \\ 0, t \ge 1 \end{cases}$ and then graph g on the interval $[0, 2]$.

```
> g:=t->piecewise(t>=0 and t<1,1-t,t<0 or t>=1,0):
  plot(g(t),t=0..2);
```

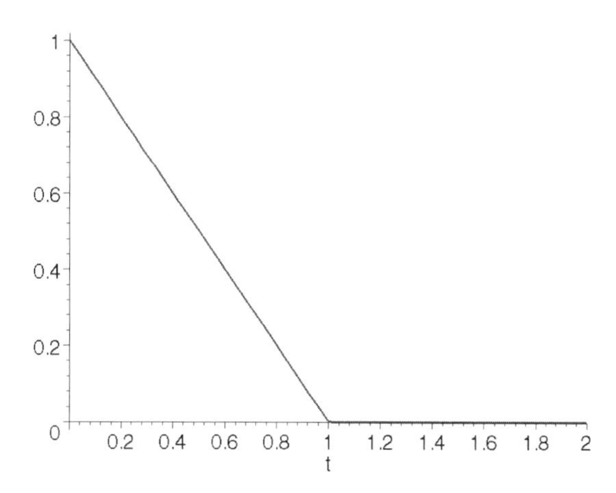

We then define $f(t) = \begin{cases} 1-t, 0 \le t < 1 \\ f(t-1), t \ge 1 \end{cases}$ using `proc`, `if` and the option `remember` in the definition of f. This causes Maple to "remember" the values of f that are computed. For example, if Maple has computed $f(2)$, then when the value of $f(3)$ is requested, Maple need not recompute $f(2)$. Note the error message that results when we first attempt to use `plot` to graph f. The error results because $f(t)$ does not make sense to Maple unless t is a (particular) number. Consequently, we use `''` to delay the evaluation of f in the second `plot` command.

```
> f:=proc(t) option remember;
       if t>=0 and t<1 then 1-t else f(t-1) fi end:
> plot(f(t),t=0..4,numpoints=100);
```

```
        Error, (in f) cannot evaluate boolean
```

```
> plot('f(t)',t=0..4,numpoints=100);
```

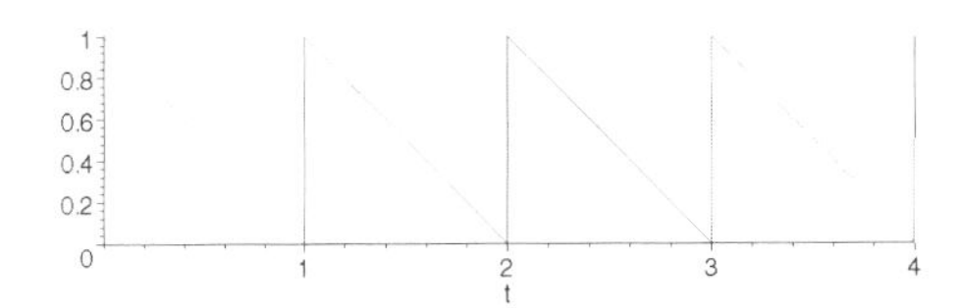

The Laplace transform of the periodic function *f* with period *P* is given by

$$\mathcal{L}\{f(t)\} = \frac{1}{1-e^{-Ps}}\int_0^P e^{-st}f(t)dt.$$

Therefore, the Laplace transform of *f* is given by $\frac{1}{1-e^{-s}}\int_0^1 e^{-st}f(t)dt$. We compute $\int_0^1 (1-t)e^{-st}dt$.

```
> stepone:=int((1-t)*exp(-s*t),t=0..1);
```

$$stepone := \frac{\mathbf{e}^{(-s)}+s-1}{s^2}$$

Therefore, the Laplace transform of $f(t)$ is $\frac{e^{-s}+s-1}{s^2(1-e^{-s})}$.

■

Example 3 illustrates the use of the command `invlaplace` to compute the inverse Laplace transform of some functions.

EXAMPLE 3: Find the inverse Laplace transform of $\frac{4}{s^2+16}$.

SOLUTION: After loading the package `inttrans`, we use `invlaplace` to see that the inverse Laplace transform of $\frac{4}{s^2+16}$ is $\sin 4t$. (If you have already loaded the `inttrans` package during your current Maple session, you do not need to reload the package by entering `with(inttrans)`.)

```
> with(inttrans):
  simplify(invlaplace(4/(s^2+16),s,t));
```

$$\sin(4t)$$

■

Solving Ordinary Differential Equations with the Laplace Transform

Laplace transforms can be used to solve a variety of differential equations. Typically, when we use Laplace transforms to solve a differential equation for a function $y(t)$, we will compute the Laplace transform of each term of the differential equation, solve the resulting algebraic equation for the Laplace transform of $y(t)$, $\mathcal{L}\{y(t)\}$, and finally determine $y(t)$ by computing the inverse Laplace transform of $\mathcal{L}\{y(t)\}$.

We instruct Maple to use Laplace transforms to solve a differential equation by including the option `method=laplace` in the `dsolve` command.

EXAMPLE 4: Let $f(t) = \begin{cases} -1, 0 \le t < 1 \\ 0, t \ge 1 \end{cases}$. Solve $\frac{dy}{dt} + 4y = f(t)$ subject to the initial condition $y(0) = 0$.

SOLUTION: To define *f*, we will make use of the command `Heaviside`: The **unit step function**, `Heaviside(x)`, is defined by

$$\mathcal{U}(x) = \begin{cases} 1, x > 0 \\ 0, x < 0 \end{cases}.$$

Thus, a piecewise-defined function like

$$f(x) = \begin{cases} g(x),\ 0 < x < a \\ h(x),\ x > a \end{cases}$$

can be defined in terms of the unit step function with $f(x) = g(x)[1 - \mathcal{U}(x-a)] + h(x)\mathcal{U}(x-a)$. In this case, we have that $f(t) = -(1 - \mathcal{U}(t-1))$.

For $t \geq 0$, Heaviside $(-t+1) = f(t)$.

```
> f:='f':y:='y':t:='t':
  f:=t->-(1-Heaviside(t-1):
  plot(f(t),t=0..2);
```

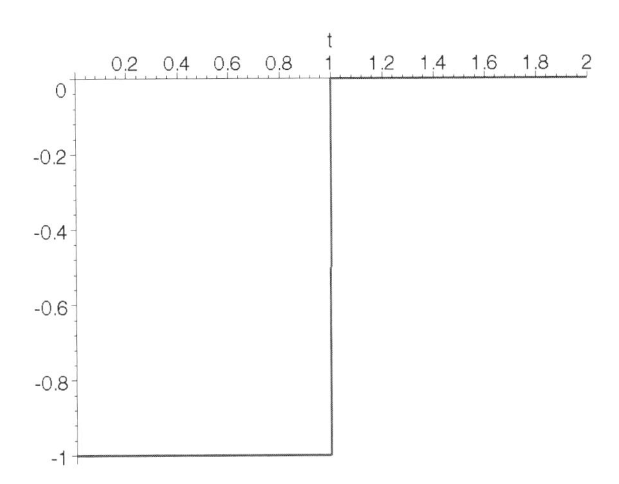

In this case, we use the command `dsolve` together with the option `method=laplace` to solve the differential equation. We proceed by defining `EqOne` to be the differential equation and then solve `EqOne` subject to the initial condition $y(0) = 0$. The resulting output is named `Sol`.

```
> EqOne:=diff(y(t),t)+4*y(t)=f(t):
  Sol:=dsolve({EqOne,y(0)=0},y(t),method=laplace);
```

$$Sol := \mathrm{y}(t) = -\frac{1}{4} + \frac{1}{4}\mathbf{e}^{(-4t)} + \frac{1}{4}\mathrm{Heaviside}(t-1)\,(1 - \mathbf{e}^{(-4t+4)})$$

The solution is then graphed on the interval $[0, 2]$ with `plot`.

```
> plot(rhs(Sol),t=0..2);
```

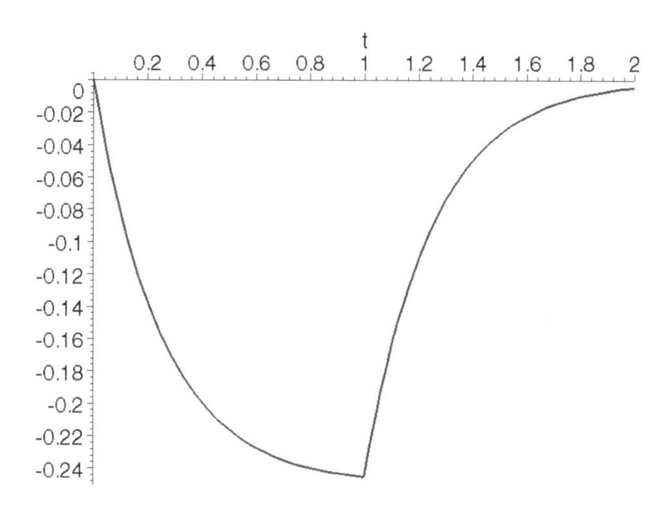

■

In cases in which `dsolve` together with the `method=laplace` option does not solve the given differential equation, Maple can often be used to perform the calculations encountered when certain differential equations are solved.

EXAMPLE 5: Let $f(t)$ be defined recursively by $f(t) = \begin{cases} 1, 0 \le t < 1 \\ -1, 1 \le t < 2 \end{cases}$ and $f(t) = f(t-2)$ if $t \ge 2$. Solve $y'' + 4y' + 20y = f(t)$.

SOLUTION: We begin by defining the piecewise defined function f. Note that `elif` is used to avoid repeated `if    fi` statements. After defining f, we graph f on the interval $[0, 6]$.

```
> f:='f':y:='y':
  f:=proc(t) option remember;
      if t<1 and t>=0 then 1 elif
        t<2 and t>=1 then -1 else
        f(t-2) fi end:
> plot('f(t)','t'=0..6,numpoints=200);
```

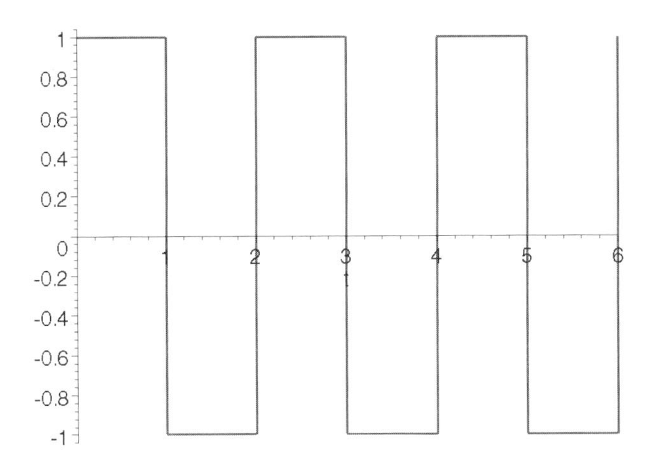

We then define LHS_Eq to be the left-hand side of the equation $y'' + 4y' + 20y = f(t)$,

```
> LHS_Eq:=diff(y(t),t$2)+4*diff(y(t),t)+20*y(t):
```

and compute the Laplace transform of each term of LHS_Eq, naming the resulting output stepone. In the result, laplace(y(t), t, s) represents $\mathcal{L}\{y\}(s)$. Note that Maple easily computes the Laplace transform of derivatives by applying the property:

$$\mathcal{L}\{f^{(n)}(t)\} = \mathcal{L}\{f\} - s^{n-1}f(0) - \cdots - sf^{(n-2)}(0) - f^{(n-1)}(0).$$

```
> stepone:=laplace(LHS_Eq,t,s);
```

$$stepone := s\,(s\,\mathrm{laplace}(\mathrm{y}(t), t, s) - \mathrm{y}(0)) - \mathrm{D}(y)(0) + 4\,s\,\mathrm{laplace}(\mathrm{y}(t), t, s)$$
$$- 4\,\mathrm{y}(0) + 20\,\mathrm{laplace}(\mathrm{y}(t), t, s)$$

Let lr denote the Laplace transform of $f(t)$, corresponding to the Laplace transform of the right-hand side of the equation $y'' + 4y' + 20y = f(t)$. We then solve the equation

$$(20 + 4s + s^2)\mathcal{L}\{y\}(s) - 4y(0) - sy(0) - y'(0) = \mathrm{lr}$$

for $\mathcal{L}\{y\}(s)$ in steptwo. To compute y, we must compute the inverse Laplace transform of $\mathcal{L}\{y\}(s)$. First in stepthree, we expand the expression obtained in steptwo.

```
> steptwo:=solve(stepone=lr,laplace(y(t),t,s));
```

$$steptwo := \frac{s\,y(0) + D(y)(0) + 4\,y(0) + lr}{s^2 + 4\,s + 20}$$

```
> stepthree:=expand(steptwo);
```

$$stepthree := \frac{s\,y(0)}{s^2 + 4\,s + 20} + \frac{D(y)(0)}{s^2 + 4\,s + 20} + 4\frac{y(0)}{s^2 + 4\,s + 20} + \frac{lr}{s^2 + 4\,s + 20}$$

The fourth term of `stepthree` is the only part of $L\{y\}(s)$ containing the term `lr`. We remove this term from `stepthree` and name the result `stepfour`. Hence, to find y, we must compute the inverse Laplace transform of `stepfour+op(4,stepthree)`.

```
> op(4,stepthree);
```

$$\frac{lr}{s^2 + 4\,s + 20}$$

```
> stepfour:=stepthree-op(4,stepthree);
```

$$stepfour := \frac{s\,y(0)}{s^2 + 4\,s + 20} + \frac{D(y)(0)}{s^2 + 4\,s + 20} + 4\frac{y(0)}{s^2 + 4\,s + 20}$$

The inverse Laplace transform of `stepfour` is computed and named `y1` with `invlaplace`. To see that the result is a real-valued function, we rewrite it in terms of trigonometric functions, using `convert` together with the `trig` option, and then using `convert` together with the `expsincos` option.

```
> y1:=simplify(invlaplace(stepfour,s,t));
```

$$\begin{aligned} y1 := \frac{1}{8}I(&-D(y)(0)\mathbf{e}^{((-2+4I)t)} + D(y)(0)\mathbf{e}^{((-2-4I)t)} \\ &-2y(0)\mathbf{e}^{((-2+4I)t)} + 2y(0)\mathbf{e}^{((-2-4I)t)} \\ &-4I\,y(0)\mathbf{e}^{((-2+4I)t)} - 4y(0)\mathbf{e}^{((-2-4I)t)}) \end{aligned}$$

```
> y1:=simplify(convert(y1,trig));
```

$$yI:=\frac{1}{4}\cosh(2t)\mathrm{D}(y)(0)\sin(4t)+\frac{1}{2}\cosh(2t)\mathrm{y}(0)\sin(4t)$$
$$+\cosh(2t)\mathrm{y}(0)\cos(4t)-\frac{1}{4}\sinh(2t)\mathrm{D}(y)(0)\sin(4t)$$
$$-\frac{1}{2}\sinh(2t)\mathrm{y}(0)\sin(4t)-\sinh(2t)\mathrm{y}(0)\cos(4t)$$

```
> y1:=simplify(convert(y1,expsincos));
```

$$yI:=\frac{1}{4}(\mathrm{D}(y)(0)\sin(4t)+2\mathrm{y}(0)\sin(4t)+4\mathrm{y}(0)\cos(4t))\mathbf{e}^{(-2t)}$$

To compute the inverse Laplace transform of $\frac{\mathrm{lr}}{s^2+4s+20}$, we begin by computing `lr`. The periodic function $f(t) = \begin{cases} 1, 0 \le t < 1 \\ -1, 1 \le t < 2 \end{cases}$ and $f(t) = f(t-2)$ if $t \ge 2$ can be written in terms of step functions as

$$f(t) = \mathcal{U}_0(t) - 2\mathcal{U}_1(t) + 2\mathcal{U}_2(t) - 2\mathcal{U}_3(t) + 2\mathcal{U}_4(t) - \cdots$$

$$= \mathcal{U}(t) - 2\mathcal{U}(t-1) + 2\mathcal{U}(t-2) - 2\mathcal{U}(t-3) + 2\mathcal{U}(t-4) - \cdots$$

$$= \mathcal{U}(t) + 2\sum_{n=1}^{\infty}(-1)^n\mathcal{U}(t-n).$$

The Laplace transform of the unit step function, $\mathcal{U}_a(t) = \mathcal{U}(t-a)$, is $\frac{e^{-as}}{s}$, $s > 0$, and the Laplace transform of $f(t-a)\mathcal{U}(t-a)$ is $e^{-as}F(s)$, where $F(s)$ denotes the Laplace transform of $f(t)$. Then,

$$\mathrm{lr} = \mathcal{L}fs = \frac{1}{s} - 2\frac{e^{-s}}{s} + 2\frac{e^{-2s}}{s} - 2\frac{e^{-3s}}{s} + \ldots = \frac{1}{\mathrm{s}}(1 - 2e^{-s} + 2e^{-2s} - 2e^{-3s} + 2e^{-4s} - \ldots)$$

and

$$\frac{\text{lr}}{s^2+4s+20} = \frac{1}{s(s^2+4s+20)}(1-2e^{-s}+2e^{-2s}-2e^{-3s}+\ldots)$$

$$= \frac{1}{s(s^2+4s+20)} + 2\sum_{n=1}^{\infty}(-1)^n \frac{e^{-ns}}{s(s^2+4s+20)}.$$

and

$$\frac{\text{lr}}{s^2+4s+20} = \frac{1}{s(s^2+4s+20)}(1-2e^{-s}+2e^{-2s}-2e^{-3s}+2e^{-4s}\ldots)$$

$$= \frac{1}{s(s^2+4s+20)} + 2\sum_{n=1}^{\infty}(-1)^n \frac{e^{-ns}}{s(s^2+4s+20)}.$$

In the following commands, we transform $\frac{\text{lr}}{s^2+4s+20}$ into $\frac{1}{s(s^2+4s+20)}$ by dividing by (`s*lr`) and name the resulting output `stepfive`. We then compute the inverse Laplace transform of `stepfive` and name the result `g`.

```
> stepfive:=simplify(op(4,stepthree)/(s*lr));
```

$$stepfive := \frac{1}{(s^2+4\,s+20)\,s}$$

```
> g:=convert(simplify(invlaplace(stepfive,s,t)),trig);
```

$$g := \left(-\frac{1}{200}+\frac{1}{400}I\right)(-8+5\,(\cosh(2\,t)-\sinh(2\,t))\,(\cos(4\,t)+I\sin(4\,t))$$
$$+4\,I\,(\cosh(2\,t)-\sinh(2\,t))\,(\cos(4\,t)-I\sin(4\,t))$$
$$+3\,(\cosh(2\,t)-\sinh(2\,t))\,(\cos(4\,t)-I\sin(4\,t))-4\,I)$$

```
> g:=simplify(convert(g,expsincos));
```

$$g := -\frac{1}{40}\mathbf{e}^{(-2\,t)}\sin(4\,t)-\frac{1}{20}\mathbf{e}^{(-2\,t)}\cos(4\,t)$$

Then, $\mathcal{L}^{-1}\left\{2(-1)^n \frac{e^{-ns}}{s(s^2+4s+20)}\right\} = 2(-1)^n g(t-n)U(t-n)$ and the inverse Laplace transform of $\frac{1}{s(s^2+4s+20)} + 2\sum_{n=1}^{\infty}(-1)^n \frac{e^{-ns}}{s(s^2+4s+20)}$ is

$y_2(t) = g(t) + 2\sum_{n=1}^{\infty}(-1)^n g(t-n)U(t-n)$. It then follows that

$$\begin{aligned} y(t) &= y_1 t + y_2(t) \\ &= y(0)e^{-2t}\cos 4t + \frac{y'(0)+2y(0)}{4}e^{-2t}\sin 4t + g(t) + 2\sum_{n=1}^{\infty}(-1)^n g(t-n)U(t-n), \end{aligned}$$

where $g(t) = \frac{1}{20} - \frac{1}{20}e^{-2t}\cos 4t - \frac{1}{40}e^{-2t}\sin 4t$.

To graph the solution for various initial conditions, we first create a table of $2(-1)^n g(t-n)H(t-n)$ for $n = 1, 2, 3,$ and 4 :

```
> array([seq([n,2*(-1)^n*subs(t=t-n,g)*
      Heaviside(t-n)],n=1..5)]);
```

$$\left[\begin{array}{ll} 1 & -2\left(-\frac{1}{40}\mathbf{e}^{(-2t+2)}\sin(4t-4) - \frac{1}{20}\mathbf{e}^{(-2t+2)}\cos(4t-4) + \frac{1}{20}\right)\text{Heaviside}(t-1) \\ 2 & 2\left(-\frac{1}{40}\mathbf{e}^{(-2t+4)}\sin(4t-8) - \frac{1}{20}\mathbf{e}^{(-2t+4)}\cos(4t-8) + \frac{1}{20}\right)\text{Heaviside}(t-2) \\ 3 & -2\left(-\frac{1}{40}\mathbf{e}^{(-2t+6)}\sin(4t-12) - \frac{1}{20}\mathbf{e}^{(-2t+6)}\cos(4t-12) + \frac{1}{20}\right)\text{Heaviside}(t-3) \\ 4 & 2\left(-\frac{1}{40}\mathbf{e}^{(-2t+8)}\sin(4t-16) - \frac{1}{20}\mathbf{e}^{(-2t+8)}\cos(4t-16) + \frac{1}{20}\right)\text{Heaviside}(t-4) \end{array}\right]$$

and then we define `y2` and `sol` to be the first five terms of the solution.

```
> y2:=g+2*sum('(-1)^n*subs(t=t-n,g)*
      Heaviside(t-n)','n'=1..4):
  sol:=y1+y2:
```

We also define two lists, `init_pos` and `init_vel`, which we will use when we graph the solutions corresponding to various initial conditions.

```
> init_pos:=[-1/2,0,1/2]:
  init_vel:=[-1/2,1/2,1]:
```

to_graph is a list of nine functions corresponding to replacing $y(0)$ by the ith number in init_pos and $y'(0)$ by the jth number in init_vel for $i=1, 2$, and 3 and $j=1, 2$, and 3.

```
> to_graph:=[seq(seq(subs({y(0)=init_pos[i],
      D(y)(0)=init_vel[j]},sol),i=1..3),j=1..3)]:
```

We then use `seq` to graph each of the functions in `to_graph` on the interval [0,5].

```
> to_show:={seq(plot(to_graph[i],t=0..5,color=BLACK),
      i=1..9)}:
```

The resulting nine graphs are displayed together as a graphics array, using `display` together with the option `insequence=true`.

```
> with(plots):
  graphics_array:=display(to_show,insequence=true):
  display(graphics_array);
```

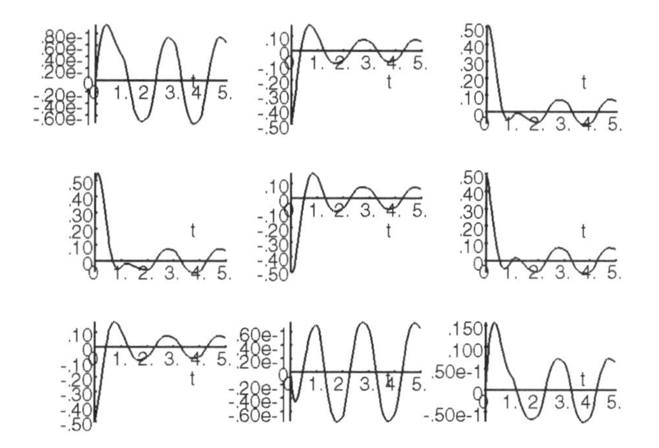

■

Application: The Convolution Theorem

In many cases, we are required to determine the inverse Laplace transform of a product of two functions. Just as in differential and integral calculus, when the derivative and integral of a product of two functions do not produce the product of the derivatives and integrals, respectively, neither does the inverse Laplace transform of the product yield the product of the inverse Laplace transforms. The **Convolution Theorem** tells us how to compute the inverse Laplace transform of a product of two functions, as follows: Suppose that $f(t)$ and $g(t)$ are piecewise continuous on $[0, +\infty]$ and both are of exponential order. Further, suppose that the Laplace transform of $f(t)$ is $F(s)$ and that of $g(t)$ is $G(s)$. Then,

$$\mathcal{L}^{-1}\{F(s)G(s)\} = \mathcal{L}^{-1}\{L\{(f*g)(t)\}\} = (f*g)(t) = \int_0^t f(t-v)g(v)dv.$$

Note that $(f*g)(t) = \int_0^t f(t-v)g(v)dv$ is called the **convolution integral**.

EXAMPLE 6: The initial-value problem used to determine the charge $q(t)$ on the capacitor in an L-R-C circuit is

$$\begin{cases} \mathcal{L}\dfrac{d^2Q}{dt^2} + R\dfrac{dQ}{dt} + \dfrac{1}{C}Q = E(t) \\ Q(0) = 0, \dfrac{dQ}{dt}(0) = 0 \end{cases},$$

where L denotes inductance, $\frac{dQ}{dt}(t) = I(t)$, $I(t)$ current, R resistance, C capacitance, and $E(t)$ voltage supply. Because $\frac{dQ}{dt}(t) = I(t)$, this differential equation can be represented as $L\frac{dI}{\mathrm{dt}} + RI + \frac{1}{C}\int_0^t I(u)\ du = E(t)$. Note also that the initial condition $Q(0) = 0$ is satisfied because $Q(0) = \frac{1}{C}\int_0^0 I(u)du = 0$. The condition $\frac{dQ}{dt}(0) = 0$ is replaced by $I(0) = 0$. (a) Solve this **integrodifferential equation**, an equation that involves a derivative as well as an integral of the unknown function, by using the Convolution Theorem. (b) Consider this example with constant values $L = C = R = 1$ and $E(t) = \begin{cases} \sin t, & 0 \le t < \pi/2 \\ 0, & t \ge \pi/2 \end{cases}$. Determine $I(t)$ and graph the solution.

SOLUTION: We proceed as in the case of a differential equation by taking the Laplace transform of both sides of the equation. The Convolution Theorem is used in determining the Laplace transform of the integral with.

$$\mathcal{L}\left\{\int_0^t I(u)du\right\} = \mathcal{L}\{1*I(t)\} = \mathcal{L}\{1\}\mathcal{L}\{I(t)\} = \frac{\mathcal{L}\{I(t)\}}{s}.$$

```
> with(inttrans):
  laplace(int(i(u),u=0..t),t,s);
```

$$\frac{\text{laplace}(\text{i}(t), t, s)}{s}$$

Therefore, application of the Laplace transform yields

$$Ls\mathcal{L}\{I(t)\} - \mathcal{L}I(0) + R\mathcal{L}\{I(t)\} + \frac{1}{C}\frac{\mathcal{L}\{I(t)\}}{s} = \mathcal{L}\{E(t)\}.$$

```
> step1:=laplace(L*diff(i(t),t)+R*i(t)+
      1/C*int(i(u),u=0..t)=E(t),t,s);
```

$$step1 := L\,(s\,\text{laplace}(\text{i}(t), t, s) - \text{i}(0)) + R\,\text{laplace}(\text{i}(t), t, s) + \frac{\text{laplace}(\text{i}(t), t, s)}{C\,s} = \text{laplace}(\text{E}(t), t, s)$$

Because $I(0) = 0$, we have

$$Ls\mathcal{L}\{I(t)\} + R\mathcal{L}\{I(t)\} + \frac{1}{C}\frac{\mathcal{L}\{I(t)\}}{s} = \mathcal{L}\{E(t)\}.$$

Simplifying and solving for $\mathcal{L}\{I(t)\}$ results in $\mathcal{L}\{I(t)\} = \dfrac{Cs\mathcal{L}\{E(t)\}}{\mathcal{L}Cs^2 + RCs + 1}$

```
> step2:=solve(subs(i(0)=0,step1),laplace(i(t),t,s));
```

$$step2 := \frac{\text{laplace}(\text{E}(t), t, s)\,C\,s}{L\,s^2\,C + R\,C\,s + 1}$$

so that

$$I(t) = L^{-1}\left\{\frac{Cs\mathcal{L}\{E(t)\}}{LCs^2 + RCs + 1}\right\}.$$

For (b), we note that $E(t) = \begin{cases} \sin t, & 0 \le t < \pi/2 \\ 0, & t \ge \pi/2 \end{cases}$ can be written as

$$(E(t) = \sin t(\mathcal{U}(t) - \mathcal{U}(t - \pi/2)) = \mathcal{U}(\pi/2 - t)\sin t).$$

Proceeding with Maple, we first define and graph E. Later we will display the graph of E along with the graph of the solution. Consequently, we name the graph of E `Plot_v` and load the `plots` package so that we may use the command `display`, which we can use to display a set of `plot` structures simultaneously. In addition, we define `cape` to be the Laplace transform of $E(t)$.

```
> E:=t->sin(t)*Heaviside(Pi/2-t):
  Plot_E:=plot(E(t),t=0..Pi):
  with(plots):
  display({Plot_E});
  cape:=laplace(E(t),t,s);
```

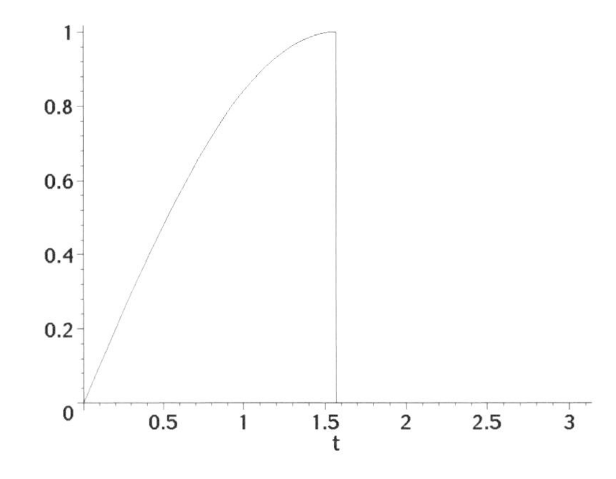

$$cape := -\frac{-1 + \mathbf{e}^{\left(-\frac{1}{2}s\pi\right)}s}{(s^2 + 1)}$$

Using (a), we define capi to be the simplified form of $\frac{\mathcal{L}(E(t))}{LC^2+RCs+1} = \frac{\text{cape}}{s^2+s+1}$ and then write capi.

```
> capi:=simplify(cape/(s^2+s+1));
```

$$capi := -\frac{-1+\mathbf{e}^{\left(-\frac{1}{2}s\pi\right)}s}{(s^2+1)(s^2+s+1)}$$

We determine $I(t)$ with invlaplace.

```
> i:=invlaplace(capi,s,t);
```

$$i: = -\cos(t) + \frac{1}{3}\mathbf{e}^{\left(-\frac{1}{2}t\right)}\sqrt{3}\sin\left(\frac{1}{2}\sqrt{3}t\right) + \mathbf{e}^{\left(-\frac{1}{2}t\right)}\cos\left(\frac{1}{2}\sqrt{3}t\right)$$
$$+\text{Heaviside}\left(t-\frac{1}{2}\pi\right)\cos(t)$$
$$+\frac{2}{3}\text{Heaviside}\left(t-\frac{1}{2}\pi\right)\mathbf{e}^{\left(-\frac{1}{2}t+\frac{1}{4}\pi\right)}\sqrt{3}\sin\left(\frac{1}{2}\sqrt{3}\left(t-\frac{1}{2}\pi\right)\right)$$

The solution is plotted in Plot_i and displayed with the forcing function in the plot that follows. Notice the effect that the forcing function has on the solution to the differential equation.

```
> Plot_i:=plot(i,t=0..10):
  display({Plot_i,Plot_E});
```

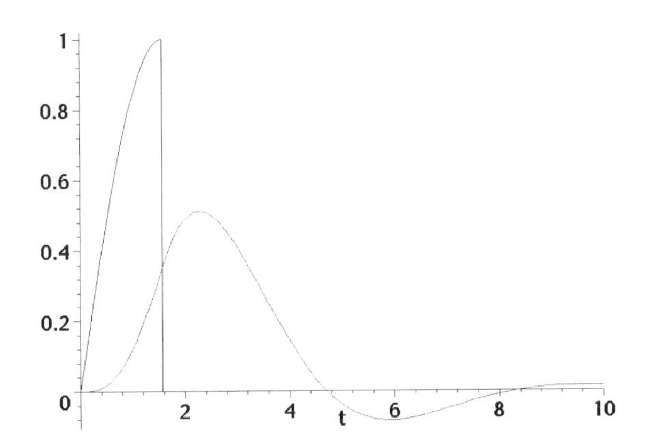

In this case, we see that we can use `dsolve` together with the option `method=laplace` to solve the initial value problem

$$\begin{cases} Q'' + Q' + Q = E(t) \\ Q(0) = 0, Q'(0) = 0 \end{cases}$$

as well.

```
>solb:=dsolve({diff(Q(t),t$2)+diff(Q(t),t)+
    Q(t)=E(t),Q(0)=0,D(Q)(0)=0},Q(t),
    method=laplace);
```

$$solb := Q(t) = -\cos(t) + \frac{1}{3} e^{(-1/2\,t)} \sqrt{3} \sin\left(\frac{1}{2}\sqrt{3}\,t\right) + e^{(-1/2\,t)} \cos\left(\frac{1}{2}\sqrt{3}\,t\right) + \text{Heaviside}\left(t - \frac{1}{2}\pi\right)\cos(t) + \frac{2}{3}\text{Heaviside}\left(t - \frac{1}{2}\pi\right) e^{(-1/2\,t + 1/4\,\pi)} \sqrt{3} \sin\left(\frac{1}{2}\sqrt{3}\left(t - \frac{1}{2}\pi\right)\right)$$

We use this result to graph $Q(t)$ and $I(t)=Q'(t)$.

```
> assign(sola):
  plot({Q(t),diff(Q(t),t)},t=0..10)
```

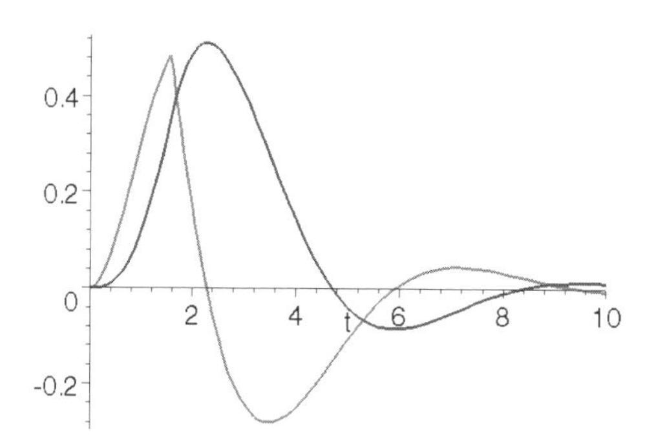

■

Application: The Dirac Delta Function

Let $\delta(t-t_0)$ denote the (generalized) function with the two properties (i) $(\delta(t-t_0)=0)$ if $t \neq t_0$; and (ii) $\int_{-\infty}^{+\infty} \delta(t-t_0)\,dt = 1$, which is called the **Dirac delta function** and is quite useful in the definition of impulse forcing functions that arise in some differential equations. The Laplace transform of $\delta(t-t_0)$ is $L\{\delta(t-t_0)\} = e^{-st_0}$. Maple contains a definition of the Dirac delta function, `Dirac`.

EXAMPLE 7: Solve $\begin{cases} x'' + x' + x = \delta(t) + \mathcal{U}(t-2\pi) \\ x(0)=0,\ x'(0)=0 \end{cases}$.

SOLUTION: We proceed by defining `Eq` to be the equation $x'' + x' + x = \delta(t) + \mathcal{U}(t-2\pi)$.

```
> Eq:=diff(x(t),t$2)+diff(x(t),t)+x(t)=
     Dirac(t)+Heaviside(t-2*Pi);
```

$$Eq := \left(\frac{\partial^2}{\partial t^2}\mathrm{x}(t)\right) + \left(\frac{\partial}{\partial t}\mathrm{x}(t)\right) + \mathrm{x}(t) = \mathrm{Dirac}(t) + \mathrm{Heaviside}(t-2\pi)$$

In this case, `dsolve`, along with the `method=laplace` option, solves the initial value problem. We name the resulting output `sola`.

```
> sola:=dsolve({Eq,x(0)=0,D(x)(0)=0},x(t),laplace);
```

$$sola := \mathrm{x}(t) = -\frac{1}{3}\sqrt{-3}\,(\mathbf{e}^{((-1/2+1/2\sqrt{-3})\,t)} - \mathbf{e}^{((-1/2-1/2\sqrt{-3})\,t)})$$
$$+\frac{2}{3}\frac{\mathrm{Heaviside}(t-2\pi)\left(\sqrt{-3}-\frac{1}{2}\mathbf{e}^{(-(1/2+1/2\sqrt{-3})(t-2\pi))}(-1+\sqrt{-3}) - \mathbf{e}^{(1/2(-1+\sqrt{-3})(t-2\pi))}\left(\frac{1}{2}+\frac{1}{2}\sqrt{-3}\right)\right)\sqrt{-3}}{\left(\frac{1}{2}+\frac{1}{2}\sqrt{-3}\right)(-1+\sqrt{-3})}$$

Of course, we can also use Maple to perform the computations encountered in the solution of this problem. We first clear all prior definitions of x and t and then define lap_Eq to be the Laplace transform of each term of the equation.

```
> lap_Eq:=laplace(Eq,t,s);
```

$$lap\_Eq := s\,(s\,\text{laplace}(\text{x}(t), t, s) - \text{x}(0)) - \text{D}(x)(0) + s\,\text{laplace}(\text{x}(t), t, s) - \text{x}(0) + \text{laplace}(\text{x}(t), t, s) = 1 + \frac{\mathbf{e}^{(-2\,s\,\pi)}}{s}$$

We then apply the initial conditions to lap_Eq by replacing each occurrence of $x(0)$ in lap_Eq by 0 and each occurrence of $x'(0)$ by 0, and we solve the resulting equation for the Laplace transform of x, naming the result lap_x. To calculate x, we must compute the inverse Laplace transform of lap_x.

```
> sub_conds:=subs({x(0)=0,D(x)(0)=0},lap_Eq);
```

$$sub\_conds := s^2\,\text{laplace}(\text{x}(t), t, s) + s\,\text{laplace}(\text{x}(t), t, s) + \text{laplace}(\text{x}(t), t, s) = 1 + \frac{\mathbf{e}^{(-2\,s\,\pi)}}{s}$$

```
> lap_x:=solve(sub_conds,laplace(x(t),t,s));
```

$$lap\_x := \frac{s + \mathbf{e}^{(-2\,s\,\pi)}}{s\,(s^2 + s + 1)}$$

We use invlaplace to compute the inverse Laplace transform of lap_x.

```
> invlaplace(lap_x,s,t);
```

$$\frac{2}{3}\mathbf{e}^{(-1/2\,t)}\sqrt{3}\,\sin\left(\frac{1}{2}\sqrt{3}\,t\right) + \text{Heaviside}(t - 2\,\pi) - \frac{1}{3}\text{Heaviside}(t - 2\,\pi)\,\mathbf{e}^{(-1/2\,t+\pi)}\sqrt{3}\,\sin\left(\frac{1}{2}\sqrt{3}\,(t - 2\,\pi)\right) - \text{Heaviside}(t - 2\,\pi)\,\mathbf{e}^{(-1/2\,t+\pi)}\cos\left(\frac{1}{2}\sqrt{3}\,(t - 2\,\pi)\right)$$

■

7.5 Systems of Ordinary Differential Equations

Homogeneous Linear Systems with Constant Coefficients

Let $\mathbf{A} = \begin{pmatrix} a_{11} & a_{12} & \cdots & a_{1n} \\ a_{21} & a_{22} & \cdots & a_{2n} \\ \vdots & \vdots & \ddots & \vdots \\ a_{n1} & a_{n1} & \cdots & a_{nn} \end{pmatrix}$ be an $n \times n$ real matrix, and let $\left\{\lambda_k\right\}_{k=1}^{n}$ be the eigenvalues and $\left\{\mathbf{v}_k\right\}_{k=1}^{n}$ the corresponding eigenvectors of **A**. Then, a general solution of the system $\mathbf{X}' = \mathbf{AX}$ is determined by the eigenvalues and corresponding eigenvectors of **A**. If the eigenvalues $\left\{\lambda_k\right\}_{k=1}^{n}$ of **A** are distinct, a general solution of $\mathbf{X}' = \mathbf{AX}$ can be written as

$$\mathbf{X}(\mathrm{t}) = \mathrm{c}_1\mathbf{v}_1\mathrm{e}^{\lambda_1 \mathrm{t}} + \mathrm{c}_2\mathbf{v}_2\mathrm{e}^{\lambda_2 \mathrm{t}} + \cdots + \mathrm{c}_\mathrm{n}\mathbf{v}_\mathrm{n}\mathrm{e}^{\lambda_\mathrm{n} \mathrm{t}},$$

where $\left\{\mathbf{v}_k\right\}_{k=1}^{n}$ are eigenvectors corresponding to $\left\{\lambda_k\right\}_{k=1}^{n}$.

If **A** has complex conjugate eigenvalues, $\lambda_1 = \alpha + \beta i$ and $\lambda_2 = \alpha - \beta i$ and corresponding eigenvectors $\mathbf{v}_1 = \mathbf{a} + \mathbf{b}i$ and $\mathbf{v}_2 = \mathbf{a} - \mathbf{b}i$. Then, one solution of $\mathbf{X}' = \mathbf{AX}$ is

$$\begin{aligned}
\mathbf{X} = \mathbf{v}_1 e^{\lambda t} &= (\mathbf{a} + \mathbf{b}i)e^{(\alpha + \beta i)t} = e^{\alpha t}(\mathbf{a} + \mathbf{b}i)e^{i\beta t} = e^{\alpha t}(\mathbf{a} + \mathbf{b}i)(\cos\beta t + \sin\beta t) \\
&= e^{\alpha t}(\mathbf{a}\cos\beta t - \mathbf{b}\sin\beta t) + ie^{\alpha t}(\mathbf{a}\cos\beta t + \mathbf{b}\sin\beta t) \\
&= \mathbf{X}_1(t) + i\mathbf{X}_2(t).
\end{aligned}$$

Now, because $\mathbf{X}$ is a solution of the system, $\mathbf{X}' = \mathbf{AX}$, we have $\mathbf{X}_1{}'(t) + i\mathbf{X}_2{}'(t) = \mathbf{AX}_1(t) + i\mathbf{AX}_2(t)$. Equating the real and imaginary parts of this equation yields $\mathbf{X}_1{}'(t) = \mathbf{AX}_1(t)$ and $i\mathbf{X}_2{}'(t) = \mathbf{AX}_2(t)$. Therefore, $\mathbf{X}_1(t)$ and $\mathbf{X}_2(t)$ are solutions of $\mathbf{X}' = \mathbf{AX}$, so any linear combination of $\mathbf{X}_1(t)$ and $\mathbf{X}_2(t)$ is also a solution. We can show that $\mathbf{X}_1(t)$ and $\mathbf{X}_2(t)$ are linearly independent, so this linear combination forms a portion of a general solution of $\mathbf{X}' = \mathbf{AX}$.

Maple can solve many systems of differential equations. To solve the 2 x 2 system with constant coefficients, $\begin{cases} dx/dt = ax + by \\ \mathrm{d}y/dt = cx + dy \end{cases}$, we enter the command

```
dsolve({diff(x(t),t)=a*x(t)+b*y(t),diff(y(t),t)=c*x(t)+d*y(t)},
                    {x(t),y(t)}).
```

EXAMPLE 1: Solve the system of equations $\begin{cases} dx/dt = -y \\ dy/dt = -x \end{cases}$.

SOLUTION: We begin by finding the eigenvalues and associated eigenvectors of the matrix of coefficients $\mathbf{A} = \begin{pmatrix} 0 & -1 \\ -1 & 0 \end{pmatrix}$ with `eigenvects`. This gives us $\lambda_1 = -1$ and $\lambda_2 = 1$. An eigenvector corresponding to $\lambda_1 = -1$ is given by $\mathbf{v}_1 = \begin{pmatrix} 1 \\ 1 \end{pmatrix}$ while that corresponding to $\lambda_2 = 1$ is $\mathbf{v}_2 = \begin{pmatrix} 1 \\ -1 \end{pmatrix}$. Therefore, a general solution is $\begin{pmatrix} x(t) \\ y(t) \end{pmatrix} = c_1 \begin{pmatrix} 1 \\ 1 \end{pmatrix} e^{-t} + c_2 \begin{pmatrix} 1 \\ -1 \end{pmatrix} e^{t}$.

```
> with(linalg):
  eigenvects([[0,-1],[-1,0]]);
```

$$[-1, 1, \{[1, 1]\}], [1, 1, \{[-1, 1]\}]$$

We can graph the solutions for various values of the constants c_1 and c_2. Notice that x and y both depend on the variable t, where $x(t)$ represents the x-coordinate at a particular time t and $y(t)$ represents the y-coordinate at a particular time t. Hence, we can plot these solutions parametrically in the xy-plane. (This type of graph is known as the **phase plane** of the system of equations.) First, the solutions are defined as `x(t)` and `y(t)`, using the formula obtained above. Several members of the family of solutions are created in `tograph` by substituting the values $-6, -4, -2, 0, 2, 4,$ and 6 for the constants `c1` and `c2`. `tograph` consists of forty-nine sets of pairs of functions, which are then plotted with `plot` in `graphone`

```
> x:=t->c1*exp(-t)+c2*exp(t):
  y:=t->c1*exp(-t)-c2*exp(t):

> vals:=[seq(2*i,i=-3..3)]:
  tograph:={seq(seq(subs({c1=i,c2=j},
```

```
    [x(t),y(t),t=-2..2]),i=vals),j=vals)}:
graphone:=plot(tograph,view=[-15..15,-15..15],
    color=BLACK):
with(plots):
display(graphone);
```

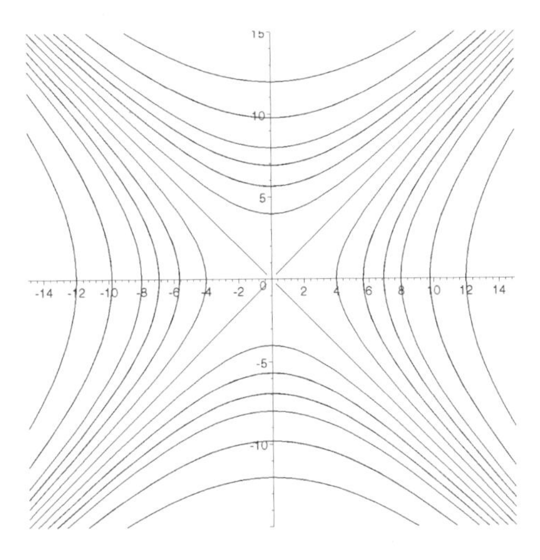

In order to determine the direction associated with these solutions, we consider the direction field (that is, the collection of vectors that represent the tangent line at points on the solutions). Note that the slope of these tangent lines is given by

$$\frac{dy}{dx} = \frac{dy/dt}{dx/dt} = \frac{cx+dy}{ax+by}$$

at each point (x, y)). The direction field is graphed with the command `PlotVectorField`, located in the DEtools package. Finally, the direction field is displayed with the solutions to illustrate the associated motion. Notice that when the associated direction field is plotted, solutions near the line in the direction of the eigenvector corresponding to the positive eigenvalue move away from the equilibrium point. On the other hand, solutions near the line in the direction of the eigenvector corresponding to the negative eigenvalue move toward the equilibrium point.

```
x:='x':y:='y':
> dirfield:=dfieldplot([diff(x(t),t)=-y(t),
      diff(y(t),t)=-x(t)],[x(t),y(t)],
```

```
    t=0..2,x=-15..15,y=-15..15):
display({graphone,dirfield});
```

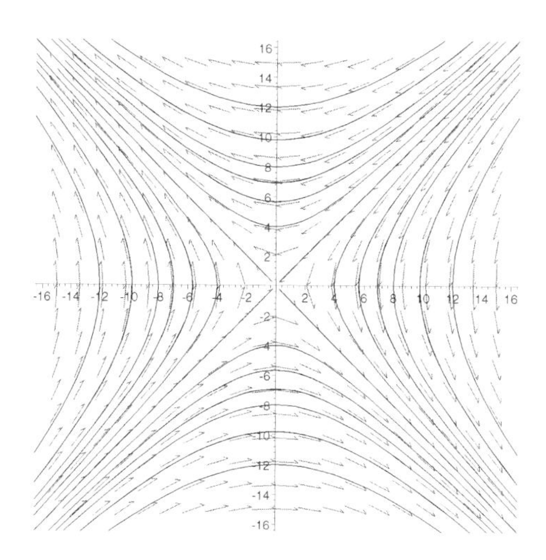

Graphs of solutions, together with direction fields, if desired, can be created without explicitly solving the system with commands like `DEplot` and `phaseportrait`, which are both contained in the `DEtools` package. For example, entering

```
> vals:=[seq(-2*i,i=-6..6)]:
  inits:=[seq([x(0)=i,y(0)=0],i=vals),
      seq([x(0)=0,y(0)=i],i=vals)];
```

inits := [[x(0) = 12, y(0) = 0], [x(0) = 10, y(0) = 0], [x(0) = 8, y(0) = 0], [x(0) = 6, y(0) = 0], [x(0) = 4, y(0) = 0], [x(0) = 2, y(0) = 0], [x(0) = 0, y(0) = 0], [x(0) = -2, y(0) = 0], [x(0) = -4, y(0) = 0], [x(0) = -6, y(0) = 0], [x(0) = -8, y(0) = 0], [x(0) = -10, y(0) = 0], [x(0) = -12, y(0) = 0], [x(0) = 0, y(0) = 12], [x(0) = 0, y(0) = 10], [x(0) = 0, y(0) = 8], [x(0) = 0, y(0) = 6], [x(0) = 0, y(0) = 4], [x(0) = 0, y(0) = 2], [x(0) = 0, y(0) = 0], [x(0) = 0, y(0) = -2], [x(0) = 0, y(0) = -4], [x(0) = 0, y(0) = -6], [x(0) = 0, y(0) = -8], [x(0) = 0, y(0) = -10], [x(0) = 0, y(0) = -12]]

defines the list of initial conditons `inits`. Then,

```
> phaseportrait([diff(x(t),t)=-y(t),
      diff(y(t),t)=-x(t)],[x(t),y(t)],
      t=-2..2,inits,x=-15..15,y=-15..15,
```

```
color=GRAY,linecolor=BLACK);
```

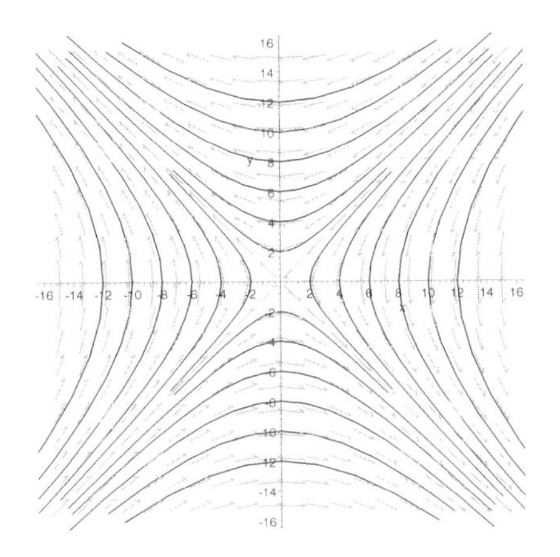

graphs (x versus y) the solutions to the system that satisfy the initial conditions specified in `inits` for $-2 \le t \le 2$. The direction field is graphed in gray, while the solution curves are graphed in black.

Last, we remark that `dsolve` is able to find a general solution of the system as well.

```
> solb:=dsolve({diff(x(t),t)=-y(t),
      diff(y(t),t)=-x(t)},{x(t),y(t)});
```

$$solb := \left\{ x(t) = \frac{1}{2}\_C1\, e^{(-t)} + \frac{1}{2}\_C1\, e^{t} - \frac{1}{2}\_C2\, e^{t} + \frac{1}{2}\_C2\, e^{(-t)}, \right.$$

$$\left. y(t) = -\frac{1}{2}\_C1\, e^{t} + \frac{1}{2}\_C1\, e^{(-t)} + \frac{1}{2}\_C2\, e^{(-t)} + \frac{1}{2}\_C2\, e^{t} \right\}$$

■

The commands `matrixDE(A,t)` and `matrixDE(A,F,t)` solve the systems $\mathbf{X}'(t) = \mathbf{A}(t)\mathbf{X}(t)$ and $\mathbf{X}'(t) = \mathbf{A}(t)\mathbf{X}(t) + \mathbf{F}(t)$, respectively. Note that `matrixDE` is contained in the `DEtools` package.

EXAMPLE 2: Find a general solution of the system of equations $\begin{cases} dx/dt = y \\ dy/dt = -x \end{cases}$.

SOLUTION: In matrix form, this system is equivalent to $\mathbf{X}' = \begin{pmatrix} 0 & 1 \\ -1 & 0 \end{pmatrix}\mathbf{X}$. The matrix of coefficients is defined in `A`, and the eigenvalues and corresponding eigenvectors are determined with `eigenvects`. The result means that the eigenvalues are $\lambda = \pm i$ and the eigenvectors are $\mathbf{v}_1 = \begin{pmatrix} i \\ 1 \end{pmatrix}$ and $\mathbf{v}_1 = \begin{pmatrix} -i \\ 1 \end{pmatrix}$. Note that the exponential function is not included because the eigenvalues are imaginary with no real part.

```
> with(linalg):
  with(DEtools):
  A:=array([[0,1],[-1,0]]):
  eigenvects(A);
```

$$[I, 1, \{[-I, 1]\}], [-I, 1, \{[I, 1]\}]$$

We find a fundamental matrix for the system using `matrixDE`.

```
> sola:=matrixDE(A,t);
```

$$sola := \left[\begin{bmatrix} \sin(t) & \cos(t) \\ \cos(t) & -\sin(t) \end{bmatrix}, [0, 0]\right]$$

This result indicates that a fundamental matrix for the system is $\Phi(t) = \begin{pmatrix} \sin t & \cos t \\ \cos t & -\sin t \end{pmatrix}$; a general solution is given by $\mathbf{X}(t) = \Phi(t)\begin{pmatrix} c_1 \\ c_2 \end{pmatrix} = \begin{pmatrix} \sin t & \cos t \\ \cos t & -\sin t \end{pmatrix}\begin{pmatrix} c_1 \\ c_2 \end{pmatrix}$.

```
> C:=matrix(2,1):
> gensol:=evalm(sola[1]&*C);
```

$$\begin{bmatrix} \sin(t)\, C_{1,1} + \cos(t)\, C_{2,1} \\ \cos(t)\, C_{1,1} - \sin(t)\, C_{2,1} \end{bmatrix}$$

In this case, we graph various solutions to the systems along with its direction field, using DEplot.

```
> inits:=[seq([x(0)=i,y(0)=i],i=-10..10)]:
  DEplot([diff(x(t),t)=y(t),
      diff(y(t),t)=-x(t)],[x(t),y(t)],
      t=-Pi..Pi,inits,x=-15..15,y=-15..15,
      color=BLACK,linecolor=GRAY);
```

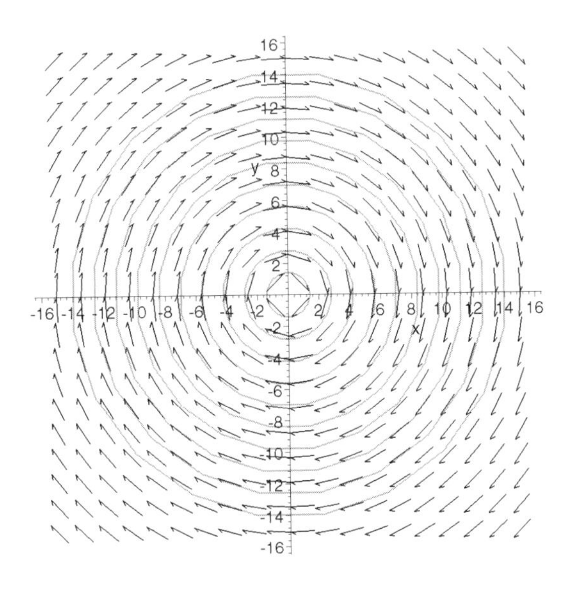

■

Initial-value problems can be solved with dsolve as well. To solve the initial-value problem

$$\begin{cases} dx/dt = ax + by \\ dy/dt = cx + dy \\ x(t_0) = x_0,\ y(t_0) = y_0 \end{cases},$$

the dsolve command is entered as

```
dsolve({diff(x(t),t)=a*x(t)+b*y(t),diff(y(t),t)=c*x(t)+d*y(t),
   x(t0)=x0,y(t0)=y0},{x(t),y(t)}).
```

EXAMPLE 3: Solve $\begin{cases} \mathbf{X}' = \begin{pmatrix} -1/4 & -4 \\ 4 & -1/4 \end{pmatrix} \mathbf{X} \\ \mathbf{X}(0) = \begin{pmatrix} 10 \\ 0 \end{pmatrix} \end{cases}$.

SOLUTION: This matrix equation is equivalent to the system

$$\begin{cases} \dfrac{dx}{\mathrm{dt}} = -\dfrac{1}{4}x - 4y \\ \dfrac{dy}{dt} = 4x - \dfrac{1}{4}y \\ x(0) = 10,\ y(0) = 0 \end{cases},$$

which we solve with `dsolve`.

```
> x:='x':y:='y':
> sol:=dsolve({diff(x(t),t)=-1/4*x(t)-4*y(t),
      diff(y(t),t)=4*x(t)-1/4*y(t),x(0)=10,y(0)=0},
      {x(t),y(t)});
```

$$sol := \{\, \mathrm{y}(t) = 10\,\mathbf{e}^{(-1/4\,t)}\sin(4\,t),\ \mathrm{x}(t) = 10\,\mathbf{e}^{(-1/4\,t)}\cos(4\,t) \,\}$$

We use `assign` to name $x(t)$ and $y(t)$ the results obtained in `sol` and then graph $x(t)$ and $y(t)$ together with `plot`. Notice that in the graph we see that $\lim_{t\to\infty} x(t) = \lim_{t\to\infty} y(t) = 0$.

```
> assign(sol):
```

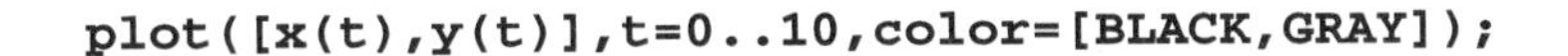

```
plot([x(t),y(t)],t=0..10,color=[BLACK,GRAY]);
```

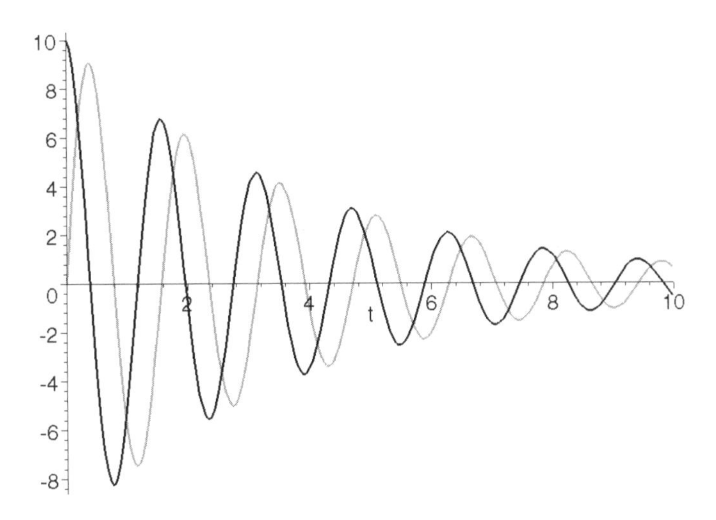

and then parametrically with `plot`, naming the result `param_plot`. We then use `dfieldplot`, which is contained in the `DEtools` package, to graph the direction field for the system for $-15 \le x \le 15$ and $-15 \le y \le 15$, naming the result `field_plot`. Both `param_plot` and `field_plot` are displayed together with `display`. In the graph, we see that the solution spirals into the origin.

```
> param_plot:=plot([x(t),y(t),t=0..10],color=BLACK):

> with(plots):
  with(DEtools):
  x:='x':y:='y':
  field_plot:=dfieldplot([diff(x(t),t)=-1/4*x(t)-4*y(t),
      diff(y(t),t)=4*x(t)-1/4*y(t)],[x(t),y(t)],
      t=-2..2,x=-15..15,y=-15..15,color=GRAY):
```

```
display({param_plot,field_plot});
```

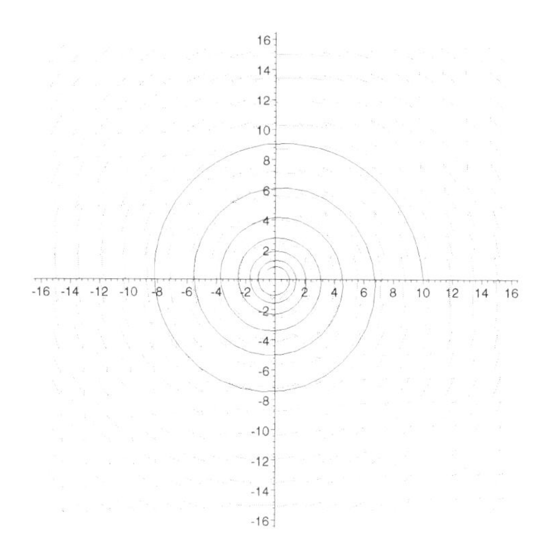

If we had not wished to generate an explicit solution to the initial-value problem, we could have used `DEplot` to generate graphical solutions. In the following `DEplot` commands, we include the options `scene=[t,x]` and `scene=[t,y]` to instruct Maple to graph t versus $x(t)$ and t versus $y(t)$, respectively.

```
> DEplot([diff(x(t),t)=-1/4*x(t)-4*y(t),
      diff(y(t),t)=4*x(t)-1/4*y(t)],[x(t),y(t)],
      t=0..10,x=-15..15,[[x(0)=10,y(0)=0]],y=-15..15,
      color=BLACK,linecolor=GRAY,stepsize=.1,
         scene=[t,x]);
  DEplot([diff(x(t),t)=-1/4*x(t)-4*y(t),
      diff(y(t),t)=4*x(t)-1/4*y(t)],[x(t),y(t)],
      t=0..10,x=-15..15,[[x(0)=10,y(0)=0]],y=-15..15,
```

```
color=BLACK,linecolor=GRAY,stepsize=.1,scene=[t,y]);
```

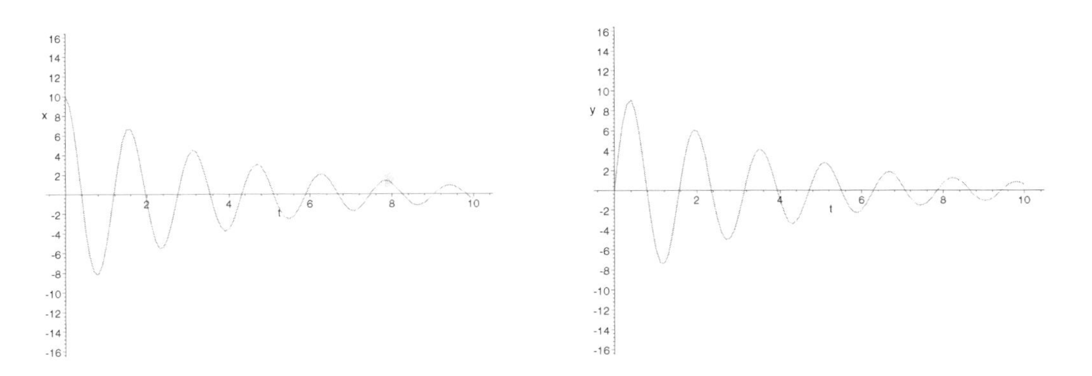

When we do not include the scene option, the parametric solution is graphed together with the direction field.

```
> DEplot([diff(x(t),t)=-1/4*x(t)-4*y(t),
      diff(y(t),t)=4*x(t)-1/4*y(t)],[x(t),y(t)],t=0..10,
      x=-15..15,[[x(0)=10,y(0)=0]],y=-15..15,
      color=BLACK,linecolor=GRAY,stepsize=.1);
```

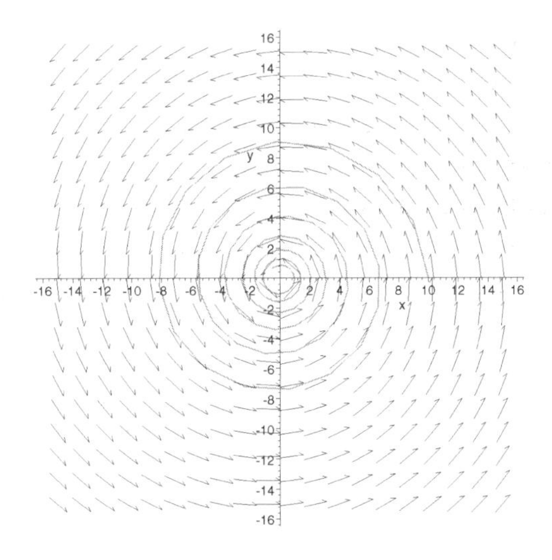

■

Variation of Parameters

We now consider nonhomogeneous systems of equations of the form $\mathbf{X}' = \mathbf{AX} + \mathbf{F}(t)$. The solution to the corresponding homogeneous system of equations $\mathbf{X}' = \mathbf{AX}$ can be represented in terms of the fundamental matrix $\Phi(t)$ and the $n \times 1$ constant vector $\mathbf{C}$ with $\mathbf{X} = \Phi(t)\mathbf{C}$. Hence, by assuming a particular solution to the nonhomogeneous system of the form $\mathbf{X}_p(t) = \Phi(t)\mathbf{V}(t)$, where $\mathbf{V}(t) = \begin{pmatrix} v_1(t) \\ v_2(t) \\ \vdots \\ v_n(t) \end{pmatrix}$, we find that a particular solution is given by $\mathbf{X}_p = \Phi(t)\int \Phi^{-1}(t)\mathbf{F}(t)dt$. Hence a general solution is determined with

$$(\mathbf{X}(t) = \Phi(t)\mathbf{C} + \mathbf{X}_p(t) = \Phi(t)\mathbf{C} + \Phi(t)\int \Phi^{-1}(t)\mathbf{F}(t)dt).$$

EXAMPLE 4: Solve $\begin{cases} x' - y = e^{-t} \\ y' + 5x + 2y = \sin 3t \\ x(0) = x_0,\ y(0) = y_0 \end{cases}$. Graph the solution for various initial conditions.

SOLUTION: We begin by clearing prior definitions of the variables `Sys`, `x`, and `y`, which were used above. We then define this nonhomogeneous system in `Sys`, and use `dsolve` together with the `method=laplace` option setting to apply the Method of Laplace Transforms to determine the solution $\{x(t), y(t)\}$ in `sol`. Note that when we use the option `method=laplace`, the solution is automatically given in terms of $x(0)$ and $y(0)$, not c_1 and c_2.

```
> Sys:='Sys':x:='x':y:='y':
  Sys:={diff(x(t),t)-y(t)=exp(-t),
      diff(y(t),t)+5*x(t)+2*y(t)=sin(3*t)}:
```

```
> sol:=dsolve(Sys,{x(t),y(t)},method=laplace);
```

$$sol := \{ x(t) = \frac{1}{4}e^{(-t)} - \frac{7}{52}e^{(-t)}\cos(2t) + e^{(-t)}x(0)\cos(2t) + \frac{35}{52}e^{(-t)}\sin(2t) + \frac{1}{2}e^{(-t)}x(0)\sin(2t) + \frac{1}{2}e^{(-t)}y(0)\sin(2t) - \frac{3}{26}\cos(3t) - \frac{1}{13}\sin(3t), y(t) = -\frac{5}{4}e^{(-t)} - \frac{3}{13}\cos(3t) + \frac{9}{26}\sin(3t) + \frac{77}{52}e^{(-t)}\cos(2t) + e^{(-t)}y(0)\cos(2t) - \frac{21}{52}e^{(-t)}\sin(2t) - \frac{5}{2}e^{(-t)}x(0)\sin(2t) - \frac{1}{2}e^{(-t)}y(0)\sin(2t) \}$$

We use `assign` to name $x(t)$ and $y(t)$ the results obtained in `sol`.

```
> assign(sol):
```

We will graph the solution for the sixteen initial conditions given in `T`.

```
> T:=[[0,0],[0,.25],[0,-.25],[.15,0],[0,1],[.5,.5],
      [-.5,.5],[-.5,-.5],[.5,-.5],[1,1],[1,0],
      [1,-1],[0,-1],[-1,-1],[-1,0],[-1,1]]:
  nops(T);
```

16

Next, we define `toplot` to be the list of ordered triples corresponding to $\{x(t), y(t), 0 \le t \le 7\}$ obtained by replacing $x(0)$ and $y(0)$ by each ordered pair given in `T`. We then generate sixteen graphs with `seq` and `plot`.

```
> toplot:={seq(subs({x(0)=T[i,1],y(0)=T[i,2]},
      [x(t),y(t),t=0..7]),i=1..16)}:
  graphs:=[seq(plot(toplot[i],color=BLACK),i=1..16)]:
```

The resulting list of graphs is displayed as an array using `display` together with the option `insequence=true`.

```
> with(plots):
  anarray:=display(graphs,insequence=true):
```

```
display(anarray);
```

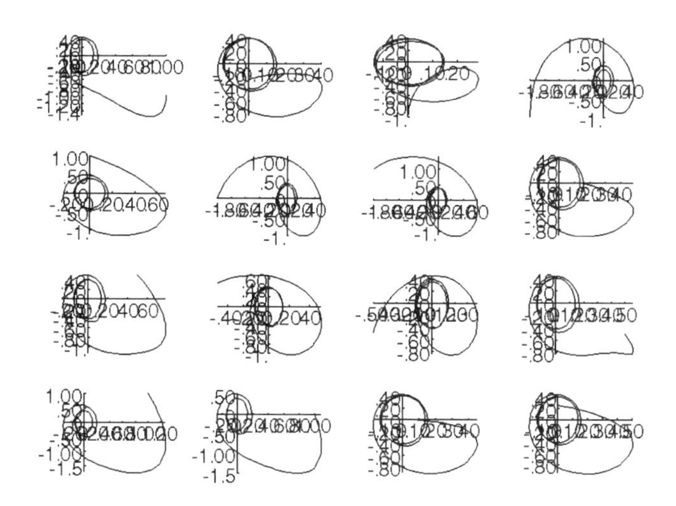

As in the previous examples, if we do not wish to generate an explicit solution, we can use some of the commands contained in the `DEtools` package to graph solutions to the system, satisfying various initial conditions. For example, next we use `DEplot` to graph the solution to the system satisfying $x(0) = -1$ and $y(0) = 2$. In the first command, we use a `stepsize` of 0.1; in the second, we use `DEplot3d`, a `stepsize` of 0.05 and the option `scene=[t,x(t),y(t)]` to generate a three-dimensional graph.

```
> x:='x':y:='y':
  with(DEtools):
  DEplot([diff(x(t),t)=y+exp(-t),
      diff(y(t),t)=-5*x-2*y+sin(3*t)],
      [x(t),y(t)],t=0..10,{[x(0)=-1,y(0)=2]},
      stepsize=0.1,linecolor=BLACK);
  DEplot3d([diff(x(t),t)=y+exp(-t),
      diff(y(t),t)=-5*x-2*y+sin(3*t)],
      [x(t),y(t)],t=0..10,{[x(0)=-1,y(0)=2]},
      stepsize=0.05,
```

```
scene=[t,x(t),y(t)],linecolor=BLACK);
```

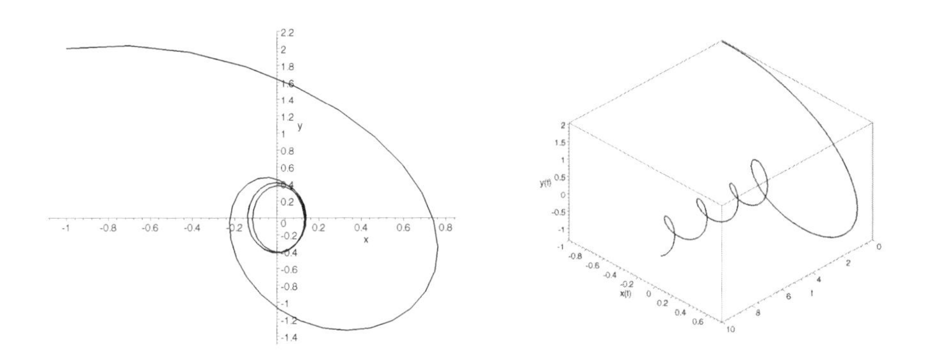

Alternatively, we can use other commands like `matrixDE` to solve the nonhomogeneous system. First, we rewrite the system in the form $\mathbf{X}' = \begin{pmatrix} 0 & 1 \\ -5 & -2 \end{pmatrix}\mathbf{X} + \begin{pmatrix} e^{-t} \\ \sin 3t \end{pmatrix}$. Next, we define $\mathbf{F}(t) = \begin{pmatrix} e^{-t} \\ \sin 3t \end{pmatrix}$ and $\mathbf{A} = \begin{pmatrix} 0 & 1 \\ -5 & -2 \end{pmatrix}$.

```
> A:=array([[0,1],[-5,-2]]);
  F:=matrix(2,1,[exp(-t),sin(3*t)]);
```

$$A := \begin{bmatrix} 0 & 1 \\ -5 & -2 \end{bmatrix}$$

$$F := \begin{bmatrix} \mathbf{e}^{(-t)} \\ \sin(3\,t) \end{bmatrix}$$

Maple returns an ordered pair when we use `matrixDE` to solve the system.

```
> solb:=matrixDE(A,F,t);
```

$$solb := \left[\begin{bmatrix} \mathbf{e}^{(-t)}\cos(2\,t) & \mathbf{e}^{(-t)}\sin(2\,t) \\ -\mathbf{e}^{(-t)}\cos(2\,t) - 2\,\mathbf{e}^{(-t)}\sin(2\,t) & -\mathbf{e}^{(-t)}\sin(2\,t) + 2\,\mathbf{e}^{(-t)}\cos(2\,t) \end{bmatrix},\right.$$
$$\left[\frac{1}{4}\mathbf{e}^{(-t)} + \frac{1}{4}\mathbf{e}^{(-t)}\cos(2\,t) - \frac{3}{26}\cos(3\,t) - \frac{1}{13}\sin(3\,t),\right.$$
$$\left.\left. -\frac{1}{52}\,\frac{13\,\mathbf{e}^{(-t)}\,\mathbf{e}^{t} + 13\,\mathbf{e}^{(-t)}\cos(2\,t)\,\mathbf{e}^{t} + 26\,\mathbf{e}^{(-t)}\sin(2\,t)\,\mathbf{e}^{t} - 18\sin(3\,t)\,\mathbf{e}^{t} + 12\cos(3\,t)\,\mathbf{e}^{t} + 52}{\mathbf{e}^{t}}\right]\right]$$

The result means that the solution to the system is $X(t) = \Phi(t)\begin{pmatrix} c_1 \\ c_2 \end{pmatrix} + \Phi_p(t)$, where $\Phi(t)$ is the first part of the solution, solb[1], and $\Phi_p(t)$ is the second, solb[2], defined next.

```
> C := matrix(2,1):
  x:=evalm(solb[1]&*C+solb[2]);
```

$$x := \left[\begin{array}{l} \mathbf{e}^{(-t)}\cos(2t)\,C_{1,1} + \mathbf{e}^{(-t)}\sin(2t)\,C_{2,1} + \frac{1}{4}\mathbf{e}^{(-t)} + \frac{1}{4}\mathbf{e}^{(-t)}\cos(2t) - \frac{3}{26}\cos(3t) - \frac{1}{13}\sin(3t) \\ (-\mathbf{e}^{(-t)}\cos(2t) - 2\,\mathbf{e}^{(-t)}\sin(2t))\,C_{1,1} + (-\mathbf{e}^{(-t)}\sin(2t) + 2\,\mathbf{e}^{(-t)}\cos(2t))\,C_{2,1} \\ \quad - \frac{1}{52}\frac{13\,\mathbf{e}^{(-t)}\,\mathbf{e}^{t} + 13\,\mathbf{e}^{(-t)}\cos(2t)\,\mathbf{e}^{t} + 26\,\mathbf{e}^{(-t)}\sin(2t)\,\mathbf{e}^{t} - 18\sin(3t)\,\mathbf{e}^{t} + 12\cos(3t)\,\mathbf{e}^{t} + 52}{\mathbf{e}^{t}} \end{array} \right]$$

The right-hand side of the equation, $\mathbf{AX}(t) + \mathbf{F}(t)$, is defined in rhs_eq.

```
> rhs_eq:=evalm(A&*x+F);
```

$$rhs\_eq := \left[\begin{array}{l} (-\mathbf{e}^{(-t)}\cos(2t) - 2\,\mathbf{e}^{(-t)}\sin(2t))\,C_{1,1} + (-\mathbf{e}^{(-t)}\sin(2t) + 2\,\mathbf{e}^{(-t)}\cos(2t))\,C_{2,1} \\ \quad - \frac{1}{52}\frac{13\,\mathbf{e}^{(-t)}\,\mathbf{e}^{t} + 13\,\mathbf{e}^{(-t)}\cos(2t)\,\mathbf{e}^{t} + 26\,\mathbf{e}^{(-t)}\sin(2t)\,\mathbf{e}^{t} - 18\sin(3t)\,\mathbf{e}^{t} + 12\cos(3t)\,\mathbf{e}^{t} + 52}{\mathbf{e}^{t}} + \mathbf{e}^{(-t)} \\ -5\,\mathbf{e}^{(-t)}\cos(2t)\,C_{1,1} - 5\,\mathbf{e}^{(-t)}\sin(2t)\,C_{2,1} - \frac{5}{4}\mathbf{e}^{(-t)} - \frac{5}{4}\mathbf{e}^{(-t)}\cos(2t) + \frac{15}{26}\cos(3t) + \frac{18}{13}\sin(3t) \\ \quad - 2\,(-\mathbf{e}^{(-t)}\cos(2t) - 2\,\mathbf{e}^{(-t)}\sin(2t))\,C_{1,1} - 2\,(-\mathbf{e}^{(-t)}\sin(2t) + 2\,\mathbf{e}^{(-t)}\cos(2t))\,C_{2,1} \\ \quad + \frac{1}{26}\frac{13\,\mathbf{e}^{(-t)}\,\mathbf{e}^{t} + 13\,\mathbf{e}^{(-t)}\cos(2t)\,\mathbf{e}^{t} + 26\,\mathbf{e}^{(-t)}\sin(2t)\,\mathbf{e}^{t} - 18\sin(3t)\,\mathbf{e}^{t} + 12\cos(3t)\,\mathbf{e}^{t} + 52}{\mathbf{e}^{t}} \end{array} \right]$$

We verify that the result obtained with matrixDE is the solution by computing $\mathbf{X}'(t) - (\mathbf{AX}(t) + \mathbf{F}(t)) = \mathbf{0}$.

```
> simplify(normal(diff(x[1,1],t))-rhs_eq[1,1],
      symbolic);
```

$$0$$

```
> simplify(normal(diff(x[2,1],t))-rhs_eq[2,1],
      symbolic);
```

0

■

Nonlinear Systems, Linearization, and Classification of Equilibrium Points

An **equilibrium point** (x_0, y_0) of the system of differential equations $\begin{cases} dx/dt = f(x, y) \\ dy/dt = g(x, y) \end{cases}$ is a point that satisfies $\begin{cases} f(x_0, y_0) = 0 \\ g(x_0, y_0) = 0 \end{cases}$. For the 2×2 system of linear differential equations $\mathbf{X}' = \mathbf{AX}$ in which the matrix of coefficients $\mathbf{A}$ has eigenvalues λ_1 and λ_2, the equilibrium point $(0, 0)$ is classified according to the following criteria:

Eigenvalues	Classification
Real: $\lambda_1 \le \lambda_2 < 0$	Stable Node
Real: $\lambda_1 \le 0 < \lambda_2$	Saddle
Real: $0 \le \lambda_1 < \lambda_2$	Unstable Node
Complex: $\lambda_1 = \overline{\lambda_2} = \alpha + \beta i,\ \beta \ne 0,\ \alpha < 0$	Stable Spiral
Complex: $\lambda_1 = \overline{\lambda_2} = \alpha + \beta i,\ \beta \ne 0,\ \alpha > 0$	Unstable Spiral
Complex: $\lambda_1 = \overline{\lambda_2} = \alpha + \beta i,\ \beta \ne 0,\ \alpha = 0$	Center

The general form of the autonomous system (in which there is no dependence on t) is $\begin{cases} dx/dt = f(x, y) \\ dy/dt = g(x, y) \end{cases}$. Approximate solutions to problems of this type can be found by considering the linearized system about each equilibrium point (x_0, y_0), which in this case satisfy $\{f(x_0, y_0) = 0, g(x_0, y_0) = 0\}$. This linearized system is given by

$$\begin{cases} dx/dt = f_x(x_0, y_0)x + f_y(x_0, y_0)y + \tilde{c}_1 \\ dy/dt = g_x(x_0, y_0)x + g_y(x_0, y_0)y + \tilde{c}_2 \end{cases},$$

where $f_x(x_0, y_0) = \frac{\partial f}{\partial x}(x_0, y_0)$, $f_y(x_0, y_0) = \frac{\partial f}{\partial y}(x_0, y_0)$, $g_x(x_0, y_0) = \frac{\partial g}{\partial x}(x_0, y_0)$, $g_y(x_0, y_0) = \frac{\partial g}{\partial y}(x_0, y_0)$, and $\tilde{c}_1$, $\tilde{c}_2$ are constants.

The matrix $J(x, y) = \begin{pmatrix} f_x(x, y) & f_y(x, y) \\ g_x(x, y) & g_y(x, y) \end{pmatrix}$ is known as the **Jacobian matrix**. Hence, the linearized system can be written as

$$\begin{pmatrix} dx/dt \\ dy/dt \end{pmatrix} = \begin{pmatrix} f_x(x_0, y_0) & f_y(x_0, y_0) \\ g_x(x_0, y_0) & g_y(x_0, y_0) \end{pmatrix} \begin{pmatrix} x - x_0 \\ y - y_0 \end{pmatrix} = J(x_0, y_0) \begin{pmatrix} x - x_0 \\ y - y_0 \end{pmatrix}.$$

Numerical Solutions of Systems of Ordinary Differential Equations

Maple's `dsolve` command with the `numeric` option setting can be used to approximate the solution of many nonlinear systems of differential equations. Of course, since this command applies a fourth-fifth order Runge-Kutta method, the correct number of initial conditions must be supplied within the `dsolve` command. Hence, the `numeric` setting enables us to use `dsolve` to find approximate solutions to initial value problems.

This command is entered in the form

```
dsolve({desys,ics},var,numeric),
```

where `desys` and `ics` represent the system of differential equations and initial conditions, respectively. The output is a procedure that represents a list of ordered pairs. The elements in the output list `(indvar,mn,[ord1,ord2,...])` represent the independent variable, a random machine number, and the order of each differential equation, respectively. Note that the result of `dsolve` with the `numeric` option setting is a procedure that can be graphed with the `odeplot` command, contained in the `plots` package. For a given value of the independent variable, the procedure returns values for the indicated independent variable and the corresponding dependent variables.

Application: Predator-Prey Model

EXAMPLE 5: The **Lotka-Volterra system** (Predator-Prey model) is the system ,

$$\begin{cases} x' = a_1 x - a_2 xy \\ y' = -b_1 y + b_2 xy \end{cases}$$

where a_1 , a_2 , b_1 , and b_2 are constants. Find and classify the equilibrium points of the Lotka-Volterra equations.

SOLUTION: We begin by defining `eq1` and `eq2` to be $x' = a_1 x - a_2 xy$ and $y' = -b_1 y + b_2 xy$, respectively, and then solving the system of equations $\begin{cases} a_1 x - a_2 xy = 0 \\ -b_1 y + b_2 xy = 0 \end{cases}$ for x and y to locate the equilibrium points.

```
> x:='x':y:='y':
  rhs_eq1:=a[1]*x-a[2]*x*y:
  rhs_eq2:=-b[1]*y+b[2]*x*y:
  cps:=solve({rhs_eq1=0,rhs_eq2=0},{x,y});
```

$$cps := \{x = 0, y = 0\}, \{x = \frac{b_1}{b_2}, y = \frac{a_1}{a_2}\}$$

To classify the equilibrium points, we first define `linmatrix` to be the matrix

$$\begin{pmatrix} \frac{d}{dx}(a_1 x - a_2 xy) & \frac{d}{dy}(a_1 x - a_2 xy) \\ \frac{d}{dx}(-b_1 y + b_2 xy) & \frac{d}{dy}(-b_1 y + b_2 xy) \end{pmatrix}.$$

```
> linmatrix:=[[diff(rhs_eq1,x),diff(rhs_eq1,y)],
      [diff(rhs_eq2,x),diff(rhs_eq2,y)]];
```

$$linmatrix := [[a_1 - (a_2 y, -a_2 x)], [b_2 y - b_1 + b_2 x]]$$

We then compute the value of `linmatrix` if $x = b_1/b_2$ and $y = a_1/a_2$,

```
> lin2:=subs(cps[2],linmatrix);
```

$$lin2 := \left[\left[0, -\frac{a_2 b_1}{b_2}\right], \left[\frac{b_2 a_1}{a_2}, 0\right]\right]$$

and the eigenvalues of this matrix. Because the eigenvalues are complex conjugates with the real part equal to 0, we conclude that the equilibrium point $(b_1/b_2, a_1/a_2)$ is a center.

```
> with(linalg):
  eigenvals(lin2);
```

$$\sqrt{-b_1 a_1}, -\sqrt{-b_1 a_1}$$

Similarly we compute the value of `linmatrix` if $x = 0$ and $y = 0$,

```
> eigenvals(subs(cps[1],linmatrix));
```

$$a_1, -b_1$$

and then the eigenvalues. Because the eigenvalues are real and have opposite signs, we conclude that the equilibrium point $(0, 0)$ is a saddle.

We use `dsolve` together with the `numeric` option to solve the system numerically if $x(0) = 1$ and $y(0) = 1$.

```
> eq1:=diff(x(t),t)=2*x(t)-x(t)*y(t):
  eq2:=diff(y(t),t)=-3*y(t)+x(t)*y(t):
  sol1:=dsolve({eq1,eq2,x(0)=1,y(0)=1},
      {x(t),y(t)},numeric);
```

$$sol1 := \textbf{proc}(rkf45\_x) \ \ldots \ \textbf{end}$$

We can evaluate this result for particular values of t. For example, entering

```
> sol1(4);
```

$$[t = 4, \mathrm{x}(t) = 5.251673011028663, \mathrm{y}(t) = .5583196721418087]$$

shows us that $x(4) \approx 5.25167$ and $y(4) \approx 0.558341$. On the other hand, entering

```
> seq(sol1(t),t=0..5);
```

$$[0=0, x(0)=1., y(0)=1.], [1=1, x(1)=4.335622930119876, y(1)=.4448252098341622],$$
$$[2=2, x(2)=2.026705097058897, y(2)=5.503181686191604],$$
$$[3=3, x(3)=1.152502750135020, y(3)=.7823187697402739],$$
$$[4=4, x(4)=5.251673019498925, y(4)=.5583196719204030],$$
$$[5=5, x(5)=1.366670016318617, y(5)=4.639806357798973]$$

creates a table of values of $x(t)$ and $y(t)$ for $t \approx 0, 1, 2, 3, 4, 5$. We then use `odeplot` to graph $x(t)$ and $y(t)$ together and then graph the solution parametrically.

```
> with(plots):
> odeplot(sol1,[[t,x(t)],[t,y(t)]],0..10,
      color=BLACK);
```

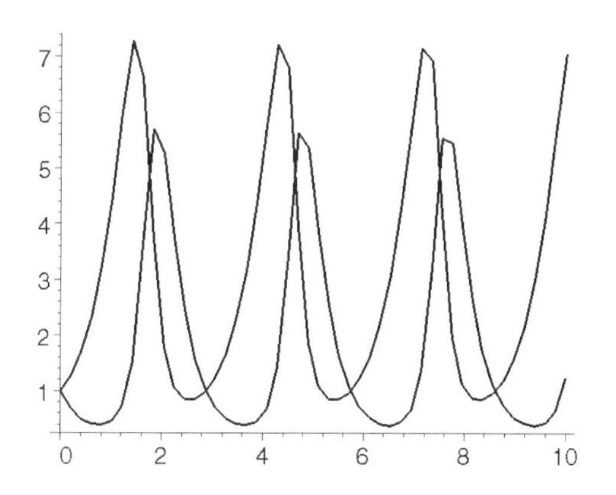


```
> odeplot(sol1,[x(t),y(t)],0..4,color=BLACK);
```

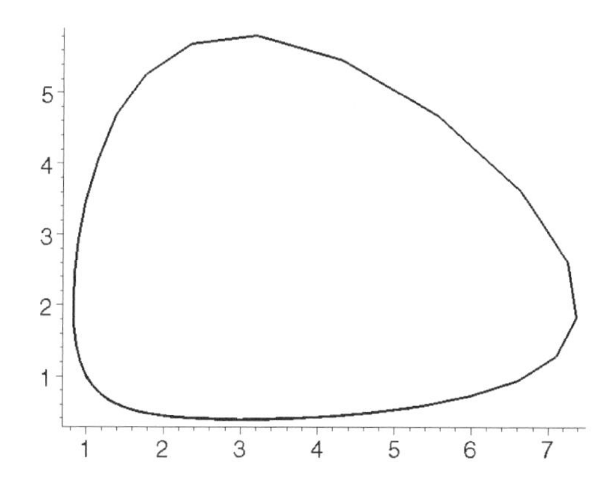

We can obtain a more accurate representation of the phase plane of the nonlinear system by graphing several solutions along with the direction field. First, we define a list of initial conditions, `inits`,

```
> inits:=[seq([x(0)=3*(1/8+3*i/40),
      y(0)=2*(1/8+3*i/40)],i=1..10)];
```

$$inits := \left[\left[x(0)=\frac{3}{5}, y(0)=\frac{2}{5}\right], \left[x(0)=\frac{33}{40}, y(0)=\frac{11}{20}\right], \left[x(0)=\frac{21}{20}, y(0)=\frac{7}{10}\right], \left[x(0)=\frac{51}{40}, y(0)=\frac{17}{20}\right], \left[x(0)=\frac{3}{2}, y(0)=1\right], \left[x(0)=\frac{69}{40}, y(0)=\frac{23}{20}\right], \left[x(0)=\frac{39}{20}, y(0)=\frac{13}{10}\right], \left[x(0)=\frac{87}{40}, y(0)=\frac{29}{20}\right], \left[x(0)=\frac{12}{5}, y(0)=\frac{8}{5}\right], \left[x(0)=\frac{21}{8}, y(0)=\frac{7}{4}\right]\right]$$

Then we use `phaseportrait` to graph the solutions that satisfy the initial conditions specified in `inits` together with the direction field.

```
> phaseportrait([eq1,eq2],[x(t),y(t)],t=0..4,
      inits,x=0..12,y=0..12,linecolor=BLACK,
      color=GRAY,stepsize=0.05);
```

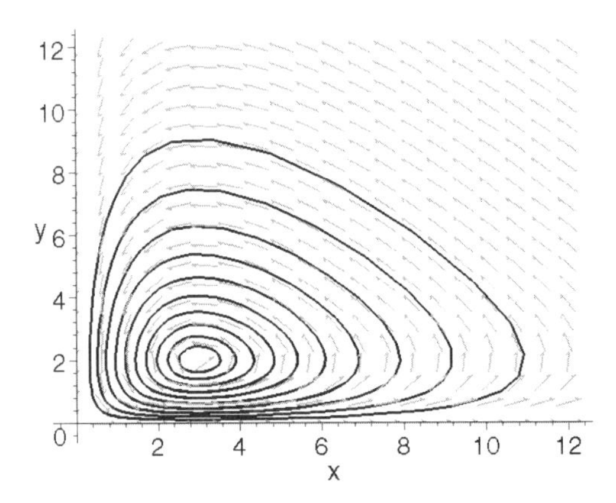

Notice that all of the solutions oscillate about the center. These solutions reveal the relationship between the two populations: prey, $x(t)$, and predator, $y(t)$. As we follow one cycle counterclockwise, beginning near the point $(2, 0)$, for example, we notice that as $x(t)$ increases, $y(t)$ increases until $y(t)$ becomes overpopulated.

Then, because the prey population is too small to supply the predator population, $y(t)$ decreases, which leads to an increase in the population of $x(t)$. Because the number of predators becomes too small to control the number in the prey population, $x(t)$ becomes overpopulated, and the cycle repeats itself.

■

Next, we consider the solution of a second-order nonlinear equation by transforming the equation into the corresponding system of equations.

EXAMPLE 6: The **Van-der-Pol equation** $x'' + \mu(x^2 - 1)x' + x = 0$ is equivalent to the system $\begin{cases} x' = y \\ y' = \mu(1 - x^2)y - x \end{cases}$. Classify the equilibrium points and plot the phase plane near the equilibrium point(s) for various values of μ.

SOLUTION: We find these equilibrium points by solving

$$\begin{cases} y = 0 \\ -x - \mu(x^2 - 1)y = 0 \end{cases}$$

From the first equation, we see that $y = 0$. Then, substitution of $y = 0$ into the second equation yields $x = 0$ as well. Therefore, the only equilibrium point is $(0, 0)$.

The Jacobian matrix for this system is

$$J(x, y) = \begin{pmatrix} 0 & 1 \\ -1 - 2\mu xy & -\mu(x^2 - 1) \end{pmatrix}$$

The eigenvalues of $J(0, 0)$ are $\lambda_{1,2} = \frac{1}{2}\left(\mu \pm \sqrt{\mu^2 - 4}\right)$.

```
> lin_mat:=array([[0,1],[-1,mu]]):
  with(linalg):
  eigs:=eigenvals(lin_mat);
```

$$eigs := \frac{1}{2}\mu + \frac{1}{2}\sqrt{\mu^2 - 4}, \frac{1}{2}\mu - \frac{1}{2}\sqrt{\mu^2 - 4}$$

Notice that if $\mu > 2$, then both eigenvalues are positive and real. Hence, we classify $(0, 0)$ as an **unstable node**. On the other hand, if $0 < \lambda < 2$, then the eigenvalues are a complex conjugate pair with a positive real part. Hence, $(0, 0)$ is an **unstable spiral**. (We omit the case $\mu = 2$ because the eigenvalues are repeated.) To graph the solutions for various values of μ and different initial conditions, we define sys. Given μ, sys(μ) returns the system $\begin{cases} x' = y \\ y' = \mu(1-x^2)y - x \end{cases}$.

```
sys:=mu->[diff(x(t),t)=y(t),
     diff(y(t),t)=mu*(1-x(t)^2)*y(t)-x(t)];
```

$$sys := \mu \rightarrow \left[\frac{\partial}{\partial t}\mathrm{x}(t) = \mathrm{y}(t), \frac{\partial}{\partial t}\mathrm{y}(t) = \mu\,(1 - \mathrm{x}(t)^2)\,\mathrm{y}(t) - \mathrm{x}(t)\right]$$

We then form various initial conditions in inits1, inits2, inits3, inits4, and inits5 and use union to join these five sets into initconds.

```
> inits1:={seq([x(0)=0.1*cos(2*Pi*i/4),
     y(0)=0.1*sin(2*Pi/4)],i=0..4)};
  inits2:={seq([x(0)=-5,y(0)=-5+10*i/9],i=0..9)};
  inits3:={seq([x(0)=5,y(0)=-5+10*i/9],i=0..9)}:
  inits4:={seq([x(0)=-5+10*i/9,y(0)=-5],i=0..9)}:
  inits5:={seq([x(0)=-5+10*i/9,y(0)=5],i=0..9)}:
  initconds:=`union`(inits1,inits2,inits3,
     inits4,inits5):
```

We then use phaseportrait to graph the solution to sys (μ) for $\mu = 1/2, 1, 3/2,$ and 3 using the initial conditions in the set initconds.

```
> with(DEtools):
  A:=array(1..2,1..2):
  A[1,1]:=phaseportrait(sys(1/2),[x(t),y(t)],
     t=0..20,initconds,x=-5..5,y=-5..5,
     arrows=NONE,linecolor=BLACK,stepsize=0.05):
  A[1,2]:=phaseportrait(sys(1),[x(t),y(t)],
     t=0..20,initconds,x=-5..5,y=-5..5,
     arrows=NONE,linecolor=BLACK,stepsize=0.05):
  A[2,1]:=phaseportrait(sys(3/2),[x(t),y(t)],
     t=0..20,initconds,x=-5..5,y=-5..5,
```

```
    arrows=NONE,linecolor=BLACK,stepsize=0.05):
A[2,2]:=phaseportrait(sys(3),[x(t),y(t)],
    t=0..20,initconds,x=-5..5,y=-5..5,
    arrows=NONE,linecolor=BLACK,stepsize=0.05):
```

These four graphics objects are displayed as an array with `display`. In each figure, we see that all of the curves approach a curve called a **limit cycle**. Physically, the fact that the system has a limit cycle indicates that for all oscillations, the motion eventually becomes periodic, which is represented by a closed curve in the phase plane.

```
> with(plots):
  display(A);
```

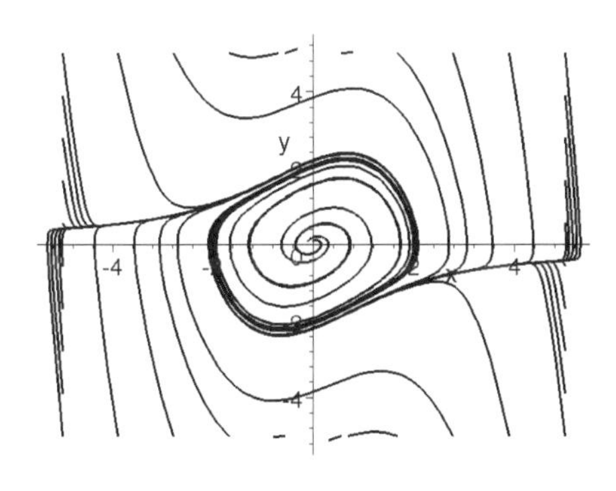

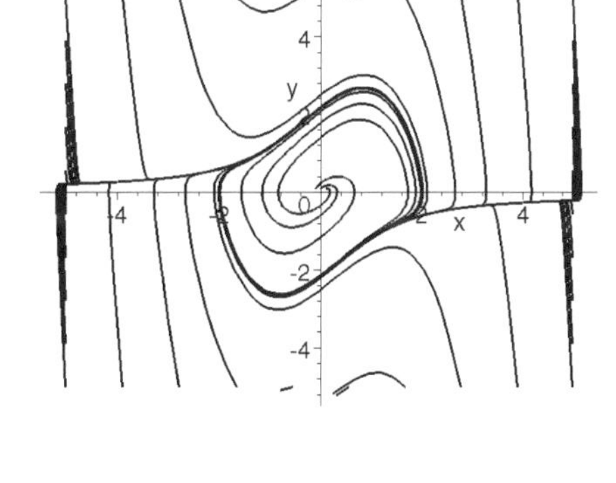

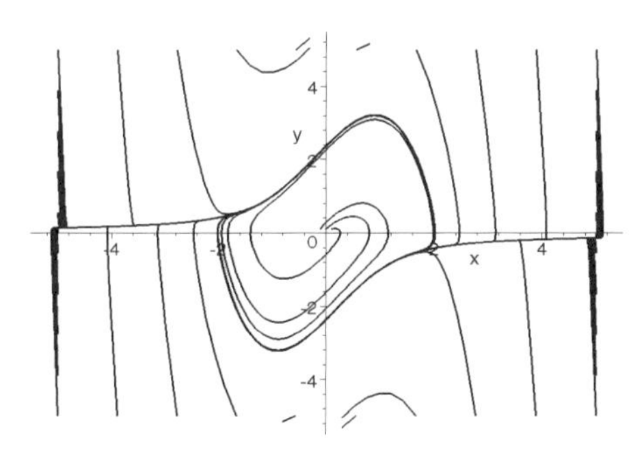

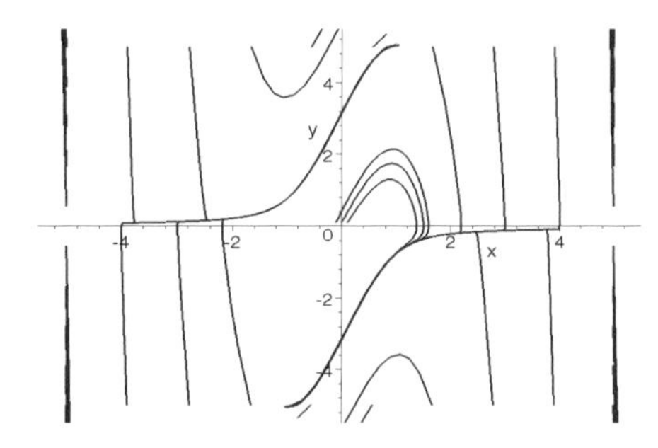

■

Application: The Double Pendulum

EXAMPLE 7: The motion of a double pendulum is modeled by the following system of equations, using the approximation $\theta \approx \theta$ for small displacements:

$$\begin{cases} (m_1 + m_2)l_1^2\theta_1'' + m_2 l_1 l_2 \theta_2'' + (m_1 + m_2)l_1 g\theta_1 = 0 \\ m_2 l_2^2 \theta_2'' + m_2 l_1 l_2 \theta_1'' + m_2 l_2 g\theta = 0 \end{cases},$$

where θ_1 represents the displacement of the upper pendulum, and θ_2 that of the lower pendulum. Also, m_1 and m_2 represent the mass attached to the upper and lower pendulums, respectively, while the length of each is given by l_1 and l_2. Suppose that $m_1 = 3$, $m_2 = 1$, and each pendulum has length 16. If $\theta_1(0) = 1$, $\theta_1'(0) = 0$, $\theta_2(0) = -1$, and $\theta_2'(0) = 0$, then solve the double pendulum problem using $g = 32$. Also, plot the solution.

SOLUTION: In this case, the system is

$$\begin{cases} 4 \cdot 16^2\theta_1'' + 16^2\theta_2'' + 4 \cdot 16 \cdot 32\theta_1 = 0 \\ 16^2\theta_1'' + 16^2\theta_2'' + 16 \cdot 32\theta_2 = 0 \end{cases},$$

which can be simplified to obtain

$$\begin{cases} 4\theta_1'' + \theta_2'' + 8\theta_1 = 0 \\ \theta_1'' + \theta_2'' + 2\theta_2 = 0 \end{cases}.$$

In the following code, we define $\theta_1(t)$ and $\theta_2(t)$ and then use `dsolve` to solve the initial-value problem. Notice that `dsolve` is unable to solve the initial-value problem unless we include the option `method=laplace`.

```
> Eq1:=4*diff(theta[1](t),t$2)+diff(theta[2](t),t$2)+
      8*theta[1](t)=0;
  Eq2:=diff(theta[1](t),t$2)+diff(theta[2](t),t$2)+
      2*theta[2](t)=0;
```

$$Eq1 := 4\left(\frac{\partial^2}{\partial t^2}\theta_1(t)\right)+\left(\frac{\partial^2}{\partial t^2}\theta_2(t)\right)+8\,\theta_1(t)=0$$

$$Eq2 := \left(\frac{\partial^2}{\partial t^2}\theta_1(t)\right)+\left(\frac{\partial^2}{\partial t^2}\theta_2(t)\right)+2\,\theta_2(t)=0$$

```
> sola:=dsolve({Eq1,Eq2,theta[1](0)=1,
      D(theta[1])(0)=0,theta[2](0)=1,D(theta[2])(0)=0},
      {theta[1](t),theta[2](t)},method=laplace);
```

$$sola := \{\,\theta_2(t)=-\frac{1}{2}\cos(2\,t)+\frac{3}{2}\cos\left(\frac{2}{3}\sqrt{3}\,t\right),\ \theta_1(t)=\frac{1}{4}\cos(2\,t)+\frac{3}{4}\cos\left(\frac{2}{3}\sqrt{3}\,t\right)\}$$

```
> solb:=dsolve({Eq1,Eq2,theta[1](0)=1,
      D(theta[1])(0)=0,theta[2](0)=1,D(theta[2])(0)=0},
      {theta[1](t),theta[2](t)});
```

$$solb :=$$

These two functions are graphed together and parametrically.

```
> assign(sola):
  plot([theta[1](t),theta[2](t)],t=0..20,
      color=[BLACK,GRAY]);
```

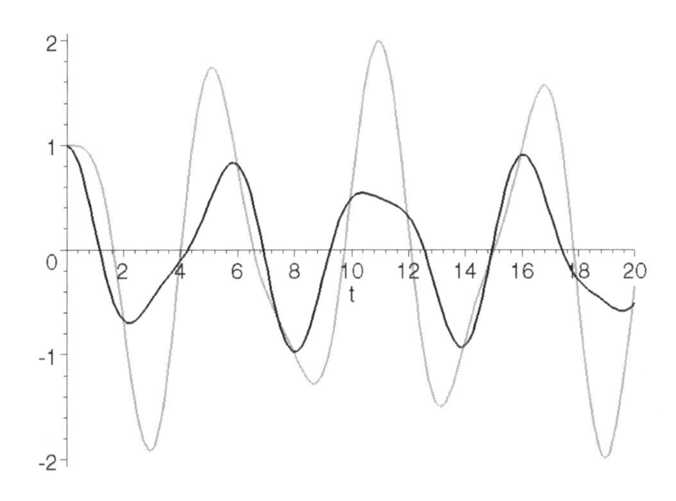

```
> plot([theta[1](t),theta[2](t),t=0..20],color=BLACK);
```

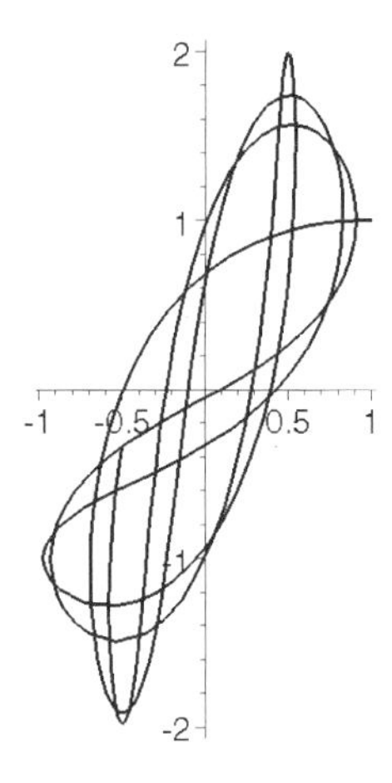

We can illustrate the motion of the pendulum by animating the motion of the pendulum with Maple's animation capabilities. First, we define the function `pen2`. The result of entering `pen2(t,l1,l2)` looks like the motion of the pendulum with lengths `l1` and `l2` at time `t`.

```
> pen2:=proc(t0,len1,len2)
      local pt1,pt2,xt0,yt0;
      xt0:=evalf(subs(t=t0,theta[1](t)));
      yt0:=evalf(subs(t=t0,theta[2](t)));
      pt1:=[len1*cos(3*Pi/2+xt0),len1*sin(3*Pi/2+xt0)];
      pt2:=[len1*cos(3*Pi/2+xt0)+len2*cos(3*Pi/2+yt0),
         len1*sin(3*Pi/2+xt0)+len2*sin(3*Pi/2+yt0)];
      plot([[0,0],pt1,pt2],xtickmarks=2,ytickmarks=2,
         view=[-32..32,-32..0]);
   end:
```

Next, we define `tvals` to be a list of sixteen evenly spaced numbers between 0 and 10. We use `seq` to calculate `pen2` for each value of t in `tvals` and name the resulting list of graphics `toshow`.

```
> with(plots):
  tvals:=[seq(10*i/15,i=0..15)]:
  toshow:=[seq(pen2(t,16,16),t=tvals)]:
```

We can animate this list of graphics using `display` together with the option `insequence=true`. One frame from the resulting animation is shown in the following screenshot:

```
> display(toshow,insequence=true);
```

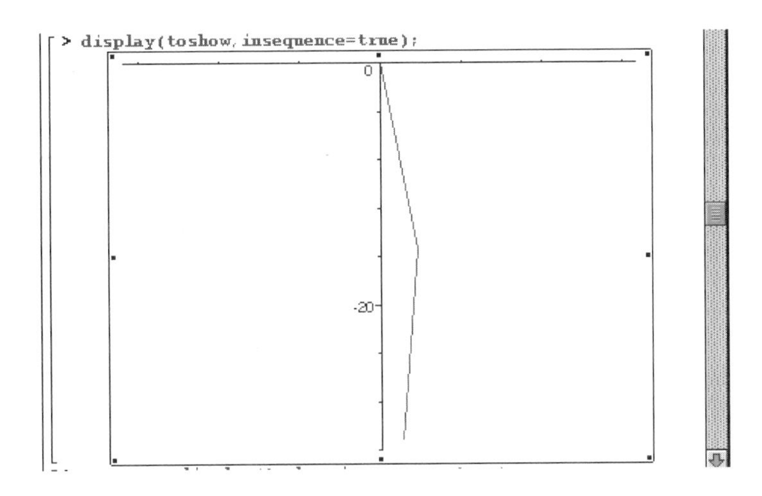

Alternatively, entering

```
> anarray:=display(toshow,insequence=true):
  display(anarray);
```

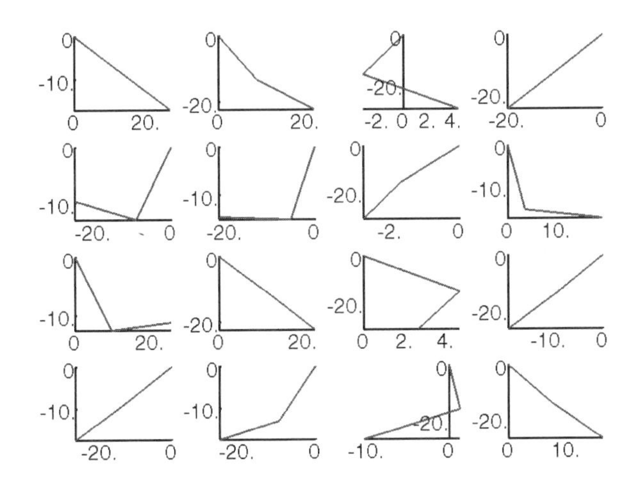

displays the list `toshow` as an array.

■

7.6 Some Partial Differential Equations

We now turn our attention to several partial differential equations. Several examples in this section will take advantage of commands contained in the `PDEtools` package. Information regarding the functions contained in the PDEtools package is obtained with `?PDEtools`.

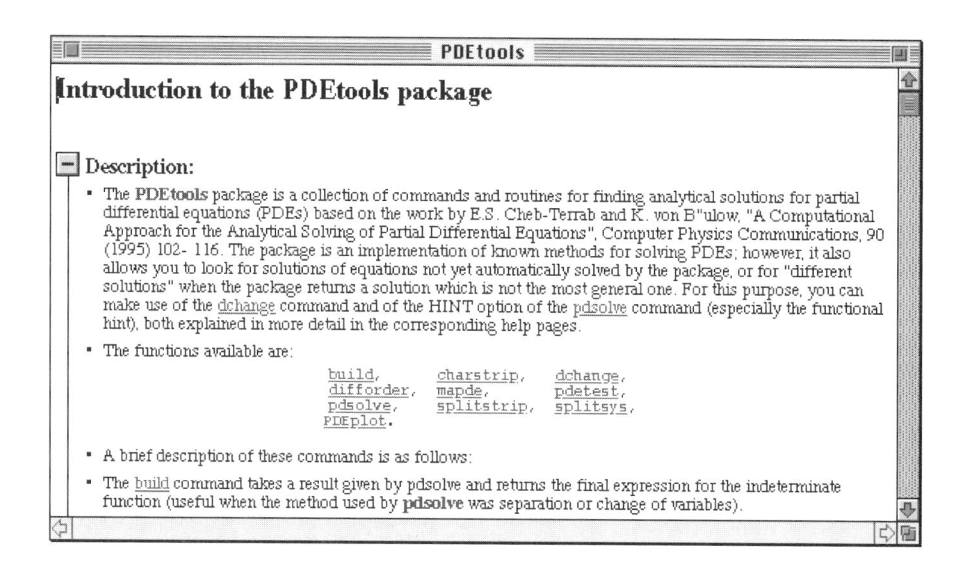

The One-Dimensional Wave Equation

Suppose that we pluck a string (like a guitar or violin string) of length p and constant mass density that is fixed at each end. A question that we might ask is: "What is the position of the string at a particular instance of time?" We answer this question by modeling the physical situation with a partial differential equation, namely the **wave equation**, in one spatial variable:

$$c^2 u_{xx} = u_{tt}.$$

In this equation, $c^2 = T/\rho$ where T is the tension of the string and ρ is the constant mass of the string per unit length. The solution $u(x, t)$ represents the displacement of the string from the x-axis at time t. In order to determine u, we must describe the boundary and initial conditions that model the physical situation. At the ends of the string, the displacement from the x-axis is fixed at zero, so we use the homogeneous boundary conditions.

$$u(0, t) = 0 \text{ and } u(p, t) = 0 \text{ for } t > 0.$$

The motion of the string also depends on the displacement and the velocity at each point of the string at $t = 0$. If the initial displacement is given by $f(x)$ and the initial velocity by $g(x)$, we have the initial conditions.

$$u(x, 0) = f(x) \text{ and } u_t(x, 0) = g(x) \text{ for } 0 < x < p.$$

Therefore, we determine the displacement of the string with the initial-boundary value problem

$$\begin{cases} c^2 u_{xx} = u_{tt}, \ 0 < x < p, t > 0 \\ u(0, t) = 0, u(p, t) = 0, t > 0 \\ u(x, 0) = f(x), u_t(x, 0) = g(x), 0 < x < p \end{cases}.$$

This problem is solved through separation of variables by assuming that $u(x, t) = X(x)T(t)$. Substitution into the wave equation yields

$$c^2 X''T = XT''$$
$$\frac{X''}{X} = \frac{T''}{c^2 T} = -\lambda'$$

so we obtain the two second-order ordinary differential equations

$$X'' + \lambda X = 0 \text{ and } T'' + c^2 \lambda T = 0.$$

At this point, we solve the equation that involves the homogeneous boundary conditions. The boundary conditions in terms of $u(x, t) = X(x)T(t)$ are $u(0, t) = X(0)T(t) = 0$ and $u(p, t) = X(p)T(t) = 0$, so we have

$$X(0) = 0 \text{ and } X(p) = 0.$$

Therefore, we determine $X_{(x)}$ by solving the eigenvalue problem

$$\begin{aligned} X'' + \lambda X &= 0, 0 < x < p \\ X(0) &= 0, X(p) = 0 \end{aligned}.$$

The eigenvalues of this problem are

$$\lambda_n = (n\pi/p)^2, n = 1, 2, \ldots$$

with corresponding eigenfunctions

$$X_n(x) = \sin(n\pi x/p), n = 1, 2, \ldots .$$

Next, we solve the equation $T'' + c^2\lambda_n T = 0$. A general solution is

$$T_n(t) = a_n \cos\left(c\sqrt{\lambda_n}t\right) + b_n \sin\left(c\sqrt{\lambda_n}t\right) = a_n \cos\frac{cn\pi t}{p} + b_n \sin\frac{cn\pi t}{p}$$

where the coefficients a_n and b_n must be determined. Putting this information together, we obtain

$$u_n(x, t) = \left(a_n \cos\frac{cn\pi t}{p} + b_n \sin\frac{cn\pi t}{p}\right)\sin\frac{n\pi x}{p},$$

so by the Principle of Superposition, we have

$$u(x, t) = \sum_{n=1}^{\infty}\left(a_n \cos\frac{cn\pi t}{p} + b_n \sin\frac{cn\pi t}{p}\right)\sin\frac{n\pi x}{p}.$$

Applying the initial displacement yields

$$u(x, 0) = \sum_{n=1}^{\infty} a_n \sin\frac{n\pi x}{p} = f(x),$$

so a_n is the Fourier sine series coefficient for $f(x)$, which is given by

$$a_n = \frac{2}{p}\int_0^p f(x)\sin\frac{n\pi x}{p}dx, n = 1, 2, \ldots$$

In order to determine b_n, we must use the initial velocity. Therefore, we compute

$$u_t(x, t) = \sum_{n=1}^{\infty}\left(-a_n\frac{cn\pi}{p}\sin\frac{cn\pi t}{p} + b_n\frac{cn\pi}{p}\cos\frac{cn\pi}{p}\right)\sin\frac{n\pi x}{p}.$$

Then,

$$u_t(x, 0) = \sum_{n=1}^{\infty} b_n\frac{cn\pi}{p}\cos\frac{cn\pi}{p} = g(x),$$

so $b_n\frac{cn\pi}{p}$ represents the Fourier sine series coefficient for $g(x)$, which means that

$$b_n = \frac{p}{cn\pi}\frac{2}{p}\int_0^p g(x)\sin\frac{n\pi x}{p}dx = \frac{2}{cn\pi}\int_0^p g(x)\sin\frac{n\pi x}{p}dx, n = 1, 2, \ldots .$$

EXAMPLE 1: Solve $\begin{cases} u_{xx} = u_{tt},\ 0 < x < 1, t > 0 \\ u(0, t) = 0, u(1, t) = 0, t > 0 \\ u(x, 0) = \sin \pi x, u_t(x, 0) = 3x + 1, 0 < x < 1 \end{cases}$.

SOLUTION: The initial displacement and velocity functions are defined first.

```
> f:=x->sin(Pi*x):
  g:=x->3*x+1:
```

Next, the functions to determine the coefficients a_n and b_n in the series approximation of the solution $u(x, t)$ are defined. Here $p = 1$ and $c = 1$.

```
> a[1]:=2*int(f(x)*sin(Pi*x),x=0..1);
```

$$a_1 := 1$$

Because n represents an integer, these results indicate that $a_n = 0$ for all $n \geq 2$.

```
> a[n]:=2*int(f(x)*sin(n*Pi*x),x=0..2);
```

$$a_n := 4 \frac{\sin(n\pi)\cos(n\pi)}{\pi(-1+n)(1+n)}$$

In fact, when we instruct Maple to assume that n is an integer with `assume`, Maple determines that $a_n = 0$.

```
> assume(n,integer):
  a[n]:=2*int(f(x)*sin(n*Pi*x),x=0..2);
```

$$a_{n\sim} := 0$$

Similarly, assuming that n is an integer, we see that $b_n = \dfrac{2(1 + (-4)^{n+1})}{\pi^2 n^2}$.

```
> b[n]:=2/(n*Pi)*int(g(x)*sin(n*Pi*x),x=0..1);
```

$$b_{n\sim} := -2 \frac{4(-1)^{n\sim} - 1}{n\sim^2 \pi^2}$$

The function u defined next computes the nth term in the series expansion. Hence, uapprox determines the approximation of order n by summing the first n terms of the expansion, as illustrated with uapprox(10). (Notice that we define uapprox(n) using the remember option so that Maple "remembers" the terms uapprox that are computed. That is, Maple needs to recompute uapprox(n-1) to compute uapprox(n) if uapprox(n-1) has already been computed.)

```
> u:=n->-2*(4*(-1)^n-
   1)/(n^2*Pi^2)*sin(n*Pi*t)*sin(n*Pi*x):
> uapprox:=proc(n) option remember;
      uapprox(n-1)+u(n);
      end:
  uapprox(0):=cos(Pi*t)*sin(Pi*x):
> uapprox(10);
```

$$\cos(\pi t)\sin(\pi x) + 10\frac{\sin(\pi t)\sin(\pi x)}{\pi^2} - \frac{3}{2}\frac{\sin(2\pi t)\sin(2\pi x)}{\pi^2} + \frac{10}{9}\frac{\sin(3\pi t)\sin(3\pi x)}{\pi^2}$$
$$-\frac{3}{8}\frac{\sin(4\pi t)\sin(4\pi x)}{\pi^2} + \frac{2}{5}\frac{\sin(5\pi t)\sin(5\pi x)}{\pi^2} - \frac{1}{6}\frac{\sin(6\pi t)\sin(6\pi x)}{\pi^2} + \frac{10}{49}\frac{\sin(7\pi t)\sin(7\pi x)}{\pi^2}$$
$$-\frac{3}{32}\frac{\sin(8\pi t)\sin(8\pi x)}{\pi^2} + \frac{10}{81}\frac{\sin(9\pi t)\sin(9\pi x)}{\pi^2} - \frac{3}{50}\frac{\sin(10\pi t)\sin(10\pi x)}{\pi^2}$$

To illustrate the motion of the string, we graph uapprox(10), the tenth partial sum of the series, on the interval [0,1] for 16 equally spaced values of t between 0 and 2 with animate. One frame from the resulting animation is displayed:

```
> with(plots):
  animate(uapprox(10),x=0..1,t=0..2,frames=16);
```

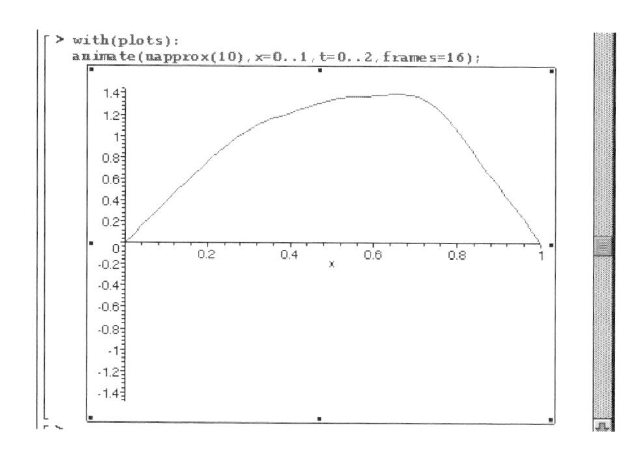

On the other hand, entering

```
> anarray:=animate(uapprox(10),x=0..1,t=0..2,
      frames=16,color=BLACK):
  display(anarray);
```

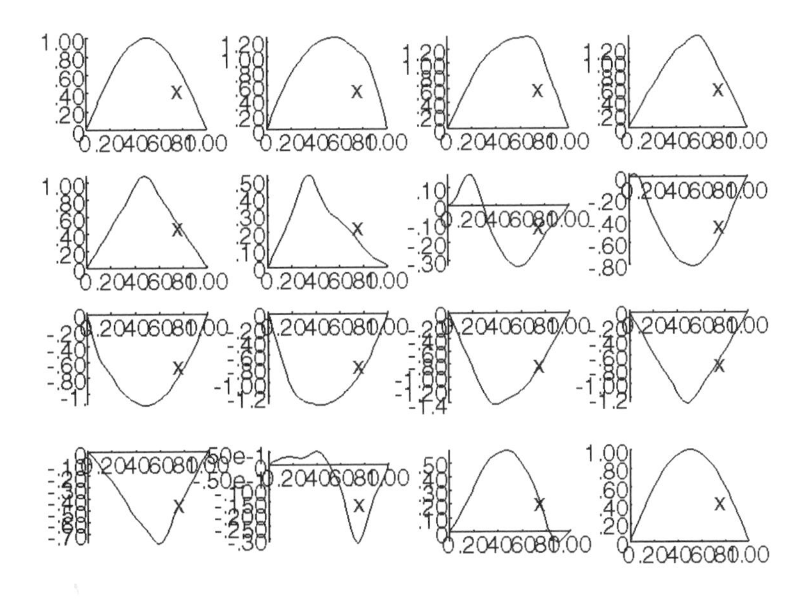

graphs `uapprox(10)` for sixteen equally spaced values of t between 0 and 2 with `animate`, and then displays the resulting graphs as an array.

Using `pdsolve`, we obtain D'Alembert's solution.

```
> pde:=diff(u(x,t),x$2)=diff(u(x,t),t$2):
  pdsolve(pde);
```

$$u(x,t) = \_F2(x+t) + \_F1\left(\frac{1}{2}x - \frac{1}{2}t\right)$$

■

The Two-Dimensional Wave Equation

One of the more interesting problems involving two spatial dimensions (x and y) is the wave equation. The two-dimensional wave equation in a circular region that is radially symmetric (not dependent on θ) with boundary and initial conditions is expressed in polar coordinates as

$$\begin{cases} c^2\left(u_{rr} + \frac{1}{r}u_r\right) = u_{tt},\ 0 < r < \rho,\ t > 0 \\ u(\rho, t) = 0,\ |u(0, t)| < \infty,\ t > 0 \\ u(r, 0) = f(r),\ u_t(r, 0) = g(r),\ 0 < r < \rho \end{cases}.$$

Notice that the boundary condition $u(\rho, t) = 0$ indicates that u is fixed at zero around the boundary; $|u(0, t)| < \infty$ indicates that the solution is bounded at the center of the circular region. Like the wave equation discussed previously, this problem is typically solved through separation of variables by assuming a solution of the form $u(r, t) = F(r)G(t)$. Applying separation of variables yields the solution

$$u(r, t) = \sum_{n=1}^{\infty} (A_n \cos ck_n t + B_n \sin ck_n t) J_0(k_n r),$$

where $\lambda_n = c\alpha_n/\rho$, $k_n = \alpha_n/\rho$ (α_n represents the nth zero of the Bessel function of the first kind of order zero), and the coefficients A_n and B_n are found through application of the initial displacement and velocity functions. With $u(r, 0) = \sum_{n=1}^{\infty} A_n J_0(k_n r) = f(r)$ and the orthogonality conditions of the Bessel functions, we find that

$$A_n = \frac{\int_0^\rho r f(r) J_0(k_n r)\, dr}{\int_0^\rho r \left[J_0(k_n r)\right]^2 dr} = \frac{2}{\left[J_1(\alpha_n)\right]^2} \int_0^\rho r f(r) J_0(k_n r)\, dr,\ n = 1, 2, \ldots,$$

Similarly, because

$$u_t(r, t) = \sum_{n=1}^{\infty} (-ck_n A_n \sin ck_n t + ck_n B_n \cos ck_n t) J_0(k_n r),$$

we have $u_t(r,0) = \sum_{n=1}^{\infty} ck_n B_n J_0(k_n r) = g(r)$. Therefore,

$$B_n = \frac{\int_0^{\rho} rg(r)J_0(k_n r)dr}{ck_n \int_0^{\rho} r\left[J_0(k_n r)\right]^2 dr} = \frac{2}{ck_n\left[J_1(\alpha_n)\right]^2}\int_0^{\rho} rg(r)J_0(k_n r)dr,\ n = 1, 2, \ldots .$$

As a practical matter, in nearly all cases, these formulas are difficult to evaluate.

EXAMPLE 2: Solve $\begin{cases} u_{rr} + \frac{1}{r}u_r = u_{tt},\ 0 < r < 1, t > 0 \\ u(1,t) = 0,\ |u(0,t)| < \infty,\ t > 0 \\ u(r,0) = r(r-1),\ u_t(r,0) = \sin \pi r,\ 0 < r < 1 \end{cases}$

SOLUTION: In this case, $\rho = 1$, $f(r) = r(r-1)$, and $g(r) = \sin \pi r$. To calculate the coefficients, we will need to have approximations of the zeros of the Bessel functions, so we load the table of zeros that were found earlier in Chapter 5 and saved as `besseltable`. Then, for $1 \le n \le 8$, α_n is the nth zero of J_0.

```
> alpha:=array([2.4048, 5.5201, 8.6537, 11.792,
       14.931, 18.071, 21.212, 24.352]):
```

Next, we define the constants ρ and c and the functions $f(r) = r(r-1)$, $g(r) = \sin \pi r$, and $k_n = \alpha_n/\rho$.

```
> c:=1:
  rho:=1:
  f:=r->r*(r-1):
  g:=r->sin(Pi*r):
  k:=n->alpha[n]/rho:
```

The formulas for the coefficients A_n and B_n are then defined so that an approximate solution may be determined. Note that we use `evalf` together with `int` to approximate the coefficients and avoid the difficulties in integration associated with the presence of the Bessel function of order zero.

```
> a:=proc(n) option remember;
       2/BesselJ(1,alpha[n])^2*
          evalf(Int(r*f(r)*BesselJ(0,k(n)*r),
```

```
            r=0..rho))
      end:

> b:=proc(n) option remember;
      2/(c*k(n)*BesselJ(1,alpha[n])^2)*
         evalf(Int(r*g(r)*BesselJ(0,k(n)*r),
            r=0..rho))
      end:
```

We now compute the first eight values of A_n and B_n. Because a and b are defined as procedures with proc using the option remember, Maple "remembers" these values for later use.

```
> array([seq([n,a(n),b(n)],n=1..8)]);
```

1	-.3235010276	.5211819702
2	.2084692034	-.1457773395
3	.007640292444	-.01342290349
4	.03838004574	-.008330225220
5	.005341000922	-.002504216150
6	.01503575901	-.002082788164
7	.003340078854	-.0008805687932
8	.007857367112	-.0008134612340

The nth term of the series solution is defined in u. Then, an approximate solution is obtained in uapprox by summing the first eight terms of u.

```
> u:='u':n:='n':
  u:=(n,r,t)->(a(n)*cos(c*k(n)*t)+b(n)*sin(c*k(n)*t))*
      BesselJ(0,k(n)*r):

> uapprox:=sum('u(n,r,t)','n'=1..8):
```

We graph uapprox for several values of t using animate3d.

```
> with(plots):
  drumhead:=animate3d([r*cos(theta),r*sin(theta),
      uapprox],r=0..1,theta=-Pi..Pi,
      t=0..1.5,frames=9):
```

```
display(drumhead);
```

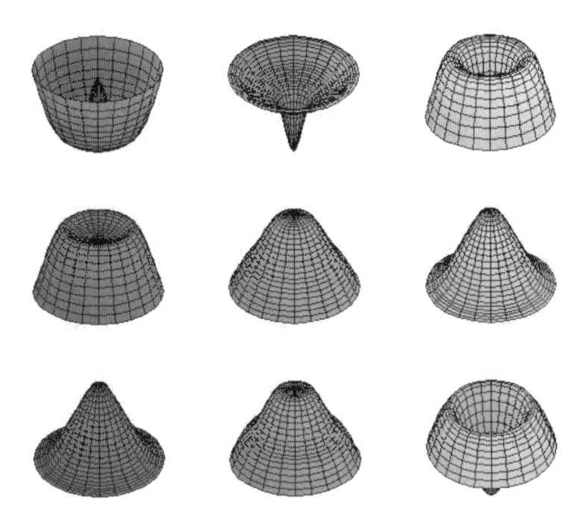

In order to actually watch the drumhead move, we can use `animate3d` to generate several graphs and animate the result. Be aware, however, that generating many three-dimensional graphics and then animating the results uses a great deal of memory and can take considerable time, even on a relatively powerful computer. We show one frame from the following animation:

```
> animate3d([r*cos(theta),r*sin(theta),
      uapprox],r=0..1,theta=-Pi..Pi,
      t=0..1.5,frames=9);
```

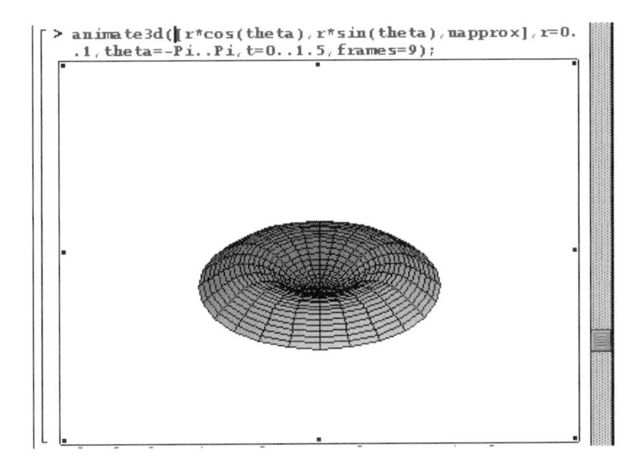

■

The problem that describes the displacement of a circular membrane in its general case is

$$\begin{cases} c^2\left(u_{rr} + \frac{1}{r}u_r + \frac{1}{r^2}u_{\theta\theta}\right) = u_{tt},\ o < r < \rho,\ -\pi < \theta < \pi,\ t > 0 \\ u(\rho, \theta, t) = 0,\ |u(0, \theta, t)| < \infty,\ -\pi < \theta < \pi,\ t > 0 \\ u(r, \pi, t) = u(r, -\pi, t),\ u_\theta(r, \pi, t) = u_\theta(r, -\pi, t),\ 0 < r < \rho,\ t > 0 \\ u(r, \theta, 0) = f(r, \theta),\ u_t(r, \theta, 0) = g(r, \theta),\ 0 < r < \rho,\ -\pi < \theta < \pi \end{cases},$$

if the displacement of the drumhead is not radially symmetric. Using separation of variables and assuming that $u(r, \theta, t) = R(r)H(\theta)T(t)$, we obtain that a general solution is given by

$$\begin{aligned} u(r, \theta, t) = {} & \sum_n a_{0n} J_0(\lambda_{0n} r)\cos(\lambda_{0n} ct) + \sum_{m,n} a_{mn} J_m(\lambda_{mn} r)\cos(m\theta)\cos(\lambda_{mn} ct) \\ & + \sum_{m,n} b_{mn} J_m(\lambda_{mn} r)\sin(m\theta)\cos(\lambda_{mn} ct) + \sum_n A_{0n} J_0(\lambda_{0n} r)\sin(\lambda_{0n} ct) \\ & + \sum_{m,n} A_{mn} J_m(\lambda_{mn} r)\cos(m\theta)\sin(\lambda_{mn} ct) \\ & + \sum_{m,n} B_{mn} J_m(\lambda_{mn} r)\sin(m\theta)\sin(\lambda_{mn} ct), \end{aligned}$$

where J_m represents the mth Bessel function of the first kind, α_{mn} denotes the nth zero of the Bessel function J_m, and $\lambda_{mn} = \alpha_{mn}/\rho$. The coefficients are given by:

$$a_{0n} = \frac{\int_0^{2\pi}\int_0^{\rho} f(r, \theta) J_0\left(\lambda_{0n} r\right) r\, dr\, d\theta}{2\pi \int_0^{\rho}\left[J_0\left(\lambda_{0n} r\right)\right]^2 r\, dr},\quad a_{mn} = \frac{\int_0^{2\pi}\int_0^{\rho} f(r, \theta) J_m\left(\lambda_{mn} r\right)\cos(m\theta) r\, dr\, d\theta}{\pi \int_0^{\rho}\left[J_m\left(\lambda_{mn} r\right)\right]^2 r\, dr},$$

$$b_{mn} = \frac{\int_0^{2\pi}\int_0^{\rho} f(r, \theta) J_m\left(\lambda_{mn} r\right)\sin(m\theta) r\, dr\, d\theta}{\pi \int_0^{\rho}\left[J_m\left(\lambda_{mn} r\right)\right]^2 r\, dr},\quad A_{0n} = \frac{\int_0^{2\pi}\int_0^{\rho} g(r, \theta) J_0\left(\lambda_{0n} r\right) r\, dr\, d\theta}{2\pi\lambda_{0n} c \int_0^{\rho}\left[J_0\left(\lambda_{0n} r\right)\right]^2 r\, dr},$$

$$A_{mn} = \frac{\int_0^{2\pi}\int_0^{\rho} g(r, \theta) J_m\left(\lambda_{mn} r\right)\cos(m\theta) r\, dr\, d\theta}{\pi\lambda_{mn} c \int_0^{\rho}\left[J_m\left(\lambda_{mn} r\right)\right]^2 r\, dr},\ \text{and}\ B_{mn} = \frac{\int_0^{2\pi}\int_0^{\rho} g(r, \theta) J_m(\lambda_{mn} r)\sin(m\theta) r dr d\theta}{\pi\lambda_{mn} c \int_0^{\rho}[J_m(\lambda_{mn} r)]^2 r dr}.$$

EXAMPLE 3: Solve

$$\begin{cases} 10^2\left(u_{rr}+\frac{1}{r}u_r+\frac{1}{r^2}u_{\theta\theta}\right)=u_{tt},\ 0<r<1,\ -\pi<\theta<\pi,\ t>0 \\ u(1,\theta,t)=0,\ |u(0,\theta,t)|<\infty,\ -\pi<\theta<\pi,\ t>0 \\ u(r,\pi,t)=u(r,-\pi,t),\ u_\theta(r,\pi,t)=u_\theta(r,-\pi,t),\ 0<r<1,\ t>0 \\ u(r,\theta,0)=\cos(\pi r/2)\sin\theta,\ u_t(r,\theta,0)=(r-1)\cos(\pi\theta/2) \end{cases}.$$

SOLUTION: The table of zeros that was found earlier in Chapter 5 and saved as `besseltable` is read in and called `getzeros`. A function `alpha` is then defined so that these zeros of the Bessel functions can be more easily obtained from the list.

```
> ALPHA:=array([
      [2.4048, 5.5201, 8.6537, 11.792,
         14.931, 18.071, 21.212, 24.352],
      [3.8317, 7.0156, 10.173, 13.324,
         16.471, 19.616, 22.760, 25.904],
      [5.1356, 8.4172, 11.620, 14.796,
         17.960, 21.117, 24.270, 27.421],
      [6.3802, 9.7610, 13.015, 16.223,
         19.409, 22.583, 25.748, 28.908],
      [7.5883, 11.065, 14.373, 17.616,
         20.827, 24.019, 27.199, 30.371],
      [8.7715, 12.339, 15.700, 18.980,
         22.218, 25.430, 28.627, 31.812],
      [9.9361, 13.589, 17.004, 20.321,
         23.586, 26.820, 30.034, 33.233]]):
```

```
> alpha:=table():
  for i from 0 to 6 do
   for j from 1 to 8 do
      alpha[i,j]:=ALPHA[i+1,j] od od:
```

The appropriate parameter values and the initial condition functions are defined as follows: Notice that the functions describing the initial displacement and velocity are defined as the product of functions. This enables the subsequent calculations to be carried out using `evalf` and `int`.

```
> c:=10:
  rho:=1:f:='f':
  f1:=r->cos(Pi*r/2):
  f2:=theta->sin(theta):
  f:=proc(r,theta) option remember;
       f1(r)*f2(theta)
       end:
```

```
> g1:=r->r-1:
  g2:=theta->cos(Pi*theta/2):
  g:=proc(r,theta) option remember;
       g1(r)*g2(theta)
       end:
```

The coefficients a_{0n} are determined with the function `a0`.

```
a0:=proc(n) option remember;
       evalf(Int(f1(r)*BesselJ(0,alpha[0,n]*r)*r,
          r=0..rho)*
       Int(f2(t),t=0..2*Pi)/
       (2*Pi*Int(r*BesselJ(0,alpha[0,n]*r)^2,
          r=0..rho)))
       end:
```

We use `seq` to generate the first five values of a0. Because `a0` was defined as a procedure using the option `remember`, Maple will not need to recompute these values when they are called later. Notice that we interpret these values to all be zero.

```
> seq(a0(n),n=1..5);
```

$.1286493367\ 10^{-16}, -.7698006025\ 10^{-18}, .2308648863\ 10^{-18}, -.1045304671\ 10^{-18}, .5708561480\ 10^{-19}$

Because the denominator of each integral formula used to find a_{mn} and b_{mn} is the same, the function bjmn, which computes this value is defined next. A table of nine values of this coefficient is then determined.

```
> bjmn:=proc(m,n) option remember;
      evalf(Int(r*BesselJ(m,alpha[m,n]*r)^2,
         r=0..rho))
      end:
> seq(seq(bjmn(m,n),m=1..3),n=1..3);
```

.08110781816, .05768792844, .04448295755. .04503456132,
.03682464114, .03110451636, .03117913218, .02701416557,
.02382360434

We also note that in evaluating the numerators of a_{mn} and b_{mn}, we must compute $\int_0^\rho r f_1(r) J_m(\alpha_{mn} r) dr$. This integral is defined in fbjmn, and the corresponding values are found for $n = 1, 2, 3$ and $m = 1, 2, 3$.

```
> fbjmn:=proc(m,n) option remember;
      evalf(Int(f1(r)*BesselJ(m,alpha[m,n]*r)*r,
         r=0..rho))
      end:
> seq(seq(fbjmn(m,n),m=1..3),n=1..3);
```

.1035741366, .07909488454, .06289255442, .02051373335,
.02755702171, .02907659382, .01039761263, .01503814332,
.01720000672

The formula to compute a_{mn} is then defined and uses the information calculated in fbjmn and bjmn. As in the previous calculation, the coefficient values for $n = 1, 2, 3$ and $m = 1, 2, 3$ are determined.

```
> a:=proc(m,n) option remember;
      evalf(fbjmn(m,n)*Int(f2(t)*cos(m*t),
         t=0..2*Pi)/(Pi*bjmn(m,n)))
      end:
```

```
> seq(seq(a(m,n),m=1..3),n=1..3);
```

$-.1891510652\ 10^{-15}, -.6261779917\ 10^{-16}, .1007528471\ 10^{-15}, -.6747128133\ 10^{-16}, -.3417654071\ 10^{-16}, .6661496811\ 10^{-16}, -.4939577444\ 10^{-16}, -.2542361469\ 10^{-16}, .5144852122\ 10^{-16}$

A similar formula is then defined for the computation of b_{mn}.

```
> b:=proc(m,n) option remember;
  evalf(fbjmn(m,n)*Int(f2(t)*sin(m*t),
         t=0..2*Pi)/(Pi*bjmn(m,n)))
      end:
```

```
> seq(seq(b(m,n),m=1..3),n=1..3);
```

$$1.276993254, .6765363938\ 10^{-16}, -.6976430811\ 10^{-16}, .4555108954, .3692508186\ 10^{-16}, -4612621176\ 10^{-16}, .3334798599, .2746822921\ 10^{-16} -.3562450680\ 10^{-16}$$

The values of A_{0n} are to be found similar to those of a_{0n}. After defining the function capa0 to calculate these coefficients, a table of values is then found.

```
> capa0:=proc(n) option remember;
  evalf(Int(g1(r)*BesselJ(0,alpha[0,n]*r)*r,
         r=0..rho)*
      Int(g2(t),t=0..2*Pi)/
      (2*Pi*c*alpha[0,n]*
         Int(r*BesselJ(0,alpha[0,n]*r)^2,
         r=0..rho)))
      end:
```

```
> seq(capa0(n),n=1..6);
```

$$.001422305064, .00005424800235, 00002675839648, 6408606429\ 10^{-16} .4959417507\ 10^{-5}, .1886364107\ 10^{-5}$$

The value of the integral of the component of g, g1, which depends on r and the appropriate Bessel functions, is defined as gbjmn.

```
>  gbjmn:=proc(m,n) option remember;
   evalf(Int(g1(r)*BesselJ(m,alpha[m,n]*r)*r,
      r=0..rho))
      end:
```

```
> seq(seq(gbjmn(m,n),m=1..3),n=1..3);
```

−.07439063140, −.05543795341, −04336144966, −.01949086495
−.02279800021, −.02267782604, −009892714342, −01303894855,
−.01416851159

Then, A_{mn} is found by taking the product of integrals, gbjmn depending on r and one depending on θ. A table of coefficient values is generated in this case as well.

```
> capa:=proc(m,n) option remember;
      evalf(gbjmn(m,n)*Int(g2(t)*cos(m*t),
         t=0..2*Pi)/(Pi*alpha[m,n]*c*bjmn(m,n)))
      end:
```

```
> seq(seq(capa(m,n),m=1..3),n=1..3);
```

.003509601162, −.002626909580, −.0005031904977, .0009045115035,
−.001032534693, −.0002460029071, .0004572947171, −.000583123673, −0001504974617

Similarly, the B_{mn} are determined.

```
> capb:=proc(m,n) option remember;
      evalf(gbjmn(m,n)*Int(g2(t)*sin(m*t),
         t=0..2*Pi)/(Pi*alpha[m,n]*c*bjmn(m,n)))
      end:
```

```
> seq(seq(capb(m,n),m=1..3),n=1..3);
```

.009879439153, −01478936902, −004249405850, .002546177172, −.005813118470,
.002077476020, .001287273144, −.003282944015, −.001270939727

Now that the necessary coefficients have been found, we must construct the approximate solution to the wave equation by using our results. Below, term1 represents the sum of the first five terms of the expansion involving a_{0n}, term2 the sum of the first nine terms involving a_{mn}, term3 the sum of the first nine terms involving b_{mn}, term4 the sum of the first five terms involving A_{0n}, term5 the sum of the first nine terms involving A_{0n}, and term6 the sum of the first nine terms involving A_{mn}. We interprety term1 and term2 to be zero.

```
> term1:=sum('a0(n)*BesselJ(0,alpha[0,n]*r)*
```

```
     cos(alpha[0,n]*c*t)',n=1..5);
```

$$term1 := .1286493367\ 10^{-16}\ \mathrm{BesselJ}(0, 2.4048\ r)\cos(24.0480\ t) - .7698006025\ 10^{-18}\ \mathrm{BesselJ}(0, 5.5201\ r)\cos(55.2010\ t) + .2308648863\ 10^{-18}\ \mathrm{BesselJ}(0, 8.6537\ r)\cos(86.5370\ t) - .1045304671\ 10^{-18}\ \mathrm{BesselJ}(0, 11.792\ r)\cos(117.920\ t) + .5708561480\ 10^{-19}\ \mathrm{BesselJ}(0, 14.931\ r)\cos(149.310\ t)$$

```
> n:='n':m:='m':
  term2:=sum('sum('a(m,n)*BesselJ(m,alpha[m,n]*r)*
     cos(m*theta)*cos(alpha[m,n]*c*t)',n=1..3)',
        m=1..3);
```

$$term2 := -.1891510652\ 10^{-15}\ \mathrm{BesselJ}(1, 3.8317\ r)\cos(\theta)\cos(38.3170\ t) - .6747128133\ 10^{-16}\ \mathrm{BesselJ}(1, 7.0156\ r)\cos(\theta)\cos(70.1560\ t) - .4939577444\ 10^{-16}\ \mathrm{BesselJ}(1, 10.173\ r)\cos(\theta)\cos(101.730\ t) - .6261779917\ 10^{-16}\ \mathrm{BesselJ}(2, 5.1356\ r)\cos(2\ \theta)\cos(51.3560\ t) - .3417654071\ 10^{-16}\ \mathrm{BesselJ}(2, 8.4172\ r)\cos(2\ \theta)\cos(84.1720\ t) - .2542361469\ 10^{-16}\ \mathrm{BesselJ}(2, 11.620\ r)\cos(2\ \theta)\cos(116.200\ t) + .1007528471\ 10^{-15}\ \mathrm{BesselJ}(3, 6.3802\ r)\cos(3\ \theta)\cos(63.8020\ t) + .6661496811\ 10^{-16}\ \mathrm{BesselJ}(3, 9.7610\ r)\cos(3\ \theta)\cos(97.6100\ t) + .5144852122\ 10^{-16}\ \mathrm{BesselJ}(3, 13.015\ r)\cos(3\ \theta)\cos(130.150\ t)$$

```
> n:='n':m:='m':
  term3:=sum('sum('b(m,n)*BesselJ(m,alpha[m,n]*r)*
     sin(m*theta)*cos(alpha[m,n]*c*t)',n=1..3)',
        m=1..3);
> n:='n':
  term4:=sum('capa0(n)*BesselJ(0,alpha[0,n]*r)*
     sin(alpha[0,n]*c*t)',n=1..5);
> n:='n':m:='m':
  term5:=sum('sum('capa(m,n)*BesselJ(m,alpha[m,n]*r)*
```

```
cos(m*theta)*sin(alpha[m,n]*c*t)',n=1..3)',m=1..3);
```

```
> n:='n':m:='m':
  term6:=sum('sum('capb(m,n)*BesselJ(m,alpha[m,n]*r)*
      sin(m*theta)*sin(alpha[m,n]*c*t)',n=1..3)',
         m=1..3);
```

An approximate solution is given as the sum of these terms as computed in u. Notice that we do not include `term1` and `term2` because they are zero.

```
> u:=term3+term4+term5+term6:
```

The position of the waves on the circular region can be viewed with Maple through the use of polar coordinates (in the xy-plane) and `animate3d`. We plot the membrane for $t = 0$ to $t = 1$ using increments of $1/8$. Hence, a sequence of nine three-dimensional plots is the result.

```
> with(plots):
  somegraphs:=animate3d([r*cos(theta),r*sin(theta),u],
      r=0..1,theta=-Pi..Pi,t=0..1,frames=9):
  display(somegraphs);
```

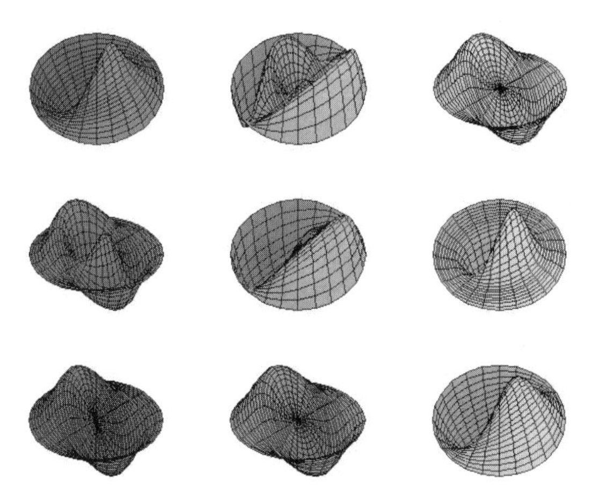

Of course, we can generate many graphs with `animate3d` and animate the result, as in Example 2. Be aware, however, that generating many three-dimensional graphics and then animating the results uses a great deal of memory and can take considerable time, even on a relatively powerful computer.

■

Other Partial Differential Equations

A partial differential equation of the form !

$$a(x, y, u)\frac{\partial u}{\partial x} + b(x, y, u)\frac{\partial u}{\partial y} = c(x, y, u)$$

is called a **first-order, quasi-linear partial differential equation**. In the case in which $c(x, y, u) = 0$, the equation is **homogeneous**; when a and b are independent of u, the equation is **almost linear**; and when $c(x, y, u)$ can be written in the form

$$c(x, y, u) = d(x, y)u + s(x, y),$$

the equation is **linear**.

Quasi-linear partial differential equations can frequently be solved using the **Method of Characteristics**. We can graph solution to quasi-linear partial differential equations with the `PDEplot` command that is contained in the `PDEtools` package. The command

```
PDEplot([a(x,y,u),b(x,y,u),c(x,y,u)],[x,y,u],
   [p(s),q(s),f(s)],s=s0..s1)
```

graphs the solution to the initial value problem ,

$$a(x, y, u)\frac{\partial u}{\partial x} + b(x, y, u)\frac{\partial u}{\partial y} = c(x, y, u),$$

where $u(x, y) = f(s)$ on the curve with parametrization $\begin{cases} x = p(s) \\ y = q(s) \end{cases}$, $s_0 \le s \le s_1$, using the method of characteristics. For a list of the options available, enter the command `?PDEplot`.

EXAMPLE 4: Use the method of characteristics to solve the initial value problem

$$\begin{cases} -3xtu_x + u_t = xt \\ u(x,0) = x \end{cases}.$$

SOLUTION: Note that pdsolve can find a general solution to the equation, but it cannot solve the initial-value problem.

```
> u:='u':x:='x':t:='t':
  pdsolve(-3*x*t*diff(u(x,t),x)+diff(u(x,t),t)=x*t);
```

$$u(x,t) = -\frac{1}{3}x + \_F1\left(\frac{2}{3}\ln(x) + t^2\right)$$

For this problem, the characteristic system is

$$\frac{\partial x}{\partial r} = -3xt \qquad x(0,s) = s$$

$$\frac{\partial t}{\partial r} = 1, \qquad t(0,s) = s$$

$$\frac{\partial t}{\partial r} = xt \qquad u(0,s) = 0$$

Since we will use the command PDEplot to graph the solution to the equation, we begin by loading the PDEtools package.

```
> with(DEtools);
```

$$[PDEplot, build, charstrip, dchange, difforder, mapde, splitstrip, splitsys]$$

We begin by using dsolve to solve $\partial t/\partial r = 1, t(0,s) = 0$

```
> d1:=dsolve({diff(t(r),r)=1,t(0)=0},t(r));
```

$$d1 := t(r) = r$$

and obtain $t = r$. Thus, $\partial u/\partial r = xt, u(0, s) = s$ which we solve next

```
> d2:=dsolve({diff(x(r),r)=-3*x(r)*r,x(0)=s},x(r));
```

$$d2 := \mathrm{x}(r) = s\,\mathbf{e}^{(-3/2\,r^2)}$$

and obtain $x = se^{-3r^2/2}$. Substituting $t = r$ and $x = se^{-3r^2/2}$ into $\partial u/\partial r = xt, u(0, s) = s$, and using dsolve to solve the resulting equation, yields the following result, named d3.

```
> d3:=dsolve({diff(u(r),r)=exp(-3/2*r^2)*s*r,
      u(0)=s},u(r));
```

$$d3 := \mathrm{u}(r) = -\frac{1}{3}\,s\,\mathbf{e}^{(-3/2\,r^2)} + \frac{4}{3}\,s$$

To find $u(x, t)$, we must solve the system of equations

$$\begin{cases} t = r \\ x = se^{-3r^2/2} \end{cases}$$

for r and s, which we do next with solve. We name the resulting output vals.

```
> vals:=solve({x=exp(-3/2*r^2)*s,t=r},{r,s});
```

$$vals := \{\, r = t,\, s = \frac{x}{\mathbf{e}^{(-3/2\,t^2)}} \}$$

Thus, the solution is given by replacing the values obtained in vals in the solution obtain in d3. We do this by using assign to assign r and s the values in vals and assign to assign $u(r)$ the value obtained in d3. We then evaluate $u(r)$. The resulting output represents the solution to the initial-value problem.

```
> assign(vals):
  assign(d3):
  u(r);
```

$$-\frac{1}{3}\,x + \frac{4}{3}\,\frac{x}{\mathbf{e}^{(-3/2\,t^2)}}$$

Finally, we verify that this result is the solution to the problem.

```
> simplify(-3*x*t*diff(u(r),x)+diff(u(r),t));
```

$$x\,t$$

The initial condition $u(x, 0) = x$ has parametrization $\begin{cases} x = s \\ t = 0 \\ u = s \end{cases}$. We use PDEplot to graph the solution for $0 \le s \le 15$. With the first command, we graph the solution drawing twenty characteristics (the default is ten), and in the second command, we draw twenty characteristics in addition to including the option basechar=true so that the base characteristics are also displayed in the resulting graph.

```
> u:='u':s:='s':
  PDEplot(-3*x*t*diff(u(x,t),x)+
      diff(u(x,t),t)=x*t,[s,0,s],s=0..15,
         numchar=20);
   PDEplot(-3*x*t*diff(u(x,t),x)+
      diff(u(x,t),t)=x*t,[s,0,s],s=0..15,
         numchar=20,basechar=true);
```

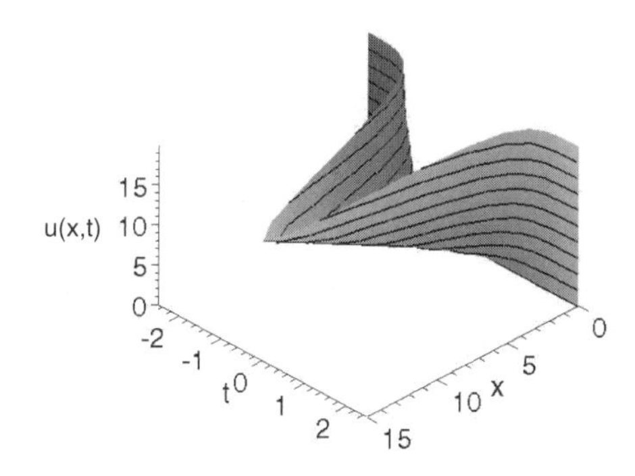

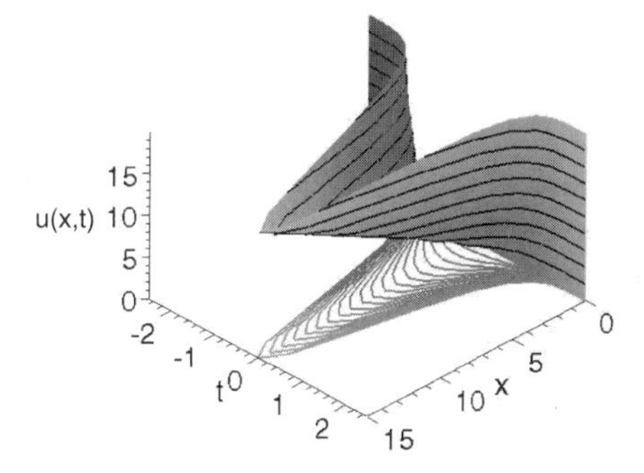

■

CHAPTER 8

Selected Graphics Topics

8.1 The plots and plottools Package

The `plots` package contains a variety of commands that allow the Maple V user to construct a variety of graphics objects. Here, we discuss several of those commands contained in the `plots` and `plottools` packages, as found in the Release 5 version of Maple V, that are most frequently used by beginning users. Some of the commands that are not discussed in this chapter have been previously discussed in the text. Every example discussed in this chapter uses commands contained in either the `plots` and/or `plottools` package, so be sure the appropriate package has been loaded prior to entering the Maple code for any example discussed here. We load the `plots` package by entering the command `with(plots)`. Any function contained in the `plots` package can be used after the `plots` package has been loaded. Remember that if a colon (`:`) is included at the end of the command instead of a semicolon, the resulting output is suppressed; the list of commands contained in the `plots` package is not displayed.

```
> with(plots);
```

[*animate, animate3d, animatecurve, changecoords, complexplot, complexplot3d, conformal, contourplot, contourplot3d, coordplot, coordplot3d, cylinderplot, densityplot, display, display3d, fieldplot, fieldplot3d, gradplot, gradplot3d, implicitplot, implicitplot3d, inequal, listcontplot, listcontplot3d, listdensityplot, listplot, listplot3d, loglogplot,*

logplot, matrixplot, odeplot, pareto, pointplot, pointplot3d, polarplot, polygonplot, polygonplot3d, polyhedra_supported, polyhedraplot, replot, rootlocus, semilogplot, setoptions, setoptions3d, spacecurve, sparsematrixplot, sphereplot, surfdata, textplot, textplot3d, tubeplot]

Detailed information regarding the plots package is obtained with `?plots`.

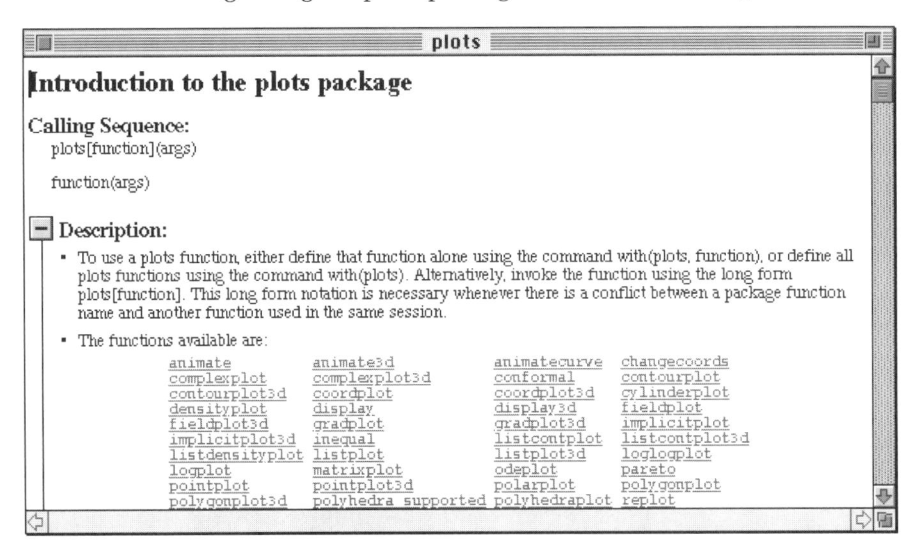

plots

Introduction to the plots package

Calling Sequence:

plots[function](args)

function(args)

Description:

- To use a plots function, either define that function alone using the command with(plots, function), or define all plots functions using the command with(plots). Alternatively, invoke the function using the long form plots[function]. This long form notation is necessary whenever there is a conflict between a package function name and another function used in the same session.
- The functions available are:

animate	animate3d	animatecurve	changecoords
complexplot	complexplot3d	conformal	contourplot
contourplot3d	coordplot	coordplot3d	cylinderplot
densityplot	display	display3d	fieldplot
fieldplot3d	gradplot	gradplot3d	implicitplot
implicitplot3d	inequal	listcontplot	listcontplot3d
listdensityplot	listplot	listplot3d	loglogplot
logplot	matrixplot	odeplot	pareto
pointplot	pointplot3d	polarplot	polygonplot
polygonplot3d	polyhedra_supported	polyhedraplot	replot

Graphing in Polar Coordinates

Loading the `plots` package enables the user to take advantage of several commands, which will improve the graphing capabilities previously available. The command discussed first, `polarplot`, allows for the graphing of functions given in polar coordinates (r, θ). The command

```
polarplot(f(θ),θ=α..β,options),
```

produces the graph of the function $r = f(\theta)$, $\alpha \le \theta \le \beta$, in polar coordinates.

More generally,

```
polarplot(expr,opt)
```

yields a two-dimensional plot in polar coordinates. Note that if `expr` is a single expression, then the given expression represents the radius r in terms of the angle θ. However, if `expr` is made up of several components, then the first component represents the radius, the second the angle, and other components the range for the indicated parameters. The options used with `polarplot` are the same as those used with `plot`.

EXAMPLE 1: Find the area of the region between the inner and outer loops of the limacon $r = 1 + 2\sin\theta$.

SOLUTION: We begin by defining $r = 1 + 2\sin\theta$ and then using the commands `polarplot` and `plot` to graph r in both polar and rectangular coordinates. The polar graph is on the right; the rectangular graph is on the left.

```
> r:='r':
  r:=theta->1+2*sin(theta):
  plot(r(theta),theta=0..2*Pi);
  polarplot(r(theta),theta=0..2*Pi);
```

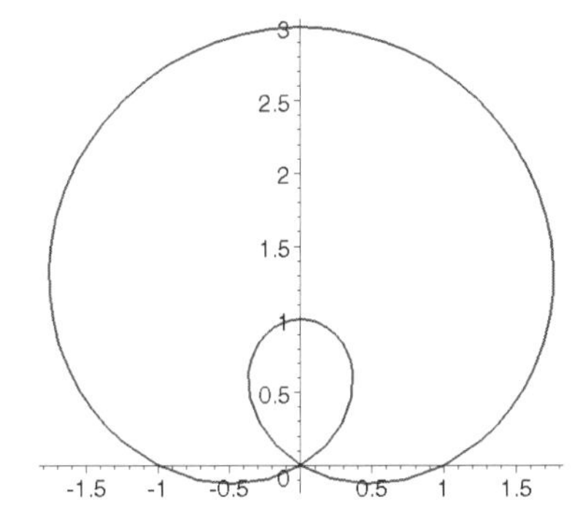

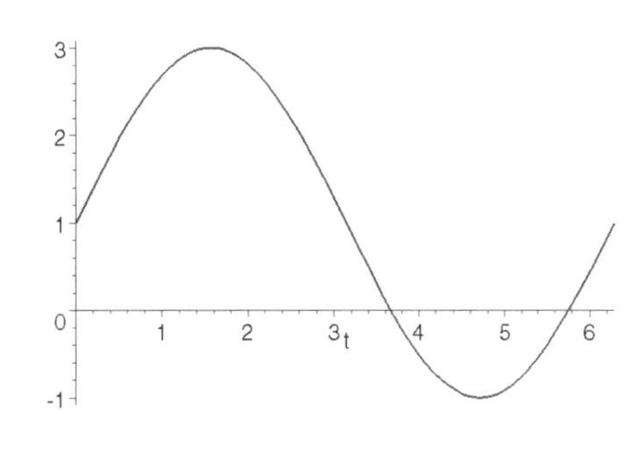

The area of the outer loop of the limacon is given by

$$\frac{1}{2}\int_{-\pi/6}^{7\pi/6}(r(\theta))^2 d\theta = \frac{1}{2}\int_{-\pi/6}^{7\pi/6}(1 + 2\sin\theta)^2 d\theta,$$

computed with `int` and named `o`.

```
> o:=1/2*int(r(t)^2,t=-Pi/6..7*Pi/6);
```

$$o := 2\pi + \frac{3}{2}\sqrt{3}$$

The area of the inner loop of the limacon is given by

$$\frac{1}{2}\int_{7\pi/6}^{11\pi/6}(r(\theta))^2 d\theta = \frac{1}{2}\int_{7\pi/6}^{11\pi/6}(1 + 2\sin\theta)^2 d\theta,$$

computed with `int` and named `i`.

```
> i:=1/2*int(r(t)^2,t=7*Pi/6..11*Pi/6);
```

$$i := \pi - \frac{3}{2}\sqrt{3}$$

Thus, the desired area is given by subtracting `i` from `o`.

```
> area:=o-i;
  evalf(area);
```

$$area := \pi + 3\sqrt{3}$$

$$8.337745078$$

We conclude that the area is $3\sqrt{3} + \pi \approx 8.33775$.

■

`polarplot`, in the same way as commands like `plot` and `plot3d`, will graph several curves. Entering

```
polarplot({r1(theta),r2(theta),...},theta=α..β)
```

graphs the curves $r_1(\theta)$, $r_2(\theta)$, ... in polar coordinates for $\alpha \leq \theta \leq \beta$.

EXAMPLE 2: Find the area inside the graph of $r = 1$ and outside the graph of $r = \cos 3\theta$.

SOLUTION: Below, we use `polarplot` and `plot` to graph the curves $r = 1$ and $r = \cos 3\theta$ in both polar and rectangular coordinates.

```
> r:='r':
  r:=theta->cos(3*theta):
  plot({r(theta),1},theta=0..2*Pi,view=[0..2*Pi,-1..1]);
  polarplot({r(theta),1},theta=0..2*Pi,
      view=[-1..1,-1..1]);
```

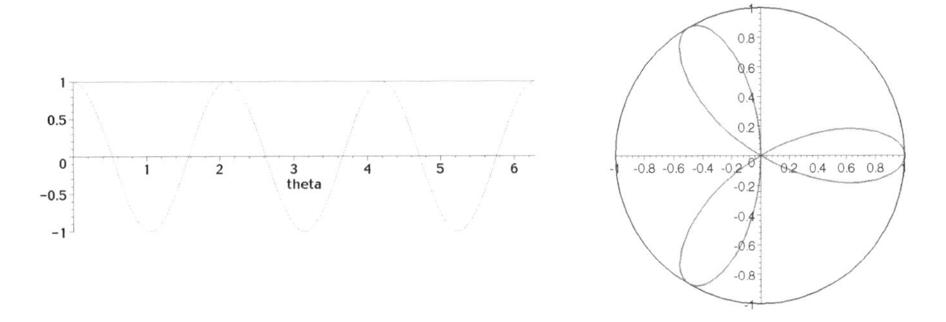

The area of the unit circle is π, while the area of the three-leafed rose is given by

$$\frac{3}{2}\int_{-\pi/6}^{\pi/6}(r(\theta))^2\,d\theta = \frac{3}{2}\int_{-\pi/6}^{\pi/6}\cos^2 3\theta\,d\theta$$

because the area of one leaf is given by $\frac{1}{2}\int_{-\pi/6}^{\pi/6}(r(\theta))^2\,d\theta = \frac{1}{2}\int_{-\pi/6}^{\pi/6}\cos^2 3\theta\,d\theta$. These values are computed in c and r, respectively. The desired area is the given by subtracting the area of the three-leafed rose from the area of the circle.

```
> c:=Pi:
  r:=3/2*int(r(theta)^2,theta=-Pi/6..Pi/6):
  area:=c-rose;r
  evalf(area);
```

$$area := \frac{3}{4}\pi$$

$$2.356194491$$

■

Conformal Mappings

A problem of interest in complex analysis is finding the image of a complex-valued function $f(z)$. The command

```
conformal(f(z),z=a+I*b..c+I*d,opts)
```

gives the image of $f(z)$ using Cartesian coordinate grid lines over the rectangular region $[a,c]\times[b,d] = \{x+iy : a\le x\le c, b\le y\le d\}$. The options given in opt may include grid=[k,m] to specify that k and m lines in the x and y directions, respectively, be mapped. Note that the default value is 11 in each case. Another option is `numxy=[k,m]`, which indicates that `k` and `m` points should be plotted along each gridline. The default setting is 15 for each of these values. The plotting options associated with `plot`, such as `style` and `tickmarks`, may be used with `conformal` as well.

EXAMPLE 3: Graph the image of the region $R = \{x + iy : 0 \le x \le 2, 0 \le y \le 2\}$ by the mapping $f(z) = \frac{z-1}{z+1}$.

SOLUTION: We use `conformal` to graph the image of the functions $id(z) = z$ and $f(z) = \frac{z-1}{z+1}$ on the domain specified. Hence, the Cartesian grid is unchanged upon application of `id(z)`. (This region can therefore be viewed as the domain of `f(z)`.) The second graph illustrates the effects that `f(z)` has on the points in the Cartesian grid. This gives the usual manner in which the domain and image of a complex-valued function are illustrated.

Because $f(z) = \frac{z-1}{z+1}$ is a Mobius transformation, $f(z)$ maps lines and circles onto lines and circles. This observation is confirmed in the graphs.

```
> id:=z->z:
  f:=z->(z-1)/(z+1):
  conformal(id(z),z=0..2+2*I);
  conformal(f(z),z=0..2+2*I);
```

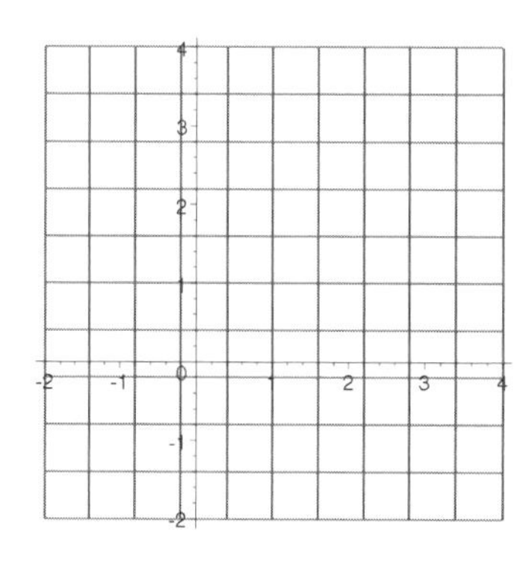

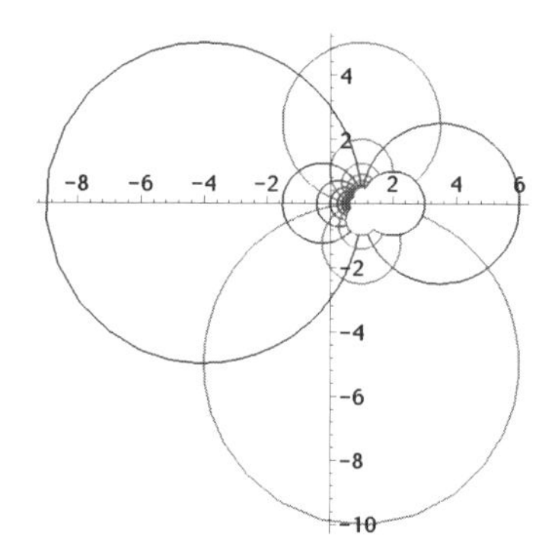

■

In addition to Cartesian coordinates, polar coordinates can also be used with

```
conformal(f(z),r=r0..r1,theta=theta0..theta1,coords=polar),
```

which produces the image of $f(z)$ over the circular region R bounded by limits placed on the polar coordinates r and θ: $R = \{re^{i\theta}: r_0 \le r \le r_1, \theta_0 \le \theta \le \theta_1\}$.

EXAMPLE 4: Graph the image of $R = \{re^{i\theta}: 0 \le r \le 2, 0 \le \theta \le 2\pi\}$ by the mapping $h(z) = \sin z$.

SOLUTION: The identity map, `id(z)=z`, is used to produce the polar grid, to be viewed as the domain of the function `h(z)`. The image of `h` is then determined with `conformal` and the two graphs are shown side by side.

```
> conformal(z,z=0..2+2*Pi*I);
  conformal(sin(z),z=0..2+2*Pi*I,coords=polar);
```

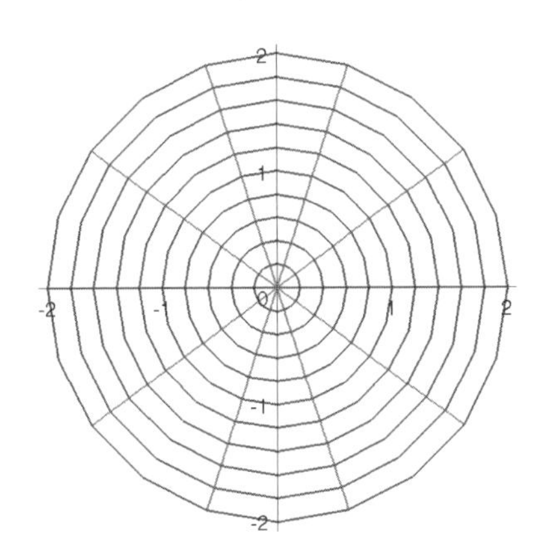

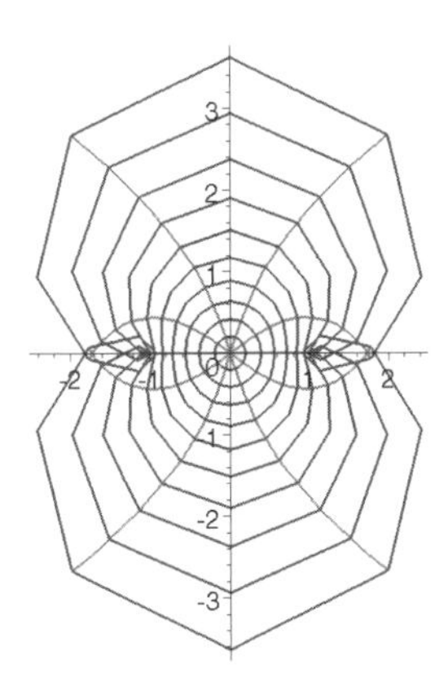

■

Two-and Three-Dimensional Graphics

As we have seen in previous examples, Maple has extensive two- and three-dimensional graphics capabilities to graph functions of one or two variables, parametric functions, and equations. We can also graph density plots of functions of two variables in addition to generating graphs in cylindrical or spherical coordinates.

The command

```
densityplot(f(x,y),x=a..b,y=c..d, options)
```

generates a density plot of $f(x,y)$ on the rectangle $[a,c] \times [b,d]$. In addition to having the same options as `plot`, `densityplot` has the option `grid=[m,n]`, which specifies the number of sample points to use in each direction when generating the graph; the default is `grid=[25,25]`, so that 625 sample points are used.

EXAMPLE 5: Generate a density plot of $f(x,y) = (x^2 + 3y^2)e^{1-(x^2+y^2)}$ on the rectangle $[-2,2] \times [-2,2]$.

SOLUTION: We use `densityplot` to generate the graph. We illustrate the difference between density and contour plots by generating a contour graph on the same rectangle with `contourplot`.

```
> densityplot((x^2+3*y^2)*exp(1-(x^2+y^2)),
      x=-2..2,y=-2..2);
  contourplot((x^2+3*y^2)*exp(1-(x^2+y^2)),
      x=-2..2,y=-2..2);
```

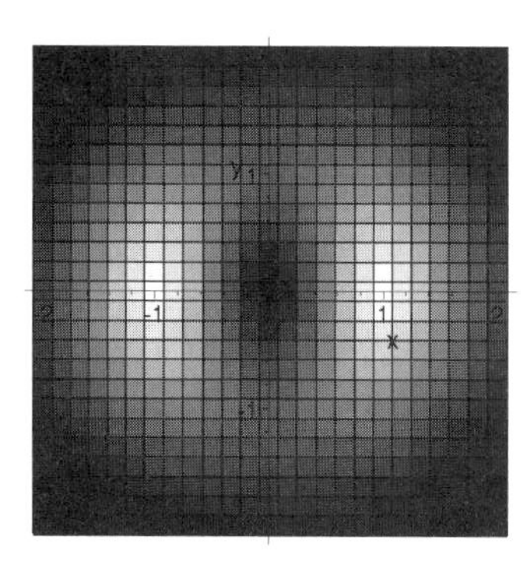

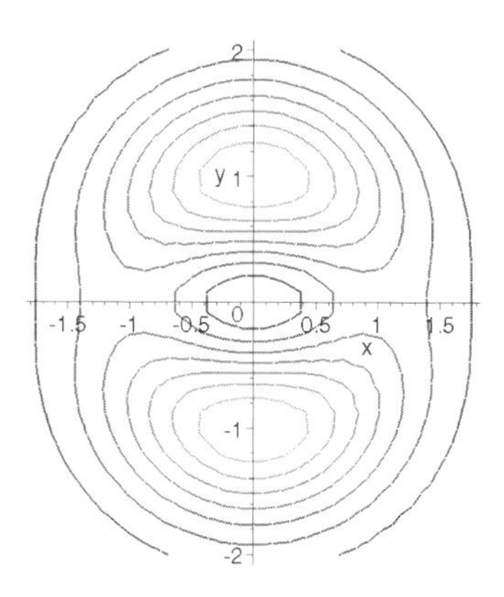

■

The command

`cylinderplot(f,range1,range2,opts)`

yields a three-dimensional plot of a surface or a parametric surface in cylindrical coordinates. If `f` is not a list, then `range1` and `range2` represent the respective ranges for `theta` and `z` while `f` represents the corresponding value of `r`. However, if `f` is a list of three components, then these components represent the coordinates `r`, `theta`, and `z` while `range1` and `range2` specify the ranges for the two parameters of the surface. The options given in `opts` are the same as those used with `plot3d`.

EXAMPLE 6: Graph $f(r,\theta) = (1+\sin\theta)\cos r$ for $0 \le \theta \le 2\pi$ and $0 \le r \le 2\pi$ in cylindrical coordinates using forty sample points on the interval $[0,2\pi]$, and fifty sample points on the interval $[0,4\pi]$. Display the resulting graph in a box.

SOLUTION: After clearing all prior definitions of f, if any, and defining *f*, we use `cylinderplot` to graph *f* for $0 \le \theta \le 2\pi$ and $0 \le r \le 2\pi$.

```
> f:='f':
  f:=(r,theta)->cos(r)*(1+sin(theta)):
  cylinderplot(f(r,theta),theta=0..2*Pi,r=0..4*Pi,
          grid=[40,50],axes=BOXED);
```

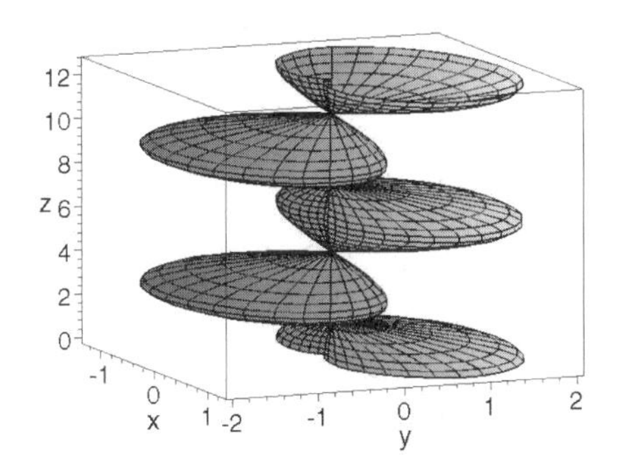

■

The command

`sphereplot(plist,range1,range2,opt)`

yields the three-dimensional plot of surface or parametric surface in spherical coordinates. If `plist` is not a list, then `plist` gives the radius as a function of the two angular coordinates ϕ and θ in spherical coordinates. However, if `plist` is made up of three components, then the components represent the coordinates r, ϕ, and θ, respectively. The same options used with `plot3d` are available for use with `sphereplot` in `opts`.

EXAMPLE 7: Use spherical coordinates to graph $f(r,z) = r + \cos r \sin z$ for $0 \le \theta \le 4\pi$ and $0 \le r \le \pi$. Use fifty sample points for the interval $[0,4\pi]$, and use thirty sample points for the interval $[0,\pi]$.

SOLUTION:

```
> sphereplot(sin(z)*cos(r)+r,r=0..4*Pi,z=0..Pi,
        grid=[50,30]);
```

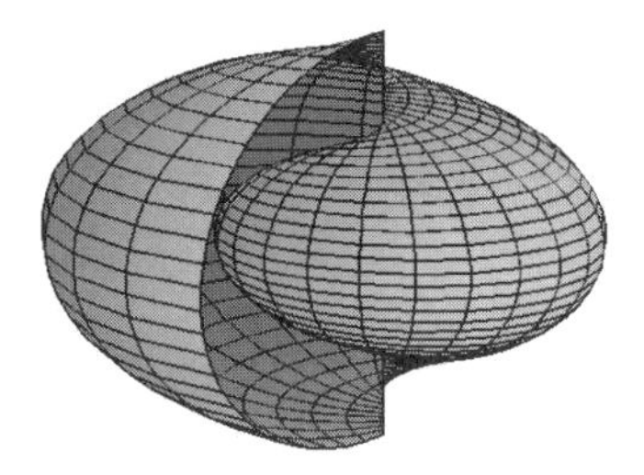

■

The command

```
spacecurve(cplist,opt)
```

yields a three-dimensional plot of the curve or curve(s) given by the components listed in `cplist`. The options that may be listed in `opt` include all of those used with `plot3d` except the component used to specify the grid size.

EXAMPLE 8: Graph the curves (a)
$\langle -10\cos(t) - 2\cos(5t) + 15\sin(2t), -15\cos(2t) + 10\sin(t) - 2\sin(5t), 10\cos(3t)\rangle$
for $0 \le t \le 2\pi$ using 200 sample points.

SOLUTION: We use the option `numpoints=200`, so that 200 sample points are used for generating the graph. Generally, a larger number of sample points results in a smoother graph. The result is a trefoil knot

```
> spacecurve([-10*cos(t)-2*cos(5*t)+15*sin(2*t),
    -15*cos(2*t)+10*sin(t)-2*sin(5*t),
      10*cos(3*t)],t=0..2*Pi,color=BLACK,
    numpoints=200,axes=BOXED);
```

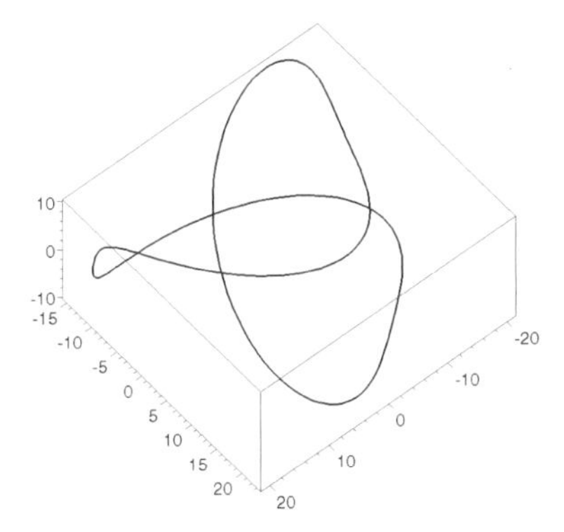

■

Multiple curves can be graphed with spacecurve as well.

EXAMPLE 9: Graph the **Borromean rings** with parametric equations

$$\begin{cases} x = 0 \\ y = r_1 \cos\theta \\ z = r_2 \sin\theta \end{cases}, \begin{cases} x = r_2 \cos\theta \\ y = 0 \\ z = r_1 \sin\theta \end{cases}, \text{and } \begin{cases} x = r_1 \cos\theta \\ y = r_2 \sin\theta, 0 \le \theta \le 2\pi, \\ z = 0 \end{cases}$$

if $r_1 = 3$ and $r_2 = 2$.

SOLUTION: All three curves are graphed together with spacecurve.

```
> r[1]:=3:
  r[2]:=2:
  spacecurve({[0,r[1]*cos(theta),r[2]*sin(theta)],
       [r[2]*cos(theta),0,r[1]*sin(theta)],
          [r[1]*cos(theta),r[2]*sin(theta),0]},
      theta=0..2*Pi,numpoints=100,color=BLACK);
```

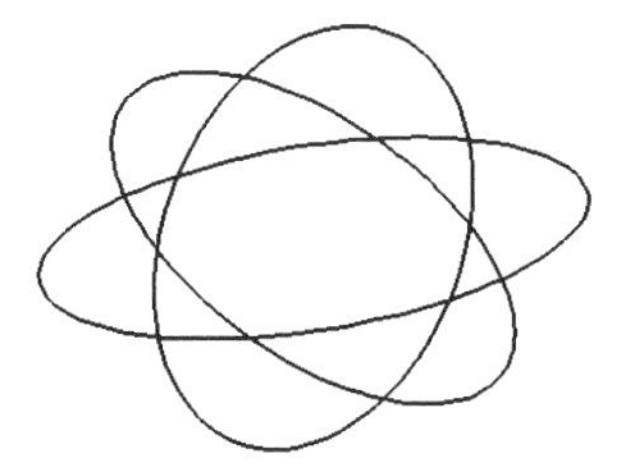

■

The command

tubeplot(clist,opt)

yields the three-dimensional tube graph about the space curve (or space curves) given in clist. Options for this command include numpoints=n, which specifies the number of points to be used to determine the space curve, radius=r, which gives the radius about the space curve, and tubepoints=m, which indicates the number of points used to plot the tube. Other specifications include t=a..b in order to specify a range on the parameter. Options available for use with spacecurve can be used with tubeplot as well.

EXAMPLE 10: Graph the **3,4 torus knot** that has parametrization

$$\begin{cases} x = \left(1+\frac{1}{2}\cos\left(\frac{n}{2m}\theta\right)\right)\cos\theta \\ y = \left(1+\frac{1}{2}\cos\left(\frac{n}{2m}\theta\right)\right)\sin\theta \\ z = \frac{1}{2}\sin\left(\frac{n}{m}\theta\right) \end{cases}$$

where $0 \leq \theta \leq 2\pi m$, $m = 3$, and $n = 4$.

SOLUTION: After defining m and n,

```
> m:=3: #  meridian
  n:=4: #  longitude
```

we define x, y, and z.

```
> x:=cos(theta)*(1+cos(n/(2*m)*theta)/2):
  y:=sin(theta)*(1+cos(n/(2*m)*theta)/2):
  z:=sin(n/m*theta)/2:
```

We then use `tubeplot` to graph the knot.

```
> knota:=tubeplot([x,y,z],theta=0..2*Pi*m,radius=0.1,
      numpoints=200):
  display(knota);
```

Using `project`, which is contained in the `plottools` package, we project the knot into the *xy*-plane. (Generally, `project(object,[[x1,y1,z1],[x2,y2,z2],[x3,y3,z3]])` projects `object` into the plane determined by (x_1,y_1,z_1), (x_2,y_2,z_2), and (x_3,y_3,z_3).)

```
> with(plottools):
  project(knota,[[0,0,0],[1,0,0],[0,1,0]]);
```

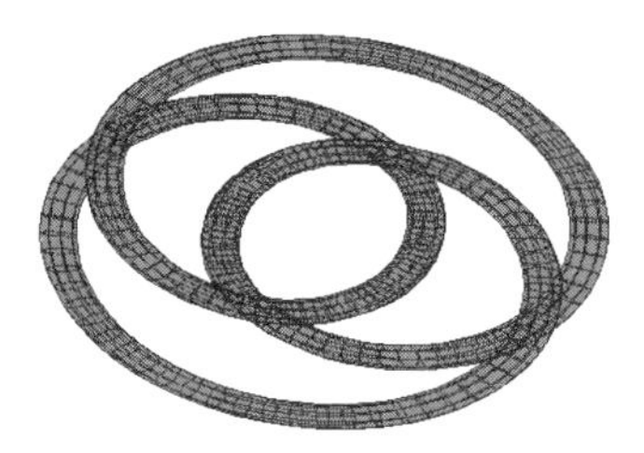

■

As with other graphics commands, multiple graphs can be generated in a single command.

EXAMPLE 11: Graph the **Borromean rings** with parametric equations

$$\begin{cases} x = 0 \\ y = r_1 \cos\theta \\ z = r_2 \sin\theta \end{cases}, \begin{cases} x = r_2 \cos\theta \\ y = 0 \\ z = r_1 \sin\theta \end{cases}, \text{and} \begin{cases} x = r_1 \cos\theta \\ y = r_2 \sin\theta, 0 \le \theta \le 2\pi, \\ z = 0 \end{cases}$$

if $r_1 = 3$ and $r_2 = 2$.

SOLUTION: The code is almost identical to the code in Example 9, except that we use `tubeplot` instead of `spacecurve`.

```
> r[1]:=3:
  r[2]:=2:
  tubeplot({[0,r[1]*cos(theta),r[2]*sin(theta)],
```

```
[r[2]*cos(theta),0,r[1]*sin(theta)],
   [r[1]*cos(theta),r[2]*sin(theta),0]},
theta=0..2*Pi,numpoints=100);
```

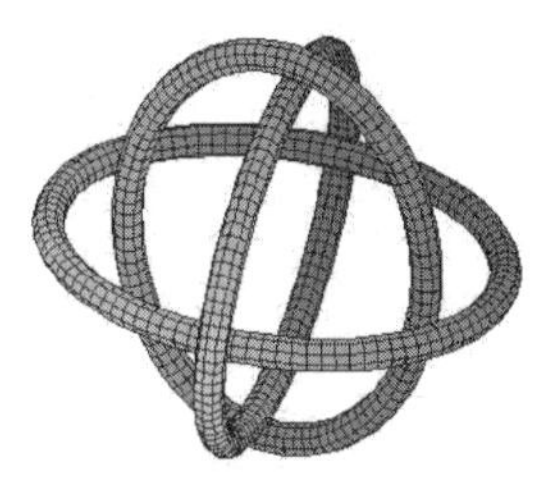

■

The `plots` package contains the command `implicitplot3d`, which can be used to graph level curves of functions of three variables and equations in three variables. The command

`implicitplot3d(eq,x=xmin..xmax,y=ymin..ymax,z=zmin..zmax)`

graphs `eq` in the parallelepiped `[xmin,xmax]×[ymin,ymax]×[zmin,zmax]`.

EXAMPLE 12: Graph the equation $-x^2-2y^2+z^2-4yz = 10$.

SOLUTION: The graph of the equation $-x^2-2y^2+z^2-4yz = 10$ in the region $[-6,6]\times[-6,6]\times[-6,6]$ is generated with `implicitplot3d`.

```
> implicitplot3d(-x^2-2*y^2+z^2-4*y*z=10,
      x=-6..6,y=-6..6,z=-6..6,axes=BOXED);
```

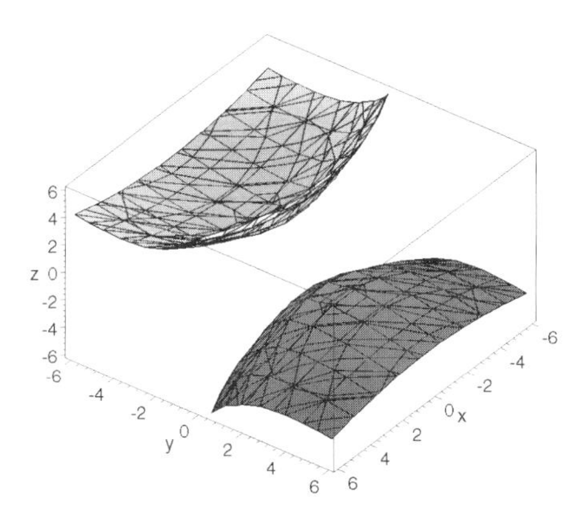

■

In addition to graphing equations, we can graph level curves of functions of three variables.

EXAMPLE 13: Sketch the level curves of $w = x^2 + z^2 - y^2$ corresponding to -1, 0, and 1.

SOLUTION: The level curves of the function $w = x^2 + z^2 - y^2$ corresponding to -1, 0, and 1 are the same as the graphs of the equations $x^2 + z^2 - y^2 = -1$, $x^2 + z^2 - y^2 = 0$, and $x^2 + z^2 - y^2 = 1$, respectively. We use `implicitplot3d` to graph each of the equations in the box $[-2,2] \times [-2,2] \times [-2,2]$.

```
> implicitplot3d(x^2+z^2-y^2=-1,
      x=-2..2,y=-2..2,z=-2..2,axes=BOXED);
  implicitplot3d(x^2+z^2-y^2=0,
      x=-2..2,y=-2..2,z=-2..2,axes=BOXED);
  implicitplot3d(x^2+z^2-y^2=1,
      x=-2..2,y=-2..2,z=-2..2,axes=BOXED);
```

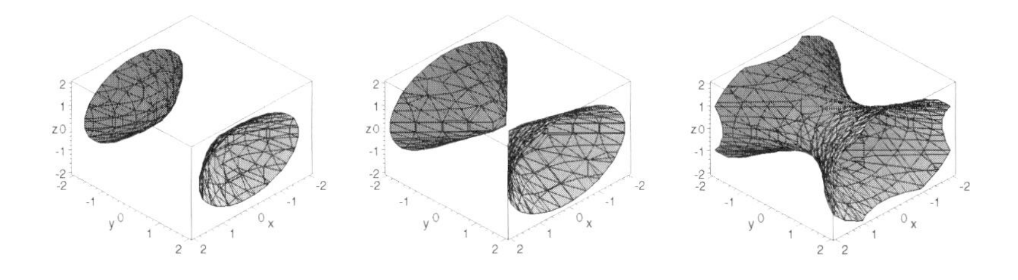

■

Graphing Vector Fields in Two and Three Dimensions

The command

```
fieldplot([f(x,y),g(x,y)],x=a..b,y=c..d)
```

graphs the vector field $\langle f(x,y), g(x,y)\rangle$ for $a \le x \le b$ and $c \le y \le d$. In addition to the options `grid=[m,n]` and `arrows=LINE`, `THIN`, `SLIM`, or `THICK` (the default is `THIN`), `fieldplot` has the same options as `plot`.

EXAMPLE 14: Graph the vector field given by the vector-valued function $\langle y,(1-x^2)y-x\rangle$ on the rectangle $[-3/2,3/2] \times [-3/2,3/2]$.

SOLUTION: We use `fieldplot` with several of its options to graph the vector field.

```
> fieldplot([y,(1-x^2)*y-x],x=-3/2..3/2,y=-3/2..3/2,
     arrows=SLIM,grid=[30,30]);
```

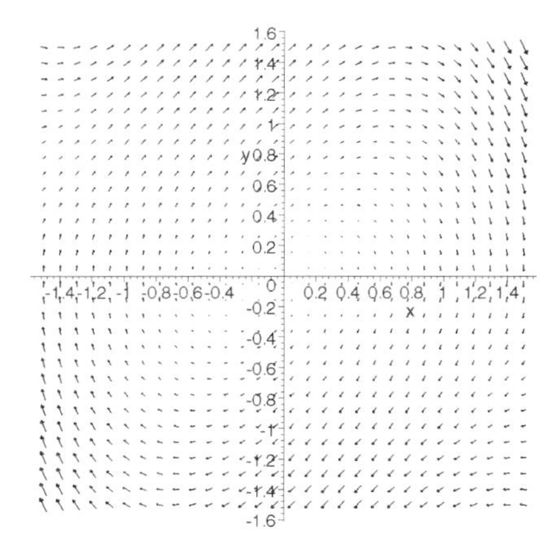

■

The command

```
gradplot(function(x,y),x=xmin..xmax,y=ymin..ymax)
```

graphs the gradient field of the function, `function(x,y)`. This is done by first computing the gradient of `function(x,y)` (which yields a vector field), and then plotting the gradient. `gradplot` has the same options as `fieldplot`.

EXAMPLE 15: Graph the gradient field of $\frac{(1-x^2)y-x}{y}$ on the rectangle $[-2,2]\times[-4,4]$.

SOLUTION:

```
> gradplot(((1-x^2)*y-x)/y,x=-2..2,y=-4..4,
      grid=[10,10]);
```

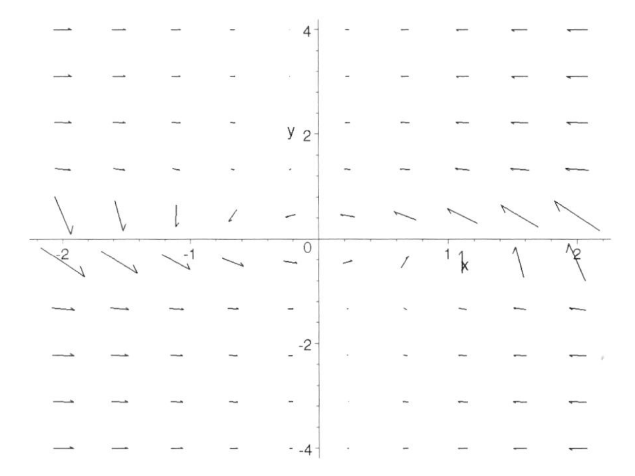

■

Vector fields can be plotted in three dimensions as well. The syntax for the `gradplot3d` and `fieldplot3d` commands are similar to those used in the two-dimensional cases discussed above with the addition of a z-component.

EXAMPLE 16: Graph the vector field $\langle -11x+4y+6z, 10x-4y+5z, 5x+8y-6z\rangle$ on the parallelepiped $[1,3]\times[0,6]\times[1,5]$.

SOLUTION: We use `fieldplot3d` to graph the vector field on $[1,3]\times[0,6]\times[1,5]$.

```
> fieldplot3d([-11*x+4*y+6*z,10*x-4*y+5*z,
      5*x+8*y-6*z],
      x=1..3,y=0..6,z=1..5,axes=BOXED,color=BLACK);
```

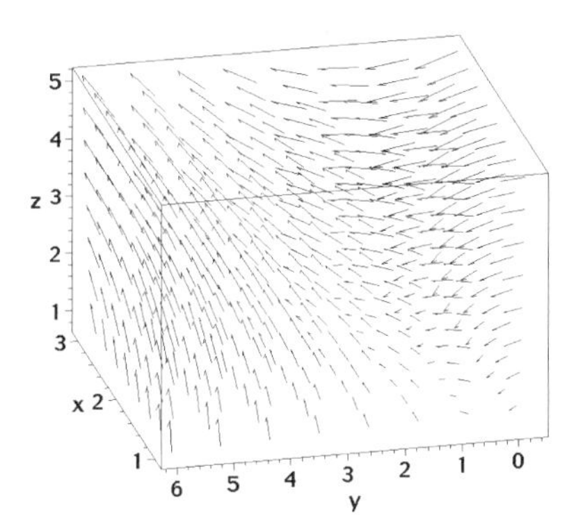

■

Example 17 illustrates the use of `gradplot3d`.

EXAMPLE 17: Graph the gradient field of $x^2 + y^2 + z^2 - 4$ on the cube $[-2,2] \times [-2,2] \times [-2,2]$.

SOLUTION: `gradplot3d` is used first to compute the gradient of $x^2 + y^2 + z^2 - 4$ and then to graph the resulting vector field on the cube $[-2,2] \times [-2,2] \times [-2,2]$.

```
> gradplot3d(x^2+y^2+z-4,x=-2..2,y=-2..2,z=-2..2,
      axes=FRAME,color=BLACK);
```

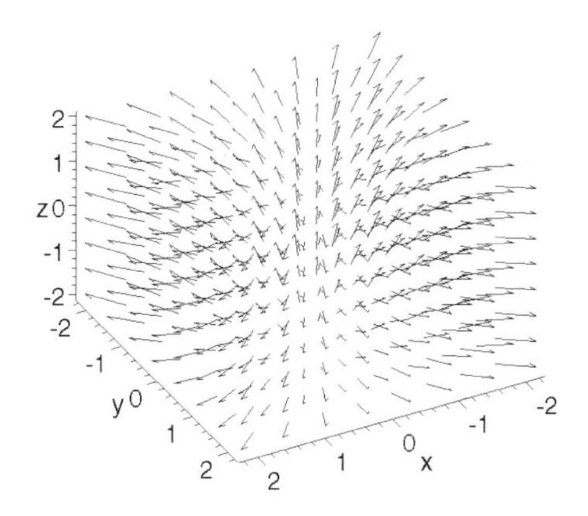

■

Graphing Lists and Arrays of Points in Three Dimensions

We have seen that `plot` can be used to graph lists of two-dimensional points. Three-dimensional points are graphed with `pointplot`, illustrated below. In addition, the command

```
matrixplot(mat,opt)
```

yields a three-dimensional graphics object by taking the x and y coordinates from the row and column indices, respectively, and the z coordinate from the corresponding matrix element of `mat`. One of the options of `matrixplot` is `heights=HISTOGRAM`, which produces a histogram in three dimensions. Other options include `axes`, `labels`, `orientation`, `projection`, `style`, and `view`.

EXAMPLE 18: Use `matrixplot` to graph the set of points determined by the matrix

$$A = \begin{pmatrix} -2 & 3 & -1 & -3 & -2 \\ 1 & -2 & -2 & -2 & 0 \\ 2 & 0 & 2 & 3 & 3 \\ 3 & -2 & 2 & 1 & 0 \end{pmatrix}.$$

SOLUTION: After defining **A**, we use `matrixplot` to graph the points determined by **A**.

```
> A:=array([[-2,3,-1,-3,-2],[1,-2,-2,-2,0],
        [2,0,2,3,3],[3,-2,2,1,0]]):
  p1:=matrixplot(A):
  display(p2);
```

After using `convert` with the `POLYGONS` option to convert `p1` to a `POLYGONS` object, we illustrate the use of the `cutin` and `cutout` commands, which are contained in the `plottools` package. `cutin` replaces each polygon with a smaller, similar polygon; `cutout` is the complementary function of `cutin`.

```
> with(plottools):
  p2:=convert(p1,POLYGONS):
  display(cutin(p2,1/2));
  display(cutout(p2,1/2));
```

■

Example 19 illustrates the option `heights=histogram`.

EXAMPLE 19: Energy consumption (in quadrillion Btu) by end-use sector for selected years is shown in the following table. Create a three-dimensional bar chart representing this data.

Year	Residential and Commercial	Industrial	Transportation
1975	24.143	31.528	18.605
1980	25.653	30.609	19.695
1985	26.682	27.200	20.067
1990	28.857	29.904	25.528

Source: *The World Almanac and Book of Facts*, 1993

SOLUTION: We first define the array `Data`, representing the energy consumption of residential and commercial users, industrial users, and transportation users, respectively.

```
> Data:=array([[24.143,25.653,26.682,28.857],
    [31.528,30.609,27.200,29.904],
    [18.605,19.695,20.067,25.528]]);
```

$$Data := \begin{bmatrix} 24.143 & 25.653 & 26.682 & 28.857 \\ 31.528 & 30.609 & 27.200 & 29.904 \\ 18.605 & 19.695 & 20.067 & 25.528 \end{bmatrix}$$

Next, we use `matrixplot` together with the options `heights=histogram` and `axes=FRAME` to generate a three-dimensional bar graph representing the data in `Data`.

```
> matrixplot(Data,heights=histogram,axes=FRAME);
```

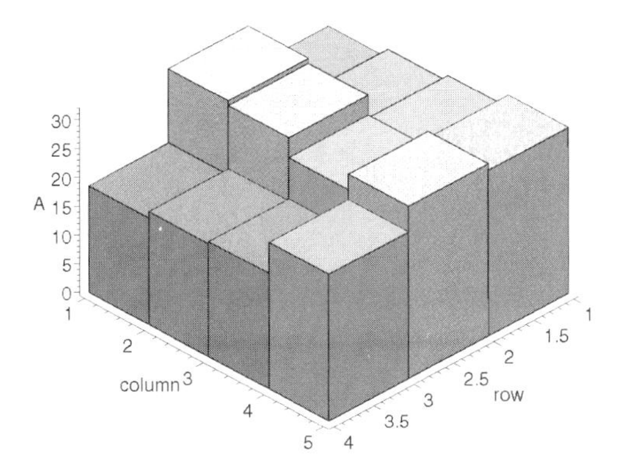

■

Three-dimensional points are graphed with pointplot. The command

`pointplot3d(ptlist,opt)`

creates a three-dimensional plot of the points listed in `ptlist`. The options, which may be entered in the option list `opt` are the same as those used with `plot3d` as well as `orientation`, `projection`, and `view`.

EXAMPLE 20: Graph the set of points (–4, 4, 2), (–1, –3, –2), (–1, –3, –2), (–5, –1, 1), (0, 0, 3), (–3, 3, 1), (–1, –5, 0), (0, 4, 5), (1, –1, 0), (–4, 0, 3).

SOLUTION: We use `pointplot3d` together with the option `axes=BOXED` to graph the set of points in space.

```
> pointplot3d({[-4,4,2],[-1,-3,-2],[-5,-1,1],
      [0,0,3],[-3,3,1],[-1,-5,0],
      [0,4,5],[1,-1,0],[-4,0,3]},
      axes=BOXED,symbol=BOX);
```

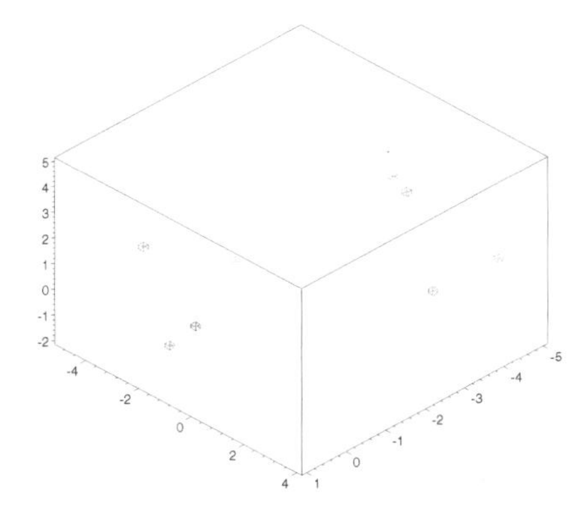

■

Graphing Geometric Objects in Two and Three Dimensions

Another useful graphics package is the `plottools` package which contains definitions of a variety of geometric objects and tools.

```
> with(plottools);
```

[*arc, arrow, circle, cone, cuboid, curve, cutin, cutout, cylinder, disk, dodecahedron, ellipse, ellipticArc, hemisphere, hexahedron, homothety, hyperbola, icosahedron, line, octahedron, pieslice, point, polygon, project, rectangle, reflect, rotate, scale, semitorus, sphere, stellate, tetrahedron, torus, transform, translate, vrml*]

As with other packages, detailed information regarding the `plottools` package is obtained with `?plottools`.

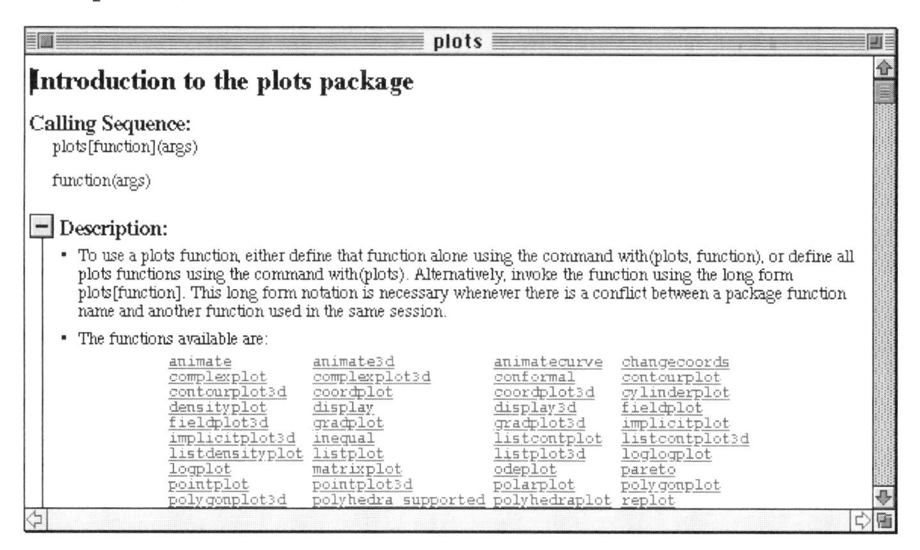

plots

Introduction to the plots package

Calling Sequence:

plots[function](args)

function(args)

Description:

- To use a plots function, either define that function alone using the command with(plots, function), or define all plots functions using the command with(plots). Alternatively, invoke the function using the long form plots[function]. This long form notation is necessary whenever there is a conflict between a package function name and another function used in the same session.
- The functions available are:

animate	animate3d	animatecurve	changecoords
complexplot	complexplot3d	conformal	contourplot
contourplot3d	coordplot	coordplot3d	cylinderplot
densityplot	display	display3d	fieldplot
fieldplot3d	gradplot	gradplot3d	implicitplot
implicitplot3d	inequal	listcontplot	listcontplot3d
listdensityplot	listplot	listplot3d	loglogplot
logplot	matrixplot	odeplot	pareto
pointplot	pointplot3d	polarplot	polygonplot
polygonplot3d	polyhedra_supported	polyhedraplot	replot

EXAMPLE 21: (a) Graph a dodecahedron. (b) Display a stellated octahedron. (c) Display a dodecahedron, octahedron, and tetrahedron in the same graph.

SOLUTION: After loading the `plottools` package, we use `dodecahedron` to define `d1`. In `d2`, we use `cutin` to replace each face of `dodecahedron([0,0,0],2)` with a similar, but smaller, face. `d1` and `d2` are displayed together with `display3d`.

```
> d1:=dodecahedron([0,0,0],1):
  d2:=cutin(dodecahedron([0,0,0],2),1/3):
  display3d({d1,d2});
```

We then use `translate`, which is contained in the `plottools` package, to translate the center of `d1` to the point (i, i, i) for $i = 0, \ldots, 4$. We display all five intersecting dodecahedra, together with `display3d`.

```
> d2:={seq(translate(d1,i,i,i),i=0..4)}:
  display3d(d2);
```

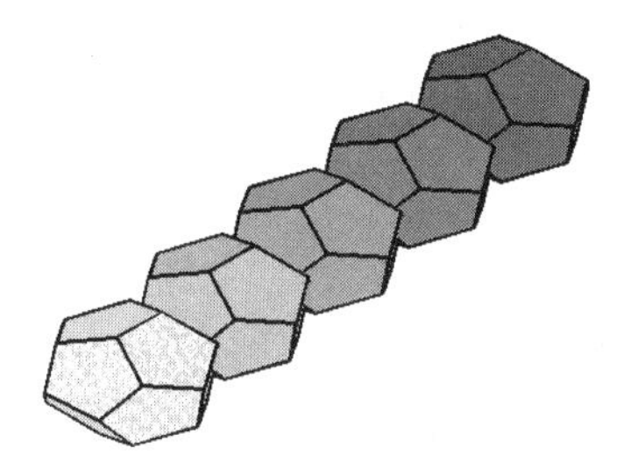

Using `cutout`, we cut out the center of each face and display the result. Notice that `cutin` and `cutout` are complementary functions.

```
> d2:={seq(cutout(translate(d1,i,i,i),4/5),i=0..4)}:
  display3d(d2,color=BLACK);
```

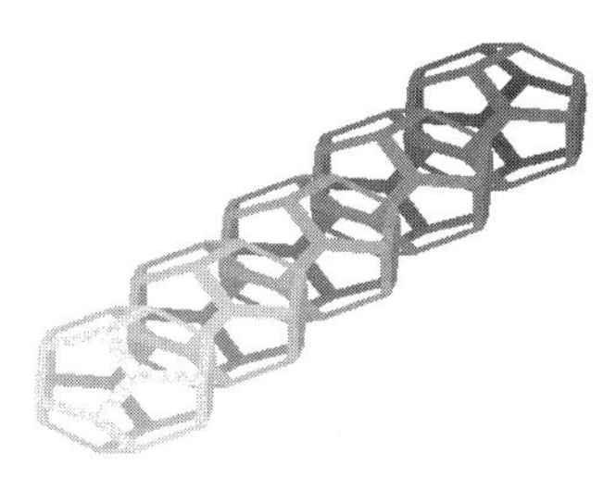

A stellated octahedron is displayed using display, stellate, and octahedron. You can make the spikes shorter by changing "6" to a number smaller than 6, or you can make the spikes longer by changing "6" to a number larger than 6. (Generally, stellate replaces each face with a stellate.)

```
> display(stellate(dodecahedron([0,0,0],1),6),
      scaling=CONSTRAINED);
```

If $0 < r < 1$ in stellate(object,r), the stellates are directed toward the center of the object.

```
> display(stellate(dodecahedron([0,0,0],1),1/8),
      scaling=CONSTRAINED);
```

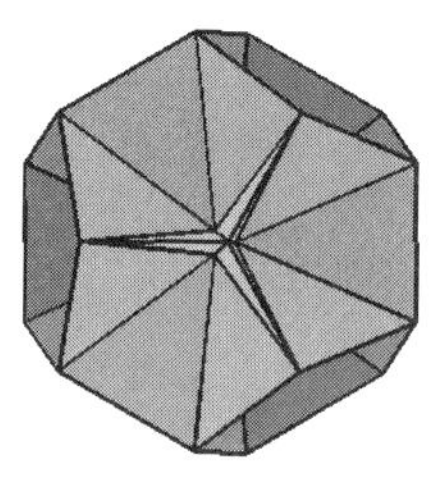

For (c), we use dodecahedron, octahedron, and tetrahedron together in the display command.

```
> display({dodecahedron([0,0,0],1/2),
      octahedron([cos(Pi/3),sin(Pi/3),0],1/3),
      tetrahedron([cos(2*Pi/3),sin(2*Pi/3),1/3],1/4)},
      scaling=CONSTRAINED);
```

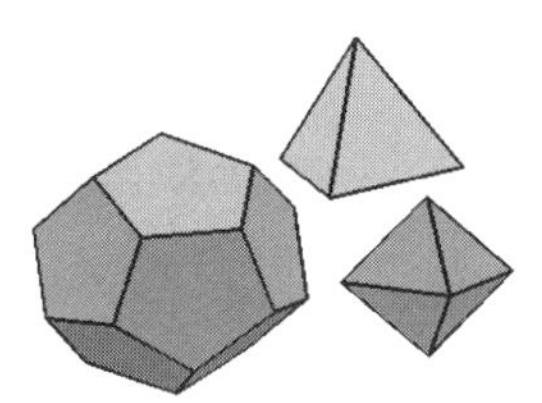

■

Tori and semitori can be generated with torus and semitorus, which are contained in the plottools package.

EXAMPLE 22: Generate a torus with meridian radius 1/2 and outer radius (distance from center of the meridian to the center of the torus) 2.

SOLUTION: The torus is defined with torus in t1. Then, a semitorus is rotated $\pi/2$ radians about the x- and y-axes using semitorus and rotate in t2. t1 and t2 are displayed together with display. (Generally rotate(object,α,β,γ) rotates object α radians counterclockwise about the x-axis, β radians counterclockwise about the y-axis, and γ radians counterclockwise about the z-axis.)

```
> t1:=torus([0,0,0],1/2,2):
  t2:=rotate(semitorus([0,0,0],Pi/2..3*Pi/2,1/2,3),
      Pi/2,-Pi/2,0):
  display({t1,t2},axes=BOXED,scaling=CONSTRAINED);
```

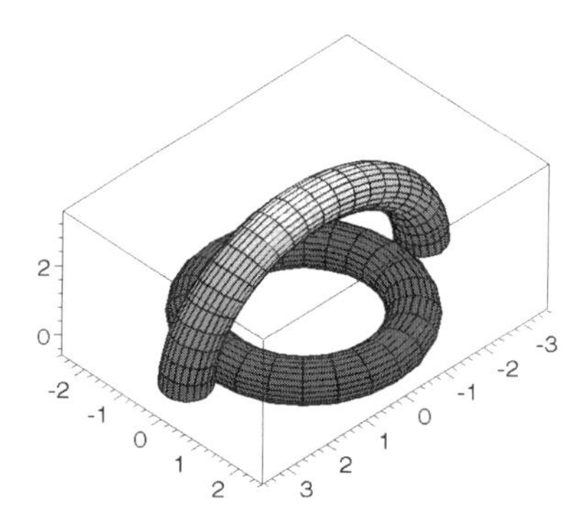

■

The command `sphere([x0,y0,z0],r)` generates a sphere with center at (x_0, y_0, z_0) and radius r.

EXAMPLE 23: Show a sphere of radius 1 surrounded by a sphere of radius 2.

SOLUTION: The sphere of radius 1 is obtained in `s2`, while the sphere of radius 2 is obtained in `s1`. The two spheres are shown together with `display`.

```
> s1:= sphere([0,0,0], 2,style=LINE,color=BLACK):
  s2:=sphere([0,0,0], 1):
  display({s1,s2},scaling=CONSTRAINED);
```

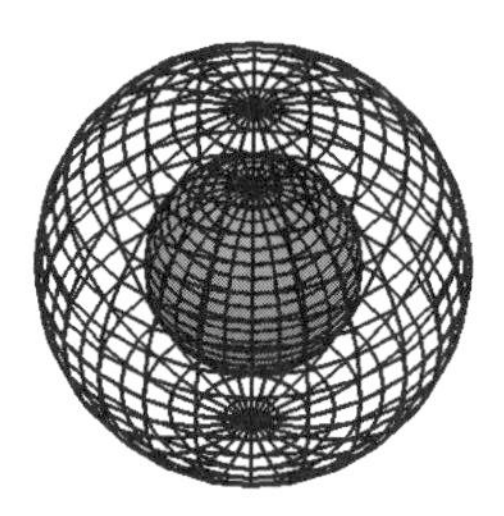

On the other hand, in the following, we define two spheres, `s1` and `s2`. The function `tr_s2` translates `s2`.

```
> s1:=sphere([0,0,0],1.5):
  s2:=sphere([0,0,0],0.5):
  tr_s2:=t->translate(s2,2.5*cos(t),
       .75+2.5*sin(t),.25*sin(t));
```

We then use `seq` to apply `tr_s2` to various values of t and display the result: a world with nine moons!

```
> tvals:=seq(2*Pi*i/8,i=0..8):
  an_array:=[seq(display({s1,tr_s2(t)},
       scaling=CONSTRAINED),t=tvals)]:
  display(an_array);
```

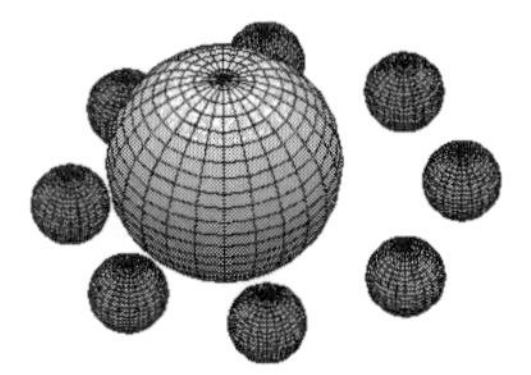

■

Selected References

Abbasian, R., and Ionescu, A. *Vector Calculus with Maple*. McGraw-Hill, 1996.

Abell, Martha L., and Braselton, James P. *Differential Equations with Maple V*, second edition. Academic Press, 1999.

——. *Modern Differential Equations: Theory, Applications, Technology*. Saunders College Publishing, 1996.

——. *Student Resource Manual for Modern Differential Equations: Theory, Applications, Technology*. Saunders College Publishing, 1996.

Adams, S. *Maple Talk*. Prentice Hall, 1996.

Anderson, P. H. *Maple V Applications to Electrical Circuit Analysis*. Self-published, 1995.

Andersson, G. *Applied Mathematics with Maple*. Chartwell-Bratt, 1997.

Articolo, G. *Partial Differential Equations and Boundary Value Problems with Maple*. Academic Press, 1998.

Bauldry, W.; Evans, B.; and Johnson, J. *Linear Algebra with Maple*. John Wiley and Sons, 1995.

Baylis, W. *Theoretical Methods in the Physical Sciences: An Introduction to Problem Solving Using Maple V*. Birkhäuser, 1994.

Blachman, N., and Mossinghoff, M. *Maple V Quick Reference*. Brooks/Cole, 1995.

Borreli, Robert L., and Coleman, Courtney S. *Differential Equations: A Modeling Perspective*, preliminary edition. John Wiley & Sons, 1996.

Carlson, J., and Johnson, J. *Multivariable Mathematics with Maple: Linear Algebra, Vector Calculus and Differential Equations*. Prentice Hall, 1997.

Cheney, Ward, and Kincaid, David. *Numerical Mathematics and Computing*, second edition. Brooks/Cole Publishing, 1985.

Churchill, Ruel V. *Operational Mathematics*, third edition. McGraw-Hill, 1972.

Coombes, K.; Hunt, B.; Lipsman, R.; Osborn, J.; and Stuck, G. *Differential Equations with Maple*, second edition. John Wiley and Sons, 1997.

Corless, R. *Essential Maple: An Introduction for Scientific Programmers*. Springer-Verlag, 1995.

Enns, R., and McGuire, G. *Nonlinear Physics with Maple for Scientists and Engineers*. Birkhäuser, 1997.

Greene, R. *Classical Mechanics with Maple*. Springer-Verlag, 1995.

Gutterman, Martin M., and Nitecki, Zbigniew H. *Differential Equations: A First Course*, third edition. Saunders College Publishing, 1991.

Heal, K. M.; Hansen, M.; and Rickard, K. *Maple V Learning Guide for Release 5*. Springer-Verlag, 1997.

Horbatsch, M. *Quantum Mechanics Using Maple*. Springer-Verlag, 1995.

Jordan, D. W., and Smith, P. *Nonlinear Ordinary Differential Equations*, second edition. Oxford University Press, 1988.

Karian, Z., and Tanis, E. *Probability and Statistics Explorations with Maple*. Prentice Hall, 1995.

Klimek, G., and Klimek, M. *Discovering Curves and Surfaces with Maple*. Springer-Verlag, 1997.

Kofler, M. *Maple: An Introduction and Reference*. Addison-Wesley, 1997.

Kreyszig, Erwin. *Advanced Engineering Mathematics*, seventh edition. John Wiley & Sons , 1993.

Marion, Jerry B., and Thornton, Stephen T. *Classical Dynamics of Particles and Systems*, fourth edition. Saunders College Publishing, 1995.

Monagan, M.; Geddes, K.; Heal, K.; Labahn, G.; and Vorkoetter, S. *Maple V Programming Guide for Release 5*. Springer-Verlag, 1997.

Powers, David L. *Boundary Value Problems*, second edition. Academic Press , 1979.

Robertson, J. *Engineering Mathematics with Maple*. McGraw-Hill, 1996.

Rosen, K. H. *Exploring Discrete Mathematics with Maple*. McGraw-Hill, 1997.

Strang, Gilbert. *Linear Algebra and its Applications*, third edition. Harcourt Brace Jovanovich, 1988.

Zwillinger, Daniel. *Handbook of Differential Equations*, second edition. Academic Press , 1992.

Note that an extensive list of books and other publications (including works in progress) devoted to Maple V can be found at

http://www.maplesoft.com/books.html

For further information, including purchasing information, about Maple V contact:

Corporate Headquarters:
Waterloo Maple, Inc.
450 Phillip Street
Waterloo, ON, Canada N2L 5J2
telephone: (519) 747-2373
fax: (519) 747-5284
email: info@maplesoft.com
web: http://www.maplesoft.com/

Index

D

G

H

N

Q

R

S

T